Lecture Notes on Data Engineering and Communications Technologies

200

Series Editor

Fatos Xhafa, *Technical University of Catalonia, Barcelona, Spain*

The aim of the book series is to present cutting edge engineering approaches to data technologies and communications. It will publish latest advances on the engineering task of building and deploying distributed, scalable and reliable data infrastructures and communication systems.

The series will have a prominent applied focus on data technologies and communications with aim to promote the bridging from fundamental research on data science and networking to data engineering and communications that lead to industry products, business knowledge and standardisation.

Indexed by SCOPUS, INSPEC, EI Compendex.

All books published in the series are submitted for consideration in Web of Science.

Leonard Barolli

Editor

Advanced Information Networking and Applications

Proceedings of the 38th International Conference on Advanced Information Networking and Applications (AINA-2024), Volume 2

 Springer

Editor
Leonard Barolli
Department of Information and Communication
Engineering
Fukuoka Institute of Technology
Fukuoka, Japan

ISSN 2367-4512 ISSN 2367-4520 (electronic)
Lecture Notes on Data Engineering and Communications Technologies
ISBN 978-3-031-57852-6 ISBN 978-3-031-57853-3 (eBook)
https://doi.org/10.1007/978-3-031-57853-3

This Springer imprint is published by the registered company Springer Nature Switzerland AG
The registered company address is: Gewerbestrasse 11, 6330 Cham, Switzerland

Paper in this product is recyclable.

Welcome Message from AINA-2024 Organizers

Welcome to the 38th International Conference on Advanced Information Networking and Applications (AINA-2024). On behalf of AINA-2024 Organizing Committee, we would like to express to all participants our cordial welcome and high respect.

AINA is an International Forum, where scientists and researchers from academia and industry working in various scientific and technical areas of networking and distributed computing systems can demonstrate new ideas and solutions in distributed computing systems. AINA is a very open society and is always welcoming international volunteers from any country and any area in the world.

AINA International Conference is a forum for sharing ideas and research work in the emerging areas of information networking and their applications. The area of advanced networking has grown very rapidly and the applications have experienced an explosive growth, especially in the area of pervasive and mobile applications, wireless sensor and ad-hoc networks, vehicular networks, multimedia computing, social networking, semantic collaborative systems, as well as IoT, big data, cloud computing, artificial intelligence, and machine learning. This advanced networking revolution is transforming the way people live, work, and interact with each other and is impacting the way business, education, entertainment, and health care are operating. The papers included in the proceedings cover theory, design and application of computer networks, distributed computing, and information systems.

Each year AINA receives a lot of paper submissions from all around the world. It has maintained high-quality accepted papers and is aspiring to be one of the main international conferences on the information networking in the world.

We are very proud and honored to have two distinguished keynote talks by Prof. Fatos Xhafa, Technical University of Catalonia, Spain, and Dr. Juggapong Natwichai, Chiang Mai University, Thailand, who will present their recent work and will give new insights and ideas to the conference participants.

An international conference of this size requires the support and help of many people. A lot of people have helped and worked hard to produce a successful AINA-2024 technical program and conference proceedings. First, we would like to thank all authors for submitting their papers. We are indebted to Program Track Co-chairs, Program Committee Members and Reviewers, who carried out the most difficult work of carefully evaluating the submitted papers.

We would like to thank AINA-2024 General Co-chairs, PC Co-chairs, Workshops Organizers for their great efforts to make AINA-2024 a very successful event. We have special thanks to the Finance Chair and Web Administrator Co-chairs.

We do hope that you will enjoy the conference proceedings and readings.

AINA-2024 Organizing Committee

Honorary Chair

Makoto Takizawa — Hosei University, Japan

General Co-chairs

Minoru Uehara — Toyo University, Japan
Euripides G. M. Petrakis — Technical University of Crete (TUC), Greece
Isaac Woungang — Toronto Metropolitan University, Canada

Program Committee Co-chairs

Tomoya Enokido — Rissho University, Japan
Mario A. R. Dantas — Federal University of Juiz de Fora, Brazil
Leonardo Mostarda — University of Perugia, Italy

International Journals Special Issues Co-chairs

Fatos Xhafa — Technical University of Catalonia, Spain
David Taniar — Monash University, Australia
Farookh Hussain — University of Technology Sydney, Australia

Award Co-chairs

Arjan Durresi — Indiana University Purdue University in Indianapolis (IUPUI), USA
Fang-Yie Leu — Tunghai University, Taiwan
Marek Ogiela — AGH University of Science and Technology, Poland
Kin Fun Li — University of Victoria, Canada

Publicity Co-chairs

Markus Aleksy	ABB Corporate Research Center, Germany
Flora Amato	University of Naples "Federico II", Italy
Lidia Ogiela	AGH University of Science and Technology, Poland
Hsing-Chung Chen	Asia University, Taiwan

International Liaison Co-chairs

Wenny Rahayu	La Trobe University, Australia
Nadeem Javaid	COMSATS University Islamabad, Pakistan
Beniamino Di Martino	University of Campania "Luigi Vanvitelli", Italy

Local Arrangement Co-chairs

Keita Matsuo	Fukuoka Institute of Technology, Japan
Tomoyuki Ishida	Fukuoka Institute of Technology, Japan

Finance Chair

Makoto Ikeda	Fukuoka Institute of Technology, Japan

Web Co-chairs

Phudit Ampririt	Fukuoka Institute of Technology, Japan
Ermioni Qafzezi	Fukuoka Institute of Technology, Japan
Shunya Higashi	Fukuoka Institute of Technology, Japan

Steering Committee Chair

Leonard Barolli	Fukuoka Institute of Technology, Japan

Tracks Co-chairs and Program Committee Members

1. Network Architectures, Protocols and Algorithms

Track Co-chairs

Spyropoulos Thrasyvoulos	Technical University of Crete (TUC), Greece
Shigetomo Kimura	University of Tsukuba, Japan
Darshika Perera	University of Colorado at Colorado Springs, USA

TPC Members

Thomas Dreibholz	Simula Metropolitan Center for Digital Engineering, Norway
Angelos Antonopoulos	Nearby Computing SL, Spain
Hatim Chergui	i2CAT Foundation, Spain
Bhed Bahadur Bista	Iwate Prefectural University, Japan
Chotipat Pornavalai	King Mongkut's Institute of Technology Ladkrabang, Thailand
Kenichi Matsui	NTT Network Innovation Center, Japan
Sho Tsugawa	University of Tsukuba, Japan
Satoshi Ohzahata	University of Electro-Communications, Japan
Haytham El Miligi	Thompson Rivers University, Canada
Watheq El-Kharashi	Ain Shams University, Egypt
Ehsan Atoofian	Lakehead University, Canada
Fayez Gebali	University of Victoria, Canada
Kin Fun Li	University of Victoria, Canada
Luis Blanco	CTTC, Spain

2. Next Generation Mobile and Wireless Networks

Track Co-chairs

Purav Shah	School of Science and Technology, Middlesex University, UK
Enver Ever	Middle East Technical University, Northern Cyprus
Evjola Spaho	Polytechnic University of Tirana, Albania

TPC Members

Burak Kizilkaya	Glasgow University, UK
Muhammad Toaha	Middle East Technical University, Turkey
Ramona Trestian	Middlesex University, UK
Andrea Marotta	University of L'Aquila, Italy
Adnan Yazici	Nazarbayev University, Kazakhstan
Orhan Gemikonakli	Final International University, Cyprus
Hrishikesh Venkataraman	Indian Institute of Information Technology, Sri City, India
Zhengjia Xu	Cranfield University, UK
Mohsen Hejazi	University of Kashan, Iran
Sabyasachi Mukhopadhyay	IIT Kharagpur, India
Ali Khoshkholghi	Middlesex University, UK
Admir Barolli	Aleksander Moisiu University of Durres, Albania
Makoto Ikeda	Fukuoka Institute of Technology, Japan
Yi Liu	Oita National College of Technology, Japan
Testuya Oda	Okayama University of Science, Japan
Ermioni Qafzezi	Fukuoka Institute of Technology, Japan

3. Multimedia Networking and Applications

Track Co-chairs

Markus Aleksy	ABB Corporate Research Center, Germany
Francesco Orciuoli	University of Salerno, Italy
Tomoyuki Ishida	Fukuoka Institute of Technology, Japan

TPC Members

Hadil Abukwaik	ABB Corporate Research Center, Germany
Thomas Preuss	Brandenburg University of Applied Sciences, Germany
Peter M. Rost	Karlsruhe Institute of Technology (KIT), Germany
Lukasz Wisniewski	inIT, Germany
Angelo Gaeta	University of Salerno, Italy
Angela Peduto	University of Salerno, Italy
Antonella Pascuzzo	University of Salerno, Italy
Roberto Abbruzzese	University of Salerno, Italy
Tetsuro Ogi	Keio University, Japan

Yasuo Ebara   Osaka Electro-Communication University, Japan
Hideo Miyachi   Tokyo City University, Japan
Kaoru Sugita   Fukuoka Institute of Technology, Japan

4. Pervasive and Ubiquitous Computing

Track Co-chairs

Vamsi Paruchuri   University of Central Arkansas, USA
Hsing-Chung Chen   Asia University, Taiwan
Shinji Sakamoto   Kanazawa Institute of Technology, Japan

TPC Members

Sriram Chellappan   University of South Florida, USA
Yu Sun   University of Central Arkansas, USA
Qiang Duan   Penn State University, USA
Han-Chieh Wei   Dallas Baptist University, USA
Ahmad Alsharif   University of Alabama, USA
Vijayasarathi Balasubramanian   Microsoft, USA
Shyi-Shiun Kuo   Nan Kai University of Technology, Taiwan
Karisma Trinanda Putra   Universitas Muhammadiyah Yogyakarta,
                                    Indonesia
Cahya Damarjati   Universitas Muhammadiyah Yogyakarta,
                                    Indonesia
Agung Mulyo Widodo   Universitas Esa Unggul Jakarta, Indonesia
Bambang Irawan   Universitas Esa Unggul Jakarta, Indonesia
Eko Prasetyo   Universitas Muhammadiyah Yogyakarta,
                                    Indonesia
Sunardi S. T.   Universitas Muhammadiyah Yogyakarta,
                                    Indonesia
Andika Wisnujati   Universitas Muhammadiyah Yogyakarta,
                                    Indonesia
Makoto Ikeda   Fukuoka Institute of Technology, Japan
Tetsuya Oda   Okayama University of Science, Japan
Evjola Spaho   Polytechnic University of Tirana, Albania
Tetsuya Shigeyasu   Hiroshima Prefectural University, Japan
Keita Matsuo   Fukuoka Institute of Technology, Japan
Admir Barolli   Aleksander Moisiu University of Durres, Albania

5. Web-Based Systems and Content Distribution

Track Co-chairs

Chrisa Tsinaraki	Technical University of Crete (TUC), Greece
Yusuke Gotoh	Okayama University, Japan
Santi Caballe	Open University of Catalonia, Spain

TPC Members

Nikos Bikakis	Hellenic Mediterranean University, Greece
Ioannis Stavrakantonakis	Ververica GmbH, Germany
Sven Schade	European Commission, Joint Research Center, Italy
Christos Papatheodorou	National and Kapodistrian University of Athens, Greece
Sarantos Kapidakis	University of West Attica, Greece
Manato Fujimoto	Osaka Metropolitan University, Japan
Kiki Adhinugraha	La Trobe University, Australia
Tomoki Yoshihisa	Shiga University, Japan
Jordi Conesa	Open University of Catalonia, Spain
Thanasis Daradoumis	Open University of Catalonia, Spain
Nicola Capuano	University of Basilicata, Italy
Victor Ströele	Federal University of Juiz de Fora, Brazil

6. Distributed Ledger Technologies and Distributed-Parallel Computing

Track Co-chairs

Alfredo Navarra	University of Perugia, Italy
Naohiro Hayashibara	Kyoto Sangyo University, Japan

TPC Members

Serafino Cicerone	University of L'Aquila, Italy
Ralf Klasing	LaBRI Bordeaux, France
Giuseppe Prencipe	University of Pisa, Italy
Roberto Tonelli	University of Cagliari, Italy
Farhan Ullah	Northwestern Polytechnical University, China

Leonardo Mostarda	University of Perugia, Italy
Qiong Huang	South China Agricultural University, China
Tomoya Enokido	Rissho University, Japan
Minoru Uehara	Toyo University, Japan
Lucian Prodan	Polytechnic University of Timisoara, Romania
Md. Abdur Razzaque	University of Dhaka, Bangladesh

7. Data Mining, Big Data Analytics and Social Networks

Track Co-chairs

Pavel Krömer	Technical University of Ostrava, Czech Republic
Alex Thomo	University of Victoria, Canada
Eric Pardede	La Trobe University, Australia

TPC Members

Sebastián Basterrech	Technical University of Denmark, Denmark
Tibebe Beshah	University of Addis Ababa, Ethiopia
Nashwa El-Bendary	Arab Academy for Science, Egypt
Petr Musilek	University of Alberta, Canada
Varun Ojha	Newcastle University, UK
Alvaro Parres	ITESO, Mexico
Nizar Rokbani	ISSAT-University of Sousse, Tunisia
Farshid Hajati	Victoria University, Australia
Ji Zhang	University of Southern Queensland, Australia
Salimur Choudhury	Lakehead University, Canada
Carson Leung	University of Manitoba, Canada
Syed Mahbub	La Trobe University, Australia
Osama Mahdi	Melbourne Institute of Technology, Australia
Choiru Zain	La Trobe University, Australia
Rajalakshmi Rajasekaran	La Trobe University, Australia
Nawfal Ali	Monash University, Australia

8. Internet of Things and Cyber-Physical Systems

Track Co-chairs

Tomoki Yoshihisa	Shiga University, Japan
Winston Seah	Victoria University of Wellington, New Zealand
Luciana Pereira Oliveira	Instituto Federal da Paraiba (IFPB), Brazil

TPC Members

Akihiro Fujimoto	Wakayama University, Japan
Akimitsu Kanzaki	Shimane University, Japan
Kazuya Tsukamoto	Kyushu Institute of Technology, Japan
Lei Shu	Nanjing Agricultural University, China
Naoyuki Morimoto	Mie University, Japan
Teruhiro Mizumoto	Chiba Institute of Technology, Japan
Tomoya Kawakami	Fukui University, Japan
Adrian Pekar	Budapest University of Technology and Economics, Hungary
Alvin Valera	Victoria University of Wellington, New Zealand
Chidchanok Choksuchat	Prince of Songkla University, Thailand
Jyoti Sahni	Victoria University of Wellington, New Zealand
Murugaraj Odiathevar	Sungkyunkwan University, South Korea
Normalia Samian	Universiti Putra Malaysia, Malaysia
Qing Gu	University of Science and Technology Beijing, China
Tao Zheng	Beijing Jiaotong University, China
Wenbin Pei	Dalian University of Technology, China
William Liu	Unitec, New Zealand
Wuyungerile Li	Inner Mongolia University, China
Peng Huang	Sichuan Agricultural University, PR China
Ruan Delgado Gomes	Instituto Federal da Paraiba (IFPB), Brazil
Glauco Estacio Goncalves	Universidade Federal do Pará (UFPA), Brazil
Eduardo Luzeiro Feitosa	Universidade Federal do Amazonas (UFAM), Brazil
Paulo Ribeiro Lins Júnior	Instituto Federal da Paraiba (IFPB), Brazil

9. Intelligent Computing and Machine Learning

Track Co-chairs

Takahiro Uchiya	Nagoya Institute of Technology, Japan
Flavius Frasincar	Erasmus University Rotterdam, The Netherlands
Miltos Alamaniotis	University of Texas at San Antonio, USA

TPC Members

Kazuto Sasai	Ibaraki University, Japan
Shigeru Fujita	Chiba Institute of Technology, Japan
Yuki Kaeri	Mejiro University, Japan
Jolanta Mizera-Pietraszko	Military University of Land Forces, Poland
Ashwin Ittoo	University of Liège, Belgium
Marco Brambilla	Politecnico di Milano, Italy
Alfredo Cuzzocrea	University of Calabria, Italy
Le Minh Nguyen	JAIST, Japan
Akiko Aizawa	National Institute of Informatics, Japan
Natthawut Kertkeidkachorn	JAIST, Japan
Georgios Karagiannis	Durham University, UK
Leonidas Akritidis	International Hellenic University, Greece
Athanasios Fevgas	University of Thessaly, Greece
Yota Tsompanopoulou	University of Thessaly, Greece
Yuvaraj Munian	Texas A&M-San Antonio, USA

10. Cloud and Services Computing

Track Co-chairs

Salvatore Venticinque	University of Campania "Luigi Vanvitelli", Italy
Shigenari Nakamura	Tokyo Denki University, Japan
Sajal Mukhopadhyay	National Institute of Technology, Durgapur, India

TPC Members

Giancarlo Fortino	University of Calabria, Italy
Massimiliano Rak	University of Campania "Luigi Vanvitelli", Italy
Jason J. Jung	Chung-Ang University, Korea

Dimosthenis Kyriazis	University of Piraeus, Greece
Geir Horn	University of Oslo, Norway
Dario Branco	University of Campania "Luigi Vanvitelli", Italy
Dilawaer Duolikun	Cognizant Technology Solutions, Hungary
Naohiro Hayashibara	Kyoto Sangyo University, Japan
Tomoya Enokido	Rissho University, Japan
Sujoy Saha	NIT Durgapur, India
Animesh Dutta	NIT Durgapur, India
Pramod Mane	IIM Rohtak, India
Nanda Dulal Jana	NIT Durgapur, India
Banhi Sanyal	NIT Kurukshetra, India

11. Security, Privacy and Trust Computing

Track Co-chairs

Ioannidis Sotirios	Technical University of Crete (TUC), Greece
Michail Alexiou	Georgia Institute of Technology, USA
Hiroaki Kikuchi	Meiji University, Japan

TPC Members

George Vasiliadis	Hellenic Mediterranean University, Greece
Antreas Dionysiou	University of Cyprus, Cyprus
Apostolos Fouranaris	Athena Research Center, Greece
Panagiotis Ilia	Technical University of Crete, Greece
George Portokalidis	IMDEA, Spain
Nikolaos Gkorgkolis	University of Crete, Greece
Zeezoo Ryu	Georgia Institute of Technology, USA
Muhammad Faraz Karim	Georgia Institute of Technology, USA
Yunjie Deng	Georgia Institute of Technology, USA
Anna Raymaker	Georgia Institute of Technology, USA
Takamichi Saito	Meiji University, Japan
Kazumasa Omote	University of Tsukuba, Japan
Masakatsu Nishigaki	Shizuoka University, Japan
Mamoru Mimura	National Defense Academy of Japan, Japan
Chun-I Fan	National Sun Yat-sen University, Taiwan
Aida Ben Chehida Douss	National School of Engineers of Tunis, ENIT Tunis, Tunisia
Davinder Kaur	IUPUI, USA

12. Software-Defined Networking and Network Virtualization

Track Co-chairs

Flavio de Oliveira Silva   Federal University of Uberlândia, Brazil
Ashutosh Bhatia   Birla Institute of Technology and Science, Pilani,
                                  India

TPC Members

Rui Luís Andrade Aguiar   Universidade de Aveiro (UA), Portugal
Ivan Vidal   Universidad Carlos III de Madrid, Spain
Eduardo Coelho Cerqueira   Federal University of Pará (UFPA), Brazil
Christos Tranoris   University of Patras (UoP), Greece
Juliano Araújo Wickboldt   Federal University of Rio Grande do Sul
                                  (UFRGS), Brazil
Haribabu K.   BITS Pilani, India
Virendra Shekhavat   BITS Pilani, India
Makoto Ikeda   Fukuoka Institute of Technology, Japan
Farookh Hussain   University of Technology Sydney, Australia
Keita Matsuo   Fukuoka Institute of Technology, Japan

AINA-2024 Reviewers

Admir Barolli   Burak Kizilkaya
Aida ben Chehida Douss   Carson Leung
Akimitsu Kanzaki   Chidchanok Choksuchat
Alba Amato   Christos Tranoris
Alberto Postiglione   Chung-Ming Huang
Alex Thomo   Dario Branco
Alfredo Navarra   David Taniar
Amani Shatnawi   Elinda Mece
Anas AlSobeh   Enver Ever
Andrea Marotta   Eric Pardede
Angela Peduto   Euripides Petrakis
Anne Kayem   Evjola Spaho
Antreas Dionysiou   Fabrizio Messina
Arjan Durresi   Feilong Tang
Ashutosh Bhatia   Flavio Silva
Beniamino Di Martino   Francesco Orciuoli
Bhed Bista   George Portokalidis

Giancarlo Fortino
Giorgos Vasiliadis
Glauco Gonçalves
Hatim Chergui
Hiroaki Kikuchi
Hiroki Sakaji
Hiroshi Maeda
Hiroyuki Fujioka
Hyunhee Park
Isaac Woungang
Jana Nowaková
Jolanta Mizera-Pietraszko
Junichi Honda
Jyoti Sahni
Kazunori Uchida
Keita Matsuo
Kenichi Matsui
Kiki Adhinugraha
Kin Fun Li
Kiyotaka Fujisaki
Leonard Barolli
Leonardo Mostarda
Leonidas Akritidis
Lidia Ogiela
Lisandro Granville
Lucian Prodan
Luciana Oliveira
Mahmoud Elkhodr
Makoto Ikeda
Mamoru Mimura
Manato Fujimoto
Marco Antonio To
Marek Ogiela
Masaki Kohana
Minoru Uehara
Muhammad Karim
Muhammad Toaha Raza Khan
Murugaraj Odiathevar
Nadeem Javaid
Naohiro Hayashibara
Nobuo Funabiki
Nour El Madhoun
Omar Darwish

Panagiotis Ilia
Petr Musilek
Philip Moore
Purav Shah
R. Madhusudhan
Raffaele Guarasci
Ralf Klasing
Roberto Tonelli
Ronald Petrlic
Sabyasachi Mukhopadhyay
Sajal Mukhopadhyay
Salvatore D'Angelo
Salvatore D'Angelo
Salvatore D'Angelo
Salvatore Venticinque
Santi Caballé
Satoshi Ohzahata
Serafino Cicerone
Shigenari Nakamura
Shinji Sakamoto
Sho Tsugawa
Sriram Chellappan
Stephane Maag
Takayuki Kushida
Tetsuya Oda
Thomas Dreibholz
Tomoki Yoshihisa
Tomoya Enokido
Tomoya Kawakami
Tomoyuki Ishida
Vamsi Paruchuri
Victor Ströele
Vikram Singh
Wei Lu
Wenny Rahayu
Winston Seah
Yong Zheng
Yoshitaka Shibata
Yusuke Gotoh
Yuvaraj Munian
Zeezoo Ryu
Zhengjia Xu

AINA-2024 Keynote Talks

Agile Edge: Harnessing the Power of the Intelligent Edge by Agile Optimization

Fatos Xhafa

Technical University of Barcelona, Barcelona, Spain

Abstract. The digital cloud ecosystem comprises various degrees of computing granularity from large cloud servers and data centers to IoT devices, leading to the cloud-to-thing continuum computing paradigm. In this context, the intelligent edge aims at placing intelligence to the end devices, at the edges of the Internet. The premise is that collective intelligence from the IoT data deluge can be achieved and used at the edges of the Internet, offloading the computation burden from the cloud systems and leveraging real-time intelligence. This, however, comes with the challenges of processing and analyzing the IoT data streams in real time. In this talk, we will address how agile optimization can be useful for harnessing the power of the intelligent edge. Agile optimization is a powerful and promising solution, which differently from traditional optimization methods, is able to find optimized and scalable solutions under real-time requirements. We will bring real-life problems and case studies from Smart City Open Data Repositories to illustrate the approach. Finally, we will discuss the research challenges and emerging vision on the agile intelligent edge.

Challenges in Entity Matching in AI Era

Juggapong Natwichai

Chiang Mai University, Chiang Mai, Thailand

Abstract. Entity matching (EM) is to identify and link entities originating from various sources that correspond to identical real-world entities, thereby constituting a foundational component within the realm of data integration. For example, in order to counter-fraud detection, the datasets from sellers, financial services providers, or even IT infrastructure service providers might be in need for data integration, and hence, the EM is highly important here. This matching process is also recognized for its pivotal role in data augmenting to improve the precision and dependability of subsequent tasks within the domain of data analytics. Traditionally, the EM procedure composes of two integral phases, namely blocking and matching. The blocking phase associates with the generation of candidate pairs and could affect the size and complexity of the data. Meanwhile, the matching phase will need to trade-off between the accuracy and the efficiency. In this talk, the challenges of both components are thoroughly explored, particularly with the aid of AI techniques. In addition, the preliminary experiment results to explore some important factors which affect the performance will be presented.

Contents

A Method for Estimating the Number of Diseases in Computed Tomography Reports of the Japanese Medical Image Database (J-MID): Variations Among Facilities

Koji Sakai[✉], Yu Ohara, Yosuke Maehara, Takeshi Takahashi, and Kei Yamada

Kyoto Prefectural University of Medicine, Kawaramachi Hirokoji Agaru Kajiicho,
Kyoto 6028566, Kyoto, Japan
`sakai3@koto.kpu-m.ac.jp`

Abstract. A medical image database is vital for research related to diagnosis, training, quality control, disease prevalence, and treatment outcomes. Accurate identification of disease names and their quantities within these databases is essential for efficient data utilization and gaining insights into disease types and distributions. It also plays a significant role as a resource for machine learning-based research. However, accurately determining the number of diseases in image databases has been a challenge in many healthcare facilities, hindering data management efficiency and subject selection for machine learning. This study aimed to analyze Japanese-language image diagnostic reports in the Japan Medical Image Database (J-MID) computed tomography image database and estimate disease counts. Disease name extraction was aided by a lexicon of affirmative and negative predicates. Since J-MID involves multiple medical institutions, we aimed to understand predicate usage and disease distribution variations among institutions. By using the lexicon, we identified differences in disease distribution among facilities and effectively narrowed down machine learning-suitable diseases within the dataset during the limited data acquisition period.

Keywords: disease name · estimation · J-MID

1 Introduction

A medical imaging database supports research related to diagnosis, training, quality management, disease prevalence, and treatment outcomes. Understanding the names and quantities of diseases within these databases is useful for efficiently utilizing the database and gaining insights from the types and distribution of diseases [1]. Additionally, it holds a significant position as a research source for studies using machine learning. There have also been studies predicting International Classification of Diseases codes from imaging reports [2, 3]. However, accurately identifying the exact number of diseases within imaging databases remains a challenge in many healthcare institutions. This impedes efficient data management and the effective selection of machine learning targets.

L. Barolli (Ed.): AINA 2024, LNDECT 200, pp. 1–8, 2024.
https://doi.org/10.1007/978-3-031-57853-3_1

This study aimed to analyze image diagnostic reports written in Japanese within the Japanese Medical Image Database (J-MID) computed tomography (CT) image database and estimate the numbers of diseases. An overview of J-MID is presented in Fig. 1 [4]. As of 2023, J-MID is a database that has collected CT and MRI images along with their radiology reports from ten facilities, and its operation began in 2018. Recently, research in lung cancer is just beginning to be reported [5]. We are involved in the research project as members of J-MID users. Participating facilities send CT and magnetic resonance images, along with their radiology reports, to a cloud-based database via SINET 6 [6]. The transmitted data is aggregated in the J-MID repository and made available to J-MID Users via SINET6.

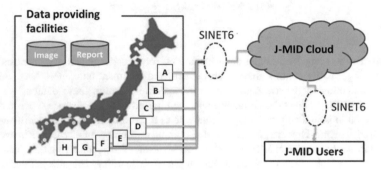

Fig. 1. The outline of J-MID project.

The extraction of disease names utilized a set of affirmative predicates related to disease names. In other words, it involved extracting affirmations of the presence of disease names. It's important to note that disease names were based on clinical diagnoses by radiologists rather than confirmed diagnoses through pathological examinations. Since J-MID involves multiple institutions, the objective was to observe the characteristics of predicates used by participating medical facilities and the distribution of diseases.

The use of predicate sets created from text within image diagnostic reports revealed differences in the distribution of diseases within each facility. Additionally, within the limited data acquisition period, it was possible to narrow down the diseases that are suitable for machine learning for each facility and for J-MID as a whole.

2 Materials and Methods

2.1 Participants

The study aimed to analyze 184,465 CT image reports stored in the image diagnostic report system at J-MID in 2022. The participating hospitals were eight facilities (A-F). The study was approved by the Ethics Committee of our facility (ERB-C-1262-1).

2.2 Data Preprocessing

We carried out a cleansing process for the image diagnostic reports in order to remove any information other than the findings and diagnoses. Then, we unified half-width and full-width characters. The disease names were extracted by the master list of disease and injuries (MLDI), which included 27,164 injuries and illnesses [7] and predicate lexicon created by an 8 years experienced radiologist (Y.O).

For the experiment, we used a desktop computer (Windows 11) equipped with an Intel® Core™ i9-10900K CPU @3.7 GHz and 64 GB of RAM. The computation was performed by NVIDIA GeForce RTX 3090 graphics card. While, for the analysis we developed an in-house program by using MATLAB R2023a.

2.3 Creation of Predicate Lexicon

For the period of first quarter in 2022, we obtained a group of chunks through morphological analysis [8] and used dependency analysis [9] to identify those that were directly related to the injury and illness names as the predicate thesaurus in J-MID CT reports. Any chunks related to injury and illness names that did not have any dependency were considered to have no predicate.

2.4 Estimation of Target Disease Count

The predicates with a positive meaning among those that directly modified the injury and illness names were selected as predicate by a radiologist (Y.O.) with 8 years of experience. The extracted combinations of injury and illness names with positive predicates received from the subject data were considered as the target disease count (Fig. 2).

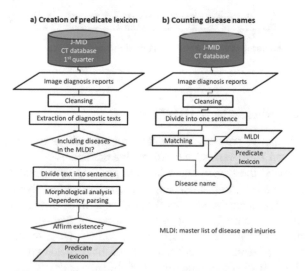

Fig. 2. Data processing procedures. a) Creation of a predicate lexicon by morphological analysis and dependency parsing. b) Counting disease names by MLDI and the created predicate lexicon.

3 Results

3.1 Number of CT Reports in J-MID 2022

Figure 3 shows the number of CT reports from eight facilities. The average number of CT reports across the eight facilities was 20,480, with the maximum at facility C being 50,264 and the minimum at facility B being 43. Due to the significant difference between the maximum and minimum, the standard deviation also became large. In terms of the overall proportion, facilities D, E, and F were roughly equal, while facilities A and G were also comparable. Therefore, it is anticipated that the contribution from facility C is substantial, while the contributions from facilities H and B are expected to be negligible.

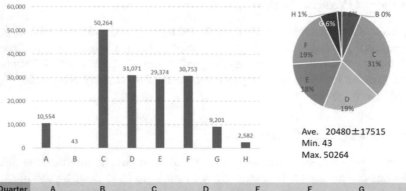

Quarter	A	B	C	D	E	F	G	H
1	263	-	9,826	106	6,961	7,871	-	-
2	3,340	-	13,707	9,927	7,511	8,056	2,149	1,389
3	3,461	-	13,248	10,438	7,394	8,151	3,502	941
4	3,490	43	13,483	10,600	7,508	6,675	3,550	252
Sum.	10,554	43	50,264	31,071	29,374	30,753	9,201	2,582

Fig. 3. The number of CT reports from eight facilities in J-MID (2022).

3.2 Affirmative Predicates in J-MID CT Report

The number of affirmative predicates extracted from the CT reports in 2022 was 1,209. Table 1 shows the top 10 representative affirmative predicates with high extraction frequency (%). It showed the percentage when the top 20 was considered as 100%. To facilitate readers' understanding, their English translations are also provided. Due to the mixed use of polite language such as "-masu" in similar expressions, Japanese has become a difficult-to-structure language with diverse expressions.

The top 5 affirmative predicates account for 76.6% of the total. It accounted for 89.3% in the top 10. In particular, there were numerous derived expressions of "doubt", which accounted for 30.2% within the top 20 of affirmative predicates.

Table 1. Representative 10 examples of affirmative predicates in radiological CT reports on J-MID

#	Affirmative predicates			
	Japanese	English	Count	%
1	Ari	there is/to exist	153,858	36.8
2	utagai-masu	I doubt	50,909	12.2
3	Utagau	to doubt	47,089	11.3
4	mitome-masu	I approve	44,556	10.7
5	kangae-masu	to think	23,445	5.6
6	Utagai	doubt/suspicion	18,634	4.5
7	tokki-subeki	remarkable	9,533	2.3
8	utagaware-masu	to be doubted	9,313	2.2
9	omoware-masu	it seems	8,281	2.0
10	Tomonau	accompany	7,279	1.7

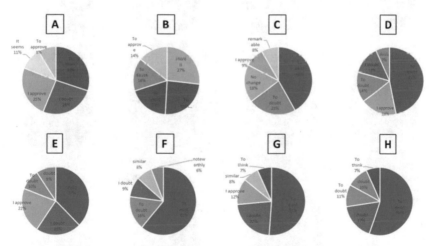

Fig. 4. The top-five of affirmative predicates on eight facilities in radiological CT reports of J-MID (2022).

3.3 Differences of Affirmative Predicates Among Facilities

The top-five of affirmative predicates on each eight facilities are shown in Fig. 4.

In each facility, expressions such as "to exist", "to doubt", "to approve", "to think", and their derivatives were observed. In particular, "to exist" was the most commonly observed in all facilities except for Facility A. "to doubt" and its derivatives were next frequently observed, followed by "I approve" and its derivatives. Facility A stood out as distinctive, with a higher occurrence of expressions like "to think" and "it seems" compared to others. Facilities C through H were characterized by the prevalence of the expression "to exist".

3.4   Top 10 Extracted Disease Names

The top 10 extracted disease names in J-MID 2022 CT reports and its contribution of each facility are shown in Fig. 5 and Table 2. The most frequent diagnosis was "inflammation", followed by "lesion", "pleural effusion", "cancer", and "ascites". In 2022, Facility F made the highest contribution to the most frequently observed diagnoses. Facility E contributed the most to "tumor", Facility D to "pneumonia", and Facility G to "colonic diverticulum". Facility C had the highest number of reports, but its contribution to the top 10 extracted diagnoses was not the greatest. Facility B had minimal original report numbers, and its contribution to the extracted diagnoses was almost non-existent except for "colonic diverticulum".

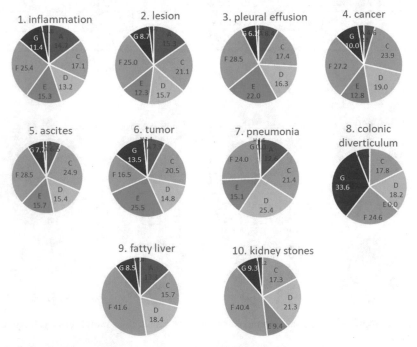

Fig. 5. The top 10 extracted disease names in J-MID 2022 CT reports and its contribution of each facility.

4   Discussions

In 2022, the J-MIT CT reports totaled 184,465, and they were derived from eight facilities, although the composition ratio was uneven. The top 10 positive predicates extracted from the diagnostic texts accounted for 89.3% of the top 20, with "to doubt" and its derivatives comprising 30%. Predicate selection varied among the facilities, but six out of the

Table 2. The total sum of top 10 extracted disease names and its contribution of each facility

#	Disease name	Sum	Contribution ratio of each facility [%]							
			A	B	C	D	E	F	G	H
1	inflammation	19,652	14.7	0.0	17.1	13.2	15.3	**25.4**	11.4	3.0
2	lesion	11,480	15.3	0.0	21.1	15.7	12.3	**25.0**	8.7	2.0
3	pleural effusion	10,124	8.4	0.0	17.4	16.3	22.0	**28.5**	6.2	1.1
4	cancer	9,853	4.6	0.0	23.9	19.0	12.8	**27.2**	10.0	2.5
5	ascites	8,667	6.5	0.0	24.9	15.4	15.7	**28.5**	7.7	1.3
6	tumor	7,326	7.7	0.0	20.5	14.8	**25.5**	16.5	13.5	1.4
7	pneumonia	7,257	12.6	0.0	21.4	**25.4**	15.1	24.0	0.0	1.5
8	colonicdiverticulum	6,146	0.0	0.0	17.8	18.2	0.0	24.6	**33.6**	5.8
9	fatty liver	6,084	13.3	0.0	15.7	18.4	0.0	**41.6**	8.5	2.4
10	kidney stone	5,781	0.0	0.0	17.3	21.3	9.4	**40.4**	9.3	2.2

The numbers written in bold are the largest among the facilities.

eight facilities made frequent use of "to exist". The major disease names extracted from the CT reports included "inflammation", "lesion", "pleural effusion", "cancer", and others. The contribution rate in the top 10 extracted disease names was biased towards one facility. Therefore, when performing machine learning using this CT data, it's important to be aware that the choice of diseases to be targeted can result in a bias in imaging techniques or radiology report content.

Generally, natural language processing is a machine learning technique used to understand the meaning of text. However, when it involves medical terminology, abbreviations, and specific cultural backgrounds of patients, it becomes complex and challenging. Japanese medical reports may contain unique disease names and symptoms, and they often use a multitude of specialized medical terms. To address these, specialized knowledge and appropriate dictionary construction are necessary for natural language processing. Furthermore, medical terminology often uses specialized abbreviations that require domain expertise for interpretation, making natural language processing more challenging. Therefore, the effective creation of predicate terminology sets is crucial.

Limitations of our method include the possibility of discrepancies between extracted diseases and the actual types of diseases, the potential for errors in cases with a small number of instances, and instances where dependency analysis may not function effectively. The proposed predicate terminology set has been verified by a single radiologist, and for broader applicability, it requires validation by radiologists from multiple facilities. Implementing machine learning for this task [10] poses the following challenge.

5   Conclusion

We proposed a method for creating a diagnostic predicate terminology set and estimating the number of disease names in CT image databases contributed by multiple facilities. This method allows for the efficient selection of disease groups suitable for machine learning within the database. Additionally, it enables an understanding of the variations between facilities.

Acknowledgement. This work was supported by JSPS KAKENHI Grant Numbers JP21K07683 and JP21K07652.

References

1. Medical image repositories. https://www.ucl.ac.uk/child-health/about-us/support-services/library/resources-z/medical-image-repositories. Accessed 25 Oct 2023
2. Krishnan, G.S., Kamath, S.S.: Ontology-driven text feature modeling for disease prediction using unstructured radiological notes. Comput. Syst. (2019). https://doi.org/10.13053/cys-23-3-3238
3. Purushotham, S., Meng, C., Che, Z., Liu, Y.: Benchmark of deep learning models on large healthcare MIMIC datasets. J. Biomed. Inform. **83**, 112–134 (2018)
4. J-MID. http://www.radiology.jp/j-mid/. Accessed 25 Oct 2023
5. Toda, N., et al.: Deep learning algorithm for fully automated detection of small (≤ 4 cm) renal cell carcinoma in contrast-enhanced computed. Invest. Radiol. **57**, 327–333 (2022)
6. SINET 6. https://www.sinet.ad.jp/en. Accessed 25 Oct 2023
7. Injury and Illness Master, Health Insurance Claims Review & Reimbursement Services, Japan. https://www.ssk.or.jp/seikyushiharai/tensuhyo/kihonmasta/kihonmasta_07.html#cms01. Accessed 25 Oct 2023
8. Kudo, T., Yamamoto, K., Matsumoto, Y.: Applying conditional random fields to Japanese morphological analysis. In: Proceedings of the 2004 Conference on Empirical Methods in Natural Language Processing (EMNLP-2004), pp. 230–237 (2004)
9. Kubo et al.: http://taku910.github.io/cabocha/. Accessed 25 Oct 2023
10. Sugimoto, K., et al.: Extracting clinical terms from radiology reports with deep learning. J. Biomed. Inform. **116**, 103729 (2021)

Enabling Dynamic Vulnerability Assessment for Multi-web Application Using Executable Directed Acyclic Graph

Jakkarin Lapmoon, Thunpisit Kosolsriwiwat, Sirinyaporn Jiraporn, and Somchart Fugkeaw[✉]

Sirindhorn International Institude of Technology, Thammasat University, Pathum Thani, Thailand

{6322772680,6322772854,6322772334}@g.siit.tu.ac.th, somchart@siit.tu.ac.th

Abstract. Implementing vulnerability assessment (VA) and penetration testing is one of the crucial methods to evaluate the security of web applications. The VA and Pentest results can be used to assess the present vulnerability of the system to help improve effective and up-to-date controls. However, using such static results for vulnerability analysis may not reflect the real situation of the organization's security policy. In addition, the integration of VA results and adaptive vulnerability calculation of vulnerabilities found in web applications are not provided by existing VA tools. Essentially, a dynamic, and interactive vulnerability assessment system for web applications is crucial due to the dynamic nature of modern web applications causing the possible new threat landscapes. In this paper, we introduce DyVAM (Dynamic Vulnerability Assessment for Multi-Web Applications), a system that integrates OWASP ZAP results and Directed Acyclic Graphs (DAGs) for supporting interactive vulnerability analysis. Our proposed system also incorporates organizational policies into the vulnerability calculation to enable a more actionable and pertinent vulnerability assessment result. Finally, we conducted the experiments to substantiate the efficiency and practicality of our proposed system. The results demonstrate that DyVAM, when implemented with our proposed multi-threading approach, significantly improves processing speed compared to traditional processing.

1 Introduction

Web applications are indispensable tools for modern organizations, enabling them to interact with customers, manage data, and offer a variety of online services. While these applications bring unparalleled convenience, they are also prime targets for cyber threats. The dynamic nature of web application ecosystems and the ever-evolving threat landscape require effective defensive approaches. One of the most palpable techniques used to prevent the attack is to continuously perform VA and Pentest and used their results to assess the security vulnerability assessment. Then, the organizations need to consider choosing the appropriate controls for mitigating the vulnerabilities. Typically, OWASP ZAP [1] is a tool that is commonly used to scan web applications to discover a list of vulnerabilities with its corresponding CVSS value. However, applying the CVSS [2] score alone obtained from the VA tool may not pertinently fit the security requirements of the organizations. This is because some organizations may have their own policies where the emphasis on severity of vulnerabilities is subject to the critical functions of the applications deployed.

Therefore, evaluating vulnerability and ensuring alignment with organizational security policies often becomes a challenging task. This requires going beyond the initial VA tool results and calculating vulnerability based on predefined policies. Moreover, vulnerability calculations are not normally integrated with existing VA tools. Generally, the vulnerability score obtained from the VA tool will be invoked into the vulnerability assessment [3]. This causes the lack of adaptive vulnerability evaluation if the vulnerability needs to be monitored through the operation time. In addition, the cost for importing vulnerability scores will be cumbersome if there are many applications engaging in the continuous assessment. Furthermore, the provision of the interactive workflow supporting dynamic calculation of the vulnerability and providing solutions to address discovered vulnerabilities become essential for the efficient handling of web application vulnerabilities. To the best of our knowledge, these features have not been provided by any VA tools.

To this end, we introduced DyVAM (Dynamic Vulnerability Assessment for Multi-Web Applications) system to fulfill the aforementioned shortfalls. DyVAM leverages the vulnerability scanning function of OWASP ZAP [1] and Directed Acyclic Graphs [4] (DAGs) for constructing the automated and interactive workflow allowing insight review of web application vulnerabilities where the standard CVSS score of the web application can be jointly calculated using the score in line with the organization policy. In addition, DyVAM can generate the executable DAG model providing interactive and adaptive vulnerability score analysis. The proposed DAG presents the association between discovered vulnerability, web application function, vulnerability score, and mitigation solution. To enable efficient processing of a huge volume of VA scanned result, we applied multi-thread programming to handle CVSS mapping and scanned result classification represented in DAG.

2   Related Work

This section discusses related works entailing web application security and vulnerability assessment.

In [2], Houmb et al. introduced a CVSS Risk Level Estimation Model, which gauges the security risk level by integrating vulnerability data, incorporating both frequency and impact estimates derived from the Common Vulnerability Scoring System (CVSS). This model applied a Bayesian Belief Network (BBN) technique to calculate CVSS for the utilization of available risk-related data. To illustrate the model's functionality, it was applied to a safety- and mission-critical system in the Measurement and Logging While Drilling (M/LWD) system.

In [3], Sethapanee et al. introduced a vulnerability and risk assessment system called AutoRAT that provides an automatic risk assessment from Nmap scanning results. The proposed risk assessment model incorporates risk scoring using the CVSS which is used as a part of final risk calculation.

In [5], the authors introduced the assessment model of the security weaknesses of the sharia-compliant crowdfunding platform, employing the Open Web Application Security Project (OWASP) methodology through the utilization of the Zed Attack Proxy (ZAP) tool. The authors proposed five stages method in analyzing the vulnerability to the security of the sharia crowdfunding website using the OWASP-ZAP framework. The results of assessment determine the level of vulnerability in the Sharia Crowdfunding Website.

In [6], the authors proposed the vulnerability assessment method using the OWASP approach with the ZAP Report. The results indicate a combination of low and moderate levels of vulnerability risk. These vulnerabilities manifest through web alerts categorized as medium, low, and informational.

In [7], the authors employed Bayesian network modeling to represent network attacks and introduced the Bayes formula for determining the likelihood of new node probabilities and attack pathways. This suggested model enables the quantification of network vulnerability assessment, with the graphical model being constructed using data from the Common Vulnerabilities and Exposures (CVE) repository.

In [8], Gu et al. proposed the approach for detecting SQL Injection Attacks (SQLIAs) and conducting vulnerability analysis within network traffic, called DIAVA. The approach is designed to issue timely alerts to users. By scrutinizing bidirectional network traffic associated with SQL operations and employing our newly devised multilevel regular expression model, DIAVA can precisely detect successful SQLIAs from the pool of potential suspects. Furthermore, DIAVA rapidly assesses the severity of these SQLIAs and evaluates the vulnerabilities of the exposed data using its GPU-based dictionary attack analysis engine.

In [9], K. Rahkema and D. Pfahl proposed a system namely SwiftDependency-Checker to enhance the accessibility of data from public vulnerability databases for Swift developers. It achieves this goal by examining dependencies declared via CocoaPods, Carthage, and Swift Package Manager (Swift PM), querying the National Vulnerability Database (NVD), and then providing warnings within Xcode regarding the usage of vulnerable library versions.

In [10], Humayun et al. conducted a systematic mapping study, analyzing 78 studies to identify prevalent cybersecurity vulnerabilities. Denial-of-service and malware were the most discussed vulnerabilities, detected using intrusion detection systems, machine learning, and algorithms. The study highlighted prominent cybersecurity research venues (IEEE, ScienceDirect) and identified the USA and India as key contributors. The absence of standardized mitigation techniques was emphasized, advocating for increased cyber awareness and training.

In [11], the authors introduced the VulDB platform, a comprehensive vulnerability management and threat intelligence platform. This platform offers detailed information about vulnerabilities, including their CVSS scores, exploit prices, and the latest exploits. Emphasizing the significance of accurate and up-to-date vulnerability data, the document highlights the platform's additional feature of informative videos discussing vulnerabilities and their exploitation.

In [12], S. Kai et al. introduced a CVSS-based vulnerability assessment method aimed at addressing practical challenges in the Common Vulnerability Scoring System. Recognizing issues with scoring variability and subjective metric weightings, the proposed method employs a decision tree-based model to reduce vulnerability metrics and minimize manual scoring errors. Experimental results showcase a reduction in scoring errors while maintaining high accuracy, emphasizing the improved precision and utility of the proposed assessment method.

Nevertheless, all the above approaches neither support adaptive vulnerability assessment nor provide real-time monitoring and analysis of vulnerabilities for multi-web applications.

3 Our Proposed Scheme

This section provides a system overview and detailed process of our DyVAM system.

3.1 System Model Overview

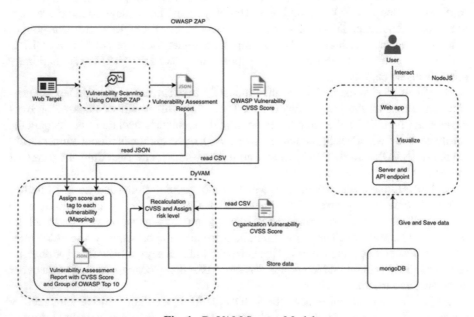

Fig. 1. DyVAM System Model

Fig. 1 presents the system model of DyVAM which includes the following components.

1. OWASP ZAP initiates the assessment by scanning target web applications, actively identifying vulnerabilities, and generating detailed JSON-based "Vulnerability Assessment Reports".
2. The Executable DAG system is responsible for extracting data from OWASP ZAP's "Vulnerability Assessment Reports". The system contains the list of Vulnerabilities and the corresponding CVSS Scores of multiple web applications. Our system classifies the vulnerabilities into ten categories based on the OWASP list for each web application. The system constructs the vulnerability analysis in a graph-based model which consists of five nodes including the web application URL, the vulnerability name classified under OWASP Top 10, the unified score calculated by the CVSS and imported organizational policy score, and the mitigation strategy. Each graph represents the vulnerability analysis plan.
3. MongoDB, an open-source database service, is dedicated to storing, managing, and accessibility. DyVAM establishes a connection to MongoDB and stores the processed vulnerability data that will be securely stored and serves as a valuable repository for future reference and easy access.
4. NodeJS (Server and API endpoint) seamlessly interacts with the MongoDB database to retrieve vulnerability data. However, the true power lies in the web application. This interface systematically presents identified vulnerabilities, offering key details such as VA program name, version, generated report timestamp, and vulnerability information.

In addition to the vulnerability identification feature, the system enriches user understanding by providing crucial evidence supporting each vulnerability, enabling precise assessments and impact analyses. Our system also acknowledges diverse security needs, allowing the customization of CVSS scores to align with specific organizational policy.

3.2 OWASP ZAP

In our Dynamic Vulnerability Assessment approach, we employ the formidable combination of OWASP ZAP (Zed Attack Proxy) version 2.13.0 and AJAX technology. This dynamic duo automates vulnerability scans with the assistance of the AJAX Spider. Utilizing the default Firefox Headless browser, our algorithm initiates a thorough assessment of web applications, particularly those built with AJAX, delving deeper than standard methods allow. The AJAX Spider enhances the scanning process by actively probing the web application's structure, enabling node-level analysis, systematic traversal, and controlled navigation to uncover potential vulnerabilities.

The dynamic vulnerability assessment process commences by selecting the specific web applications to scan and generate the vulnerability JSON report. The chosen web applications undergo in-depth scrutiny using OWASP ZAP to uncover potential security vulnerabilities.

3.3   Our Proposed Dynamic Vulnerability Assessment Based on Executable DAG System

Our proposed vulnerability assessment scheme was designed and developed to deliver interactive and actionable analysis of the scanned VA results through the executable DAG system. The system is modeled in a workflow-based orchestrated by DAGs. In essence, our proposed DyVAM system consists of two major functional modules where their details are described as follows.

Executable DAG Generation with Multi Data Mapping and Classification: Data mapping and classification module is responsible for mapping VA scanned results and CVSS score and classifying the results into OWASP Top10 category for each web application. During this phase, the system retrieves web vulnerability information from JSON files generated by OWASP ZAP. Given that our system involves analyzing vulnerability assessment results for numerous web applications, the volume of scanned results can be substantial. Consequently, the system must handle a considerable amount of data for the purposes of mapping and classification.

Therefore, we applied DAG and multithread programming, to parallelize the tasks by generating threads to execute the data mapping and classification in each order of such tasks for multi-web applications. We also designed the programming for achieving optimal number of threads for supporting tasks to avoid too many threads as they could hurt the performance. Algorithm 1 presents how this system module is executed.

Algorithm 1 Data Mapping and Classification

```
Input Normalized Vulnerability Assessment JSON report
(vaJsonReport),    OWASP    Vulnerability    CVSS    Score
(owarpCvssScore),   Vulnerability   List   OWASP   Top   10
(owarpTop10),  Organization  Vulnerability  Weight  Score
(orgWeightScore)
Result = new Map<SiteName, List<mappingData>>
Def mappingAndClassification(vaRecord, owarpCvssScore,
owarpTop10, orgWeightScore):
  vulKey = vaRecord.key
  If vulKey in owarpTop10:
    siteName = vaRecord.site
    cvssScore = owarpCvssScore.get(vulKey)
    weightScore = orgWeightScore.get(vulKey)
    remediation = vaRecord.remediation
    If Result.get(siteName) exits:
      List = Result.get(siteName)
      List.add({vulKey, cvssScore, weightScore,
remediation})
    Else:
      List = new List<mappingData>
      List.add({vulKey, cvssScore, weightScore,
remediation})
      Result.add(siteName, List)
    End If
  End If
End Def
N = Total Vulnerabilities
TN = Thread number
numPerThread = N/ TN
For tn in TN:
 tn.execute(mappingAndClassification(vaJsonReport.getRe
 cord(tn.num:  tn.num + numPerThread), owarpCvssScore,
 owarpTop10, orgWeightScore)
End for
Output  Result  Vulnerability  Assessment  Report  With
Mapped Classification and Score (vaReportScoreMapped)
```

The "Algorithm 1: Data Mapping and Classification" is designed to enrich a normalized vulnerability assessment report with additional data. It takes as inputs the Vulnerability Assessment Report in JSON format (vaJsonReport), OWASP Vulnerability CVSS Score (owarpCvssScore), a list of vulnerabilities from OWASP Top 10 (owarp-Top10), and Organization Vulnerability Weight Score (orgWeightScore). The algorithm works by defining mappingAndClassification function, which processes each vulnerability record. It checks if the vulnerability key is listed in the OWASP Top 10. If so, it maps the vulnerability to its corresponding CVSS score and organizational weight score and retrieves any remediation information. This data is then aggregated under the corresponding site name. The algorithm executes this function in parallel across multiple threads to efficiently process the data. Each thread handles a portion of the total vulnerabilities, distributing the workload evenly. The final output of the algorithm is a comprehensive vulnerability assessment report that includes a detailed mapping and classification of each vulnerability, augmented with CVSS scores, organizational weight scores, and remediation strategies. Algorithm 2 presents how the vulnerability score, and level are calculated, based on the invocation of organizations.

Algorithm 2 Calculate Vulnerability Score and Level

```
Input  Vulnerability  Assessment  Report  With  Mapped
Classification and Score (vaReportScoreMapped)
Def calVulnerabilityScoreAndLevel(vaReportScoreMapped):
  For vaReportSite in vaReportScoreMapped:
    For vul in vaReportSite.vulnerabilityList:
        newCvssScore = vul.cvssScore + vul.weightScore)
/ 2
      If newCvssScore >= 9.0 AND newCvssScore <= 10:
        risk = Critical
      Else If newCvssScore >=7.0 AND newCvssScore <
9.0:
        risk = High
      Else If newCvssScore >= 4.0 AND newCvssScore <
7.0:
        risk = Medium
      Else:
        risk = Low
      End If
      vul.cvssScore = newCvssScore
      vul.risk = risk
    End For
  End For
End Def
Output  Vulnerability  Assessment  Report  With  Mapped
Classification and Score with New CVSS score and Risk
Levels for each site (vaReportScoreMappedFinal)
```

As shown in Algorithm 2, the algorithm takes an input of the Vulnerability Assessment Report with Mapped Classification and Score (vaReportScoreMapped). The core function is the calVulnerabilityScoreAndLevel function, which iterates through each site in the report. For every site, the function further iterates through its list of vulnerabilities.

Within this iteration, the algorithm calculates a new CVSS score for each vulnerability. This new score is an average of the existing CVSS score and an organizational weight score assigned to the vulnerability. The calculation is as follows: newCvssScore = (vul.cvssScore + vul.weightScore)/2. Based on this newly computed score, the algorithm then determines the risk level of each vulnerability using a set of conditional checks as defined in the algorithm.

After computing the new CVSS score and determining the risk level, these values are then updated in the vulnerability record. Algorithm 3 presents how our executable DAG is generated.

Algorithm 3 Executable DAG Generation

```
Input Vulnerability Assessment Report with Mapped
Classification and Score with New CVSS and Risk Levels
for each site (vaReportScoreMappedFinal)
Def DAG = {
   webAppId: string
   vulnerabilities [vulnerability: {
      vulName: string
      cvssScore: number
      remediation: string
   }]
}
For vaReportSite in vaReportScoreMappedFinal:
   Result = new DAG
   Result.webAppId = vaReportSite
   For vul in vaReportSite.vulnerabilityList:
      vulnerability.vulName = vul.vulKey
      vulnerability.cvssScore = vul.cvssScore
      vulnerability.remediation = vul.remediation
      Result.vulnerabilities.add(vulnerability)
   End For
   Results.add(Result)
End For
Output Results: Efficiently processed data using DAG and
Multi-Processing
```

As illustrated in Algorithm 3, the Executable DAG focuses on structuring vulnerability assessment data into an executable Directed Acyclic Graph (DAG) format, particularly beneficial for parallel processing. The input for this algorithm is a vulnerability assessment report, already enhanced with mapped classifications, new CVSS scores, and risk levels for each site. The algorithm defines a DAG structure for each web application, encapsulating essential vulnerability information like names, CVSS scores, and remediation strategies.

The end product is a list of executable DAGs, one for each web application, making the data ready for efficient, multi-threaded processing which improves the overall efficiency of the vulnerability assessment process. Figure 2 and Fig. 3 below show the general model of DAG and its detailed activity as depicted in Algorithm 3.

Fig. 2. DAG High Level Model outlining the workflow of a vulnerability assessment process to the assignment of remediation tasks.

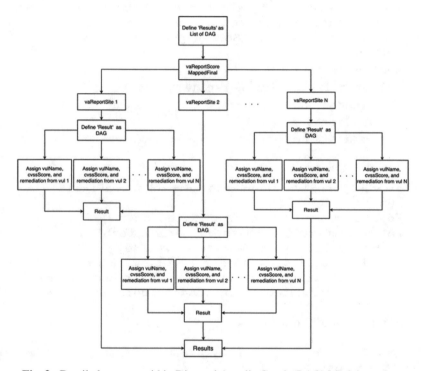

Fig. 3. Detailed process within Directed Acyclic Graph (DAG) Model structure

4   Evaluation

In this section, we present the evaluation of the DyVAM system, focusing on its performance in scanning and identifying vulnerabilities across a variety of web applications. Our evaluation is based on comprehensive scans of ten distinct websites, each chosen for their unique characteristics and security postures. Our DyVAM system successfully scanned these websites and generated a comprehensive dataset comprising 161 identified vulnerabilities. These findings are thoroughly documented in a JSON report, which contains a total of 395,371 lines.

4.1   Functionality Comparison

Table 1 presents the functionality comparison of DyVAM and related works including schemes [3, 7, 8].

Table 1. Comparison of DyVAM and related works

Features/Tools	[3]	[7]	[8]	DyVAM
VA Analysis for Multi-Web app	No	No	No	Yes
CVSS Scoring	Yes	No	No	Yes
Classification of Vulnerability	Yes	Yes	Limited	Yes
Executable VA Analytical Results	Yes	Yes	No	Yes
Provision of Mitigation Solution	Yes	No	No	Yes

The comparison table clearly demonstrates that DyVAM encompasses all the features essential for a dynamic and thorough vulnerability assessment, setting it apart from the works referenced in [3, 7, 8]. DyVAM is the only work that facilitates vulnerability analysis for multiple web applications. Moreover, it stands out by offering CVSS scoring, capabilities for classifying vulnerabilities, and providing executable vulnerability assessment analytical results, features that are lacking in [7, 8]. Additionally, DyVAM is unique in its provision of mitigation solutions.

4.2 Performance Evaluation

To substantiate the performance of our proposed system, we conducted experiments to measure the performance of the data mapping and classification algorithm by comparing the processing time used by our proposed multithread processing algorithm and single processing. The experiments were conducted on a high-performance PC equipped with an AMD Ryzen 9 5900HX processor with Radeon Graphics, running at 3.30 GHz. The system boasts 16.0 GB of installed RAM, with 15.4 GB available for optimal performance.

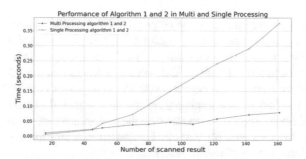

Fig. 4. Processing time between multi-processing and single-processing modes.

As shown in Fig. 4, as the number of scanned applications increases, a divergence emerges, most notably beyond the 50-application mark, where the multi-processing line remains relatively flat, and the single-processing line begins a steep ascent. The graph clearly illustrates the superiority of the multiprocessing approach within the DyVAM system at point 161 on the x-axis, as evidenced by its significantly lower execution times. In addition, we conducted the experiment to measure the processing time for creating DAG by varying the number of websites where the OWAS Zap was used to scan the vulnerability (Fig. 5).

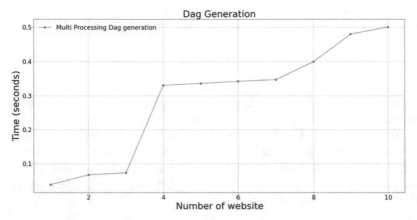

Fig. 5. DAG Execution Time

The graph shows that when the number of websites increases, the time taken for DAG generation escalates gradually, showcasing the scalability of the DyVAM system. At 4 number of websites, it took around 0.33 s, where the 10 number of websites it took just 0.5 s. The graph delineates a consistent increase in processing time, which is indicative of the system's capability to handle a growing workload efficiently.

5   Conclusion

In this paper, we proposed the DyVAM (Dynamic Vulnerability Assessment for Multi-Web Applications) system as an automated OWASP Top10 vulnerability analysis and classification. Our proposed parallel vulnerability data mapping and classification is efficiently provable for processing substantial volume of VA scanned results from multiple web applications. In addition, we introduced Directed Acyclic Graphs (DAGs) to graphically represent the flow of web applications and their detected vulnerabilities clusters. Finally, we conducted experiments to measure the performance of our proposed system. The experimental results confirm that our DyVAM is efficient thanks to the proposed design and implementation of multithread programming.

For future works, we will perform larger scale of the experiments by engaging more transactions and implementing our system in real cloud environment. In addition, we will investigate the technique to enable DyVAM to cooperatively work with other vulnerability analysis and detection tools such as SIEM, threat intelligence applications.

References

1. Makino, Y., Klyuev, V.: Evaluation of web vulnerability scanners. In: 2015 IEEE 8th International Conference on Intelligent Data Acquisition and Advanced Computing Systems: Technology and Applications (IDAACS), vol. 1, pp. 399–402. IEEE (2015)
2. Houmb, S.H., Franqueira, V.N., Engum, E.A.: Quantifying security risk level from cvss estimates of frequency and impact. J. Syst. Softw. **83**(9), 1622–1634 (2010)
3. Sethapanee, A., Nimitrchai, T., Fugkeaw, S.: Autorat: automated risk assessment tool for network mapper scanning. In: Meesad, P., Sodsee, S., Jitsakul, W., Tangwannawit, S. (eds.) IC2IT 2022. LNNS, vol. 453, pp. 99–110. Springer, Cham (2022). https://doi.org/10.1007/978-3-030-99948-3_10
4. Prostov, A., Amfiteatrova, S.S., Butakova, N.G.: Construction and security analysis of private directed acyclic graph based systems for internet of things. In: 2021 IEEE Conference of Russian Young Researchers in Electrical and Electronic Engineering (ElConRus), St. Petersburg, Moscow, Russia, pp. 2394–2398 (2021)
5. Lathifah, F.B.A., Rosidah, A., et al.: Security vulnerability analysis of the sharia crowdfunding website using owasp-zap. In: 2022 10th International Conference on Cyber and IT Service Management (CITSM), pp. 1–5. IEEE (2022)
6. Cahyani, D.D., Dewi, L.P.W.P., Suryadi, K.D.R., Listartha, I.M.E.: Analisis kerentanan website smp negeri 3 semarapura menggunakan metode pengujian rate limiting dan owasp. INSERT: Inf. Syst. Emerg. Technol. J. **2**(2), 106–112 (2021)
7. Huang, L., Chen, X., Lai, X.: Computer network vulnerability assessment and safety evaluation application based on bayesian theory. Int. J. Secur. Appl. **10**(12), 359–368 (2016)
8. Gu, H., et al.: Diava: a traffic-based framework for detection of SQL injection attacks and vulnerability analysis of leaked data. IEEE Trans. Reliab. **69**(1), 188–202 (2019)
9. Rahkema, K., Pfahl, D.: Swiftdependencychecker: detecting vulnerable dependencies declared through cocoapods, carthage and swift pm. In: Proceedings of the 9th IEEE/ACM International Conference on Mobile Software Engineering and Systems, pp. 107–111 (2022)
10. Humayun, M., Niazi, M., Jhanjhi, N., Alshayeb, M., Mahmood, S.: Cyber security threats and vulnerabilities: a systematic mapping study. Arab. J. Sci. Eng. **45**, 3171–3189 (2020)
11. [Online]. Available: https://vuldb.com/
12. Kai, S., Zheng, J., Shi, F., Lu, Z.: A CVSS-based vulnerability assessment method for reducing scoring error. In: 2021 2nd International Conference on Electronics, Communications and Information Technology (2021)

Discovery of RESTful Web Services Based on the OpenAPI 3.0 Standard with Semantic Annotations

Alberto Tuti Soki and Frank Siqueira(✉)

Federal University of Santa Catarina, Florianópolis, Brazil
alberto.soki@posgrad.ufsc.br, frank.siqueira@ufsc.br

Abstract. Due to the large quantity and diversity of computational services currently available, there is a need for mechanisms to automatically discover services taking into account the requirements of potential customers. This work is focused on the discovery of RESTful web services described using the OpenAPI/Swagger language extended through semantic annotations. Service discovery is based on functional and non-functional requirements, which define the profile of the RESTful service sought by the customer. In order to enable semantic discovery, a profile ontology for RESTful services was created. Service descriptions were stored in a database, and mappings between ontology elements and database elements were defined. As a result, service discovery can be performed by executing SPARQL queries on the database. For experimental validation, we added a set of REST descriptions to the database and executed SPARQL queries and similarity algorithms to locate the desired services.

Keywords: SOA · Web Services · Semantic Web · Service Discovery · REST · OpenAPI · Swagger

1 Introduction

The web has become the most widely used medium for consuming services such as banking and payments, geolocation, e-commerce, and data retrieval. For this purpose, Application Programming Interfaces (APIs) exist not only for consumption but also for the integration of distributed systems [10]. Distributed systems enable interoperability among themselves using standardized protocols and languages without the need for direct human intervention. Communication protocols (e.g., SOAP and HTTP), service description languages (WSDL and OpenAPI/Swagger), publication, and location specifications are fundamental elements of a Service-Oriented Architecture (SOA). Due to their interoperability capabilities, APIs based on web services have become the central focus for various technological and industrial actors in different fields [17].

According to [18], APIs have the ability to specify how consumers (both humans and machines) can access services and data, especially when they are built based on SOA. Services described using a syntactic language – such as WSDL and OpenAPI/Swagger – are, in general, understood only by human consumers and not by machines. Systems that

© The Author(s), under exclusive license to Springer Nature Switzerland AG 2024
L. Barolli (Ed.): AINA 2024, LNDECT 200, pp. 22–34, 2024.
https://doi.org/10.1007/978-3-031-57853-3_3

only allow human comprehension of service descriptions tend to exhibit low performance in the discovery of such services. This is due to lack of understanding of the semantic meaning of service descriptions and user queries [9].

To enhance understanding in the service discovery process, especially by machines/systems, it is strongly advisable to include semantic descriptions or annotations [9, 19]. This can be done in both the description and discovery phases of the discovery process. User queries can specify functional and non-functional requirements (e.g., response time). In the literature, most of the approaches for service discovery and composition are focused on services based on the SOAP protocol and on the WSDL description language [19]. In this paper, our goal is the discovery of RESTful services described by the OpenAPI/Swagger language, based on functional and non-functional requirements, as well as semantic annotations.

In the remainder of this paper, after presenting an overview of the related work (Sect. 2), we introduce a discovery system for REST services (Sect. 3), followed by an experimental evaluation of the system (Sect. 4) and by a summary of the contributions of this work (Sect. 5).

2 Related Work

This section briefly surveys the existing literature on REST (REpresentational State Transfer) web service discovery, drawing comparisons with our approach.

The Bidirectional Sentence-Word Topic Model (bi-SWTM) [9] is a Bayesian model that incorporates semantics into user descriptions and queries for efficient service-query similarity calculations. In our proposal, we have adopted the strategy of reducing computational costs by selectively applying semantic annotation only to input and output parameters in REST service descriptions.

The work presented in [19] introduces a model with semantic annotations for describing web services, requests, and an algorithm for service discovery and composition. It uses OWL-S to annotate WSDL+ service functionalities and OWL to semantically describe web service requests. However, these technologies are not applicable to REST services. In our proposal, we present a profile ontology that enhances service discovery, including user-specific information and service details.

The authors of [11] introduce Transformation-Annotation-Discovery (TAD) for discovering and composing RESTful services. This method transforms OpenAPI/ Swagger descriptions into a Neo4J graph, automatically annotating semantic concepts using Latent Dirichlet Allocation (LDA) and expanding terms with WordNet. In our proposal, semantic annotations are included directly in the OpenAPI/Swagger descriptions, aiming to simplify user interaction and reduce computational costs.

A model to semantically annotate entities in the OpenAPI/Swagger specification, associating them with entities in an ontology, is described in [12]. It is noteworthy that, in our proposed approach, the inclusion of the "x-otherTerms" and "x-kindOf" extensions stands out, introducing additional semantic concepts for each attribute of "parameter" and "response" objects. This semantic enrichment is a distinctive feature contributing to a more expressive and contextualized representation of the specification, setting our proposal apart from the approach presented in [12].

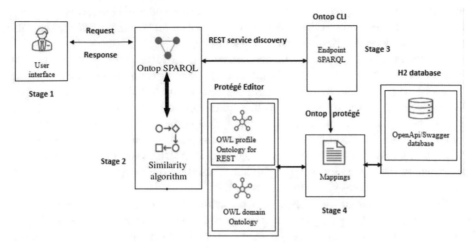

Fig. 1. System for REST web services discovery.

3 Discovery System for REST Services

This chapter discusses a system developed to simplify the identification and utilization of RESTful web services. The system has been designed to meet the growing demand for REST services in an ever-evolving technological environment. The primary purpose of this system is to enable users to find and access relevant RESTful web services based on their specific needs. This is particularly important in a scenario where many REST web services are available, and users need to locate those that meet their requirements.

3.1 System Overview

The system functions in several stages, as shown in Fig. 1, beginning with user interaction, where details such as service name, parameters, and response time are provided. Internal processes then search and retrieve relevant information about services matching these criteria. The system maps the service profile ontology to data in an OpenAPI/Swagger specification, ensuring alignment between user needs and service capabilities. This is done without requiring in-depth knowledge of the service structure, making REST web service discovery accessible to a wider audience.

The stages of the system can be described as follows:

- Stage 1: This stage involves a user interface with four fields: in the first one, the user must enter the desired REST Web service name; in the second, the input parameter(s); in the third, the output parameter(s); and in the fourth field, the maximum desired response time. A desired input is associated with a desired output; if the user does not provide an input, the system should return the contact and server information for each found Web service with the desired output. Server information is related to the base URL where the service is hosted, possible endpoints, and HTTP methods (such as POST, GET, DELETE, among others). The user-provided information is processed using Ontop SPARQL.

- Stage 2: is responsible for searching for the desired Web service based on the user-provided profile (desired service name, input, and output). The search is performed through a SPARQL endpoint, where Ontop SPARQL queries are executed. The query can verify if a Web service with the profile provided by the user exists in the OpenAPI/Swagger database. At this stage, the algorithm for sorting the results returned by the query and the similarity calculation algorithm are also incorporated.
- Stage 3: provides a SPARQL endpoint to Stage 2. This endpoint is created using the Ontop CLI tool, provided by the Protégé ontology editor. To generate an endpoint, this tool requires mapping between a specific ontology and a database.
- Stage 4: provides Stage 3 with the mapping between the OWL REST profile ontology and the OpenAPI/Swagger database, as well as the mapping between the OWL domain ontology and the domain database. All mappings were created using the Ontop Protégé tool. Section 3.2 describes the mapping between the profile ontology and the OpenAPI/Swagger database.

3.2 OWL Ontology for REST Profile

The modeling of the profile ontology involved a detailed analysis of the elements of the OpenAPI/Swagger Specification, with a focus on the "info", "servers", and "paths" objects, which provide essential information for accessing the resources of a specific REST web service. After modeling, the Protégé ontology editor was used for ontology development, as shown in the ontoGraf image in Fig. 2.

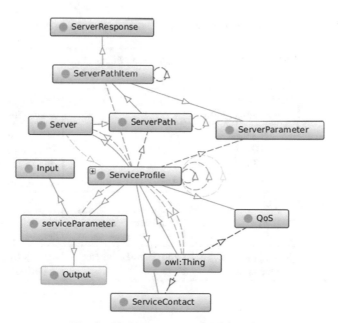

Fig. 2. OWL ontology for REST profile.

The ServiceParameter class has subclasses designated to receive information from users, which may include input and output parameters. Additionally, these subclasses also handle possible Quality of Service (QoS) parameters, represented by the QoS sub-class. An example of a QoS parameter is the desired 'response time' for each request. All other classes have been directly linked to the elements found in the OpenAPI/Swagger specification ("info", "servers", and "paths").

3.3 Mapping Between the Profile Ontology and OpenApi/Swagger

Table 1 shows the correspondence of each element (classes and data properties) contained in the REST profile ontology with their respective elements in the OpenAPI/Swagger specification. The "ServiceProfile" class has the following object properties: "hasServer" (which relates it to the "Server" class), "hasPath" (which relates it to the "Server-Path" class), and "hasContact" (which relates it to the "ServiceContact" subclass). The "Server" class has the object property "hasPath" that relates it to the "Paths" class. Finally, the "ServerPath" class is composed of the following object properties: "hasPathItem", which, in turn, has the properties "hasResponse" and "hasParameter".

To semantically enrich the ontology, some class, object, and data restrictions were defined. In this approach, the connection between the REST profile ontology and descriptions in the OpenAPI/Swagger format will be established through a relational database, such as H2 [7]. This database serves as a repository capable of storing mul-tiple descriptions of REST services as per the OpenAPI/Swagger specification, which will be described in the sequence.

Table 1. Mapping between the Profile Ontology and OpenAPI/Swagger.

OWL ontology for REST profile		OpenAPI/Swagger	
Classes	Data property	Objects	Fields/Objects
ServiceProfile	serviceName, serviceDescription	Info	title (string-required) description (string)
ServiceContact	contactName, contactUrl, contactEmail	Info	name (string) url (string) email (string)
Server	url	Server	url (string)
ServerPath	endpoint	Paths	-
ServerPathItem	method	Path Item	get, put, post, delete, options
ServerParameter	parameterName, parameterIn	Path Item/Parameter	name(string-req.) in (string)

3.4 OpenApi/Swagger Database

Relational databases can be interpreted as RDF graphs through mapping techniques and ontologies [3, 6]. This mapping process is facilitated by technologies such as Ontop, which rely on the R2RML language standard established by the W3C for mapping purposes [4, 13].

By adopting this approach, we chose to use a specific relational database, named OpenApi/Swagger database, to store REST service descriptions. The decision to use a dedicated database, rather than extracting OpenAPI/Swagger descriptions from repositories such as ProgrammableWeb, provides a controlled testing environment. This choice is particularly valuable for experimental evaluation, as it allows the insertion of real or simulated OpenAPI/Swagger descriptions.

In constructing the OpenAPI/Swagger database, a similar procedure was followed to the one used in modeling the REST profile ontology. This procedure involved using specific elements from the OpenAPI specification (such as "info", "servers", and "paths") to create elements in the profile ontology and fields in the database. Essentially, the specification elements "info-object", "contact-object", "server-object", "path-object", "path-item-object", "parameter-object", and "response-object" were transformed into tables in the OpenAPI/Swagger database.

Figure 3 illustrates the model of the OpenAPI/Swagger database, intended to store a wide variety of REST web service descriptions.

Fig. 3. OpenAPI/Swagger database.

The "security" element of the specification was not included in this approach, meaning that the services are public and do not require prior authentication. After modeling and creating the REST profile ontology and the OpenAPI/Swagger database, a mapping between them was created to enable the execution of potential queries for REST web services using Ontop SPARQL.

3.5 Mapping Using Ontop Protégé/CLI

A mapping is responsible for establishing the correspondence between data models from a data source and an RDF (Resource Description Framework) graph [15]. In this context, the mapping details the process of translating data from the source into RDF representation. Here, the data source in question is the OpenAPI/Swagger database, while the profile ontology is represented as an RDF/OWL graph. To perform this mapping, the Ontop Protégé plugin can be used, which is integrated into the Protégé ontology editor and management system [14].

To enable the use of the mapping outside of Protégé, the Ontop CLI tool can be employed. This tool takes as input the credentials for accessing the database, the ontology, and the corresponding mappings [16]. The Ontop CLI, in turn, generates a link (a SPARQL endpoint) that can be used in any other tool, allowing the application of the mapping in different contexts.

In this approach, both Ontop Protégé and Ontop CLI were applied to establish an effective connection between the REST profile ontology and the OpenAPI/Swagger database. This connection enables the efficient retrieval and manipulation of service descriptions. Figure 4 illustrates an example mapping, providing a more concrete view of the mentioned process. In this mapping, several details are specified to define how database information relates to the ontology:

- The Target part (triples template) in Fig. 4 receives the mapping, and the Source part (SQL query) receives the resource source.
- In the mapping, the part ":uni2/info_object/{i_id}" specifies the directory (uni2) where the database is located, the name of the database (info_object), and its primary key (i_id).
- The part ":ServiceProfile;" indicates that the mapping is with the ServiceProfile class of the profile ontology.

Target (Triples Template):
```
<http://example.org/voc#uni2/info_object/{i_id}> a :ServiceProfile ; :serviceName {title}^^xsd:string ;
:serviceDescription {description}^^xsd:string ; :hasContact <http://example.org/voc#uni2/contact_object/{c_id}> ;
:hasServer <http://example.org/voc#uni2/server_object/{s_id}>.
```
Source (SQL Query):
```
SELECT * FROM "uni2"."info_object" AS "info"  JOIN "uni2"."contact_object" AS "contact" ON
("contact"."info_object_id" = "info"."i_id") JOIN "uni2"."server_object" AS "server" ON
("server"."info_object_id" = "info"."i_id")
```

Fig. 4. Mapping between the ServiceProfile class and the info_object table

- The part ":serviceTitle {title}^^xsd:string;:serviceDescription {description}^^ xsd:string;" informs that the data property serviceTitle belonging to the Service-Profile class is mapped to the title field in the info_object table. The same applies to the serviceDescription data property.
- Finally, the part ":hasContact:uni2/contact_object/{c_id};:hasServer:uni2/server_object/{s_id}" indicates that the hasContact object property, which is an instance of the Servi-ceProfile class, is mapped to the contact_object table. In this case, a one-to-one

relationship is being created between the info_object table and the contact_object table. The same mapping applies to the server_object table.

- The "SELECT * FROM "uni2"."info_object" in the source (SQL Query) indicates the intention to manipulate/access any information in the info_object table and other tables related to it through the profile ontology.

3.6 Ontop SPARQL Query

SPARQL, recommended by the W3C for RDF/OWL graphs [21], is employed in this approach, specifically Ontop SPARQL in conjunction with Ontop Protégé/CLI tools for queries [16]. Ontop SPARQL is applied to query and discover service profiles in the OpenAPI/Swagger database based on user inputs/outputs and through the RESTful profile ontology. It retrieves crucial information when a service is identified, including the service name, contact details, server access details (endpoints, HTTP methods), input parameters (with URL locations), and response parameters. This information allows the user to invoke the desired service. In turn, Ontop CLI provides an Ontop SPARQL endpoint for the query to be executed in any environment and programming language. In this approach, the PHP/Laravel programming language/environment was used.

Figure 5 presents the Ontop SPARQL query. At line 2, the variable "sparqlQuery" stores the query result, while "prefix" at line 3 represents the profile ontology. Between lines 5 and 7, the desired data is specified, and from line 8, the source of this data is defined. At line 29, a filter is applied to instruct the query to retrieve services containing the term provided by the user, represented by the variable "userInputServiceName". Next, a similarity algorithm is employed to find services that meet user needs. This algorithm semantically compares user inputs with OpenAPI/Swagger descriptions (related to the info, parameter, and response objects), analyzing each character of the terms/strings. This provides a more detailed approach compared to "FILTER regex", which cannot perform this semantic analysis inherent to the similarity algorithm used.

3.7 Function for Similarity Calculation

The algorithm depicted in Fig. 6 is the Levenshtein distance algorithm, widely employed to assess semantic similarity between two strings based on their character sequences [2, 8].

In this approach, the algorithm is invoked after the execution of the SPARQL query presented in Fig. 5, which retrieves all descriptions stored in the OpenAPI/Swagger database based on the service intended by the user. In line 1, the variable "userData" receives data provided by the user, such as the desired service and intended inputs/outputs. Simultaneously, the variable "openApiDescription" receives titles and input/output parameters of services obtained through the Ontop SPARQL query. In line 6, the similarity between input parameters is calculated, resulting in a degree of similarity ranging from 0 to 1, where values closer to 1 indicate higher semantic similarity between terms. This function is invoked in different instances. First, to calculate the similarity between the user-desired service (title) and the services (titles) contained in the OpenAPI/Swagger database. Next, the process is repeated for the user-desired input

```
1 function sparqlQuery(userInputServiceName):
2  sparqlQuery = '
3   PREFIX : <http://example.org/serviceProfile#>
4    SELECT
5     ?title ?contact_name ?contact_email ?contact_url ?rKindOf
6     ?base_path ?end_point ?method ?parameter_name ?pKindOf
7     ?parameter_location ?pOtherTerms ?properties ?rOtherTerms
8     WHERE {
9      ?x a :ServiceProfile ; :serviceTitle  ?title;
10      :hasContact ?contact; :hasServer ?server.
11      ?contact a :ServiceContact; :contactName ?contact_name;
12      :contactEmail ?contact_email; :contactUrl ?contact_url.
13      ?server a :Server; :url ?base_path;
14      :hasServerPath ?serverPath.
15      ?serverPath a :ServerPath; :endPoints ?end_point;
16      :hasServerPathItem ?path_item.
17      ?path_item a :ServerPathItem; :method ?method;
18      :hasServerParameter ?parameter.
19      ?parameter a :ServerParameter; :requestParameterName ?parameter_name;
20      :parameterIn ?parameter_location; :hasAtherTermsParameter
21      ?parameterTerms; :hasResponse ?response.
22      ?parameterTerms a :OtherTerms; :xOtherTerm ?pOtherTerms; :xKindOf
23      ?pKindOf.
24      ?response a :ServerResponse; :hasSchema ?schema.
25      ?schema a :Schema; :properties ?properties; :hasAtherTermsResponse
26      ?responseTerms.
27      ?responseTerms a :OtherTerms; :xOtherTerm ?rOtherTerms; :xKindOf
28      ?rKindOf.
29      FILTER regex(?title, "'.userInputServiceName.'", "i").
30        }'
31     return sparqlQuery
```

Fig. 5. Ontop SPARQL Query for REST services discovery.

```
1 function calculationOfSimilarityDegree(userData,
   openApiDescription):
2    userData = Str(userData)
3    openApiDescription = Str(openApiDescription)
4    userData,openApiDescription = splitWordsByUppercase(userData,
     openApiDescription)
5    levenDistance = levenshtein(userData, openApiDescription)
6    similarity = 1 - (levenDistance / max(len(ucwords(userData)),
     len(ucwords(openApiDescription))))
7    return similarity
```

Fig. 6. Function for calculating similarity between terms (strings).

and the parameters described in each service. Finally, the same approach is applied for the user-desired output.

This function can be called repeatedly due to the "x-Other Terms" extension added to the OpenAPI/Swagger descriptions. The similarity threshold defined for the first instance is greater than or equal to 0.30 (30%), while for the second and third instances, it is greater than or equal to 0.60 (60%) for input/output parameters. Services that meet these criteria are considered as candidate services. For example, the 30% threshold was defined through the following test: if there is a service "Parcel Tracker" in the database, and the user inputs "Tracker Parcel", the similarity level may result in less than 50%. In the conducted test, a result of 30% was obtained.

3.8 Semantic Annotation

Approaches such as [1, 5, 20] apply extensions to objects or attributes in the OpenAPI specification to add additional semantics to these elements. The "x-refersTo" extension is used to link properties to entities in a semantic model, such as an ontology. In turn, "x-kinOf" defines a specialization between a property and a semantic model, such as a class/subclass in an ontology.

In our approach, the "x-kindOf" extension has been applied to the properties of "parameter" and "response" objects. In addition to the "x-kinOf" extension, a new extension called "x-otherTerms" was created. This extension adds to the objects a set of terms equivalent to the main property (e.g., "name" or "properties" in Table 1). The main properties and their corresponding terms included in the "x-otherTerms" extension point to a semantic concept through the "x-kinOf" extension.

Let us assume that a specific user is looking for a service to track maritime shipments and wants a service that has "Seal Number" as the input parameter and provides the estimated arrival date (ETA) of the ship as the output parameter. Table 2 illustrates the OpenAPI/Swagger description with the inclusion of extensions for semantic annotation, "x-kindOf" and "x-otherTerms", aiming to semantically enrich the description of these objects and meet user requirements.

Table 2. Semantic Annotation Extensions for Parameter and Response Objects.

Parameter Object	Schema for Response Object
{ "parameters":	{ "components": {
{	"schemas": {
"name": "containerId",	"Response": {
"x-otherTerms": {	"type": "object",
"boxNumber": "Box Number",	"properties": {
"isoNumber": "ISO Number",	"id": {
...	"type": "integer",
}	"description": "The user ID.",
"x-kindOf":	**"x-otherTerms"**: {
"https://www.wikidata.org/wiki/Q1959563"	"additionalProperty": ""
"in": "path",	},
"description": "Container ID",	**"x-kindOf"**:
"required": true,	"https://www.wikidata.org/wiki/Q205892"
"schema": {	},
type: "string" } }}	...
	} }},

The user entered "Seal Number" as input, but the input parameter for this service is "ContainerID." When the similarity function is triggered to check the degree of semantic similarity between the two terms, it will result in a low score, and the service will be discarded. However, before discarding it, the function checks the terms in "x-otherTerms" and finds that it contains a semantically similar or identical term to what the user entered, which is precisely "Seal Number". Furthermore, the semantic concept is represented in a domain ontology or knowledge base, indicated by the "x-kindOf" extension. As a result, the service is no longer discarded. Then, the same procedure is carried out for the "Response object". If the service does not provide the desired output based on the input, then it is discarded.

4   Experimental Results

Some service descriptions were added to the OpenAPI/Swagger database, aimed at testing two specific cargo/package tracking services: the first being the "1- Tracking Management System" and the second the "2- Inventory Management Tracker." The descriptions related to the parameter and response objects for these two services are in Table 3. The semantic concepts related to the parameters may point to a domain ontology or directly to a knowledge base such as "Wikidata" and "DBpedia".

Table 3. Descriptions of the parameter and response objects.

Service	Parameter Object			Response/Schema Objeto		
	name	xOtherTerm	xKindOf	propertie	xOtherTerm	xKindOf
1	containerId	container number, box number, ...	Wikidata	etd	expected departure date	Wikidata
2	containerNumber	container id, booking number, ...	Wikidata	etd	planned departure date	Wikidata

In the first test, the system was instructed to search for a "Tracking Management System" service with the input parameter "container id" and the output parameter "etd". The obtained results are shown in Table 4. The specified degree of similarity between the desired service title by the user and those present in the OpenAPI/ Swagger database should be equal to or greater than 0.30 (30%). Similarly, for the input and output parameters, the degree of semantic similarity between the terms should be equal to or greater than 0.60 (60%) on a scale ranging from 0 to 1.

The Ontop SPARQL query (Fig. 5, line 29) retrieved services 1 and 2, containing terms similar to the service entered by the user (Tracking Management System) and are recorded in the OpenAPI/Swagger database. The input parameter (containerNumber) of service 2 differs from the one specified by the user (container id), and the degree of semantic similarity between these two parameters was only 46% (Table 4), which is below the established threshold (60%). Through the "xOtherTerm" of service 2, the similarity degree calculation function (Fig. 6) checks for other terms semantically similar

Table 4. Results obtained in the first test.

User data	OpenApi/Swagger database	Similarity degree	xOtherTerms	Similarity degree
service: Tracking Management System	service (1)	1.00 (100%)		
	service (2)	0.46 (46%)		
input: container id	containerId (1)	1.00 (100%)		
	containerNumber (2)	0.46 (46%)	container id (2)	1.00 (100%)
Output: etd	etd (1)	1.00 (100%)		
	etd (2)	1.00 (100%)		

to the one specified by the user. In this case, the term "container id" was found with 100% similarity (Table 4), but any other term with a degree equal or greater than 60% would also be accepted.

5 Conclusions

The initial experiments confirm that the discovery of REST web services using a REST profile ontology is highly facilitating and flexible. The application of post-SPARQL query similarity algorithms has proven valuable in locating specific services. Significantly, relational web databases, such as H2, have been crucial, acting as repositories for REST web service descriptions.

The subsequent steps should encompass all possible scenarios related to input and output parameters, along with incorporating non-functional requirements, such as the desired response time, during the service search. Simultaneously, it is essential to assess the scalability of the system, considering the inclusion of more web descriptions in the OpenAPI/Swagger database, along with additional criteria, such as computational efficiency and usability.

Acknowledgement. This study was financed in part by CAPES (*Coordenação de Aperfeiçoamento de Pessoal de Nível Superior*) - Brazil, Finance Code 001.

References

1. Apostolakis, I., Mainas, N., Petrakis, G.M.: Simple querying service for OpenAPI descriptions with semantic extensions. Inf. Syst. 102241 (2023)
2. Behara, K.N., Bhaskar, A., Chung, E.: A novel approach for the structural comparison of origin-destination matrices: Levenshtein distance. Transp. Res. Part C: Emerg. Technol. **111**, 513–530 (2020)
3. Bereta, K., Xiao, G., Koubarakis, M.: Ontop-spatial: ontop of geospatial databases. J. Web Semant. **58**, 100514 (2019)

4. Calvanese, D., et al.: Ontop: answering SPARQL queries over relational databases. Semant. Web **8**(3), 471–487 (2017)
5. Grünewald, E., Wille, P., Pallas, F., Borges, M.C., Ulbricht, M.R.: TIRA: an OpenAPI extension and toolbox for GDPR transparency in RESTful architectures. In: 2021 IEEE European Symposium on Security and Privacy Workshops, pp. 312–319 (2021)
6. Galgonek, J., Vondrášek, J.: A comparison of approaches to accessing existing biological and chemical relational databases via SPARQL. J. Cheminform. **15**(1), 1–14 (2023)
7. H2 Database: Main Page (2023). http://www.h2database.com/html/main.html
8. Jalal, S., Yadav, D.K., Negi, C.S.: Web service discovery with incorporation of web services clustering. Int. J. Comput. Appl. **45**(1), 51–62 (2023)
9. Li, S., et al.: Bi-directional Bayesian probabilistic model based hybrid grained semantic matchmaking for Web service discovery. World Wide Web **25**(2), 445–470 (2022)
10. Martin-Lopez, A.: Automated analysis of inter-parameter dependencies in web APIs. In: Proceedings of ACM/IEEE 42nd International Conference on Software Engineering: Companion Proceedings, pp. 40–142 (2020)
11. Ma, S.-P., et al.: Real-world RESTful service composition: a transformation-annotation - discovery approach. In: 2017 IEEE 10th Conference on Service-Oriented Computing and Applications (SOCA). IEEE (2017)
12. Mainas, N., Petrakis, E.G.M., Sotiriadis, S.: Semantically enriched open API service descriptions in the cloud. In: 2017 8th IEEE International Conference on Software Engineering and Service Science (ICSESS). IEEE (2017)
13. Ontop VKG:. Introduction (2023). https://ontop-vkg.org/guide/
14. Ontop VKG: Database and Ontop Setup (2023). https://ontop-vkg.org/tutorial/basic/setup.html#database-setup
15. Ontop VKG: Key concepts (2023). https://ontop-vkg.org/guide/concepts.html#virtual-knowledge-graph-vkg
16. Ontop VKG: Setting up an Ontop SPARQL endpoint with Ontop CLI (2023). https://ontop-vkg.org/tutorial/endpoint/endpoint-cli.html
17. Papazoglu, M.P.: Service-oriented computing: concepts, characteristics and directions. In: Proceedings of 4th International Conference on Web Information Systems Engineering (WISE), pp. 3–12. IEEE (2003)
18. Segura, S., et al.: Metamorphic testing of RESTful web APIs. In: Proceedings of the 40th International Conference on Software Engineering, p. 882 (2018)
19. Talantikite, H.N., Aissani, D., Boudjlida, N.: Semantic annotations for web services discovery and composition. Comput. Stand. Interfaces **31**(6), 1108–1117 (2009)
20. Tzavaras, A., Mainas, N., Petrakis, E.G.M.: OpenAPI framework for the Web of Things. Internet Things, 100675 (2023)
21. W3C: SPARQL Query Language for RDF (2008). https://www.w3.org/TR/rdf-sparql-query/

PerfResolv: A Geo-Distributed Approach for Performance Analysis of Public DNS Resolvers Based on Domain Popularity

Marcelo Almeida Silva, Muriel Figueredo Franco[✉], Eder John Scheid, Luciano Zembruzki, and Lisandro Zambenedetti Granville

Institute of Informatics (INF) – Federal University of Rio Grande do Sul (UFRGS), Porto Alegre, Brazil
{marceloalmeida.silva,mffranco,ejscheid,lzembruzki, granville}@inf.ufrgs.br

Abstract. The Domain Name System (DNS) represents one of the cornerstones of the World Wide Web and plays an indispensable role in its operation. DNS is an extensive distributed database structured to resolve readable domain names for people, companies, and institutions into corresponding and reliable IP addresses. This paper presents PerfResolv, an approach for performance analysis on public DNS resolver servers (e.g., Google, Cloudflare, OpenDNS, Quad9, and ComodoDNS). The analysis was performed with PerfResolv located at geographically distributed points in three different countries: Brazil, Switzerland, and Australia. The results were obtained considering the response time for resolving domain names with different popularity levels to verify if and how geolocation, domain name popularity, week, day, and time affect the performance of DNS resolver servers. The results show considerable fluctuations in the response time of some DNS resolvers, with a variation of up to 40% in response time across different hours of the day. Further, there are differences between the resolution time of popular and unpopular domains, which are also influenced by the geolocation of the measurement monitors.

1 Introduction

The Internet has become indispensable due to the importance of its applications and possibilities of use in different sectors of industry, government, and entertainment. The Domain Name System (DNS) represents one of the foundations of the Internet, playing an indispensable role in its operation. The DNS works a large, distributed, and structured database for resolving readable domain names into corresponding and valid Internet Protocol (IP) addresses [16]. The creation of DNS began in the 1980s, with the emergence of the Advanced Research Projects Agency Network (ARPANET) [14]. The DNS concepts and standards were published in Request for Comments (RFC) 882 and 883, updated later in RFCs 1034 and 1035. These standards are still relevant today and widely used [12].

Different DNS protocol approaches and implementations make it possible to offer users varying levels of performance and security. DNS is also used as

L. Barolli (Ed.): AINA 2024, LNDECT 200, pp. 35–47, 2024.
https://doi.org/10.1007/978-3-031-57853-3_4

an instrument for doing business, with various organizations offering domain resolution as a service (*i.e.*, DNS resolvers) focusing on providing better performance or security for users and services. Private DNS servers can offer services that guarantee security, privacy, and performance for users willing to pay for such services [9]. In addition, there is also the possibility of local DNS servers, which can be installed in the user's infrastructure (*i.e.*, companies) and provide more agility in name resolution. However, in both scenarios, there are trade-offs between costs, complexity of operation, and security [8].

Public DNS servers have emerged as an option to meet this demand efficiently, reliably, and at no cost. These servers are free and reliable DNS resolvers offered by large companies that maintain Internet services (*e.g.*, Google, Cloudflare, and Cisco). Public DNS servers can also provide security-related features, specifically for identifying malicious domains [1]. However, public DNS servers are a growing concern in academia and industry due to the centralization of services, network flows, and infrastructure [19]. In addition, outsourcing services to public DNS resolution companies that do not adopt adequate security measures can create risks for users of the service since name resolution can be negatively affected in cases of malicious attacks, such as Amplification, Spoofing, and Denial-of-Service (DoS) [7,20].

In this context, the current literature focuses on approaches related to identifying and quantifying DNS centralization [3,15]. However, most research does not analyze the performance of public DNS resolver servers in depth (*e.g.*, considering the popularity of domains and times of use), nor does it place the necessary emphasis on the steps to define approaches that measure such performance [6]. Therefore, there is an opportunity and the necessary motivation for approaches that analyze the performance of public DNS resolvers.

In this work, we propose PerfResolv, an approach to evaluate the performance of public DNS resolver servers in different locations worldwide using domain names with varying levels of popularity taken from various domain lists. To this end, the performance of the different Public DNS resolvers (*e.g.*, Google, Cloudflare, and Quad9) is collected and measured, and the measurements obtained are analyzed and compared according to different criteria, such as time of day, day of the week, and geolocation. The Tranco list [13] is used as the list of domains to be used and analyzed based on its popularity. The results provide an overview of the factors affecting their performance, including security, domain popularity, and geolocation as crucial factors.

The rest of this work is organized as follows. Section 2 examines the related work. Section 3 describes the PerfResolv approach, while Sect. 4 reports the results obtained using the PerfResolv and discusses the findings. Finally, Sect. 5 concludes the paper and presents opportunities for future work.

2   Related Work

The performance analysis of DNS resolvers is essential since users look for servers offering a better trade-off between security and performance. A literature review was conducted to understand the performance measurement scenario of DNS resolvers and their different security levels over the last five years. Based on this, eight studies were selected and compared with the proposed PerfResolv. A summary of the analysis carried out is shown in Table 1.

The columns (*cf.* Table 1) represent the different attributes analyzed, including which DNS protocols were used and which metrics were collected for evaluation. The number of resolvers used and whether works considered domain popularity were also considered in the analysis performed. In addition, it was checked whether datasets with the results of the measurements are available, as well as whether the measurements were carried out in a geo-distributed manner (*i.e.*, different global collection points). Finally, the list of domains used as a basis for the work is listed when available.

Table 1. Comparison of Related Work on Analysis of DNS Performance

Work	Description	DNS Protocol	Metrics	Number of Resolvers	Popularity of Domains	Distributed Measurements	List of Domains
[4]	Centralized analysis of DoH and local DNS performance in a university network	Do53, DoT, DoH	DNS query time, Page load time	4	No	No	No
[10]	Analysis of the performance of DNS protocols and the impacts of encryption through Web page loading time	Do53, DoT, DoH	Page load time, DNS lookup time	3	No	Yes	Tranco 1M
[6]	Performance analysis of DNS servers using RIPE probes	Do53	DNS lookup time	10	No	Yes	Alexa Top 1M
[5]	Performance analysis of different DNS implementations on the BrightData network	DoH, Do53, DNSSEC	Page load time	4	No	Yes	No
[17]	Web tool that tests the performance of more than 200 global servers, supporting dozens of different public and private resolvers.	Do53	DNS lookup time	Variable	No	Yes	No
[1]	Analysis of the performance of local and public DNS servers in relation to malicious domains and their impacts on Web page loading time	Do53, DoH	DNS lookup time, Page load time	5	No	No	Cisco Top 1M Cisco umbrella
[11]	Analysis of the performance of DNS protocols in Web page loading time	DoT, DoH, Do53, DoTCP, DoQ	Page load time	313	No	Yes	Tranco 1M
[2]	Performance analysis of DNS servers violating TTL in the BrightData network	DNSSEC, DoH	DNS lookup time	27,000	No	Yes	Tranco 1M
This work (PerfResolv)	Analysis of the performance of public DNS resolvers concerning the popularity of domain names	Do53	DNS lookup time	5	Yes	Yes	Tranco 1M and new (unpopular) domains

Different studies have analyzed the performance of the traditional DNS (Do53), DNS over TLS (DoT), and DNS over HTTPS (DoH) protocols. [4] investigated the exploration of the political implications of DoH by measuring its performance compared with conventional DNS. In [10], the authors present measurements of how encryption using DNS Security Extensions (DNSSEC) affects the end-user experience in web browsers. In another paper, [6] checked the impact on reliability, security, and response time caused by DNS centralization.

The work [5] presented an approach to measuring DoH and Do53 performance called BrightData. In this work, the authors measured the protocols' performance on 22,052 customers at 224 collection points in different countries. Similarly, the DNSPerf tool [17] provides latency test results from more global collection points every minute. During the tests, DNSPerf uses a variable number (from dozens to hundreds) of different public and private resolvers. Also focusing on performance, [1] analyzed the DNS resolvers used by an Italian Internet Service Provider (ISP) to compare them with the public resolvers provided by Google and Cisco.

In [11], the authors analyzed the performance of the DNS protocol over the Quick UDP Internet Connection (QUIC) protocol, which Google proposed as a protocol alternative for the transport layer. In this work, the authors compared the response times of DNS-over-Quic (DoQ) compared to Do53 (both UDP and TCP), DoT, and DoH for single queries and evaluated their impact on Web performance and performed on DNS resolvers around the world. Finally, [2] presented an approach that measures the performance of DNS resolvers that violate Time-to-Live (TTL) using an HTTPS proxy service at five different collection points.

Thus, based on the literature review, it is possible to identify several studies focusing on performance and security aspects related to the DNS protocol and associated services. However, none of the studies analyzed used domain popularity as a metric for evaluation. Therefore, the PerfResolv approach is proposed throughout this article to measure how domain popularity can impact the response time of different public DNS resolver servers.

3   PerfResolv Approach

PerfResolv is proposed as an approach to automate (i) the process of measuring the performance of public DNS resolver servers and (ii) the analysis of domains with different levels of popularity. The measurements and analyses are done in a geo-distributed manner using monitors managed by PerfResolv.

In this way, PerfResolv allows a refined analysis of public DNS resolution servers and their performance levels (i.e., average response time) for domains with different levels of popularity, such as frequently accessed (more popular) and new domains (less popular). The days of the week and times of use can also be analyzed to understand the Internet usage profile and its impact on the performance of public DNS resolution servers. GMT (Greenwich Mean Time) is used to help analyze measurements from different countries with different time zones.

The approach proposed and implemented by PerfResolv is shown in Fig. 1, including the flows and components required for all phases. These components work from monitoring and collection to analysis of the data obtained.

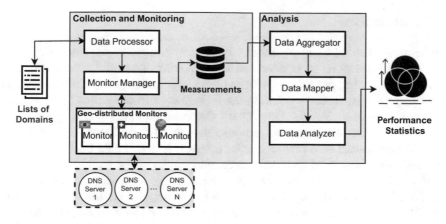

Fig. 1. PerfResolv's Architecture

During the monitoring and collection phase, the list of domains is initially defined for analysis and processed by the *Data Processor*. This processing separates domains according to their popularity (*i.e.*, based on their position in rankings and number of accesses). The domains are then forwarded to the *Monitor Manager*, which controls and communicates with geo-distributed monitors in different regions of the world (*e.g.*, specific countries or cities). The monitors run periodic queries (*c.g.*, per hour or days) for all the defined domains. These parameters and configurations can be adjusted according to experiment requirements to encompass different needs that network operators or companies might have.

For example, the monitors can be strategically distributed to observe the behavior of public DNS resolvers, which can vary in performance and security according to the domain, resolver, and region. During monitoring, measurements are taken for each domain for every query requested. All the measurements taken are forwarded to the *Monitor Manager*, which centralizes the measurements and stores them in a local database for later analysis.

During the analysis phase, the measurements are aggregated using the average of each domain's measurements per hour. Criteria for analysis are then defined and used to create subsets of the data, for example, filtered by country, DNS resolver, and domain type (popular and non-popular).

The implementation of PerfResolv and the data used is publicly available[1]. The PerfResolv monitoring and collection components were implemented using Python 3.11. The monitors use the *dnspython 2.3* library to resolve the domains in different public DNS resolvers and collect the performance measurements. The measurements are aggregated (*i.e.*, the average of the measurements taken for a domain at a given time) and mapped to Comma-Separated Values (CSV). The aggregated results are then analyzed using the *pandas 2.0.2* library. Finally, visualizations are generated using the *Vega-Altair 5.0.1* library.

---

[1] https://github.com/ComputerNetworks-UFRGS/PerfResolv.

4   Evaluation

The results obtained by applying the PerfResolv approach are discussed in this section. This allows us to understand the behavior of the different public DNS resolvers concerning domain popularity, time, and geolocation. The experiments were carried out by initially defining the IP addresses of the public DNS resolvers to be analyzed (*i.e.*, Google, Cloudflare, Comodo, OpenDNS, and Quad9), and the responses obtained were the response time to perform the query and receive the DNS response. These results can be accessed in the public PerfResolv repository and were processed to obtain aggregate statistics for analysis.

In total, 60 domains were selected, considering their popularity, for analysis: 20 of which were popular, 20 medium in popularity, and 20 completely new. Popular domains are extracted from the top of the Tranco list, where the medium are extracted from the position 300,000 (*i.e.*, still in the top 30% of the Tranco list). All the new domains were generated within a Brazilian academic and research network for this work only, thus ensuring they have little access or are permanently added to DNS caches worldwide. The experiments were conducted on Microsoft Azure Virtual Machines (VM) running in 3 countries (Australia, Brazil, and Switzerland), with 30 repetitions every hour for each domain and resolver. The period they covered was five days (Monday to Friday). In this way, it is possible to obtain the average of each execution cycle to improve the accuracy of our analysis.

Error bars were added for all plots in order to show the variability of the measurements collected. However, the bars are not prominent in the plots due to the slight error variance. Based on that, it is possible to conclude that the variance of the data is insignificant for our experiments.

4.1   Popular Domains

The first analysis checks the response time for popular and medium domains. Thus, the performance of the resolvers in milliseconds (ms) is obtained for the resolution of domains with the most hits and known (*i.e.*, 20 most popular domains) and also for domains that are not so popular but are still known (*i.e.*, 20 domains with medium popularity). These domains were selected from the Tranco list.

The results are shown in Figs. 2 (a), (b), and (c). Figure 2 (a) compares resolvers in Australia. The CloudFlare DNS resolver server performed best for popular and medium domains in this analysis. In contrast, the OpenDNS server performed worst for popular domains, and Comodo DNS performed worst for medium domains. On the other hand, the Google and Quad9 servers obtained similar performances for medium domains.

Figure 2 (b) compares the monitors in Brazil, where the Quad9 resolver have a better average performance in popular and medium domains, and Comodo DNS performed worse in medium domains. On the other hand, the Quad9 and OpenDNS resolvers have a similar performance in popular domains. However, when analyzing Cloudflare, it is possible to observe a discrepancy in performance

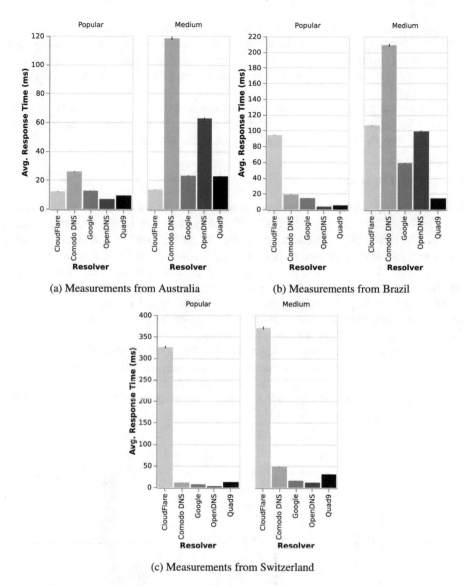

(a) Measurements from Australia   (b) Measurements from Brazil

(c) Measurements from Switzerland

Fig. 2. Performance Analysis of Resolution Time for Popular and Medium Domains Observed from Different Countries

for both popular and medium domains. As Cloudflare is known for having one of the best overall performances of public DNS resolvers, a more thorough analysis of which routing and security policies are implemented by the resolver specifically is necessary, as these policies can directly impact Cloudflare-related experiments.

Figure 2 (c) compares the resolvers through measurements taken by monitors located in Switzerland. The OpenDNS server performed best for popular and medium domains. Again, the CloudFlare server performed poorly on popular and medium domains. According to massive statistics from companies specializing in DNS monitoring and discussions with experts, an adequate domain resolution time should be no more than 100 ms and, in everyday situations, remain close to 20 ms [17].

In addition to domains with relevant popularity, new domains with few hits should also be analyzed to see how much the popularity impacts the resolution time. The results of the analysis of new domains are shown below.

4.2   New Domains

After analyzing popular domains, a list of new domains was also defined to check the impact of little-known and accessed domains on the resolution time of each server. For that, 20 *.br* domains were created in a Brazilian academic network, and the resolution time was compared with the other levels of popularity.

Figures 3, 4 and 5 show comparisons of resolver performance for popular, medium, and new domains for each of the countries used in the collection. The results show an impact on response times for new domains. For example, while the average response time for popular and medium domains was under 50 ms, the average for new domains was over 100 ms.

The measurements taken in Australia (*cf.* Fig. 3) show that the performance of all resolvers was worse for new domains. This analysis indicates that OpenDNS performed much less well than the others for new domains and is the second worst for medium domains, even though it was a resolver that performed very well for popular domains.

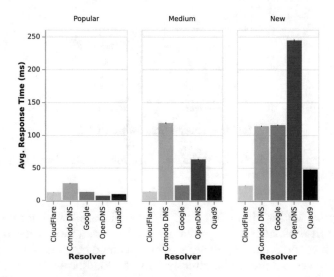

Fig. 3. Performance Analysis of New Brazilian's Domains Measured from Australia

In the measurements carried out in Brazil for new domains (*cf.* Fig. 4), the Quad9 server obtained the best performance, while CloudFlare had the worst performance. It is important to note that all new domains have *.br* as their Country Code Top Level Domain (ccTLD). Therefore, it is expected that all DNS resolvers would have the best performance for new domains for the Brazil data collection. However, this was not true for Cloudflare, which performed worse in Brazil than in Australia for such domains. Again, this may be because Cloudflare implements security policies that interfere with the experiments carried out.

In Fig. 5, it can be seen that the OpenDNS server performed best for new domains when monitored from Switzerland, while Cloudflare performed worst for new domains.

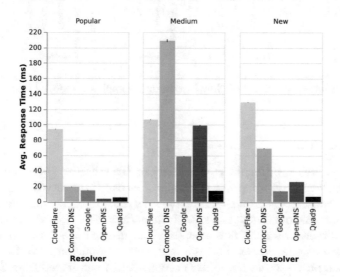

Fig. 4. Performance Analysis of New Brazilian Domains Measured from Brazil

Based on the results obtained, it is possible to gain insights into the differences in domain resolution according to their popularity and verify that some DNS providers handle requests differently, such as by implementing different security policies. The experiments conducted in Switzerland and Brazil were executed with varying implementations of DNS lookup tools (*e.g.*, dnspython and dig) [18] and using other machines in order to see if the reason for the behavior seen in the Cloudflare resolver could be related to implementation flaws or network limitations. However, the results showed no considerable variation. Therefore, it is important to carry out a thorough analysis not only of performance but also of other policies involving DNS resolution.

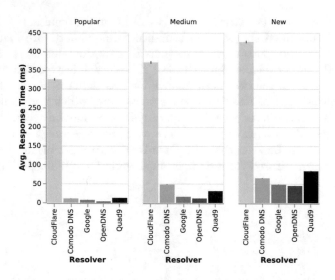

Fig. 5. Performance Analysis of New Brazilian's Domains Measured from Switzerland

4.3   Discussion

The quantitative experiments show a considerable difference between the resolutions of popular vs. non-popular domains. This can be observed empirically and was proven valid - based on quantitative analysis - by following the PerfResolv approach. The geolocation of the measurement monitors (*i.e.*, where the requests are made) directly impacts new domains that use ccTLDs from countries with unusual international access. For example, new *.br* domains have the worst performance compared to more popular domains when monitored from Switzerland and Australia.

In addition, it was possible to see that domains considered medium (in terms of popularity) can perform considerably poorly in some cases. This is because they are rarely accessed in specific locations. Therefore, domains that are popular on the Tranco list tend to be popular in all the locations analyzed (*e.g.*, facebook.com, google.com, and instagram.com). In contrast, medium domains (*e.g.*, shopiro.ca, freestart.hu, and tabsite.com) can be popular in specific locations and unpopular in others.

Furthermore, based on daily analysis, the days of the week did not impact the performance of DNS resolvers since all resolvers maintain a similar performance pattern from Monday to Friday. However, the time of day does generate some variations in performance. For example, the range from 10:00–12:00 and 14:00–16:00 h have shown longer response times than others. The Cloudflare, Quad9, and Google resolvers remained stable but with some peaks during these periods. On the other hand, the OpenDNS and Comodo resolvers had much more constant fluctuations during the experiments, with response times varying by up to 40% between different times of the day.

5   Conclusions and Future Work

In summary, this paper presented the PerfResolv approach, a geo-distributed approach for analyzing the performance of public DNS resolvers based on the popularity of domains. The proposed approach allowed performance metrics to be collected, taking into account not only the performance of the resolvers but also the type and popularity of the domains. Using monitors located in three different countries (*i.e.*, Brazil, Switzerland, and Australia) and a list of 60 domains classified according to their popularity, experiments were conducted to evaluate the response time in resolving domain names with different levels of popularity (*i.e.*, popular, medium and new). The results made it possible to analyze how popularity affects the performance of DNS resolver servers.

The experiments indicated a considerable difference between the resolution time of popular and unpopular domains, impacted by the location of the measurement monitors. In addition, it was found that domains of medium popularity can perform poorly in some locations due to low access. Popular domains on the Tranco list are popular (*i.e.*, accessed frequently) in all the places analyzed, while medium domains are only popular in specific locations. Regarding the possible influence of the days of the week on the performance of DNS resolvers, there was no significant impact, thus maintaining a similar pattern from Monday to Friday. However, the time of day generated variations in performance, with mid-morning and mid-afternoon being the periods with the most extended response times. The oscillations identified in some DNS resolvers indicate a variation of up to 40% in response times over the different hours of the day.

Future work includes an analysis of the resolution time of unpopular domains in different resolvers, countries, and ccTLDs. This allows for a more comprehensive understanding of the impact of domain popularity on DNS resolver performance. In addition, it is important to investigate the trade-offs between security and performance to explore measures to improve resolver efficiency without compromising user security. Additional performance metrics, such as RTT and TTL, can be explored by PerfResolv to obtain a more complete and detailed view of DNS resolver performance. Finally, in-depth investigations of each DNS resolver analyzed should be conducted to mitigate the different oscillations and behaviors observed.

Acknowledgements. This work was supported by The São Paulo Research Foundation (FAPESP) under the grant number 2020/05152-7, the PROFISSA project.

References

1. Affinito, A., Botta, A., Ventre, G.: Local and public DNS resolvers: do you trade off performance against security? In: IFIP Networking Conference (Networking 2022), Catania, Italy, pp. 1–9 (2022)
2. Bhowmick, P., Ashiq, M.I., Deccio, C., Chung, T.: TTL violation of DNS resolvers in the wild. In: Brunstrom, A., Flores, M., Fiore, M. (eds.) PAM 2023. LNCS, vol. 13882, pp. 550–563. Springer, Cham (2023). https://doi.org/10.1007/978-3-031-28486-1_23

3. Boeira, D.F., Scheid, E.J., Franco, M.F., Zembruzki, L., Granville, L.Z.: Traffic centralization and digital sovereignty: an analysis under the lens of DNS servers. In: 37th IEEE/IFIP Network Operations and Management Symposium (NOMS 2024), Seoul, South Korea, pp. 1–9 (2024)
4. Borgolte, K., et al.: How DNS over HTTPS is reshaping privacy, performance, and policy in the internet ecosystem. In: 47th Research Conference on Communication, Information and Internet Policy (TPRC 47), Washington, D.C, USA, pp. 1–9 (2019)
5. Chhabra, R., Murley, P., Kumar, D., Bailey, M., Wang, G.: Measuring DNS-over-HTTPS performance around the world. In: 21st ACM Internet Measurement Conference (IMC 2021), New York, USA, pp. 351–365 (2021)
6. Doan, T.V., Fries, J., Bajpai, V.: Evaluating public DNS services in the wake of increasing centralization of DNS. In: IFIP Networking Conference (Networking 2021), Espoo and Helsinki, Finland, pp. 1–9 (2021)
7. Franco, M., et al.: SecGrid: a visual system for the analysis and ML-based classification of cyberattack traffic. In: 46th IEEE Conference on Local Computer Networks (LCN 2021), Edmonton, Canada, pp. 1–8 (2021)
8. Hao, S., Wang, H., Stavrou, A., Smirni, E.: On the DNS deployment of modern web services. In: 31st IEEE International Conference on Network Protocols (ICNP 2015), San Francisco, USA, pp. 100–110 (2015)
9. Hounsel, A., Borgolte, K., Schmitt, P., Holland, J., Feamster, N.: Analyzing the costs (and benefits) of DNS, DoT, and DoH for the modern web. In: ACM/IRTF Applied Networking Research Workshop (ANRW 2019), Montreal Quebec, Canada, pp. 20–22 (2019)
10. Hounsel, A., Borgolte, K., Schmitt, P., Holland, J., Feamster, N.: Comparing the effects of DNS, DoT, and DoH on web performance. In: The Web Conference 2020 (WWW 2020), Taipei, Taiwan, pp. 562–572 (2020)
11. Kosek, M., Schumann, L., Marx, R., Doan, T.V., Bajpai, V.: DNS privacy with speed? Evaluating DNS over QUIC and its impact on web performance. In: 22nd ACM Internet Measurement Conference (IMC 2022), Nice, France, pp. 44–50 (2022)
12. Lai, T.L., Tsai, M.H.: Design and implementation of a DNS server with geolocation capability. In: 22nd Asia-Pacific Network Operations and Management Symposium (APNOMS 2021), Tainan, Taiwan, pp. 370–373 (2021)
13. Le Pochat, V., Van, T., Tajalizadehkhoob, S., Korczyński, M., Joosen, W.: Tranco: a research-oriented top sites ranking hardened against manipulation. In: 26th Network and Distributed System Security Symposium (NDSS 2019), San Diego, USA, pp. 1–15 (2019)
14. Leiner, B.M., et al.: A brief history of the internet. Comput. Commun. Rev. **39**(5), 22–31 (2009)
15. Moura, G.C., Castro, S., Hardaker, W., Wullink, M., Hesselman, C.: Clouding up the internet: how centralized is DNS traffic becoming? In: 20th ACM Internet Measurement Conference (IMC 2020), Pittsburgh, USA, pp. 42–49 (2020)
16. Park, J., Khormali, A., Mohaisen, M., Mohaisen, A.: Where are you taking me? Behavioral analysis of open DNS resolvers. In: 49th IEEE/IFIP International Conference on Dependable Systems and Networks (DSN 2019), Portland, USA, pp. 493–504 (2019)
17. PerfOps: DNSPerf - DNS Performance Analytics and Comparison (2023). https://www.dnsperf.com

18. Pinto, J., Scheid, E., Franco, M., Granville, L.: Analyzing and comparing DNS lookup tools in Python. In: XX Escola Regional de Redes de Computadores (ERRC 2023), Porto Alegre, Brazil, pp. 49–54 (2023)
19. Zembruzki, L., Jacobs, A.S., Granville, L.Z.: On the consolidation of the internet domain name system. In: IEEE Global Communications Conference (GLOBECOM 2022), Rio de Janeiro, Brazil, pp. 2122–2127 (2022)
20. Zembruzki, L., Jacobs, A.S., Landtreter, G.S., Granville, L.Z., Moura, G.C.M.: measuring centralization of DNS infrastructure in the wild. In: Barolli, L., Amato, F., Moscato, F., Enokido, T., Takizawa, M. (eds.) AINA 2020. AISC, vol. 1151, pp. 871–882. Springer, Cham (2020). https://doi.org/10.1007/978-3-030-44041-1_76

DARS: Empowering Trust in Blockchain-Based Real-World Applications with a Decentralized Anonymous Reputation System

Mouhamed Amine Bouchiha[✉], Yacine Ghamri-Doudane, Mourad Rabah, and Ronan Champagnat

L3i Laboratory, La Rochelle University, La Rochelle, France
{mouhamed.bouchiha,yacine.ghamri,mourad.rabah,
ronan.champagnat}@univ-lr.fr

Abstract. Blockchain-based Reputation Systems (BRS) are a recent and essential development in decentralized trust and reputation management. The decentralization, transparency, and efficiency brought by Blockchain (BC) are clearly what we always hoped for to build effective trustless reputation systems. Despite these promising attributes, existing BRS face a critical challenge in countering common reputation attacks, including whitewashing, self-promotion, and bad-collision attacks. Currently, BRS rely on reputation scores or tokens linked to the same address or public key, which severely limits their widespread adoption as it raises concerns about possible retaliation, hence the reluctance of users to engage and provide feedback. In this work, we propose and develop a Decentralized Anonymous Reputation System (DARS) for trust-related applications. In DARS, users can use different pseudonyms when interacting with each other to hide their digital identities. In our system design, all pseudonyms of a specific user, yet, are cryptographically linked to the same access token, allowing honest users to maintain their reputation and preventing malicious ones from starting over. This is achieved through the use of zkSNARK proofs for set membership via Merkle trees over commitments. We extended our framework with an efficient reputation model that respects all the security and privacy properties of our formal model. Finally, we developed a prototype of the proposed framework using emerging technologies and cryptographic tools. The evaluation results demonstrate the feasibility and effectiveness of DARS.

1 Introduction

Reputation, referencing the overall opinion towards a user or an entity, gained widespread adoption since its inception [1]. Reputation Systems (RS) aim to hold users accountable for their behaviors [2]. While traditional models present the central server as semi-honest, maintaining user privacy becomes uncertain if it becomes malicious due to external or internal compromises [3]. Despite cryptographic approaches addressing some security issues, the single point of failure and the lack of transparency remain significant challenges for centralized reputation systems [4].

Reputation systems often involve gathering and analyzing user data to determine and display reputation scores. This aspect can give rise to privacy concerns, particularly

© The Author(s), under exclusive license to Springer Nature Switzerland AG 2024
L. Barolli (Ed.): AINA 2024, LNDECT 200, pp. 48–61, 2024.
https://doi.org/10.1007/978-3-031-57853-3_5

if users feel uncomfortable about sharing personal information or activity data [6]. Addressing worries of potential retaliation, which often deter users from engaging and offering feedback, can involve employing feedback-independent reputation models [5]. Nevertheless, the persisting issue of linking reputation scores and tokens to a single master key remains a cause for concern regarding potential tracking. A recent approach to address this issue involves the development of decentralized privacy-preserving reputation models, allowing users to interact and share feedback confidentially and seamlessly. This advancement stands as a significant leap in trust and reputation management, offering both robust reputation management and user privacy preservation. The use of cryptographic techniques with decentralized systems such as Blockchain [10, 11] can help reputation systems guarantee user privacy without compromising transparency and efficiency. However, existing solutions fall short of being entirely decentralized since they depend either on a centralized entity or a group of trusted peers to handle identities, credentials, and security parameters. Additionally, despite the use of BC technologies in numerous research efforts [13–15] to develop decentralized and privacy-centric reputation systems, these proposed solutions fail to sufficiently tackle the efficiency challenges of on-chain reputation management. Furthermore, the issues associated with the implementation of real-world blockchain-based reputation frameworks, particularly the Oracle problem [18], have not undergone thorough examination. Therefore, to overcome all these issues, we propose in this paper, "DARS", a fully Decentralized Anonymous Reputation System for real-world applications. The main contributions of this research include:

- A decentralized reputation system that is constructed on top of two distinct ledgers to separate identity management from business activities.
- A system that relies on Decentralized Oracle Networks not only to automate smart contracts execution but also to import credentials from existing systems to prevent Sybil attacks.
- The use of zkSNARK proofs for Set Membership over commitments, allowing DARS users to gain the ability to generate and use numerous pseudonyms to safeguard their digital identity and ensure anonymity.
- A design of a reputation model that achieves all the security and privacy properties of our formal model.
- A proof-of-concept for the proposed framework, leveraging emerging technologies and cryptographic tools is developed. This allows for a more meaningful assessment of DARS's capabilities.

The remainder of this paper is structured as follows: Sect. 2 describes related work. The security model is presented in Sect. 3. Section 4 deals with cryptographic building blocks. Section 5 is devoted to the construction of the proposed Decentralized Anonymous Reputation System (DARS). Section 6 is dedicated to the performance evaluation of the proposal and Sect. 7 to the security analysis.

2 Related Work

Blockchain (BC)-based reputation systems are now essential for trust-based applications such as retail marketing, mobile crowdsensing, and decentralized markets. Considerable research has focused on developing anonymous and privacy-preserving reputation systems for these areas. These systems make it challenging to link a user's actions/feedback with their true identity. However, maintaining an accurate reputation without any ties to identity raises difficulties, as the reputation score should reflect precisely the user's activity within the system. In an attempt to bring an answer to the above concern Liu et al. [10] propose an anonymous reputation system for retail marketing in the industrial IoT environment. The system utilizes smart contracts (SCs) on a PoS-based BC, ensuring transparency and public verifiability even against malicious attacks. It prioritizes anonymity through randomizable signatures and zero-knowledge proofs (ZKPs). However, the system's reliance on centralized IDentity Management (IDM) for managing identities, credentials, and security parameters creates a potential security risk and vulnerability to a single point of failure. [11] introduces a privacy-focused reputation system using BC in mobile crowdsensing with limited resources. Reputation scores are updated globally using SCs through feedback averages. The system employs additive secret sharing and delegation sets for privacy in a dynamic setting. However, like prior research, it operates under a semi-honest model, leading to potential security concerns. BPRF [12] presents a BC-based privacy-preserving reputation framework for participatory sensing systems. The system uses SCs to manage the reputation scores of participants based on their sensing data and corresponding feedback. The solution employs group signatures and a partially blind signature algorithm to protect user information. To achieve greater transparency, Schaub et al. [13] proposed a fully decentralized reputation system on top of a public BC with a blind signature to guarantee consumer anonymity. Soska et al. [14] proposed an anonymous reputation system based on a ring signature, which resulted in linear overhead when generating the anonymous review proof. Although, [12–15] have made significant efforts to investigate the use of BC technology in constructing robust privacy-preserving reputation systems, the proposed solutions have not fully examined the efficiency and scalability issues associated with BC-based reputation management. Moreover, the implementation challenges specific to BC-based solutions like the Oracle problem [18], have not been explored in depth in these studies. To overcome all the above issues, we propose in this paper, "DARS", a decentralized anonymous reputation system for BC-based real-world applications.

3 Security Model

In this section, we introduce the adversarial model and explore the security features of the proposed DARS.

3.1 Adversarial Model

For our adversarial model, we borrow the assumptions of [7]. Therefore, an adversary is able to statically and actively corrupt up to t of the n nodes in the committee, for $t < n/3$. In addition, the adversary can corrupt any number of external entities, such as users and applications.

3.2 Security Properties

Under the above assumptions, we outline the security properties and objectives of DARS as follows:

- **Sybil resistance:** A user cannot have any credentials other than his/her own.
- **Unforgeability:** An adversary cannot forge the credentials of honest users or impersonate them.
- **User privacy**: It is infeasible for an adversary to ascertain a user's attributes through the examination of issued identification information, the analysis of transaction data during interactions with other users, or the observation of the ongoing evaluation of interactions.
- **Reputation binding:** The user's reputation is unique and stored publicly in the BC. Although users can generate as many pseudonyms as they wish, all are cryptographically linked to the same access token.
- **Forward Reputation binding:** No user should be able to mint/use a reputation token with a reputation score higher than that linked to his/her last token.

4 DARS Building Blocks

In this section, the main cryptographic building blocks upon which the DARS system is built are presented.

4.1 Commitment Scheme

A commitment scheme is a cryptographic protocol that allows a party, referred to as the committer, to commit to a chosen value without revealing it, while still being able to prove its validity later on [19]. It is designed to fulfill two crucial security properties: (*i*) Hiding Property: Given $COMM(x)$, it should be computationally infeasible to determine the original value x. (*ii*) Binding Property: It should be computationally infeasible to find two distinct values x_1 and x_2 such that $COMM(x_1) = COMM(x_2)$.

4.2 *zkSNARKs*

Zero-Knowledge Succinct Non-Interactive Argument of Knowledge (zkSNARKs) are an advanced variant of Zero-Knowledge Proofs (ZKPs). More precisely, a zkSNARK scheme is a Non-Interactive Zero-Knowledge (NIZK) scheme [8], wherein the proof itself is a self-contained data block that can be verified without requiring any interaction from the prover [20,21]. A zkSNARK construction consists of three algorithms (Gen, Prov, Verif) defined as follows:

- The key generator G takes a secret parameter λ and a program C, and generates two publicly available keys: a proving key pk, and a verification key vk. These keys are public parameters that need to be generated only once for a given program C.
- The prover P takes as input the proving key pk, a public input t, and a private witness w. The algorithm generates a proof $\pi = Prov(pk, t, w)$ that the prover knows a witness w and that the witness satisfies the program C.

– The verifier V computes $\text{Verif}(vk, t, \pi)$ which returns true if the proof is correct, and false otherwise. Thus, this function returns true if the prover knows a witness w satisfying C.

4.3   *zkSNARKs* for Set Membership via Merkle Trees

The set membership problem via Merkle trees involves proving that an element belongs to a set using the Merkle tree data structure. More formally, Given a set S containing n elements and a Merkle tree constructed from the hash values of these elements, the set membership problem is to prove that a specific element x belongs to the set S without revealing any other elements in S [9]. Merkle trees alone do not provide ZK property. To achieve this property, we need to combine Merkle trees with additional techniques such as zkSNARK or other cryptographic primitives [16].

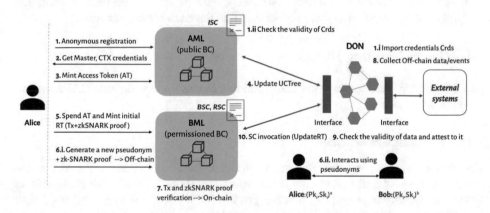

Fig. 1. Overview of DARS Framework, comprising an Access Management Ledger (AML), a Business Management Ledger (BML), and a Decentralized Oracle Network (DON).

5   DARS Framework

In this section, we describe our proposed Decentralized Anonymous Reputation System (DARS). An overview of DARS is shown in Fig. 1. We use the previously mentioned building blocks on top of two separate ledgers, *Access Management Ledger (AML)* and *Business Management Ledger (BML)*, to decouple identity management from business operations. AML is a public permissionless BC responsible for managing identities and access, while BML is a permissioned BC that implements the overall business logic of a real-world application. Our construction aims to ensure robust and efficient reputation management over different pseudonyms. We build our framework over four main phases:

Phase1-Registration. The registration process takes place between a user u and the AML committee. The AML relies on a Decentralized Oracle Network (DON) to ensure uniqueness while issuing master credentials M_{cred} for any valid user u. As in [7], we make use of DON to import identities from existing systems. For example, Alice can use her credentials on her Social Security Administration (SSA) account to generate a credential certifying her Social Security Number (SSN). Our DON uses the DECO protocol [17] to provide privacy for user data. DECO is a three-party protocol involving a prover denoted as P, a verifier denoted as V," and a TLS server denoted as S. The protocol enables P to persuade V that a data item, which may be private to P, obtained from S, meets a specified Predicate. DECO relies on multiparty computation (MPC) to protect the confidentiality and authenticity of the data, and on zero-knowledge proofs (ZKPs) to prove that a predicate is satisfied.

Once the identity of the user u is verified by the committee nodes following the DECO protocol, the corresponding user will be able to post his set of claims on AML and get access on BML using context-based credentials. To obtain a new credential for the ctx context, *e.g.* "a trading or crowdsourcing activity", u must submit to the committee $(pk_{ctx}, M_{cred}, C_{ctx}^u)$: a new identifier to be used in the *ctx* context, master credentials and a set of pre-credentials with the claims C_{ctx}^u required by *ctx*. The committee upholds a set of $Granted_{ctx}$ identifiers denoting those that have already obtained a credential within this specific context. If M_{pk^u} of M_{cred} is not part of this set, a credential is issued. Finally, (M_{pk^u}, pk_{ctx}^u) is added to $Granted_{ctx}$. For more details on issuing context-based credentials, see [7].

In our construction, master credentials are purposely excluded from any interactions with the BML to prevent linking them to the user's real identity. On the other hand, contextual credentials are used exclusively on the AML and remain cryptographically hidden on the BML to separate identity management from business operations. To guarantee these properties, the committee nodes maintain a CRH-based Merkle tree with root rt called *UCTree*, which contains all the user access commitments. When a user u is successfully registered with the committee nodes he/she must provide a commitment to his/her credentials. The user will use this commitment to access the BML without revealing any information that could be linked to his/her identity. To do that the user proceeds as follows: u generates an address key pair (apk, ask), the address public key and private key, respectively; u samples a random a and computes $cm_u = \text{COMM}_a(pk_{ctx}^u)$, then computes $cm_A := \text{COMM}_b(cm_u \| apk)$ for a secret b, and defines $AT := (pk_{ctx}^u, a, b, cm_A)$. A corresponding mint access token transaction, $tx_{AM} := (apk, cm_u, cm_A)$, is added to the AML (accepted only if pk_{ctx}^u is known to the committee). The *UCTree* is then updated with a new leaf (cm_A). We use DON to synchronize any changes made to the *UCTree* on the AML with the BML. This construction allows the user to prove to BML validators that he/she has valid credentials on AML efficiently and anonymously, *i.e.* the time and space complexity is logarithmic to the size of *UCTree*, and pk_{ctx}^u remains cryptographically hidden in cm_u.

Phase2-User Anonymity with a Reputation Token. The second phase in our construction is the mint of a reputation token that is cryptographically linked to the user's contextual credentials. This phase aims to hide the user's digital identity pk_{ctx}^u

while ensuring robust reputation management through the utilization of zkSNARK proofs and a commitment scheme.

To interact with other users and post transactions on the BML, the user must spend their access tokens and mint their initial reputation token RT_{init}. This is equivalent to adding a new leaf to the $RCTree$ (similar to $UCTree$) containing a commitment to its initial reputation score. To achieve the property of forward reputation binding the user must sample a random serial number S_n for each new reputation token. S_n is then released when using the token. This is realized as follows, u first generates a new key pair (pk, sk), then samples a random s and computes $S_n := \text{PRF}_{sk}(s)$ using a Pseudo-Random Function (PRF), and commits to the tuple (pk, R_{init}, s) in two steps: $cm_P := \text{COMM}_r(pk \| s)$ for a random r, and then $cm_R := \text{COMM}_{r'}(R_{init} \| cm_P)$ for another random r'. The outcomes comprise: (i) a reputation token $RT_{init} := (pk, R_{init}, s, r, r', cm_R)$ and (ii) a RT mint transaction $tx_{RM} := (R_{init}, cm_P, r', cm_R)$. However, this alone does not fulfill the criteria for the transaction to gain acceptance on the BML. The user must provide a zkSNARK proof π_A of the NP statement *"I know a secret b such that $\text{COMM}_b(cm_u \| apk)$ appears as a leaf in a CRH-based Merkle tree $UCTree$ whose root is rt"*. This prerequisite permits access to the BML exclusively for authorized users. With this in mind, we edit the tx_{RM} transaction to $tx_{RM} := (R_{init}, cm_P, r', cm_R, \pi_A)$ which is submitted to the BML. The tx_{RM} is accepted if and only if the π_A and cm_R are valid. Because of commitment nesting, anyone can verify that cm_R in tx_{RM} is a commitment of a token of value R_{init} (by checking that $\text{COMM}_{r'}(R_{init} \| cm_P)$ equals cm_R), but is unable to identify the owner through the knowledge of the address key pk or the serial number S_n (derived from s), as these are hidden in cm_P. Finally, user anonymity is achieved because the proof π_A is zero-knowledge: while cm_u and apk are revealed, no information about b is revealed, and finding which of the many commitments in $UCTree$ corresponds to tx_{RM} is equivalent to inverting $f(b) := \text{COMM}_b(X)$, which is assumed to be infeasible [16].

Phase3-Reputation Token Use/Spending. So far, user u has minted his initial reputation token RT_{init}. He can therefore interact with any user v on the BML by submitting transactions. Within DARS, users' reputation scores are tied to their most recent reputation commitment cm_R. As a result, for a user u to engage with other users, they must reveal the nested commitment to display their reputation score. Since only the user possessing the secret r' can unveil it, there's no susceptibility to forgery. Additionally, tx_{RM} is submitted using a pseudonym different from pk_{ctx}^u, and since u can generate numerous pseudonyms (ideally, a new pseudonym apk^{new} for each new interaction), the likelihood of disclosing its actual identity is effectively eliminated.

Let's delve deeper into the details. Users within the BML can utilize their reputation token through the submission of a reputation spending transaction, denoted as tx_{spend}. This transaction enables them to create a new token of identical value to the current one. Consider a scenario where a user u possesses a pair of address keys (apk^{old}, ask^{old}), wishes to consume its current token $RT^{old} := (apk^{old}, R^{old}, s^{old}, r^{old}, r'^{old}, cm_R^{old})$ and produce a new one RT^{new}, targeted at the public address key apk^{new}. The user u proceeds as follows, (i) u samples serial number randomness s^{new}; (ii) u computes $cm_P^{new} := \text{COMM}_r^{new}(apk^{new} \| s^{new})$ for a random r^{new}; and then computes (iii) $cm_R^{new} := \text{COMM}_{r^{new}}(R^{new}||cm_P^{new})$ for a random r^{new}. This yields the token $RT^{new} := (apk^{new}, R^{new}, s^{new}, r^{new}, r'^{new}, cm_R^{new})$. Next, u generates a zkSNARK proof π_{RS} for the following NP statement:

"Given the RCTree root rt_R, serial number S_n^{old}, and token commitment cm_R^{new}, I know a token RT^{old}, RT^{new}, and address secret key sk^{old} such that:

(i) The tokens are well-formed: $cm_P^{old} := \text{COMM}_{r^{old}}(apk^{old}||s^{old})$ and $cm_R^{old} := \text{COMM}_{r'^{old}}(R^{old}||cm_P^{old})$ for cm_R^{old} and similarly for cm_R^{new}.

(ii) The secret key matches the public key: $apk^{old} = \text{PRF}_{ask^{old}}(0)$.

(iii) The serial number is calculated correctly: $S_n^{old} = \text{PRF}_{ask^{old}}(s^{old})$.

(iv) The commitment cm_R^{old} appears as a leaf in $RCTree$ whose root is rt_R.

(v) The reputation values are equal $R^{new} = R^{old}$."

A resulting spend transaction $tx_{RS} := (rt_R, S_n^{old}, cm_R^{new}, \pi_{RS})$ is sent to the BML. tx_{RS} gets rejected if S_n^{old} appears in a prior transaction. Thus, the user is forced to use his/her most recent reputation token for each new interaction.

Phase4-Reputation Update. This process is done automatically using the DON and smart contracts. We use the DON to collect the off-chain data needed to evaluate the interaction, and then trigger the reputation module implemented using smart contracts to update the reputation scores of the users involved in the interaction using their pseudonyms. The update process takes place once the interaction is over. For a more comprehensive explanation, let's consider a scenario within a marketplace. Let's suppose that user u wants to introduce a new product into the system. In this case, u is faced with two choices: use the existing reputation token or mint a new one, as detailed earlier. Then, u can simply add the new product to the system by posting the corresponding transaction $tx_{newProd} := (prodID, price, DescriptionpathonIPFS, ...)$. Once the product is listed for sale, when another user v, intends to purchase this item from u, v has the option to either utilize their existing reputation token or spend it to generate a new one, retaining the same reputation score and ensuring ongoing anonymity. Additionally, v is required to provide a zkSNARK proof demonstrating the absence of a shared secret key with u. This condition is necessary for our construction, as it prevents self-promotion attacks. To do this, v computes $H_i^v := H(prodID||sk_{ctx}^v)$. Then produces a zkSNARK proof π_i^v for the following NP statement, **"Given the product identifier prodID, I know sk_{ctx}^v, cm_R, and RT_i such that, H_i^v is computed correctly and shares the same sk_{ctx}^v with apk_i".** The proof is then sent to the BML as part of the new order transaction $tx_{newOrd} := (ordID, info, H_i^v, \pi_i)$. Like v, if u chooses to approve v's order, u is required to compute the interaction hash $H_i^u := H(prodID||sk_{ctx}^u)$ using its own sk_{ctx}^u and provide the corresponding zkSNARK proof. The proof is then sent to BML as part of the order acceptance transaction $tx_{accOrd} := (ordID, info, H_i^u, \pi_i^u)$. The transaction is rejected if H_i^u and H_i^v are identical or if the proof π_i^u is invalid.

Once the off-chain interaction is over, users u and v must transmit the data needed to evaluate the interaction. In our system design, we use DON to collect data from external systems, verify it, and calculate the required values of all the metrics used in the reputation model. Then, the Reputation Smart Contract (RSC) is triggered to perform the evaluation and update the global reputation scores. Both u and v have shown their last reputation commitments using their pseudonyms apk_i^u

and apk_i^y, respectively. Consequently, the RSC will update the reputation scores of u and v automatically and transparently using the revealed information. It's crucial to note that the evaluation of the interaction itself should prioritize privacy, ensuring that our reputation model doesn't utilize or disclose any details regarding the users' identities engaged in the interaction. To achieve this, we propose employing the following formula for interaction evaluation:

$$\begin{cases} \mathbf{T}_i = \mathscr{P} \left[\omega_p + \omega_t F_t + \omega_a F_a \right] \\ \omega_p, \omega_t, \omega_a \in [0,1] \; ; \; \omega_p + \omega_t + \omega_a = 1 \end{cases} \tag{1}$$

where $\mathscr{P}$ is a Boolean which refers to the presence of the proof "1" or not "0", *i.e.* whether the interaction outside the chain has actually taken place or not. This could be proof of delivering a "product" or completing a "task". ω_p is the weight of the proof itself. ω_t and ω_a are the weights of the time t and the amount a of the interaction, respectively. F_t and F_a are the functions that normalize t and a, respectively ($F_a, F_t \in [0,1]$). The value and timing metrics are implemented to thwart coordinated attacks. These attacks occur when users collude to boost each other's reputation by engaging in multiple low-cost interactions within a brief timeframe. The formula can be extended with additional contextual factors (feedback, data quality...), provided they do not reveal any information about the user's digital or physical identity. The global reputation update is performed using the following formula [5]:

$$R_{new} = \begin{cases} (1 - \mathcal{W}_f)R_{old} + \mathcal{W}_f \mathbf{T}_i \; ; \mathbf{T}_i \geq T_{min} \\ \mathcal{W}_f R_{old} + (1 - \mathcal{W}_f)\mathbf{T}_i \; ; \mathbf{T}_i < T_{min} \end{cases} \tag{2}$$

where R_{old} is the old reputation, $\mathbf{T}_i$ is the value of the interaction, T_{min} is the trust threshold, and $\mathcal{W}_f \in [0,1]$ is a weighting function that gives more or less relevance to $\mathbf{T}_i$, depending on the role played by the user in the interaction, *e.g.* "seller" or "buyer", and the value of the interaction itself (*i.e.* positive or negative interaction).

6   Evaluation and Results

In this section, we first introduce the evaluation environment and experimental setup for the proposed DARS. We then discuss the on/off-chain evaluation results.

6.1   Evaluation Environment

We carried out the benchmarks on our local BC platform. The platform is a cluster of two HPE ProLiant XL225n Gen10 Plus servers dedicated to the experimentation and evaluation of BC solutions. Each server features two AMD EPYC 7713 64-Core 2 GHz processors and 2×256 GB RAM.

6.2   Experimental Setup

To evaluate the proposed solution, we developed a proof of concept for the Decentralized Anonymous Reputation System (DARS) by leveraging cutting-edge technologies and cryptographic tools. The circuits employed in DARS are implemented using the circom programming language and the circomlib library[1]. We utilized the snarkjs library[2] to compile the circuits and perform the powers of tau ceremony for the trusted setup. Additionally, we developed the smart contracts of DARS using the Solidity programming language[3] and established a local network consisting of twelve validators using Hyperledger Besu[4] as BC client with Proof of Authority (PoA) as consensus protocol. We utilized Web3js library[5] for developing the client side and deploying the system's smart contracts. Lastly, for conducting benchmarking tests, we utilized Hyperledger Caliper[6].

6.3 Performance Evaluation

Three metrics are considered for DARS performance evaluation:

- **Time overhead:** refers to the processing time for the proving and verification operations. This time is measured off-chain for the proving operation; or from when a specific transaction that contains a zkSNARK proof is received at the smart contract (on-chain) for verification until the appropriate response is sent back to the prover.
- **Throughput:** refers to the number of successful transactions per second (TPS).
- **Latency:** is the time difference in seconds between the submission and completion of a transaction.

Table 1. Time overhead measurements for the zkSNARK proofs generation and verification using the Groth16 proving system.

Tx type	Proving (ms)	Verification (ms)	Overall Time (ms)	Call Data size
spendAT (π_A)	2400	730	3130	705B
spendRT (π_{RS})	2900	950	3850	705B
spendRT (π_{RS})	480	640	1120	705B

Table 2. Time overhead measurements for zkSNARKS proofs generation and verification using the PlonK proving system.

Tx type	Proving (ms)	Verification (ms)	Overall Time (ms)	Call Data size
spendAT (π_A)	67000	760	67760	1750B
spendRT (π_{RS})	79500	935	80435	1750B
newOrd (π_i)	3400	670	4070	1750B

---

[1] https://github.com/iden3/circomlib.
[2] https://github.com/iden3/snarkjs.
[3] https://docs.soliditylang.org.
[4] https://besu.hyperledger.org.
[5] https://web3js.readthedocs.io.
[6] https://github.com/hyperledger/caliper-benchmarks.

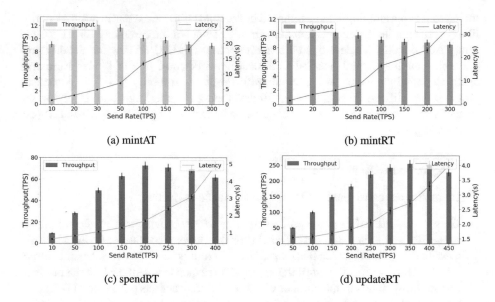

(a) mintAT

(b) mintRT

(c) spendRT

(d) updateRT

Fig. 2. Latency and Throughput of DARS under different send rates

A. Time Overhead. We employed two distinct proving systems, namely zkSNARK Groth16 and Plonk to evaluate the time overhead of our circuits. Groth16 is a circuit-specific preprocessing general-purpose zkSNARK construction that has become a standard choice in various BC projects [20]. This popularity is owed to its proofs' constant size and efficient verifier time. However, Groth16 necessitates a circuit-specific trusted setup during its preprocessing phase, which could be considered a drawback. On the other hand, PlonK represents a universal preprocessing general-purpose zkSNARK construction [21]. This proving scheme features an updatable preprocessing phase and boasts a short and constant verification time. Nevertheless, PlonK proofs tend to be larger and take more time (60–80 s) to generate compared to Groth16. Table 1 presents the timing and memory-related measurements for the Groth16 ZK-proof components, namely π_A, π_{RS}, and π_i. The π_A allows proving the existence of valid AML credentials efficiently and anonymously, π_{RS} proves the validity of RT, and π_i attests the validity of the interaction, preventing self-promotion attacks (see Sect. 5). Additionally, Table 2 displays the corresponding measurements utilizing the PlonK construction. Experimental results show that proof generation and verification take only a few milliseconds (480–2900 ms) when using the Groth16 scheme, whereas proof generation with the PlonK system takes a relatively longer time (3.4–80 s). Compared to verification with Groth16, no significant difference is observed for the proof verification using PlonK.

B. Throughput and Latency. We conducted a series of experiments using Caliper to qualitatively evaluate DARS' performance. The experiments involved changing the Tx sending rate (ranging from 10 to 500 TPS) using a consistent network configuration for the four main operations performed within our system. The results are illustrated in Fig. 2. As the Tx sending rate increases for each operation, the throughput also increases accordingly. Regarding the updateRT Tx, it reaches a peak of 255 TPS with a sending rate of 350 TPS, and then experiences a decline, indicating an overloaded system. On the other hand, system latency for the updateRT Tx remains relatively small and stable (less than 3 s), as long as the system is not overloaded. The remaining operations exhibit similar behavior, but their performances are comparatively lower. Notably, the mintAT and mintRT operations stand out as the most computationally intensive due to the substantial amount of computation needed to insert the commitments cm_A and cm_R into the $UCTree$ and $RCTree$ structures, respectively. This heightened computational intensity directly contributes to the heavier workload experienced by these operations. It is essential to highlight that all operations are currently sent directly to the main chain without employing any scaling solution. Indeed, high scalability was not our primary objective in this paper. Though the achieved scalability remains competitive.

7 Security Analysis

In this section, we examine the key security risks and the measures implemented by DARS to counter these threats.

- **Sybil Attacks:** involve creating multiple pseudonymous to manipulate the reputation system. In DARS, each legitimate user is granted only one valid credential for each specific context. Specifically, the AML committee maintains a set of $Granted_{ctx}$ identifiers, representing those that have already received a credential within that context. If M_{pk^u} in M_{cred} is not part of this set, a credential is granted; otherwise, no additional credential is issued for that user. This design ensures that a user cannot generate and utilize more than one valid access token per context. Consequently, DARS effectively guards against Sybil attacks.
- **Unforgeability:** Identity Theft is mitigated in the AML subsystem as users' keys remain in their wallets. These keys are utilized only for signing challenges during the protocol as part of credential verification. Consequently, the assurance of unforgeability within this subsystem is a direct consequence of the overall unforgeability of signatures.
- **User Privacy:** Regarding the privacy of credential issuance, it's important to note that generating a pre-credential for a claim within the Oracle protocol does not disclose any information about the user. Furthermore, given the commitment's hiding property and the privacy guarantees provided by the Secure Multi-Party Computation (SMPC) evaluation, there is no opportunity for an attacker to gain knowledge about the user during the issuance process. In addition, since no personal information is used when evaluating interaction within BML, there is no risk of de-anonymization or leakage of information about interacting parties.
- **Reputation Binding:** Our DARS is based on the forgery-proof nature of the cryptographic signatures used to create contextual credentials and submit access tokens. This ensures that the reputation score remains cryptographically linked to the original user or entity.
- **Forward Reputation Binding:** DARS satisfies this property if the signature scheme prevents forgery, the commitment scheme maintains the hiding and binding properties, and the zkSNARK scheme ensures soundness and ZK properties. These com-

bined properties help ensure that the new reputation score is reliable, private (if not shown) and consistently linked to the entity (access token) it represents.

8 Conclusion

In this work, we have proposed a decentralized anonymous reputation system that combines Blockchain and zkSNARKs. The system is built on top of two separate ledgers to decouple identity management from business activities. We make use of Decentralized Oracle Networks not only to automate SCs execution as it is traditionally used but also to import credentials from external systems to prevent sybil attacks without compromising user privacy. DARS users can generate/use as many pseudonyms as they wish on the BC to protect their digital identity and guarantee continued anonymity. The proposed framework relies on two Collision-Resistant Hash-based Merkle trees, UCTree and RCTree, over a list of access and reputation commitments, respectively, to guarantee anonymity while maintaining effective reputation management. We also designed a general reputation model that achieves the security and privacy properties of our formal model. Our design is suitable for all trust-based applications, such as decentralized marketplaces and crowdsourcing platforms. We implemented and tested a prototype of the proposed framework using cutting-edge technologies and cryptographic tools. The results of this evaluation demonstrate the feasibility and effectiveness of DARS. Thus, achieving its objective for practical and realistic usage.

Ongoing research in the fusion of ZKPs and BC shows promising potential. Specifically, advancements in L2 scaling solutions like zkRollups[7] offer hope for substantial improvements. Consequently, our upcoming focus will center on improving the scalability of DARS.

Acknowledgements. This work received support from the Nouvelle Aquitaine region through funding from the B4IoT project.

References

1. Resnick, P., Kuwabara, K., Zeckhauser, R., Friedman, E.: Reputation systems. Commun. ACM **43**(12), 45–48 (2000)
2. Hasan, O., Brunie, L., Bertino, E.: Privacy-preserving reputation systems based on blockchain and other cryptographic building blocks: a survey. ACM Comput. Surv. (CSUR) **55**(2), 1–37 (2022)
3. Zheng, Y., Duan, H., Wang, C.: Learning the truth privately and confidently: encrypted confidence-aware truth discovery in mobile crowdsensing. IEEE Trans. Inf. Forensics Secur. **13**(10), 2475–2489 (2018)
4. Zhang, Y., Deng, R.H., Zheng, D., Li, J., Wu, P., Cao, J.: Efficient and robust certificateless signature for data crowdsensing in cloud-assisted industrial IoT. IEEE Trans. Industr. Inf. **15**(9), 5099–5108 (2019)
5. Bouchiha, M.A., Ghamri-Doudane, Y., Rabah, M., Champagnat, R.: GuRuChain: guarantee and reputation-based blockchain service trading platform. In: IFIP Networking Conference (IFIP Networking), Barcelona, Spain, pp. 1–9 (2023)

---

[7] https://docs.zksync.io/userdocs/intro/.

6. Malik, S., Gupta, N., Dedeoglu, V., Kanhere, S., Jurdak, R.: TradeChain: decoupling traceability and identity in blockchain enabled supply chains. In: IEEE 20th (TrustCom), Shenyang, China, pp. 1141–1152 (2021)
7. Maram, D., et al.: CanDID: can-do decentralized identity with legacy compatibility, Sybil-resistance, and accountability. In: IEEE Symposium on Security and Privacy, SP 2021, San Francisco, CA, USA, pp. 1348–1366 (2021)
8. Abe, M., Fehr, S.: Perfect NIZK with adaptive soundness. In: Vadhan, S.P. (ed.) TCC 2007. LNCS, vol. 4392, pp. 118–136. Springer, Heidelberg (2007). https://doi.org/10.1007/978-3-540-70936-7_7
9. Benarroch, D., Campanelli, M., Fiore, D., Gurkan, K., Kolonelos, D.: Zero-knowledge proofs for set membership: efficient, succinct, modular. Des. Codes Crypt. **91**, 3457–3525 (2023)
10. Liu, D., Alahmadi, A., Ni, J., Lin, X., Shen, X.: Anonymous reputation system for IIoT-enabled retail marketing atop PoS blockchain. IEEE Trans. Industr. Inf. **15**(6), 3527–3537 (2019)
11. Zhao, K., Tang, S., Zhao, B., Wu, Y.: Dynamic and privacy-preserving reputation management for blockchain-based mobile crowdsensing. IEEE Access **7**, 74694–74710 (2019)
12. Jo, H.J., Choi, W.: BPRF: blockchain-based privacy-preserving reputation framework for participatory sensing systems. PLoS ONE **14**(12), e0225688 (2019)
13. Schaub, A., Bazin, R., Hasan, O., Brunie, L.: A trustless privacy-preserving reputation system. In: 31st IFIP International Information Security and Privacy Conference (SEC), Belgium, pp. 398–411 (2016)
14. Soska, K., Kwon, A., Christin, N., Devadas, S.: Beaver: a decentralized anonymous marketplace with secure reputation. Cryptology ePrint Archive, 2016/464 (2016)
15. Dimitriou, T.: Decentralized reputation. In: 11th ACM Conference on Data and Application Security and Privacy, CODASPY 2021, New York, NY, USA, pp. 119–130 (2021)
16. Ben-Sasson, E., et al.: Zerocash: decentralized anonymous payments from Bitcoin. In: IEEE Symposium on Security and Privacy, pp. 459–474, Berkeley, CA, USA (2014)
17. Zhang, F., Maram, D., Malvai, H., Goldfeder, S., Juels, A.: DECO: liberating web data using Decentralized Oracles for TLS. In: ACM SIGSAC Conference on Computer and Communications Security, USA, pp. 1919–1938 (2020)
18. Breidenbach, L., et al.: Chainlink 2.0: Next Steps in the Evolution of Decentralized Oracle Networks (2021)
19. Gennaro, R.: Multi-trapdoor commitments and their applications to proofs of knowledge secure under concurrent man-in-the-middle attacks. In: Franklin, M. (ed.) CRYPTO 2004. LNCS, vol. 3152, pp. 220–236. Springer, Heidelberg (2004). https://doi.org/10.1007/978-3-540-28628-8_14
20. Groth, J.: On the size of pairing-based non-interactive arguments. In: Fischlin, M., Coron, J.-S. (eds.) EUROCRYPT 2016. LNCS, vol. 9666, pp. 305–326. Springer, Heidelberg (2016). https://doi.org/10.1007/978-3-662-49896-5_11
21. Ariel, G., Williamson, Z.J., Ciobotaru, O.: PlonK: permutations over Lagrange-bases for oecumenical noninteractive arguments of knowledge. Cryptology ePrint Archive, 2019/953 (2019)

An Empirical Comparison of Outlier Detection Methods for Identifying Grey-Sheep Users in Recommender Systems

Yong Zheng[✉]

Center for Decision Making and Optimization, Illinois Institute of Technology,
Chicago, IL, USA
yzheng66@iit.edu

Abstract. Collaborative filtering is a popular recommendation technique predicting user preferences through the analysis of similar users' historical behaviors, offering personalized recommendations based on shared interests. While collaborative filtering algorithms are widely used, they face well-known challenges like rating sparsity, cold-start problems, and the presence of grey-sheep users. The grey-sheep users are the users with uncommon item preferences, and they can be treated as outliers. It is surprising that the outlier detection technologies were not fully examined to identify the grey-sheep users. In this paper, our study addresses this gap by empirically comparing multiple state-of-the-art outlier detection methods and also introducing novel user representations to enhance the outlier detection process in recommender systems.

1 Introduction

Recommender systems (RSs), as a type of information systems to alleviate overloaded information, can generate personalized lists of recommended items based on user preferences. These recommendation algorithms can be categorized into content-based approaches (e.g., approaches using semantic ontologies or other natural language processing techniques) [27], collaborative filtering (e.g., neighborhood-based approaches, matrix factorization and deep neural network based methodologies) [11,35], and hybrid models (e.g., the switching and cascading methods) [8].

Collaborative filtering (CF) techniques have gained widespread popularity as algorithms, due to their ease of development and implementation, minimal data requirements, flexible extensions, and the benefits they offer in explaining recommendations. During the past decades, there were tons of recommendation algorithms proposed and developed based on the CF approaches, e.g., the neighborhood-based CF [31,32], matrix factorization [24], and neural matrix factorization [19], etc.

© The Author(s), under exclusive license to Springer Nature Switzerland AG 2024
L. Barolli (Ed.): AINA 2024, LNDECT 200, pp. 62–73, 2024.
https://doi.org/10.1007/978-3-031-57853-3_6

Though CF algorithms are popular in RSs, they suffer from well-known challenges. These include but are not limited to

- the *sparsity issue* [18,29] in rating preferences characterized by limited or insufficient user-item interactions or ratings;
- the *cold-start problems* [5,37] with the introduction of new users or items;
- the presence of the *grey-sheep users (GS users)* [3,17,41] who have unusual preferences that are not aligned with the majority of users.

There are several solutions proposed to alleviate the sparsity issue and the cold-start problems. However, tackling the presence of grey-sheep (GS) users remains an area of ongoing investigation. Existing research [17,41] universally acknowledges that GS users should be treated as outliers within user groups. Most researchers proposed to separate these GS users from the regular users, and other recommendation algorithms, such as the content-based approaches, should be applied for these GS users [13,17,36]. Despite the extensive study of outlier detection [9,21,39] in data mining and machine learning, it is surprising that no research has specifically compared these methods for identifying GS users in recommendation systems. This paper aims to fill this gap by presenting an empirical study of state-of-the-art outlier detection techniques for identifying GS users in recommendation systems.

Our contributions in this paper can be summarized in two folds – on one hand, we proposed to represent each user by using the latent-factor vectors which were demonstrated as being more effective than other representations. On the other hand, our empirical comparison of outlier detection approaches discovered more effective techniques for identifying GS users, in comparison with existing research. And we also delivered more insights about the relationships between user representations and outlier detection methods.

2 Related Work

In this section, we discuss related work in CF, GS users and outlier detection.

2.1 Collaborative Filtering

CF is a key technique employed in recommender systems, where it harnesses the collective preferences and behaviors of users to offer tailored recommendations. The traditional CF relies on the neighborhood-based approaches, such as user-based k-nearest neighbor (UserKNN) CF [31] and item-based KNN CF [32]. Take the UserKNN for example, it was built based on the principle that users who share similar tastes or have exhibited similar interactions in the past are likely to appreciate comparable items. The corresponding rating prediction function in UserKNN can be shown in Eq. 1, where a is a user, i is an item, and N_a is a user neighborhood of the target user a (i.e., k users similar to a). $r_{u,i}$ refers to the neighbor u's rating on the item i, and the *sim* function (e.g., Pearson correlation, cosine similarity or other similarity measures) is used to calculate the similarity between two users based on their co-rated items. The algorithm finally calculates $P_{a,i}$, which is the predicted rating that user a is expected to assign to item i.

$$P_{a,i} = \bar{r}_a + \frac{\sum\limits_{u \in N_a} (r_{u,i} - \bar{r}_u) \times sim(a,u)}{\sum\limits_{u \in N_a} sim(a,u)} \tag{1}$$

With the development of latent-factor models, matrix factorization [24] became a popular CF in RSs, due to its capability of alleviating the rating sparsity problems. In matrix factorization, each user or item can be represented by a vector associated with latent factors. The latent factors are used to represent underlying factors or reasons why a user likes items in a specific domain, e.g., movie genre, movie actors or directors in a movie domain. The user vector, therefore, captures the weights that refer to the importance on these latent factors from the perspective of a specific user. Recently, neural matrix factorization and neural collaborative filtering models [19] were proposed in order to take advantage of the form of generalized matrix factorization in the neural network structure.

2.2 Grey-Sheep Users

The *sparsity issue* [18,29], *cold-start problems* [5,37], and the presence of *GS users* [17,41] are three major challenges in CF. Though there are several existing solutions to alleviate the sparsity issue and cold-start problems, the challenge of GS users is still under investigation. At the early stage, researchers devoted their efforts in the definition of GS users. At the beginning, researchers believe that GS users have unusual tastes, and they do not agree or disagree with other users [13,28].

Later, J. McCrae et al. classified users within recommender systems into three distinct categories [28]. Firstly, the majority of users fall into the class of *White Sheep*, characterized by high rating correlations with several other users. On the other hand, *Black Sheep* users typically exhibit very few or even no correlating users, which is deemed acceptable given the context of rating sparsity. The more significant challenge arises with *Grey Sheep* users, who possess different opinions or uncommon tastes, leading to low correlations with many users and resulting in unconventional recommendations for their correlated peers. Consequently, a GS user is often described as "a small number of individuals who would not benefit from pure collaborative filtering systems because their opinions do not consistently agree or disagree with any group of people" [10]. Researchers universally agreed that GS users should be separated from the regular user groups, and other recommendation techniques, such as content-based approaches, should be applied to serve these GS users [13,17,36]. Therefore, the most important research problems here is the identification of GS users in recommender systems.

Different clustering [13,14,23] technologies were made to help identify GS users. Ghazanfar et al. [13,14] made the first attempt to identify GS users, separate them from regular users, and apply content-based recommendation algorithms to these GS users in order to provide better recommendations in comparison to the ones by UserKNN. In their approach, they believed that GS users may fall on the boundary of user clusters, therefore they modified the KMeans++ clustering technique and proposed a new centroid selection approach to help identify users on the cluster boundaries. The primary limitation of their approach lies in the challenge of determining the optimal number of clusters and the considerable computational cost required for convergence in the clustering process. Additionally, the technique is susceptible to unpredictable variations stemming from initial settings and other parameters. In their experiments, they illustrate that content-based recommendation algorithms can enhance recommendation performance for GS users. In contrast, Gras et al. [17] adopt outlier detection based on the distribution of user ratings, incorporating consideration for the imprecision of ratings, i.e., prediction errors. However, while the rating prediction error can assess whether a user is a GS user, it may not be suitable for specifically identifying GS users. This is because a GS user is not the sole factor leading to significant prediction errors.

According to the definitions of White, Black and GS users, we later proposed to use the distribution statistics of user-user similarities to represent each user [40,41], and utilized the local outlier factor (LOF) [6] to help identify the GS users. It is worth noting that we also proposed a methodology to validate the identified GS users. Namely, the prediction errors by using UserKNN for these GS users should be significantly higher than the errors for the non-GS users (i.e., other regular users excluding the identified GS users).

2.3 Outlier Detection

Despite the extensive study of outlier detection [9,21,39] in data mining and machine learning, it is surprising that no research has specifically compared these outlier detection methods for identifying GS users in recommendation systems. Existing research either developed their own approaches [13,14] or utilized limited or single outlier detection approach (e.g., LOF) [40,41] to identify GS users in RSs.

Anomaly detection or outlier detection offers approaches to identify unusual patterns, anomalies, or observations that deviate significantly from the norm within a dataset [9,21]. The goal of outlier detection is to sift through large datasets and pinpoint observations that exhibit unique characteristics, allowing analysts and systems to give special attention to these exceptional cases. Outlier detection methods can be broadly categorized into three main types: supervised [2], unsupervised [7], and semi-supervised [12] techniques. PyOD [39], an open-source toolbox for scalable outlier detection, was developed and released in 2019, where the state-of-art outlier detection approaches were implemented. These implemented methods can be classified into four categories:

- *Proximity-based approaches* rely on the neighborhood estimation (e.g., the KNN method [4]) or density of data clusters (e.g., the LOF approach [6]).
- *Linear models* involve constructing boundaries or representations of normal data patterns, such as the One-Class SVM [33] and principal component analysis (PCA) [34] methods.
- *Neural network based methods*, exemplified by AutoEncoder variants [1], leverage complex architectures to reconstruct input data and detect outliers based on discrepancies between the original and reconstructed data.
- *Ensembling approaches* combine the strengths of multiple base models to enhance outlier detection robustness. Isolation forest [26], for instance, creates an ensemble of isolation trees, efficiently isolating outliers.

It is worth noting that most of the outlier detection techniques in PyOD supports unlabeled data, since these approaches can produce outlier scores where we can rank data points by these scores to pick up the top ones as outliers. In our scenario of GS user identifications, we do not have any known labels, therefore we use these techniques in an unsupervised way.

3   Experimental Design and Workflow

We plan to utilize the PyOD library to perform an empirical comparison among the state-of-art outlier detection techniques for the purpose of identifying GS users in RSs. First of all, we selected two data sets for benchmark. One is the popular MovieLens data[1] where there are 100,000 ratings given by 943 users on 1682 movies, and the rating sparsity is 93.7%. Another one is the Yahoo!Movie data [22], where we used a subset of this data to select the overall ratings associated with users who gave at least 20 ratings. There are 450,067 ratings given by 811 users on 3066 movies, and the rating sparsity is 98.2% in this Yahoo!Movie data. In these two data sets, we make sure that each user has at least 20 ratings on the items, in order to avoid the case of black sheep users. Afterwards, we execute the following workflow:

1) Run UserKNN and matrix factorization models on these two data sets, tune up hyperparameters to get the best models. More specifically, we tuned the number of neighbors and similarity measures (i.e., Pearson correlation, cosine similarity, mean squared difference) in the UserKNN model, and we used a grid search to find the optimal number of latent factors, the number of max learning iterations, the learning and regularization rates in matrix factorization.
2) We save the prediction errors for each user based on the best performing UserKNN model. These prediction errors will be used to validate the identified GS users. As pointed out in our previous research [40,41], UserKNN suffers from the presence of GS users significantly. Therefore, the prediction errors by the identified GS users should be significant higher than the errors by

---

[1] https://grouplens.org/datasets/movielens/100k/.

other regular users. Matrix factorization may alleviate the issue of GS users slightly. In order to better detect GS users, we use the prediction errors from UserKNN for validation purpose only.

3) According to our optimal UserKNN and matrix factorization models, we are able to represent each user in two ways.

- The first method is representing a user by the distribution statistics of user-user similarities, and this method was demonstrated to work better than the clustering-based identification approaches [41]. More specifically, we use the optimal similarity measure (i.e., mean squared difference in our experiments for these two data sets) to calculate the user-user similarities between each two users. Given a user, we can acquire the similarity between this user and all other users. We finally represent this user by using a vector of distribution statistics of the similarities, including min, max, mean, q_1, q_2, q_3, standard deviation and skewness.

- We can also represent each user by using the user vector generated from matrix factorization. Each latent factor refers to the underlying reason why a user likes the items. The values in the vector denote the importance on these factors from the perspective of a user. In our experiments, the optimal number of latent factors was 100 in these two data sets. Therefore, each user was represented by a latent vector with length of 100. This representation was never utilized in existing methods of GS user identifications.

4) Due to that our previous research has demonstrated the effectiveness of using LOF on the user representations by similarity distributions over other existing approaches, we use LOF as the baseline approach and perform an empirical comparison of outlier detection methods in the PyOD library to identify GS users in these two data sets.

5) We utilize the same approach [41] to valid the identified GS users. More specifically, we use the Mann-Whitney U test to perform two independent sample hypothesis testing at 99% confidence level, based on the prediction errors by two groups of users – the identified GS users and the other regular users. If the test shows a significant result indicating the prediction errors by UserKNN for GS users are larger than the errors by other regular users, we consider the identified outliers as GS users. Note that, we also tune up hyper-parameters for each outlier detection, in order to find more outliers that can be validated as GS users by the Mann-Whitney U test.

The outlier detection techniques used in our experiments can be listed as follows:

- Six proximity-based approaches, including LOF [6] which evaluates the local density deviation of data points and identifies outliers according to the relative density compared with their neighbors, angle-based outlier detection (ABOD) [25] that assesses the angles between data points to identify outliers based on their distinct angular patterns, cluster-based LOF (CBLOF) [20] which leverages clustering and density-based methods to identify outliers based on the local outlier factor within clusters, histogram-based outlier score (HBOS) [15] that uses histogram-based analysis to score data points based on their deviation from the expected distribution, and two KNN based methods [30], largest KNN (LgKNN) and average KNN (AvgKNN) that utilize k-nearest neighbors to identify outliers based on the sizes of their local neighborhoods based on the largest and average distance, respectively.

- Two linear models, including one-class SVM (OCSVM) [33] that models normal data instances using a hyperplane and identifies outliers as instances lying beyond this learned boundary, and PCA [34] which identifies anomalies by analyzing the principal components of the data, effectively capturing variations and patterns indicative of outliers.
- Two neural network based methods, including AutoEncoders (AE) [1] which uses neural nets for outlier detection by reconstructing inputs and identifying anomalies through reconstruction errors, and LUNAR [16] that utilizes autoencoder and graph-based methods to rank anomalies by local behaviors in the feature space.
- Two ensembling approaches, including isolation forest (IFOREST) [26] which creates an ensemble of isolation trees to efficiently isolate outliers, and LSCP [38] which is an unsupervised parallel outlier detection ensemble which selects competent detectors in the local region of a test instance.

4   Experimental Results and Findings

As indicated by the workflow above, we first run UserKNN and matrix factorization on the two data sets. In the MovieLens data, matrix factorization outperformed UserKNN with a mean absolute error (MAE) of 0.719, surpassing the 0.732 MAE achieved by UserKNN. However, in the Yahoo! Movie dataset, UserKNN exhibited slightly superior performance, with a MAE of 0.641 compared to 0.644 by matrix factorization. Regarding optimal settings, mean squared difference proved to be the ideal similarity measure for UserKNN in both datasets, with the optimal number of user neighbors being 70 for MovieLens and 50 for Yahoo! Movie. For matrix factorization, the optimal number of factors was 100 in both datasets, with consistent learning and regularization rates of 0.01 and 0.1, respectively. The optimal maximal learning iteration was 50 for MovieLens and 150 for Yahoo! Movie.

Regarding the process of outlier detection, the experimental results on the MovieLens data can be shown in Table 1. There are results associated with 12 outlier detection approaches, where the LOF method using the distribution of similarities as the user representation is considered as the baseline approach [41]. We use the columns in yellow background to indicate the results using the distribution of similarities as the user representations, while the columns in green tell the results using the latent-factor vectors from matrix factorization as the user representations.

Table 1. Identification of GS users for the MovieLens Data

Models	Representation: Distribution of Similarities					Representation: Vector of Latent factors				
	# of GS	MAE (GS)	MAE (Non-GS)	Diff	p-value	# of GS	MAE (GS)	MAE (Non-GS)	Diff	p-value
LOF	N/A	N/A	N/A	N/A	N/A	89	1.101	0.722	0.379	1.25E-32
ABOD	N/A	N/A	N/A	N/A	N/A	100	1.038	0.725	0.313	3.39E-27
CBLOF	95	0.842	0.749	0.093	5.29E-03	95	1.098	0.72	0.378	1.27E-35
HBOS	95	0.91	0.741	0.169	1.63E-07	95	1.086	0.721	0.365	1.72E-33
LgKNN	93	0.841	0.749	0.092	9.06E-03	95	1.095	0.72	0.375	3.25E-35
AvgKNN	N/A	N/A	N/A	N/A	N/A	50	1.106	0.724	0.382	1.83E-31
OCSVM	95	0.959	0.736	0.223	2.21E-12	95	1.092	0.721	0.371	1.22E-34
PCA	N/A	N/A	N/A	N/A	N/A	95	1.091	0.721	0.37	5.01E-35
AE	95	0.871	0.745	0.126	2.83E-04	95	1.094	0.72	0.374	2.34E-35
LUNAR	95	0.848	0.748	0.1	4.44E-03	95	1.094	0.72	0.374	5.26E-34
IForest	N/A	N/A	N/A	N/A	N/A	95	1.083	0.722	0.361	3.39E-34
LSCP	95	1.018	0.729	0.289	1.35E-21	81	1.144	0.619	0.525	7.57E-28

In terms of columns indicating the results of GS user identifications, we list the number of successful GS users, the MAE associated with the identified GS users and non-GS users from UserKNN, the differences between these two MAEs, and the p-value in the Mann-Whitney U test which is a non-parametric two independent sample test. Be aware that we employ "NA" to signify unsuccessful GS user identifications, indicating that the identified users or outliers did not satisfy the Mann-Whitney U test at a 99% confidence level.

By using the distribution of similarities as the user representations on the MovieLens data, we can observe that we can find up to 95 valid GS users. If several outlier detection methods can find the same amount of the outliers, we further use the column "Diff" to figure out the best one. The method with largest number of identified GS users and also the largest difference in MAEs became the optimal approach in our experiments. Namely, LSCP presents the best in the MovieLens data if we the distribution of similarities as the user representations. Note that the LSCP uses a combination of outlier detection methods. In our experiments, we selected the top two best-performing methods (highlighted by using bold fonts in Table 1) to be integrated in the LSCP approach. Note that the baseline LOF failed to identify GS users if we use 99% as confidence level in the Mann-Whitney U test.

Alternatively, we can observe that the results by using the latent-factor vectors as user representations are better than using the distribution of similarities to represent users. On one hand, all outlier detection approaches can identify valid GS users, which indicates that the representation by using latent-factor vectors is more robust and effective. On the other hand, we can identify more valid GS users by using the ABOD approach. In addition, the differences in MAEs (in the green columns) are significantly higher than the differences shown in the yellow columns. The optimal model in this scenario is the ABOD approach, where the ensembling method, LSCP, did not achieve better results.

Table 2. Identification of GS users for the Yahoo!Movie Data

Models	# of GS	MAE (GS)	MAE (Non-GS)	Diff	p-value
LOF	76	1.116	0.625	0.491	6.74E-24
ABOD	**90**	**1.03**	**0.627**	**0.403**	**3.32E-19**
CBLOF	81	1.142	0.619	0.523	7.85E-27
HBOS	**81**	**1.147**	**0.619**	**0.528**	**8.34E-30**
LgKNN	81	1.102	0.624	0.478	1.02E-22
AvgKNN	66	1.165	0.628	0.537	4.20E-23
OCSVM	81	1.109	0.623	0.486	4.97E-25
PCA	81	1.095	0.624	0.471	3.17E-22
AE	81	1.095	0.624	0.471	3.17E-22
LUNAR	81	1.131	0.62	0.511	9.98E-25
IForest	81	1.134	0.62	0.514	1.17E-27
LSCP	81	1.14	0.619	0.521	1.06E-28

Table 2 presents the results of outlier detection methods by using the latent-factor vectors as user representations in the Yahoo!Movie data. Note that we did not present the results by using the distribution of similarities to represent users, since all outlier detection approaches cannot find any valid GS users (i.e., identified GS users that have significant higher MAEs than the ones by other users at 99%). If we lower the confidence level to 95%, ABOD and LUNAR approaches can find a pool of valid GS users. This observation, again, demonstrates the effectiveness and robustness by using the latent-factor vectors as user representations in the process of outlier detection for GS user identifications. Furthermore, there are two possible reasons which may lead to zero valid GS users identified if we use the distribution of similarities to represent users – one is the limited number of features in this representation methods, where we only have 8 features, including the min, max, mean, q_1, q_2, q_3, standard deviation and skewness from the distribution of similarities; another possible factor is the sparsity in the data. Note that the rating sparsity in the Yahoo!Movie data is 98.2%. With sparser ratings, the similarities calculated from users' co-rated items will not be reliable – this is also the reason why the user representation by using latent-factor vectors is better than the representation by using the distribution of user-user similarities.

According to the results in Table 2, we can observe that ABOD is the optimal outlier detection method, since it can identify the most number of valid GS users. The LSCP approach which combines ABOD and HBOS in this Yahoo!Movie data failed to offer further improvements or identify more valid GS users. Note that this ABOD approach also performs the best in the MovieLens data, if we use the latent-factor vectors as user representations.

5   Conclusions and Future Work

In this paper, we propose two approaches for representing users in RSs: one involves utilizing the distribution statistics of user-user similarities derived from UserKNN, and the other employs latent-factor vectors generated through matrix factorization. We conduct an empirical comparison of state-of-the-art outlier detection methods to effectively identify genuine GS users in RSs. Two key findings have emerged from our study. Firstly, using latent-factor vectors as user representations yields superior results in outlier detection compared to using the distribution of similarities to represent users. The weakness in the representation of using similarity distributions lie in the limited number of features and the unreliability of similarity values in the distribution if rating data is sparse. Additionally, we observe that ABOD stands out as the overall winner among all outlier detection approaches when latent-factor vectors represent users. In our future work, we plan to extend our studies beyond movie datasets and explore datasets in other domains. Furthermore, we aim to delve deeper into the distribution of unusual preferences over items (e.g., preferences over different item categories or content features) to gain a better understanding of identified GS users. This knowledge will enable us to develop more effective recommendation models tailored to the preferences of these GS users.

References

1. Aggarwal, C.C.: An Introduction to Outlier Analysis. Springer, Cham (2017)
2. Aggarwal, C.C.: Supervised outlier detection. In: Outlier Analysis, pp. 219–248 (2017)
3. Alabdulrahman, R., Viktor, H.: Catering for unique tastes: targeting grey-sheep users recommender systems through one-class machine learning. Expert Syst. Appl. **166**, 114061 (2021)
4. Angiulli, F., Pizzuti, C.: Fast outlier detection in high dimensional spaces. In: Elomaa, T., Mannila, H., Toivonen, H. (eds.) PKDD 2002. LNCS, vol. 2431, pp. 15–27. Springer, Heidelberg (2002). https://doi.org/10.1007/3-540-45681-3_2
5. Bobadilla, J., Ortega, F., Hernando, A., Bernal, J.: A collaborative filtering approach to mitigate the new user cold start problem. Knowl.-Based Syst. **26**, 225–238 (2012)
6. Breunig, M.M., Kriegel, H.P., Ng, R.T., Sander, J.: LoF: identifying density-based local outliers. In: ACM Sigmod Record, vol. 29, pp. 93–104. ACM (2000)
7. Campos, G.O., Zimek, A., Sander, J., Campello, R.J., Micenková, B., Schubert, E., Assent, I., Houle, M.E.: On the evaluation of unsupervised outlier detection: measures, datasets, and an empirical study. Data Min. Knowl. Disc. **30**, 891–927 (2016)
8. Çano, E., Morisio, M.: Hybrid recommender systems: a systematic literature review. Intell. Data Anal. **21**(6), 1487–1524 (2017)
9. Chandola, V., Banerjee, A., Kumar, V.: Anomaly detection: a survey. ACM Comput. Surv. (CSUR) **41**(3), 15 (2009)
10. Claypool, M., Gokhale, A., Miranda, T., Murnikov, P., Netes, D., Sartin, M.: Combining content-based and collaborative filters in an online newspaper. In: Proceedings of ACM SIGIR Workshop on Recommender Systems, vol. 60 (1999)

11. Ekstrand, M.D., Riedl, J.T., Konstan, J.A., et al.: Collaborative filtering recommender systems. Found. Trends® Hum.-Comput. Interact. **4**(2), 81–173 (2011)
12. Gao, J., Cheng, H., Tan, P.N.: Semi-supervised outlier detection. In: Proceedings of the 2006 ACM Symposium on Applied Computing, pp. 635–636 (2006)
13. Ghazanfar, M., Prugel-Bennett, A.: Fulfilling the needs of gray-sheep users in recommender systems, a clustering solution. In: Proceedings of the 2011 International Conference on Information Systems and Computational Intelligence, pp. 18–20 (2011)
14. Ghazanfar, M.A., Prügel-Bennett, A.: Leveraging clustering approaches to solve the gray-sheep users problem in recommender systems. Expert Syst. Appl. **41**(7), 3261–3275 (2014)
15. Goldstein, M., Dengel, A.: Histogram-based outlier score (HBOS): a fast unsupervised anomaly detection algorithm. In: KI-2012: Poster and Demo Track, vol. 1, pp. 59–63 (2012)
16. Goodge, A., Hooi, B., Ng, S.K., Ng, W.S.: Lunar: unifying local outlier detection methods via graph neural networks. In: Proceedings of the AAAI Conference on Artificial Intelligence, vol. 36, pp. 6737–6745 (2022)
17. Gras, B., Brun, A., Boyer, A.: Identifying grey sheep users in collaborative filtering: a distribution-based technique. In: Proceedings of the 2016 Conference on User Modeling Adaptation and Personalization, pp. 17–26. ACM (2016)
18. Grčar, M., Mladenič, D., Fortuna, B., Grobelnik, M.: Data sparsity issues in the collaborative filtering framework. In: Nasraoui, O., Zaïane, O., Spiliopoulou, M., Mobasher, B., Masand, B., Yu, P.S. (eds.) WebKDD 2005. LNCS, vol. 4198, pp. 58–76. Springer, Heidelberg (2006). https://doi.org/10.1007/11891321_4
19. He, X., Liao, L., Zhang, H., Nie, L., Hu, X., Chua, T.S.: Neural collaborative filtering. In: Proceedings of the 26th International Conference on World Wide Web, pp. 173–182 (2017)
20. He, Z., Xu, X., Deng, S.: Discovering cluster-based local outliers. Pattern Recogn. Lett. **24**(9–10), 1641–1650 (2003)
21. Hodge, V., Austin, J.: A survey of outlier detection methodologies. Artif. Intell. Rev. **22**(2), 85–126 (2004)
22. Jannach, D., Zanker, M., Fuchs, M.: Leveraging multi-criteria customer feedback for satisfaction analysis and improved recommendations. Inf. Technol. Tourism **14**, 119–149 (2014)
23. Kaur, B., Rani, S.: Identification of gray sheep using different clustering algorithms. In: Goyal, D., Gupta, A.K., Piuri, V., Ganzha, M., Paprzycki, M. (eds.) ICIMMI 2020. LNNS, vol. 166, pp. 211–217. Springer, Singapore (2021). https://doi.org/10.1007/978-981-15-9689-6_24
24. Koren, Y., Bell, R., Volinsky, C.: Matrix factorization techniques for recommender systems. Computer **42**(8), 30–37 (2009)
25. Kriegel, H.P., Schubert, M., Zimek, A.: Angle-based outlier detection in high-dimensional data. In: Proceedings of the 14th ACM SIGKDD International Conference on Knowledge Discovery and Data Mining, pp. 444–452 (2008)
26. Liu, F.T., Ting, K.M., Zhou, Z.H.: Isolation forest. In: 2008 Eighth IEEE International Conference on Data Mining, pp. 413–422. IEEE (2008)
27. Lops, P., De Gemmis, M., Semeraro, G.: Content-based recommender systems: state of the art and trends. In: Recommender Systems Handbook, pp. 73–105 (2011)
28. McCrae, J., Piatek, A., Langley, A.: Collaborative filtering (2004). http://www.imperialviolet.org

29. Pan, W., Xiang, E., Liu, N., Yang, Q.: Transfer learning in collaborative filtering for sparsity reduction. In: Proceedings of the AAAI Conference on Artificial Intelligence, vol. 24, pp. 230–235 (2010)
30. Ramaswamy, S., Rastogi, R., Shim, K.: Efficient algorithms for mining outliers from large data sets. In: Proceedings of the 2000 ACM SIGMOD International Conference on Management of Data, pp. 427–438 (2000)
31. Resnick, P., Iacovou, N., Suchak, M., Bergstrom, P., Riedl, J.: Grouplens: an open architecture for collaborative filtering of netnews. In: Proceedings of the 1994 ACM Conference on Computer Supported Cooperative Work, pp. 175–186. ACM (1994)
32. Sarwar, B., Karypis, G., Konstan, J., Riedl, J.: Item-based collaborative filtering recommendation algorithms. In: Proceedings of the 10th International Conference on World Wide Web, pp. 285–295 (2001)
33. Schölkopf, B., Platt, J.C., Shawe-Taylor, J., Smola, A.J., Williamson, R.C.: Estimating the support of a high-dimensional distribution. Neural Comput. **13**(7), 1443–1471 (2001)
34. Shyu, M.L., Chen, S.C., Sarinnapakorn, K., Chang, L.: A novel anomaly detection scheme based on principal component classifier. In: Proceedings of the IEEE Foundations and New Directions of Data Mining Workshop, pp. 172–179. IEEE Press (2003)
35. Su, X., Khoshgoftaar, T.M.: A survey of collaborative filtering techniques. In: Advances in Artificial Intelligence, vol. 2009, p. 4 (2009)
36. Tennakoon, A., Gamlath, N., Kirindage, G., Ranatunga, J., Haddela, P., Kaveendri, D.: Hybrid recommender for condensed sinhala news with grey sheep user identification. In: 2020 2nd International Conference on Advancements in Computing (ICAC), vol. 1, pp. 228–233. IEEE (2020)
37. Wei, J., He, J., Chen, K., Zhou, Y., Tang, Z.: Collaborative filtering and deep learning based recommendation system for cold start items. Expert Syst. Appl. **69**, 29–39 (2017)
38. Zhao, Y., Nasrullah, Z., Hryniewicki, M.K., Li, Z.: LSCP: locally selective combination in parallel outlier ensembles. In: Proceedings of the 2019 SIAM International Conference on Data Mining, pp. 585–593. SIAM (2019)
39. Zhao, Y., Nasrullah, Z., Li, Z.: PyOD: a python toolbox for scalable outlier detection. J. Mach. Learn. Res. **20**(96), 1–7 (2019). http://jmlr.org/papers/v20/19-011.html
40. Zheng, Y., Agnani, M., Singh, M.: Identification of grey sheep users by histogram intersection in recommender systems. In: Cong, G., Peng, W.C., Zhang, W., Li, C., Sun, A. (eds.) ADMA 2017. LNCS, vol. 10604, pp. 148–161. Springer, Cham (2017). https://doi.org/10.1007/978-3-319-69179-4_11
41. Zheng, Y., Agnani, M., Singh, M.: Identifying grey sheep users by the distribution of user similarities in collaborative filtering. In: Proceedings of The 6th ACM Conference on Research in Information Technology. ACM (2017)

Adaptive Consensus: Enhancing Robustness in Dynamic Environments

Kshitij Mandyal$^{(\boxtimes)}$ and Dharmendra Prasad Mahato

Department of Computer Science and Engineering, National Institute of Technology Hamirpur, Hamirpur 177005, Himachal Pradesh, India
{22mcs104,dpm}@nith.ac.in

Abstract. Despite being one of the most researched subjects, there are still decades-long gaps in understanding regarding consensus. In particular, a basic concern about the communication complexity of fast randomised Consensus against a (strong) adaptive adversary who crashes processes arbitrarily online remains unresolved in the classic message-passing situation where processes fail. It dates back to the groundbreaking works of B. Joseph and Ben-Or [PODC 1998] and Aspnes and Waarts [JACM 1998, SICOMP 1996]. Later, Hajiaghayi, Kowalski, and Olkowski created an algorithm against adaptive adversary that maintains nearly optimal (up to factor $O(log^3 n)$) time complexity $O(\sqrt{n} \times log^{5/2} n)$ while reducing the communication gap a nearly linear factor to $O(\sqrt{n} \times polylogn)$ bits per process. The algorithm worked in three phases namely Phase 1, Phase 2 and Phase 3. In this paper we will focus mainly on the Phase 1 part of the algorithm. The previous algorithm utilized a fixed threshold value to determine whether to broadcast a message containing 1 or to stay silent. We propose a new algorithm that adapts to this threshold dynamically based on the current state of the system to improve responsiveness and efficiency.

Keywords: parameterised consensus · distributed consensus

1 Introduction

Fault-tolerant Consensus is one of the fundamental issues in distributed computing, which occurs when several independent processes wish to concur on a common value among the original ones despite process or communication medium failures [23].

In distributed systems, faulty components or malicious nodes looking to cause havoc-make it difficult to come to an agreement. These "adversaries" can manipulate data or crash systems, depending on their abilities, knowledge, and goals. We create algorithms that anticipate their movements and maintain resilience to keep the agreement in sync. This study dives into the world of fault-tolerant consensus, where reaching a consensus becomes a strategic battle against disruptions, maintaining system integrity even in risky situations.

Since its introduction Pease, Shostak, and Lamport [28,29], a great deal of algorithms and impossibility results have been created and examined. These have

L. Barolli (Ed.): AINA 2024, LNDECT 200, pp. 74–84, 2024.
https://doi.org/10.1007/978-3-031-57853-3_7

been used to address various distributed computing and system problems as well as to uncover a number of new, significant issues and their solutions [8]. In the majority of the classical distributed models, we are still far from achieving even asymptotically optimal solutions, despite our continuous efforts in this direction.

In particular, since the fundamental works of Bar-Joseph and Ben-Or [9] and Aspnes and Waarts [5,6] in the previous century in the classical message-passing setting with crashes of processes, there is still an open topic concerning the communication complexity of randomised consensus. A polylogarithmic amortised amount of communication bits can be found in a deterministic approach; however, deterministic solutions are at least linearly slow, as demonstrated by Fisher and Lynch [18]. Additionally, randomised techniques with amortised communication efficiency and speed are employed against weak opponents; as demonstrated by Gilbert and Kowalski [22], both formulas are $O(\log n)$ or better.

In this situation, there is no time-and-communication optimal technique, as Hajiaghayi, Kowalski, and Olkowski [26] show. Moreover, a parameterized algorithm improves amortised communication by almost a linear factor.

Consensus Problem:

Consensus, defined by the following three conditions, is the process by which all non-crashed processes come to an agreement on some of the input values:

Validity: A decision may be made based on only one of the starting values.

Consensus: No two processes make distinct value decisions.

Termination: At some point, every alive process makes a decision.

1.1 Our Contributions

The Parameterized Consensus algorithm proposed by M.T. Hajiaghayi, D.R. Kowalski, and J. Olkowski [26] shows incredibly good results in terms of time and space complexity. The algorithm utilizes a three-phase approach to reach a consensus decision among a set of processes. The first phase involves calculating local values and selecting a candidate value. The second phase confirms the candidate value through a consensus protocol. Finally, the third phase broadcasts the confirmed decision value to all processes.

In this research paper, we will mainly focus on the Phase 1 of the Parameterized Consensus algorithm in which It calculates local values, selects a candidate value, and communicates with neighboring super-processes. We will focus on the dynamic threshold adjustment in this algorithm. The algorithm utilizes a fixed threshold value to determine whether to broadcast a message containing 1 or stay silent. We are proposing a new algorithmic approach to adapt this threshold dynamically based on the current state of the system to improve responsiveness and efficiency.

If we compare the complexity of the initial algorithm with the proposed algorithm then we come to an interesting observation.

In static environments, where the network topology and the values of the super-processes remain constant, the initial algorithm generally has better time complexity. This is due to the fact that the suggested algorithm's time complexity increases logarithmically with the number of processes (n), yet it grows linearly with the number of iterations (x). The original technique is more efficient in static situations since the number of iterations (x) is usually significantly fewer than the number of processes (n).

Nonetheless, the suggested technique can be advantageous in dynamic contexts where the values of the super-processes or the network architecture can fluctuate regularly. This is a result of the suggested algorithm's increased resistance to modifications in the super-process variables and network topology. The convergence of the method may be impacted by changes in the super-process values and network architecture, which are more susceptible to changes in the original algorithm. Also, the modified algorithm does not change the behavior of the overall algorithm. The modified algorithm only adapts the threshold in the dynamic environments and does not affect the stability of the overall algorithm.

Therefore, the choice of which algorithm to use depends on the specific requirements of the application. If the application requires a fast and efficient algorithm in static environments, then the initial algorithm is a good choice. If the application requires a more resilient algorithm in dynamic environments, then the proposed algorithm is a better choice.

2 Related Work

Early Work on Consensus. Pease, Shostak, and Lamport [28,29] introduced the Consensus problem. Deterministic solutions were the subject of early work. Fisher, Lynch, and Paterson [19] demonstrated that, even in the event that one process fails, the problem remains unsolvable in an asynchronous environment. Fisher and Lynch [18] demonstrated that if up to f processes crash, a synchronous solution needs $f + 1$ rounds.

When taking into account each of these performance indicators independently, the optimal complexity of consensus with crashes is known in terms of both time and message count (or communication bits). It was demonstrated by Amdur, Weber, and Hadzilacos [4,24] that even in certain failure-free executions, the amortised quantity of messages per process remains at least constant. The optimal deterministic algorithm, presented by Chlebus, Kowalski, and Strojnowski [11–13] in, resolves consensus in an amortised amount of communication bits per process $O(polylogn)$ and an asymptotically optimal time $\theta(n)$.

Efficient Randomized Solutions for Weak Adversaries. It turned out that randomness could break through a linear time barrier for time complexity. But whenever randomization is taken into consideration, many adversary-generating failure scenarios could be taken into consideration. Consensus algorithms for oblivious adversaries were created by Chor, Merritt, and Shmoys [15]. In other words, the adversary that knows the algorithm but must determine which process fails and when before the execution begins. A randomised consensus algorithm was proposed by Gilbert and Kowalski [22]. It uses $O(1)$ amortised communication bits per process to achieve optimal communication complexity.

It terminates in $O(\log n)$ time with high probability, allowing crash failures of up to $f < n/2$.

Randomized Solutions Against (Strong) Adaptive Adversary. It is well known that reaching consensus against an adaptive adversary is more expensive than reaching consensus against weaker adversaries. Bar-Joseph and Ben-Or [9] provided the time-optimal randomised solution to the consensus problem. Their technique utilises $O(\frac{n^{3/2}}{\log n})$ amortised communications bits per process, and it operates in $O(\frac{\sqrt{n}}{\log n})$ anticipated time. Additionally, they demonstrated that their conclusion was ideal in terms of time complexity, but in this case, the communication is significantly improved.

Impact of Asynchrony on Consensus Complexity. Dolev and Reischuk [16] and Hadzilacos et al. [17,25] established an amortized message complexity lower bound for deterministic consensus in Byzantine failures [20]. King and Saia [27] demonstrated sublinear expected communication complexity under certain adversary limitations. Abraham et al. [1] emphasized the necessity of limitations for achieving subquadratic time complexity in Byzantine failures. Alistarh et al. [2,8] showed nearly optimal communication complexity ($O(n \log n)$) in asynchronous scenarios with fewer than $n/2$ possible process failures, improving upon Aspnes and Waarts' [6] earlier $O(n \log^2 n)$ result. The asymptotic almost optimality is supported by Attiya and Censor-Hillel's [7] $\Omega(n)$ lower bound, while Aspnes [5] provided a $\Omega(n/\log^2 n)$ lower bound on the expected coin flips.

Chlebus and Kowalski [11–14] introduced Fault-tolerant Gossip, presenting a deterministic algorithm solving Gossip in $O(\log^2 f)$ time with $O(\log^2 f)$ amortized messages per process, given $n - f = \Omega(n)$.[16] They established a lower bound of $\Omega(\frac{\log n}{\log(n \log n) - \log f})$ on the rounds for $O(polylogn)$ amortized messages. Subsequent papers [10,11,21] achieved $O(polylogn)$ message complexity per process for $f < n$, maintaining polylogarithmic time complexity. General Gossip, however, necessitates $\Omega(n)$ communication bits per process. For randomized gossip against an adaptive adversary, $O(n \log^2 n)$ rounds with $O(n \log^3 n)$ communication bits per process are achievable for a constant number of rumors of constant size and $f < \frac{n}{3}$ [3,8].

From the perspectives of both time and communication optimality, M.T. Hajiaghayi, D.R. Kowalski, and J. Olkowski [26] investigated the Consensus problem in the traditional message-passing paradigm with processes crashing. Between these two complexity measurements, they found an intriguing trade-off:

$$Time \times Amortized\_Communication = \Omega(n).$$

3   Methodology

In the realm of distributed computing, consensus algorithms play a crucial role in ensuring consistent data across multiple nodes. However, traditional consensus algorithms, such as ParameterizedConsensus:Phase_1, face challenges in maintaining consensus in dynamic environments where network topology and node values are prone to frequent changes. To address these challenges, this research proposes utilizing ParameterizedConsensus:Phase_1 (Dynamic Adjustment) to enhance the capability of consensus algorithms in dynamic environments.

Overview and Operation of ParameterizedConsensus:Phase_1

ParameterizedConsensus:Phase_1, is a probabilistic consensus algorithm that relies on a gossip-based approach to achieve consensus. It operates by iteratively updating the candidate value based on the opinions of neighboring nodes. During each iteration, nodes broadcast their current candidate value to their neighbors and then update their own candidate value based on the majority opinion among their neighbors.

ParameterizedConsensus:Phase_1 is particularly well-suited for static environments due to its efficient convergence properties and low communication overhead. However, in dynamic environments, its performance can degrade due to network disruptions and changes in node values. The constant broadcasting of messages can become inefficient in dynamic networks, and changes in node values can lead to oscillations and delays in reaching consensus.

Advantages of the Proposed Algorithm in Dynamic Environments

In dynamic situations, ParameterizedConsensus:Phase_1 (Dynamic Adjustment) has a number of advantages over it. It makes use of a parameter-based strategy that adjusts to the network's dynamic characteristics. The algorithm can adapt to changes in network topology and node values by dynamically adjusting its parameters based on the status of the system at that moment.

Reducing communication overhead in dynamic contexts is one of ParameterizedConsensus:Phase_1 (Dynamic Adjustment)'s main advantages. It minimises pointless message transfers, saving bandwidth and lessening network congestion by adaptively changing its parameters. Furthermore, due to its parameter-based design, ParameterizedConsensus:Phase_1 (Dynamic Adjustment) is more resistant to variations in node values, guaranteeing that consensus is upheld even in the event of node failures or shifts in node opinions.

Applications of the Proposed Algorithm. ParameterizedConsensus: Phase_1 (Dynamic Adjustment) works very effectively in dynamic settings with fluctuating node values and erratic network conditions. It is especially helpful in the following applications:

1. **Distributed sensor networks:** In sensor networks, nodes may become unavailable due to battery depletion or environmental factors. The proposed algorithm can maintain consensus even when node membership changes dynamically.
2. **Blockchain systems:** In blockchain networks, nodes may join or leave the network frequently, and block propagation can be delayed due to network congestion. The proposed algorithm can ensure consensus among nodes despite these challenges.
3. **Distributed data storage:** In distributed data storage systems, replicating data across multiple nodes requires consistent consensus algorithms. The

proposed algorithm can maintain data consistency even in dynamic environments.

3.1 Proposed Algorithm

We have modified the existing algorithm proposed by Mohammad et al. [26] as below. The Algorithm 1 proposed by Mohammad et al. [26], is designed to particularly be well-suited for static environments due to its efficient convergence properties and low communication overhead.

In our paper, we have modified the algorithm by using a different approach to dynamically adjust the threshold. For instance, when the system is experiencing high network congestion or a large number of processes are inactive, the threshold could be increased to reduce unnecessary communication and improve efficiency. Conversely, when the system is operating smoothly and a large number of processes are active, the threshold could be decreased to facilitate faster convergence and improved responsiveness. This will ultimately help the system to become more robust in changing environments.

The following presumptions regarding the threat and fault models are necessary for the algorithm to be robust:

1. **Fault Model:** While crashing and Byzantine failures can occur in processes, the overall number of faulty processes is limited to f, a known system parameter.
2. **Threat Model:** Adversaries may try to disrupt communication, introduce misleading data, or even take the identity of authorized participants. They cannot, however, take control over more than f processes.

3.2 Working

We can understand the proposed algorithm in four simple steps:

1. **Monitor System State:** Continuously monitor the state of the system, including metrics such as the number of active processes, the rate of message exchanges, and the average message propagation delay.
2. **Adapt Threshold Dynamically:** Based on the monitored system state, dynamically adjust the threshold value used to determine whether to broadcast a message containing 1 or stay silent.
3. **Periodic Threshold Updates:** Implement a mechanism to periodically update the threshold value based on the latest system state information. This ensures that the threshold remains relevant and adaptable to changing system conditions.
4. **Hysteresis Mechanism:** Incorporate a hysteresis mechanism to prevent the threshold from fluctuating too rapidly. This can help stabilize the behavior of the algorithm and avoid unnecessary threshold adjustments.

INPUT:

1. P: The set of all processes in the system.
2. $\{SP_1, ..., SP_x\}$: A list of all super-processes in the system.
3. H: The communication graph representing the connections between super-processes.
4. x: The number of super-processes in the system.
5. p: The identifier of the current process.
6. b_p: The bias parameter used in the BiasedConsensus function.

OUTPUT:

1. Candidate value

Algorithm 1: *ParameterizedConsensus:Phase_1 (Dynamic Adjustment)*

---

1. Initialize `is_active` to true.
2. Calculate the initial candidate value using `BiasedConsensus`(p, SP[p], bp).
3. Define a function `getThreshold()` that dynamically determines the threshold based on the current system state.
4. Enter the main loop:
   (a) If the candidate value is 1, update it using `BiasedConsensus`(p, SP[p], candidate_value).
   (b) Calculate the current threshold using the `getThreshold()` function.
   (c) If `is_active` is true and the candidate value is 1, broadcast a message containing 1 to every member of the neighboring super-processes in $SE(SP[p], SPj)$ using a gossip-based approach.
   (d) Stay silent for a random number of rounds determined by the current threshold.
   (e) If `is_active` is true and the candidate value is 1, check if any message containing 1 was received in the previous round.
       i. If a message containing 1 was received, set `is_active` to false.
   (f) If the number of rounds since the last broadcast exceeds a pre-defined limit, set `is_active` to false.
5. Return the final candidate value.

3.3 Comparison and Analysis

Figure 1 shows a graphical representation of how our proposed algorithm performs against the other existing algorithm in a static environment and Fig. 2 shows a graphical representation of how our proposed algorithm performs against the other existing algorithm in a dynamic environment.

The analysis done on the algorithm required a simulation of a dynamic environment where the system state is evolving and the external factors are influencing the system behavior. The experiment also includes the scenario of node failures. Thus, proving the robustness of the proposed algorithm in dynamic environments.

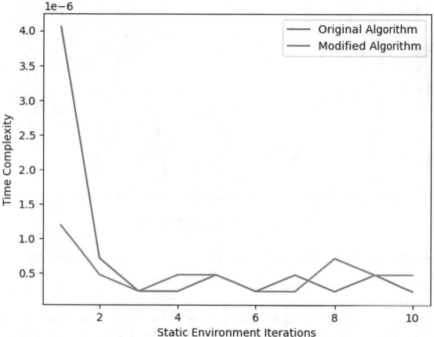

Fig. 1. Comparison of Algorithm Time Complexity in Static Environments

Over multiple iterations, we can see that on average, the modified algorithm is performing better than the original algorithm. The modification in the proposed algorithm is specifically geared towards adapting to dynamic environments by introducing a dynamic threshold ('getThreshold()' function). This dynamic threshold allows the algorithm to adjust its behavior based on the current system state, potentially making it more responsive to changes. Also, the space complexity of the modified algorithm remains the same as the modified algorithm does not store any additional data structure which could increase the memory usage. Hence, providing benefits in dynamic environments without increasing the space complexity.

In dynamic environments, where conditions may change rapidly, having adaptive features can be advantageous. Here are some reasons why the modified version might perform better in dynamic environments:

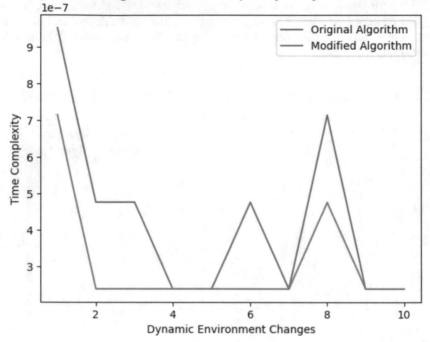

Fig. 2. Comparison of Algorithm Time Complexity in Static Environments

1. **Dynamic Threshold:**
   The 'getThreshold()' function has been added to enable the algorithm to dynamically decide the threshold depending on the status of the system at that moment. This flexibility can be quite important for managing changes and uncertainties in a changing setting.
2. **Flexible Silence Periods:** The modified version incorporates a dynamic silence period based on the current threshold. This flexibility allows the algorithm to adjust the duration of silent periods, potentially improving responsiveness to changes.
3. **Adaptive Communication:**
   For the exchange of information, both versions rely heavily on gossip-based communication. With its dynamic threshold, the updated version may be able to change the communication frequency in response to changes in the system's observed dynamics.

It's crucial to remember that real performance depends on the dynamic environment's unique traits as well as the actions of functions such as Gossip and BiasedConsensus. Flexibility may result in extra costs, and a set approach may work better.

4   Conclusion and Future Directions

The modified algorithm with dynamic threshold adjustment is an efficient and responsive approach for reaching consensus in dynamic environments. By adapting its behavior to the current system state, the modified algorithm reduces unnecessary communication overhead and improves responsiveness compared to the original algorithm. This makes it a suitable choice for systems that experience varying network conditions or process activity levels.

Future studies can be conducted on this proposed algorithm to develop more sophisticated threshold functions. One can even try hybrid approaches to get better results and conduct theoretical analysis to formally characterize the performance of the modified algorithm in different dynamic environments.

References

1. Abraham, I., et al.: Communication complexity of byzantine agreement, revisited. In: Proceedings of the 2019 ACM Symposium on Principles of Distributed Computing, pp. 317–326 (2019)
2. Alistarh, D., Aspnes, J., King, V., Saia, J.: Communication-efficient randomized consensus. Distrib. Comput. **31**, 489–501 (2018)
3. Alistarh, D., Gilbert, S., Guerraoui, R., Zadimoghaddam, M.: How efficient can gossip be? (on the cost of resilient information exchange). In: Abramsky, S., Gavoille, C., Kirchner, C., Meyer auf der Heide, F., Spirakis, P.G. (eds.) ICALP 2010. LNCS, vol. 6199, pp. 115–126. Springer, Heidelberg (2010). https://doi.org/10.1007/978-3-642-14162-1_10
4. Amdur, E.S., Weber, S.M., Hadzilacos, V.: On the message complexity of binary byzantine agreement under crash failures. Distrib. Comput. **5**, 175–186 (1992)
5. Aspnes, J.: Lower bounds for distributed coin-flipping and randomized consensus. J. ACM (JACM) **45**(3), 415–450 (1998)
6. Aspnes, J., Waarts, O.: Randomized consensus in expected $o(n\log^2 n)$ operations per processor. SIAM J. Comput. **25**(5), 1024–1044 (1996)
7. Attiya, H., Censor, K.: Lower bounds for randomized consensus under a weak adversary. In: Proceedings of the Twenty-Seventh ACM Symposium on Principles of Distributed Computing, pp. 315–324 (2008)
8. Attiya, H., Welch, J.: Distributed Computing: Fundamentals, Simulations, and Advanced Topics, vol. 19. Wiley, Hoboken (2004)
9. Bar-Joseph, Z., Ben-Or, M.: A tight lower bound for randomized synchronous consensus. In: Proceedings of the Seventeenth Annual ACM Symposium on Principles of Distributed Computing, pp. 193–199 (1998)
10. Chlebus, B.S., Kowalski, D.R.: Robust gossiping with an application to consensus. J. Comput. Syst. Sci. **72**(8), 1262–1281 (2006)
11. Chlebus, B.S., Kowalski, D.R.: Time and communication efficient consensus for crash failures. In: Dolev, S. (ed.) DISC 2006. LNCS, vol. 4167, pp. 314–328. Springer, Heidelberg (2006). https://doi.org/10.1007/11864219_22
12. Chlebus, B.S., Kowalski, D.R.: Locally scalable randomized consensus for synchronous crash failures. In: Proceedings of the Twenty-First Annual Symposium on Parallelism in Algorithms and Architectures, pp. 290–299 (2009)

13. Chlebus, B.S., Kowalski, D.R., Olkowski, J.: Fast agreement in networks with byzantine nodes. In: 34th International Symposium on Distributed Computing (DISC 2020). Schloss Dagstuhl-Leibniz-Zentrum für Informatik (2020)
14. Chlebus, B.S., Kowalski, D.R., Strojnowski, M.: Fast scalable deterministic consensus for crash failures. In: Proceedings of the 28th ACM Symposium on Principles of Distributed Computing, pp. 111–120 (2009)
15. Chor, B., Merritt, M., Shmoys, D.B.: Simple constant-time consensus protocols in realistic failure models. J. ACM (JACM) **36**(3), 591–614 (1989)
16. Dolev, D., Reischuk, R.: Bounds on information exchange for byzantine agreement. J. ACM (JACM) **32**(1), 191–204 (1985)
17. Dwork, C., Halpern, J.Y., Waarts, O.: Performing work efficiently in the presence of faults. In: Proceedings of the Eleventh Annual ACM Symposium on Principles of Distributed Computing, pp. 91–102 (1992)
18. Fischer, M.J., Lynch, N.A.: A lower bound for the time to assure interactive consistency. Inf. Process. Lett. **14**(4), 183–186 (1982)
19. Fischer, M.J., Lynch, N.A., Paterson, M.S.: Impossibility of distributed consensus with one faulty process. J. ACM (JACM) **32**(2), 374–382 (1985)
20. Galil, Z., Mayer, A., Yung, M.: Resolving message complexity of byzantine agreement and beyond. In: Proceedings of IEEE 36th Annual Foundations of Computer Science, pp. 724–733. IEEE (1995)
21. Georgiou, C., Kowalski, D.R., Shvartsman, A.A.: Efficient gossip and robust distributed computation. Theor. Comput. Sci. **347**(1–2), 130–166 (2005)
22. Gilbert, S., Kowalski, D.R.: Distributed agreement with optimal communication complexity. In: Proceedings of the Twenty-First Annual ACM-SIAM Symposium on Discrete Algorithms, pp. 965–977. SIAM (2010)
23. Gupta, I., Van Renesse, R., Birman, K.P.: Scalable fault-tolerant aggregation in large process groups. In: 2001 International Conference on Dependable Systems and Networks, pp. 433–442. IEEE (2001)
24. Hadzilacos, V.: Fault-tolerant broadcasts and related problems. In: Distributed Systems, pp. 97–145 (1993)
25. Hadzilacos, V., Halpern, J.Y.: Message-optimal protocols for byzantine agreement. In: Proceedings of the Tenth Annual ACM Symposium on Principles of Distributed Computing, pp. 309–323 (1991)
26. Hajiaghayi, M.T., Kowalski, D.R., Olkowski, J.: Improved communication complexity of fault-tolerant consensus. In: Proceedings of the 54th Annual ACM SIGACT Symposium on Theory of Computing, STOC 2022, pp. 488–501. Association for Computing Machinery, New York (2022)
27. King, V., Saia, J.: Breaking the o (n 2) bit barrier: scalable byzantine agreement with an adaptive adversary. J. ACM (JACM) **58**(4), 1–24 (2011)
28. Lamport, L., Shostak, R., Pease, M.: The byzantine generals problem. In: Concurrency: The Works of Leslie Lamport, pp. 203–226 (2019)
29. Pease, M., Shostak, R., Lamport, L.: Reaching agreement in the presence of faults. J. ACM **27**(2), 228–234 (1980)

COTTONTRUST: Reliability and Traceability in Cotton Supply Chain Using Self-sovereign Identity

Janaína F. B. Duarte, Gilson S. Junior, Gabriel F. C. da Silva,
Maurício Pillon$^{(\boxtimes)}$, Guilherme P. Koslovski, and Charles C. Miers

Graduate Program in Applied Computing (PPGCAP), Santa Catarina State
University (UDESC), Joinville, Brazil
{janaina.duarte,gilson.sj,gabriel.silva111}@edu.udesc.br,
{mauricio.pillon,guilherme.koslovski,charles.miers}@udesc.br

Abstract. A unique digital identity in cyberspace is essential for identifying devices, individuals, or objects. Associating a reliable digital identity to a virtual entity ensures the identification of those involved in transactions and data manipulations. In this regard, the Self-Sovereign Identity (SSI) is a new paradigm that delegates data management to the entity to which the identity belongs. SSI is decentralized and independent of any certifying organization, thus naturally more scalable. This paper explores the use of SSI model in the ecosystem involving the cotton supply chain, proposing a decentralized architecture for the verification of certification seals and cotton auditing. The architecture establishes trust in the information and transactions carried out among the various participants in this chain, characterized by its complexity and dispersed geography. By instantiating the platform in a proof-of-concept scenario, the goal is to analyze how SSI can be effectively used to promote trust and transparency in the chain, with the potential to address challenges related to the authenticity, certification, and traceability of cotton. The COTTONTRUST prototype is currently in operation, and our results are promising, benefiting both cotton producers and gins in obtaining fair prices, as well as buyers and final consumers in ensuring quality assurance.

1 Introduction

The cotton culture has become one of Brazil's main commodities, gaining significant prominence due to the growth and development of production systems, which contributed to Brazil's position as the fifth-largest global exporter of cotton [5]. The cotton supply chain is known for its length and complexity, involving numerous stages from seed production to the final consumer [3].

The cotton production process begins with agricultural inputs such as seeds, fertilizers, pesticides, and implements. On the farm, production includes harvesting lint, seed, and linters, with cotton processing being vital in mechanically separating fibers from seeds, resulting in bales of lint cotton. These bales are then traded both domestically and internationally. Proper identification of produced bales is crucial, achieved through labels containing essential information

© The Author(s), under exclusive license to Springer Nature Switzerland AG 2024
L. Barolli (Ed.): AINA 2024, LNDECT 200, pp. 85–97, 2024.
https://doi.org/10.1007/978-3-031-57853-3_8

about origin, quality, and specific characteristics. Additionally, visual classification reports and analyses assess cotton quality [20]. Certified cotton receives accompanying certification seals, providing reliable external recognition that the cotton meets specific quality, sustainability, or origin standards established by certifying bodies. Following farm production, lint cotton bales are directed to textile industries, including spinning, weaving, and knitting, or are sold on the international market. *In this context, the reliability of information plays a fundamental role. Information recorded on labels, classification reports, and certification seals must be accurate and trustworthy for producers, buyers, and other stakeholders to make informed decisions, ensuring transparency and the quality of the final product [4,8]. Through reliable information, cotton traceability from its origin is guaranteed, essential for meeting customer demands, complying with international regulations, and ensuring fair and equitable pricing through transparent negotiations between producers and buyers.*

The complexity and geographical dispersion of the chain hinder reliability due to the absence of a central authority in which all involved parties can unanimously place trust. The distributed and competitive nature of the chain involves independent entities that may lack mutual trust. *In this scenario, transparency becomes crucial as there is no central entity holding universal reliability [2]. Each party, due to autonomy and competition, seeks tangible guarantees and clear information to ensure quality and authenticity throughout the entire cotton supply chain [11].*

The cotton industry faces significant challenges in adopting sustainable production methods, a demand that has grown considerably due to both environmental regulations and increasing consumer awareness. The push for sustainable cotton responds to a rising desire for agricultural practices that minimize adverse environmental impacts, promote soil and water conservation, and reduce chemical usage [15]. Additionally, the increasing preference for organic cotton underscores consumer concerns about products cultivated without the use of pesticides and chemical fertilizers. This trend reflects not only a demand for more ethical agricultural practices but also a general awareness of the need for sustainable approaches in the fashion and apparel industry. Sustainable and organic cotton not only meets ethical standards but also sustainability requirements, becoming an increasingly relevant and attractive choice for conscientious consumers [7]. *From the consumer's perspective, certification seals are crucial for increasing transparency and informing about the quality, safety, and sustainability of agricultural products, including cotton [9]. However, their vulnerability to counterfeiting due to being printed on products compromises reliability. Additionally, verifying this information is not easy for consumers or other chain stakeholders, emphasizing the need to provide a reliable means of verifying seal claims, offering verifiability tools to consumers.*

Therefore, all these factors highlight the need for reliable and traceable chains with verifiability. Blockchain technology is studied in this regard, improving integrity and traceability in management systems [6]. However, challenges remain, especially for small organizations, regarding costs, digital gaps, and compliance with data privacy regulations. When it comes to trust, information decentralization, privacy, security, and transparency, SSI has emerged as a new paradigm, decentralizing control and ownership of personal information. Based on verifiable credentials and decentralized identifiers, it arose from the need for users to have absolute and exclusive control over their data [14,17]. However, the most common application of SSI focuses primarily on individual identities and Internet of Things (IoT) devices. While the literature emphasizes the absence of restrictions to extend SSI application beyond personal domains or IoT [12,14,18,19], with potential benefits for other relationships requiring trust, a non-exhaustive search did not reveal authors exploring its use in other scenarios.

Thus, our main contributions concern about trust, traceability, verifiability, and auditability inherent in the cotton production chain require an innovative approach to meet desired requirements. Therefore, the primary question this research answers is: *How to achieve traceability and reliability in the cotton production chain, considering challenges related to geographical dispersion, chain complexity, and diversity of involved parties, employing SSI model concepts?* Therefore, we developed a transparent and decentralized system based on SSI concepts, designed to provide tools to verify certification seals, as well as to trace and audit the cotton production chain. Additionally, it aims to build trust in information and transactions among chain participants, characterized by complexity and geographical dispersion. The COTTONTRUST prototype is currently in operation. Since no personal data is stored on the blockchain, the system adheres strictly to privacy and data protection regulations. As data is stored in credentials that are cryptographically verifiable in real-time, the authenticity of data and traceability information can be digitally verified in fractions of a second, providing an enhanced user experience. Furthermore, the decentralized method of information processing adopted by COTTONTRUST contributes to process transparency, consolidating reliability and efficiency in the management of the cotton production chain.

This paper is organized as follows. Section 2 addresses background, prior concepts of SSI and a set of related works found in literature. Section 3 presents COTTONTRUST, while the experimental analysis is presented in Sect. 4.

2   Related Works

Researchers have been concerned with developing various approaches to achieve traceability and auditability in supply chains in general, each applying different methods. Most works identified adopt either a centralized or decentralized approach with the application of blockchain technology to achieve supply chain traceability [1,4,10,13,16]. Specifically, Cocco et al. [4] present a system based on Self- Sovereign Identity (SSI) to support food supply chains, focusing on the visibility of food certifications. The system is implemented by combining SSI, Ethereum blockchain, and the Interplanetary File System (IPFS). It provides visibility into process and food certifications, which are currently issued by approved certification bodies and delivered and stored in paper form by various participants in the food supply chain. However, the system in question does not utilize digital wallets and verifiable credentials to store certifications; instead, it opts for IPFS.

Agrawal et al. [1] investigate and propose a blockchain-based traceability framework for the textile and apparel supply chain. The authors employ smart contracts to capture interactions among stakeholders in the chain and distributed ledger technology to store and authenticate transactions, with a primary outcome being a simulation demonstrating the achievement of traceability. It is noteworthy that, while this work proposes a traceability solution for the textile and apparel supply chain, this solution does not incorporate SSI. In turn, Hader et al. [10] propose an integrated Big Data framework with blockchain for traceability in the textile supply chain, which provides an information platform for all chain stakeholders, emphasizing transparency and information sharing. However, the work does not address aspects related to information privacy, and the solution does not incorporate SSI.

Sezer et al. [16] introduce a framework for supply chain traceability that preserves the privacy of third parties through the use of smart contracts and a permissioned blockchain. The framework promises traceability and auditability in transactions; however, it does not based on SSI. In turn, Malik et al. [13] propose a privacy-preserving framework called TradeChain, which separates the identity and business events of supply chain actors by managing two separate blockchains. One blockchain utilizes a public permissionless structure, leveraging privacy preservation and decentralized features of SSI, while the other is a permissioned blockchain specifically for recording supply chain events. In transactions recorded on the supply chain events blockchain, actors use Decentralized Identifiers (DIDs) to prove their true identity. Despite employing SSI to verify the identities of actors, the concept is not applied to supply chain events, as proposed in COTTONTRUST.

While blockchain technology offers greater benefits for supply chain traceability compared to centralized approaches, it also presents some challenges and issues that must be considered, including: (i) Scalability: the increase in transactions and participants can congest the network, impairing performance and widespread adoption. (ii) Costs: maintaining the network, validating transactions, and storing data require substantial resources, posing difficulties for smaller companies. (iii) Privacy and data protection: the immutable transparency of the blockchain raises concerns about the security of sensitive data, despite the use of encryption. (iv) Compliance with general data protection laws: the distributed and immutable nature of blockchain makes it challenging to comply with data protection laws. These issues underscore the importance of exploring other decentralized alternatives, such as the use of SSI, to overcome the limitations of centralized approaches or those using blockchain technology, and to promote more reliable, transparent, and fault-resistant traceability in the cotton supply chain.

The Table 1 presents a comparison among works related to COTTON-TRUST, addressing crucial aspects related to the autonomy and control of entities in managing their personal data, the ability to provide real-time data

Table 1. Evaluation Criteria - Comparison between related works.

	Cocco et al. [4]	Agrawal et al. [1]	Hader et al. [10]	Sezer et al. [16]	Malik et al. [13]	COTTON TRUST
Autonomy and control of entities in the management of their data.	No	No	No	No	No	Yes
Ability to provide real-time data traceability.	No	Yes	Yes	Yes	No	Yes
Ability to verify the authenticity of data in real-time.	Yes	No	No	No	Yes	Yes
Compliance with general data protection laws.	No	No	No	No	Yes	Yes

traceability, the capability to verify the authenticity of data in real-time, and compliance with data protection regulations. Upon analyzing this comparison, it is noteworthy that COTTONTRUST emerges as a comprehensive solution, effectively addressing all the mentioned points.

3 COTTONTRUST Architecture

The COTTONTRUST is a decentralized architecture grounded in the SSI model, implemented through the Hyperledger Indy blockchain, with the aim of achieving traceability, reliability, auditability, and verifiability in the cotton production chain. We adopted a system's design approach that conceptualizes the cotton production chain as an interconnected network of nodes (entities participating in the chain). For exemplifying, Fig. 1 covers the cotton lint bale production and commercialization. Although the underlying processes may vary, the system's operability remains consistent throughout the entire extent of the production chain. The entities participating in the chain, the focus of this work, are: *Seed Producer*: Responsible for producing cotton seeds used in the agricultural process. *Farm*: Location where cotton cultivation takes place. *Ginner*: Conducts the processing of harvested cotton, separating fibers from seeds and assembling cotton lint bales for commercialization. *Trader*: Engages in the negotiation and trade of cotton, acting as an intermediary between producers, processing plants, and clients. *International Market*: Represents entities or companies located outside the country that acquire cotton for various purposes, such as textile production. *Domestic Market*: Represents entities or companies within the country that acquire cotton for various purposes, including textile production or other industrial uses. *Certification Bodies*: Responsible for certifying that cotton meets specific standards of quality, sustainability, or origin. *Cotton Graders*: Accredited experts who assess and classify cotton based on characteristics such as quality, fiber type, and other relevant specifications.

By nature, the cotton production chain is inherently decentralized, as evidenced in Fig. 1, aligning with the principles of SSI. Each entity in the chain may be located in geographically distant regions, even in different countries, emphasizing the importance of decentralization and highlighting the potential benefits of employing SSI to promote transparency and reliability in a scenario characterized by extensive geographical dispersion. Although each entity in the

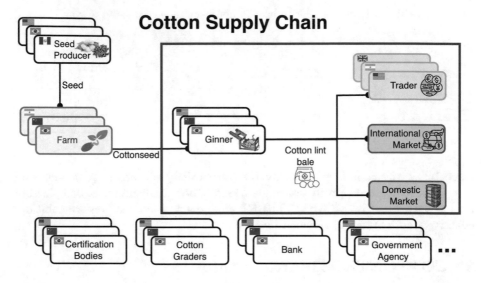

Fig. 1. The chain covering the phases of cotton lint bale production and commercialization.

chain has its specific characteristics, from a technical standpoint, they can be abstracted in a similar manner. In this work, these operational units are designated as COTTON-CELL.

3.1   COTTON-CELL Components and Structure

Each entity in the chain possesses a structure called COTTON-CELL, whose structural perspective is exemplified in Fig. 2 and incorporates the following fundamental key components: (a) Verifiable Data Registry, implemented through the Hyperledger Indy blockchain; (b) Digital Agents and Wallets; (c) DIDs; and (d) Verifiable Credentials. Each entity is uniquely identified by a DID, which

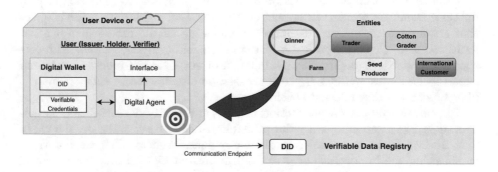

Fig. 2. COTTON-CELL components and structure.

is immutably recorded on the blockchain. Additionally, each entity can possess multiple Verifiable Credentials associated with its DID. DIDs function as unique identifiers, ensuring a unique digital representation for each entity in the cotton chain.

Within the context of COTTON-CELL, entities accumulate various Verifiable Credentials associated with their DID. These credentials encompass a variety of relevant data, including, for example, certification seals, reports, compliance with specific standards, proofs, as well as other essential documents that circulate among the chain entities. The relevance of verifiable credentials to the real world lies in their ability to provide a reliable and verifiable digital representation of tangible information. For instance, a *Sustainable Cotton Certification Seal* credential, issued by a competent authority, can be associated with a specific DID on the blockchain. This credential not only attests to the product's quality but can also be verified by other parties along the chain. Finally, by employing Verifiable Credentials, COTTON-CELL establishes a layer of trust and transparency in the cotton production chain, offering the ability to verify the authenticity and validity of information in real-time without the need for contact with the issuing entity. This not only strengthens the integrity of the chain but also facilitates well-informed decision-making by producers, buyers, regulators, and other involved participants.

3.2   Operation of COTTONTRUST

Throughout its existence, each entity accumulates various credentials, which are stored in its respective digital wallet. These credentials play a crucial role, serving as evidence and enabling the entity to validate specific information when necessary, providing concrete support in situations where it is necessary to prove certain facts or aspects of its identity. In the operational context of COTTON-TRUST, the functional dynamics are represented in Fig. 3(a) and outlined as follows:

1. An entity in the chain requests (or receives an offer for) a verifiable credential from an issuing entity.
2. The issuing entity, upon receiving and accepting the request, issues the credential, digitally signing it with its DID and private key, and then forwarding it to the requesting entity, which stores it in its digital wallet.
3. At any time, any entity in the chain has the ability to request information and, if authorized by the credential holder, access the credentials of each participant. In this process, the requester directs the request to the digital wallet of the credential holder. The digital wallet, in turn, generates proof, which incorporates essential information, including the DID of the issuer. This DID enables authenticity verification through a blockchain query to confirm its legitimacy.

To facilitate understanding, an example related to the Sustainable Certification Seal *SouABR* will be employed. This seal, in Brazil, is conferred by the

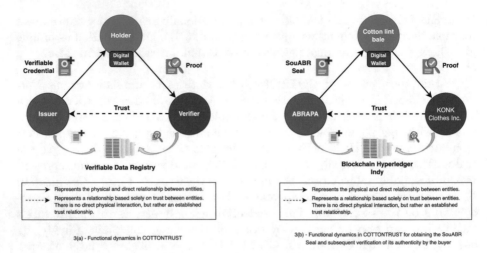

3(a) - Functional dynamics in COTTONTRUST

3(b) - Functional dynamics in COTTONTRUST for obtaining the SouABR Seal and subsequent verification of its authenticity by the buyer

Fig. 3. Functional dynamic in COTTONTRUST - Trust Triangle.

certifying body ABRAPA. The procedure to obtain the *SouABR* seal begins with a process external to the chain, in which the producing farm requests an audit of its processes from ABRAPA, in accordance with the requirements established by this particular certification. The cotton produced by the certified farm, therefore, will carry the *SouABR* certification seal, and when sold to a buyer, they can verify if the seal is indeed authentic and issued by a trustworthy entity. For the sake of simplification, it is presumed that the following prerequisites have been met: (i) The entities relevant to the scenario in question are duly registered on the blockchain, possessing DIDs. (ii) The external-to-the-chain process, which involves the analysis for granting the Seal, has already been conducted and approved. The functional dynamics in COTTONTRUST, for obtaining the *SouABR* Seal and subsequent verification of its authenticity by the buyer, are represented in Fig. 3(b) and outlined as follows:

1. ABRAPA issues a verifiable credential related to the *SouABR* Certification Seal for the cotton produced by the audited farm.
2. The *SouABR* Seal credential is stored in the digital wallet of the cotton lint bale.
3. The buyer, KONK Clothes Inc., wishes to verify if the cotton they are intending to purchase has the *SouABR* seal and if it is authentic. Therefore, they request proof from the digital wallet of the cotton bale. If the cotton bale consents, its digital wallet generates and returns the proofs to the verifier. The proof contains the decentralized identifier (DID) of the issuer (ABRAPA), which the buyer uses to confirm its legitimacy on the blockchain. This operation is done in real-time and requires no contact with the issuing entity.

3.3   COTTON-TRANSACTIONS

This prototype portrays a simplified scenario of cotton lint bale production and sale, originating from a farm, processed by a Ginner, and traded to an international market customer. In practice, various entities may engage in each stage of the supply chain. Each entity initiates its involvement by registering on the blockchain, creating its DID, and obtaining its digital wallet. This marks the initial phase of REG_ENTITY transactions, involving ledger write operations. Once registered, entities receive verifiable credentials stored in their digital wallets. These credentials serve as evidence that can be employed to validate various facts throughout the production chain. Following, during the sale of cotton lint bales, the buyer acquires products from the processing plant, triggering a new transaction known as SELL_COTTON. This transaction incorporates a blockchain query for proof verification. The buyer can thus confirm the authenticity of certification seals on the cotton bales, verify the origin of the producing farm, and ensure that the seeds come from reliable producers.

4   Experimental Analysis

The COTTONTRUST prototype was deployed on a physical machine with the following specifications: 148GB of RAM and an Intel(R) Xeon(R) CPU E5-2620 v2 @ 2.10GHz processor, featuring 4 physical cores and a total of 8 logical cores. The operating system used was Linux Ubuntu Focal Fossa 20.04.6 LTS, with Kernel Linux 5.4.0-164-generic. The employed tools included Docker version 24.0.6 (build ed223bc), Docker-Compose 1.25.0, Docker-py 4.1.0, and Python3 3.8.10. Finally, the foundational infrastructure used in conjunction with the prototype is accessible through the following link: https://github.com/mauriciopillon/cottontrust.

4.1   Test Plan

In the performance evaluation of the prototype, we have adopted the total transaction time as the primary metric to measure efficiency in REG_ENTITY and SELL_COTTON transactions, which were chosen for modeling in the initial scenario. This choice is grounded in the critical importance of these processes for the overall functionality of the platform. CPU and Memory Consumption for each type of transaction will also be analyzed, providing a comprehensive view of the system's performance in terms of computational resources. The tests aim to assess the scalability of the system and its behavior under different transaction loads, with expected results to gain insights into the efficiency, scalability, and the ability of the COTTONTRUST system to effectively manage its resources. This data is essential for optimization and validation in various operational contexts. All results are calculated based on the average of 10 executions.

Modeled Operations:

- REG_ENTITY: This experimental scenario comprises the execution of the following operations: (i) Registration of Ginner, Cotton lint bale, and International Market customer entities. (ii) Creation of 100 entities of each type. (iii) Identification and registration on the blockchain by DID, with corresponding digital wallets.
- SELL_COTTON: This scenario represents the simulation of 10000 cotton bale sales by the Cotton Beneficiary Plant to international market buyers.

Test Scenario:

- Blockchain Node Variation: the application is hosted by 4, 8, 16, and 32 nodes on the blockchain. For each scenario, we accounted the average total transaction time for REG_ENTITY and SELL_COTTON transactions.

4.2   Results and Discussions

Initially, the Total Transaction Time for the REG_ENTITY type (Fig. 4), summarized by the number of nodes in the blockchain, indicated a variation in transaction time as the number of nodes increases. Despite the results for 4 and 8 nodes showing relatively constant stability, where we observed consistent transaction times around 3ms, the slight increase in transaction time when moving from 8 to 16 nodes may indicate a transition point where network complexity begins to marginally impact performance. Upon expanding from 16 to 32 nodes, we observed an increase in transaction time, suggesting a potential influence of increased network complexity. The increase in average transaction time as the number of nodes on the blockchain increases is a behavior partly expected in distributed systems. This trend is related to the need for consensus among nodes to validate and record transactions in the shared ledger because, as the number of nodes increases, the complexity of consensus also increases since more entities need to agree on the validity of transactions.

Figure 5 summarized the Total Transaction Time for SELL_COTTON in relation to the number of nodes in the blockchain. The results show very similar transaction times, indicating a relatively constant stability or efficiency, with a slight increase in transaction time with 32 nodes. The SELL_COTTON transactions have significantly lower times compared to REG_ENTITY transactions. The REG_ENTITY transactions, involving blockchain consensus and ledger recording, have inherently higher processing times. The SELL_COTTON transactions, primarily queries, have lower times as they do not involve consensus for updating the ledger. The observed difference aligns with expectations, given the nature of the transactions. It is common for write-intensive operations (like REG_ENTITY) to take more time due to the consensus mechanism and ledger update.

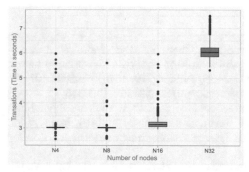
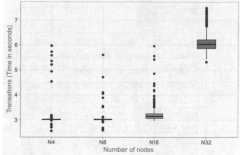

Fig. 4. Registration transactions   **Fig. 5.** Cotton bale sales transactions

5 Considerations and Future Work

This project proposes an architecture and a transparent, decentralized system designed to provide tools for verifying certification seals and tracking and auditing the cotton supply chain. Additionally, it aims to build trust in information and transactions among the chain participants, characterized by complexity and geographical dispersion. The implementation of the system based on the principles of Self-Sovereign Identity (SSI) offers distinct benefits compared to alternative models. These include efficiency and scalability by eliminating the need for a central authority to verify credentials, enhanced authenticity and integrity through SSI for maintaining verifiable credentials, reduced reliance on third parties compared to blockchain-based models, increased user privacy and control, adaptability, and ease of implementation due to decentralization, and lower operational costs by eliminating centralized intermediaries and simplifying verification processes. Comparing these results with blockchain-based and centralized models, COTTONTRUST stands out in terms of efficiency, user control, privacy, and adaptability in the cotton supply chain. As a long-term future direction, usability tests are essential to enhance, eliminate, or add features based on analysis. Additionally, exploring the applicability of this model in other areas and investigating governance mechanisms for legal acceptance are areas of future research.

Acknowledgements. This work was funding by the National Council for Scientific and Technological Development (CNPq), the Santa Catarina State Research and Innovation Support Foundation (FAPESC), UDESC, and developed at LabP2D. This work received financial support from the Coordination for the Improvement of Higher Education Personnel - CAPES - Brazil.

References

1. Agrawal, T.K., Kumar, V., Pal, R., Wang, L., Chen, Y.: Blockchain-based framework for supply chain traceability: a case example of textile and clothing industry. Comput. Ind. Eng. **154**, 107130 (2021). https://doi.org/10.1016/j.cie.2021.107130. https://www.sciencedirect.com/science/article/pii/S0360835221000346

2. Astill, J., et al.: Transparency in food supply chains: a review of enabling technology solutions. Trends Food Sci. Technol. **91**, 240–247 (2019). https://doi.org/10.1016/j.tifs.2019.07.024. https://www.sciencedirect.com/science/article/pii/S0924224418309178

3. Buainain, A.M., et al.: Cadeia produtiva do algodão, vol. 4. Bib. Orton IICA/CATIE (2007)

4. Cocco, L., Tonelli, R., Marchesi, M.: Blockchain and self sovereign identity to support quality in the food supply chain. Future Internet **13**(12), 301 (2021)

5. International Cotton Advisory Committee: World cotton statistics report (2023). https://icac.shinyapps.io/ICAC_Open_Data_Dashboaard/#. Accessed 20 Nov 2023

6. Dasaklis, T.K., Voutsinas, T.G., Tsoulfas, G.T., Casino, F.: A systematic literature review of blockchain-enabled supply chain traceability implementations. Sustainability **14**(4), 2439 (2022)

7. Duarte, L.O., Vasques, R.A., Fonseca Filho, H., Baruque-Ramos, J., Nakano, D.: From fashion to farm: green marketing innovation strategies in the Brazilian organic cotton ecosystem. J. Clean. Prod. **360**, 132196 (2022)

8. GlobalFoodSafetyResource: The importance of food safety certification (2023). https://www.identityblog.com/stories/2005/05/13/TheLawsOfIdentity.pdf. Accessed 25 Nov 2023

9. Goyal, A., Parashar, M.: Organic cotton and BCI-certified cotton fibres. In: Sustainable Fibres for Fashion and Textile Manufacturing, pp. 51–74. Elsevier (2023)

10. Hader, M., Tchoffa, D., Mhamedi, A.E., Ghodous, P., Dolgui, A., Abouabdellah, A.: Applying integrated blockchain and big data technologies to improve supply chain traceability and information sharing in the textile sector. J. Ind. Inf. Integr. **28**, 100345 (2022). https://doi.org/10.1016/j.jii.2022.100345. https://www.sciencedirect.com/science/article/pii/S2452414X22000176

11. Khan, M., Parvaiz, G.S., Dedahanov, A.T., Abdurazzakov, O.S., Rakhmonov, D.A.: The impact of technologies of traceability and transparency in supply chains. Sustainability **14**(24), 16336 (2022)

12. López, M.A.: Self-sovereign identity: the future of identity: self-sovereignity, digital wallets, and blockchain. Inter-American Development Bank, vol. 10, p. 0002635 (2020)

13. Malik, S., Gupta, N., Dedeoglu, V., Kanhere, S.S., Jurdak, R.: Tradechain: decoupling traceability and identity in blockchain enabled supply chains. In: 2021 IEEE 20th International Conference on Trust, Security and Privacy in Computing and Communications (TrustCom), pp. 1141–1152. IEEE (2021)

14. Preukschat, A., Reed, D.: Self-Sovereign Identity. Manning Publications, New York (2021)

15. Seitz, P., Strong, R., Hague, S., Murphrey, T.P.: Evaluating agricultural extension agent's sustainable cotton land production competencies: subject matter discrepancies restricting farmers' information adoption. Land **11**(11), 2075 (2022)

16. Sezer, B.B., Topal, S., Nuriyev, U.: Tppsupply: a traceable and privacy-preserving blockchain system architecture for the supply chain. J. Inf. Secur.

Appl. **66**, 103116 (2022). https://doi.org/10.1016/j.jisa.2022.103116. https://www.sciencedirect.com/science/article/pii/S2214212622000096

17. Soltani, R., Nguyen, U.T., An, A.: A survey of self-sovereign identity ecosystem. Secur. Commun. Netw. **2021** (2021). https://doi.org/10.1155/2021/8873429
18. Sporny, M., Longley, D., Sabadello, M., Reed, D., Steele, O., Allen, C.: Decentralized identifiers (DIDs) v1.0 (2021). https://www.w3.org/TR/did-core/. Accessed 20 Nov 2023
19. Strüker, J., et al.: Self-sovereign identity-foundations, applications, and potentials of portable digital identities. Project Group Business & Information Systems Engineering of the Fraunhofer Institute for Applied Information Technology FIT, Bayreuth (2021)
20. Tesema, A.F., Sayeed, M.A., Turner, C., Kelly, B.R., Hequet, E.F.: An approach for obtaining stable, reproducible, and accurate fibrogram measurements from high volume instruments. Agronomy **12**(5) (2022). https://doi.org/10.3390/agronomy12051120. https://www.mdpi.com/2073-4395/12/5/1120

Interoperability Between EVM-Based Blockchains

Alessandro Bigiotti[1(✉)], Leonardo Mostarda[1], Alfredo Navarra[2], Andrea Pinna[3], Roberto Tonelli[3], and Matteo Vaccargiu[3]

[1] Division of Computer Science, 9 Via Madonna delle Carceri, 62032 Camerino, Italy
`{alessandro.bigiotti,leonardo.mostarda}@unicam.it`
[2] Department of Mathematics and Computer Science, 1 Via Vanvitelli, 06123 Perugia, Italy
`alfredo.navarra@unipg.it`
[3] Department of Mathematics and Computer Science, 72 Via Ospedale, 09124 Cagliari, Italy
`{pinna.andrea,roberto.tonelli,matteo.vaccargiu}@unica.it`

Abstract. Blockchain is a disruptive technology that is changing the dynamics of numerous societal contexts. Interest in this technology is growing in both the academic and social spheres. Different blockchains usually work as isolated worlds that cannot communicate. The research and implementation of efficient interoperability protocols between blockchains should increase their expressiveness and make them more versatile and applicable in many real-world contexts. To this end, we investigate a possible interoperability protocol that aims at connecting Ethereum-based blockchains. By exploiting the properties of Ethereum Virtual Machines, we propose an interoperability protocol that works at the application level and makes use of off-chain processes and events to finalise an inter-chain transaction. Use cases such as synchronisation or movement of data, transfer of fungible and non-fungible tokens and cross-chain smart contract execution are addressed. Finally, we show the event semantics needed to drive inter-chain transactions.

Keywords: EVM blockchain · interoperability · inter-chain · smart contract · semantic

1 Introduction

The advent of blockchain is revolutionising numerous sectors of society. The main properties on which this technology is based concern the presence of distributed algorithms combined with cryptographic techniques. These properties allow blockchain participants to exchange information securely without the help of trusted third parties. The first blockchain was Bitcoin [13] which currently allows a distributed payment system by implementing the bitcoin cryptocurrency. This technology has continued to evolve, introducing the possibility of developing smart contracts, digital contracts that are executed by the blockchain [1]. The first blockchain with these characteristics was Ethereum [24], which gives the possibility of developing smart contracts in Solidity [27], a Turing complete language executed by the Ethereum Virtual Machine. With the advent of smart contracts, the blockchain has increased its expressive capabilities,

L. Barolli (Ed.): AINA 2024, LNDECT 200, pp. 98–109, 2024.
https://doi.org/10.1007/978-3-031-57853-3_9

becoming a distributed, self-secure and programmable environment suitable for numerous use cases. Nowadays, the blockchain offer is vast and there are numerous projects that are very different from each other. In fact, from a technical point of view, they differ in the strategies adopted in distributed protocols and cryptography. From a governance point of view [22], there are public blockchains in which everyone can participate and private blockchains in which only a pre-established set of participants is allowed.

However, blockchains, due to their characteristics, are closed environments and do not have the ability to communicate with each other. This constitutes a barrier to the application of blockchain in social contexts where communication between different parties is desirable. To increase the expressiveness and application possibilities of this technology, it is essential to look for efficient communication protocols and techniques. In this paper, we investigate an interoperability protocol between EVM-based blockchains. Today, there are numerous public blockchains based on EVM such as Polygon (MATIC) [5], Binance Smart Chain (BNB) [4], Avalanche (AVAX) [15,17], and, at the same time, there are frameworks to build private EVM-based blockchains such as Hyperledger Besu [10].

By exploiting the properties of EVMs, we propose an *ad hoc* protocol to enable interoperability among EVM-based blockchains, both public and private, which requires only two transactions to finalise an inter-chain transaction, that is, a transaction tx_a on a source blockchain B_A and a transaction tx_b on a destination blockchain B_B. The protocol aims to cover numerous use cases, such as asset transfer, token transfer (fungible and non-fungible), data synchronisation and cross-chain execution of smart contracts.

The paper is structured as follows: Sect. 2 contains an introduction to the interoperability problem and the use cases addressed, Sect. 3 contains the proposed protocol, Sect. 4 contains the semantics of the events and the results obtained, Sect. 5 contains some useful discussions and Sect. 6 contains the conclusions.

2 Interoperability Problem and Use Cases

Interoperability between blockchains refers to the possibility of two blockchains communicating. The purpose of communication can be to share some data, move some resources, or interact with some smart contracts stored in other blockchains [7]. The main barrier to blockchain interoperability lies in the constraints that make up the blockchain itself. Each blockchain adopts a consensus algorithm with specific rules in charge of validating transactions and adding them to the validated ledger; the rules followed by the consensus algorithm determine the throughput of the blockchain. Each blockchain has its own ledger that tracks the global status and transaction history. Each blockchain makes use of different cryptographic primitives to manage authentication and integrity. Finally, each blockchain makes use of a different language for the development of smart contracts. These characteristics make the blockchains isolated worlds unable to access data external to the blockchain itself. One way to access data from the outside world is to insert it into smart contracts via transactions. This type of interoperability appears at the application layer.

In general, blockchains need trusted third parties to communicate [25] and interoperability between blockchains requires that the presence of a transaction tx_a on the source blockchain B_A implies the presence of a mirror transaction tx_b in the destination blockchain B_B. We consider the following use cases:

- **Asset and Token transfer**: Asset transfer was one of the first use cases proposed. It contemplates the possibility of moving an asset (cryptocurrency or token) from a source blockchain to a destination blockchain [9,20,21]. In general, in this use case, there is an asset A on a source blockchain B_A owned by an account a. The goal is to move the asset to a destination blockchain B_B and associate it with a precise account b. This use case involves deleting the asset from the source blockchain (known as *burn*) and creating the corresponding asset on the destination blockchain (known as *mint*).
- **Data synchronisation and copy**: With the introduction of smart contracts, the blockchain becomes a programmable system increasing its potential and characteristics. In particular, smart contracts allow the saving of generic data within the blockchain. In this use case, we have two blockchains B_A and B_B in which two smart contracts with two different data are stored, respectively. The objective of this use case may concern the duplication of data between B_A and B_B [8], the synchronisation of data between B_A and B_B [23], or simply the movement of information between the two blockchains.
- **Cross-chain execution of smart contracts**: This use case aims to extend the interaction between blockchains to generic interactions between smart contracts [20]. The generic interaction between smart contracts allows the implementation of more complex logic, extending the applicability of the blockchain to numerous real-world contexts. In this use case, we have two blockchains B_A and B_B in which two smart contracts S_A and S_B are maintained, respectively. The goal is to allow the blockchain B_A to execute the functions of the smart contract S_B.

We note that the data synchronisation or data copy use cases could be seen as a particular case of the more general inter-chain execution of smart contracts. While asset and token transfers do not, as they require precise symmetry in the operations to be carried out between the two blockchains (i.e., *burn* on the source blockchain and *mint* on the destination blockchain).

3   EVM-Based Interoperability

In this section, we present an interoperability protocol at the application level that allows the execution of generic cross-chain transactions using off-chain monitoring processes called watchtowers [11,12,26], which have the task of monitoring the events emitted by the smart contracts deployed on the various blockchains. EVM-based blockchains leverage the versatility that comes from Solidity, a Turing complete programming language. Solidity allows the development of complex logic and the issuing of events recorded in the blockchain at the time of transaction validation. Every time a transaction is sent on the EVM blockchain, it is evaluated and validated by the blockchain validator nodes. If the transaction is validated, it is added to the distributed ledger and induces a change of state in the blockchain, for example, modifying the account balance or the state of a smart contract. EVM-based blockchains may share a set of standards for defining struc-tured objects present within the blockchain as fungible and non-fungible tokens. Fun-gible tokens in EVM blockchains follow the ERC20 standard [14], while non-fungible tokens follow the ERC721 standard [6]. Another property that EVM blockchains share is the ability to emit events that are recorded as logs within validated transactions. Tak-ing advantage of these properties, in the next section, a protocol will be presented that allows EVM-based blockchains to communicate and covers the use cases presented in Sect. 2.

3.1 Events and Watchtowers Interoperability Protocol

The main idea of the protocol is based on two basic concepts: the presence of an off-chain process that acts as a watchtower and the presence of interface smart contracts that, by generating particular events, will guide the watchtowers in inter-chain transac-tions. The main features are:

- **Watchtowers**: The *watchtower* (W) is an off-chain process linked to a blockchain that has the ability to read the events produced by smart contracts present on other blockchains. Each blockchain has its own watchtower that can send transactions to the blockchain it is tied to, but cannot send transactions to other blockchains. Watch-towers are responsible for generating cryptographic proof for users who want to ini-tiate an inter-chain transaction. Watchtowers can be single or distributed processes depending on trust, security and requirements needed.
- **Smart Contracts**: Two interface smart contracts *Initiator* (I) and *Finaliser* (F) are deployed on each blockchain. These smart contracts contain the functions needed to initiate and finalise inter-chain transactions. Each function present in these smart contracts will have to emit different events that are used by the watchtowers to oper-ate inter-chain transactions.

Each blockchain has its own watchtower that can read events from other blockchains and write to its own blockchain. Figure 1 shows how two blockchains B_B and B_C are connected to a blockchain B_A. The watchtowers W_B and W_C don't communicate directly with W_A, but indirectly, reading the events produced by the smart contracts *Initiator* and *Finaliser* stored on the involved blockchains. Each watchtower can access only the data related to the target smart contract. This means that the blockchain B_B can only access data relating to the events of the smart contracts responsible for managing communica-tion with B_B. This guarantees the isolation of communication and also allows privacy between the various blockchains to be maintained (as only the data involved in inter-chain transactions can be read by the various watchtowers). In Fig. 1, the blockchain B_B cannot access the data coming from the blockchain B_C and vice versa. Each inter-chain transaction has an associated validity time Δ within which the transaction must be finalised or rejected. In the case of rejection, the rollback of the changes made to the source blockchain must be envisaged. In particular, if we want to make an inter-chain transaction from a source blockchain B_A to a destination blockchain B_B, the watchtower W_A must monitor an interval of Δ_A, while the watchtower W_B must monitor an interval of $\Delta_B + c\lambda_b$; where c is a multiplication factor and λ_B is the maximum time needed to

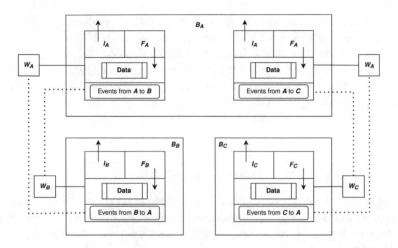

Fig. 1. The architecture that allows communications between three different blockchains B_A, B_B and B_C. The watchtowers W_i don't communicate directly, but read the events emitted by target smart contracts *Initiator* (I_i) and *Finaliser* (F_i).

produce a block on the destination blockchain. Δ_A indicates the amount of time the inter-chain transaction is valid on the source blockchain and $\Delta_B + c\lambda_B$ indicates the amount of time allowed to finalise the inter-chain transaction on the destination blockchain. The following constraint must hold:

$$\Delta_A > \Delta_B + c\lambda_B \tag{1}$$

Suppose we want to perform an inter-chain transaction from a source blockchain B_A to a destination blockchain B_B. The protocol involves the following steps:

1 An account a that wants to initiate an inter-chain transaction must request a cryptographic proof from the watchtower W_A;

2 The account a, using the proof, sends a transaction to the *Initiator* smart contract $tx_a(a, I_A)$ on B_A. After the transaction $tx_a(a, I_A)$ is validated, on B_A there is an *Inter-chain Transaction Request Event* (e_a);

3 The watchtower W_A, reading the event e_a, becomes aware of the initiation of an inter-chain transaction and has to start the validity time Δ_A; if Δ_A expires, W_A makes a rollback on B_A;

4 The watchtower W_B, reading the *Inter-chain Transaction Request Event* e_a from the source blockchain, becomes aware that an inter-chain transaction has begun. W_B, using the data contained in the event e_a, prepares a transaction $tx_b(W_B, F_B)$ directed to the *Finaliser* smart contract of the destination blockchain B_B;

5 After W_B sends the transaction, the validity time $\Delta_B + c\lambda_B$ must start;

success If the transaction tx_b is validated by the destination blockchain within $\Delta_B + c\lambda_B$, an event *Inter-chain Transaction Completed* (e_b) is emitted that finalises the inter-chain transaction. The watchtower W_A, reading the event e_b, becomes aware of the completion on the inter-chain transaction and stops the counter Δ_A. It ends in a *success* state;

abort If the transaction tx_b is not validated on the destination blockchain within $\Delta_B + c\lambda_B$, W_B goes in timeout and deletes the transaction from the waiting pool of the destination blockchain; W_A times out as no events were issued by the destination blockchain. W_A, using the data in e_a, rolls back the state of the source blockchain. It ends in an *abort* state.

The steps involved by the protocol are shown in Fig. 2. The events *Inter-chain Transaction Request* and *Inter-chain Transaction Completed* indicated in Fig. 2 contain all information necessary for the watchtowers to execute the relevant transactions on their respective blockchains. In order for watchtowers to operate correctly, they must share the same semantics to interpret the data contained in the events in order to reconstruct the respective transactions. To this end, in the next section we show the event format for the use cases examined in Sect. 2.

4   Event Semantics

The interoperability semantics are contained within the events emitted by the *Initiator* and *Finaliser* smart contracts. The idea is to use a data format shared by the watchtowers following a strategy similar to the restful pattern [16,18] with the difference that the response is dichotomous: *success* or *abort*.

The events emitted by the smart contracts *Initiator* and *Finaliser* contain particular messages in JSON format. The general structure of a message contained in an event is expressed as follows:

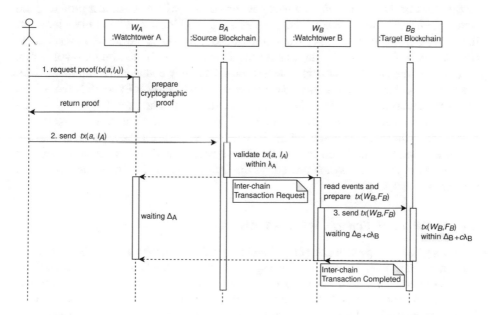

Fig. 2. The sequence diagram of the steps followed by the inter-chain protocol.

```
{"header": {
        "source_chain": <chain id>,
        "tx_type": <inter-chain type>,
        "block_number": <block number>,
        "timestamp": <current timestamp>,
        "sender": <address from source chain>,
        "receiver": <address to destination chain>
    },
    "body": {
        "instructions": {
            "service": <service name>,
            "function": <function name>
            ... other if needed ...
        },
        "content": {
            "key_1": value_1,
                ...      ...
            "key_n": value_n,
        }
    }
}
```

The *header* contains the following data: *source_chain* is the id of the chain where the inter-chain transaction started; *tx_type* determines the type of the inter-chain transaction (asset or token transfer, data synchronisation or copy, cross-chain execution of smart contracts); *service* is the name of the service (for example, it could be the name of the asset or the token); *block_number* is the number of the block that produced the event; *timestamp* is the time of the generation of the block; *sender* is the address of the account that started the inter-chain transaction; *receiver* is the address to which the transaction is directed in the destination blockchain (could be the future owner of an asset or a token in the destination blockchain). The *body* contains the following data: *instructions* contains some indications useful to the watchtower to conduct the related inter-chain transaction, such as the name of the function to be executed in cross-chain execution of smart contracts, other data could be introduced to allow complex operations, and *content* contains the actual data needed to perform the related transaction, like the amount of the token to transfer (in the case of an asset or token transfer) or the data to be copied or synchronised, or the data to pass to the target function in the destination blockchain.

4.1   Initiator and Finaliser Smart Contracts

In this section, we present the general structure of the function responsible for carrying out an inter-chain transaction. Listing 1 shows the general structure of the *Initiator* smart contract. This must have a particular inter-chain request function for each use case. The function (lines 3–4) requires the presence of a virtual nonce stored within the smart contract and a string that links the invoked service to a specific smart contract deployed on the source blockchain. An account that wants to initiate an inter-chain transaction must provide the correct nonce to the smart contract. The interface in line 8 allows other smart contracts to be modified via internal transactions. The variable `mappingServicesToContracts` is a map `string => address` which allows to trace the address of a smart contract via the name of the service. This is a fundamental part of the smart contract as it allows to make changes to the various smart contracts linked by the *Initiator* in order to implement the desired logic (for example, they will be ERC20 or ERC721 smart contracts for token transfer, or generic smart contracts for data synchronisation or cross-chain execution of smart contracts). After the internal transactions are completed, we must construct the message (lines 10–29) to insert into the event that will be read by the watchtower of the destination blockchain. Depending on the type of inter-chain transaction, the content of the message can change and take on different meanings. For example, in the case of a token transfer, line 13 could be ERC20Transfer or ERC721Transfer, line 20 could be the name of the token and the content will be just the amount of the transfer; line 21 must not be specified as the function in the target blockchain will always be *mint*. In the case of data synchronisation or execution of cross-chain transactions, the type (line 13) could be DATA_SYNC or CCESC (Cross-Chain Execution Smart Contract), the receiver (line 17) may probably not be necessary and the name of the function (line 21) will indicate the particular function to be performed on the *Finaliser* smart contract of the destination blockchain. After the message has been constructed, an event will be emitted (line 32) indicating the start of an inter-chain transaction which will be read by the watchtower of the destination blockchain. In the smart contract, verification of the cryptographic proof provided by the listener was omitted to allow the user to perform an inter-chain transaction.

```
1  contract Initiator {
2      // list of variables and data structs
3      function interChainTransactionRequest(
4          uint nonce, string calldata service, ...) public {
5          require(nonceSet[msg.sender][service]+1 == nonce);
6          nonceSet[msg.sender][service] = nonce;
7
8          Interface(mappingServicesToContracts[service]).targetFunction(...);
9
10         string memory message = string(string.concat(
11             '{"header":{',
12                 '"source_chain":"',Strings.toString(block.chainid),'",',
13                 '"type":"InterChainType",',
14                 '"block_number":"' ,Strings.toString(block.number);,'",',
15                 '"timestamp":"', Strings.toString(block.timestamp),'",',
16                 '"sender":"', Strings.toHexString(msg.sender),'"},',
17                 '"receiver":"', receiver,'"},',
18             '"body":{',
19                 '"instructions":{',
20                     '"service":"', service,'",',
21                     '"function":"functionOnDestBlockchain"',
22                 '},'
23                 '"content":{'
24                     '"param1":"', val1String, '",',
25                     '"param2":"', val2String, '"',
26                     ...',
27                 '},',
28             '},',
29         '}'
30     ));
31
```

```
32          emit  interChainTransactionRequestEvent (... ,  message );
33    }
34 }
```

Listing 1. General structure of the smart contract *Initiator.sol*.

The smart contract *Finaliser* has a similar structure, with the difference that the generation of events contains only one value in the body, which indicates the success of the inter-chain transaction (in the case of failure no event is generated).

The *Initiator* and the *Finaliser* smart contracts may undergo numerous changes due to the addition or removal of various inter-chain functions. Therefore, access to such smart contracts will be granted via a proxy smart contract responsible for keeping track of the updates of the aforementioned smart contracts and making them invisible to users. The full version of the smart contracts *Initiator* and *Finaliser* can be inspected in [2], where we have implemented the use cases: token transfer (ERC20 and ERC721), data synchronisation and cross-chain execution of smart contracts.

Table 1 shows the gas consumption of each inter-chain transaction and the related costs on different blockchains[1]. The function cross-chain execution of smart contracts (CCESC) seems the most expensive as the modification can be arbitrary complex. The costs on Hyperledger Besu depend on how it is configured (gas-free or not) and the token used for payments.

Table 1. Gas consumption and costs for each function presents within the smart contracts *Initiator* and *Finaliser*. The stats refer to the smart contracts in [2].

Smart Contract	Function	Gas Consumed	ETH	AVAX	BNB	MATIC
Initiator	Deploy	2475375 gas	313.23$	15.627$	1.882$	0.258$
	Add Service	45175 gas	5.712$	0.285$	0.034$	0.0047$
	SYNC_DATA	149735 gas	18.93$	0.945$	0.113$	0.0156$
	ERC20_TRANSFER	97816 gas	12.36$	0.62$	0.074$	0.010$
	ERC721_TRANSFER	90521 gas	11.44$	0.57$	0.068$	0.009$
	CCESC	262575 gas	33.2$	1.65$	0.19$	0.02$
Finaliser	Deploy	1937814 gas	245.04$	12.23$	1.47$	0.20$
	Add Service	45206 gas	5.716$	0.285$	0.034$	0.0047$
	SYNC_DATA	73325 gas	9.27$	0.46$	0.055$	0.007$
	ERC20_TRANSFER	87395 gas	11.05$	0.55$	0.06$	0.009$
	ERC721_TRANSFER	127579 gas	16.13$	0.80$	0.09$	0.013$
	CCESC	196353 gas	24.82$	1.239$	0.149$	0.02$

The protocol could be applied in some Internet of Things [19] and Industry 4.0 contexts, such as the use case presented by the authors in [3], where they propose a

---

[1] The costs refer to the prices as of December 14, 2023, and could change depending on numerous factors, such as the congestion of the particular blockchain, changes in the prices of the underlying cryptocurrency and so on.

technique to store pollution data on a private Hyperledger Besu blockchain. This protocol could allow the payment of ETH (or AVAX, MATIC, ...) depending on the amount of pollutants registered.

5 Discussion

The presented interoperability protocol can be applied to any EVM blockchain, public or private. However, attention must always be paid to numerous factors, such as the block production time of the blockchains under consideration and the transaction costs. If the two blockchains have quite different block production times, the Inequality 1 constraint must be carefully configured. The virtual nonce allows to prevent the fast blockchain from saturating the slow one, preventing an account from sending an inter-chain transaction if there is a transaction pending from its account in the destination chain. In this case, the watchtower will not provide cryptographic proof. As regards transaction costs, if in private blockchains we have wide freedom of configuration and in general transactions are cheap, on the contrary, in public blockchains there are costs that can vary significantly from blockchain to blockchain for the same use case.

Finally, when applying the protocol between a private and a public blockchain, the main attention must be devoted to the data contained in the events to avoid violating the privacy of the private chains. In the case of an inter-chain transaction from a private blockchain to a public one, it is advisable to include only the hashes of private blockchain transactions in the event content rather than the data themselves.

Given the versatility of the event semantics, the protocol could easily be extended to other blockchains that involve the development of smart contracts using Turing complete programming languages and the emission of events or log systems that can be queried externally.

6 Conclusion

In this paper, we have shown an inter-chain protocol that allows covering the use cases of asset and token transfer, data synchronisation and data copy, and smart contract executions between EVM-based blockchains. The protocol works at the application level and uses off-chain processes that behave like watchtowers and events emitted by particular interface smart contracts. This could help increase applicability within the Ethereum ecosystem, but not only, as it could easily be extended to numerous blockchains that implement event or log systems and support a Turing-complete programming language.

The problem of interoperability, however, remains an open challenge and new proposals could also emerge thanks to the development of new blockchains or the evolution of existing blockchains.

Acknowledgements. This research was funded by Ministero dell'Università e della Ricerca (MUR), issue D.M. 351/2022 "Borse di Dottorato" - Dottorato di Ricerca di Interesse Nazionale in "Blockchain & Distributed Ledger Technology", under the National Recovery and Resilience Plan (NRRP), and by the Italian National Group for Scientific Computation GNCS-INdAM. This work was partially supported by project SERICS (PE00000014) under the MUR National Recovery and Resilience Plan funded by the European Union-NextGenerationEU.

References

1. Alharby, M., Aldweesh, A., Moorsel, A.V.: Blockchain-based smart contracts: a systematic mapping study of academic research. In: 2018 International Conference on Cloud Computing, Big Data and Blockchain (ICCBB), pp. 1–6 (2018). https://doi.org/10.1109/ICCBB.2018.8756390
2. Bigiotti, A.: Github repo for initiator and finlizer smart contracts (2023). https://github.com/alessandrobigiotti/evm-based-bc-interoperability-semantics
3. Bigiotti, A., Mostarda, L., Navarra, A.: Blockchain and IoT integration for air pollution control. In: Barolli, L. (ed.) Advances on P2P, Parallel, Grid, Cloud and Internet Computing, pp. 27–38. Springer, Cham (2024)
4. Binance: BNB smart chain whitepaper (2019). https://github.com/bnb-chain/whitepaper/blob/master/WHITEPAPER.md
5. Bjelic, M., Nailwal, S., Chaudhary, A., Deng, W.: POL: one token for all polygon chains, pp. 1–25 (2017). https://polygon.technology/papers/pol-whitepaper
6. Bradić, S., Delija, D., Sirovatka, G., Žagar, M.: Creating own NFT token using ERC721 standard and solidity programming language. In: 2022 45th Jubilee International Convention on Information, Communication and Electronic Technology (MIPRO), pp. 1053–1056 (2022). https://doi.org/10.23919/MIPRO55190.2022.9803593
7. Falazi, G., Breitenbücher, U., Leymann, F., Schulte, S., Yussupov, V.: Transactional cross-chain smart contract invocations. Distrib. Ledger Technol. (2023). https://doi.org/10.1145/3616023
8. Fynn, E., Bessani, A., Pedone, F.: Smart contracts on the move (2020)
9. Hope-Bailie, A., Thomas, S.: Interledger: creating a standard for payments. In: Proceedings of the 25th International Conference Companion on World Wide Web, pp. 281-282. International World Wide Web Conferences Steering Committee (2016). https://doi.org/10.1145/2872518.2889307
10. Hyperledger: Hyperledger besu (2022). https://www.hyperledger.org/use/besu
11. Khabbazian, M., Nadahalli, T., Wattenhofer, R.: Outpost: a responsive lightweight watchtower. In: Proceedings of the 1st ACM Conference on Advances in Financial Technologies, pp. 31–40. Association for Computing Machinery (2019). https://doi.org/10.1145/3318041.3355464
12. Liu, B., Szalachowski, P., Sun, S.: Fail-safe watchtowers and short-lived assertions for payment channels. In: Proceedings of the 15th ACM Asia Conference on Computer and Communications Security, pp. 506–518. Association for Computing Machinery (2020). https://doi.org/10.1145/3320269.3384716
13. Nakamoto, S.: Bitcoin: a peer-to-peer electronic cash system (2008)
14. Norvill, R., Fiz, B., State, R., Cullen, A.: Standardising smart contracts: automatically inferring ERC standards. In: 2019 IEEE International Conference on Blockchain and Cryptocurrency (ICBC), pp. 192–195 (2019). https://doi.org/10.1109/BLOC.2019.8751350
15. van Renesse, R., Emin Gün Sirer, E.A.: Avalanche withepaper (2019). https://www.avalabs.org/whitepapers
16. Richardson, L., Ruby, S.: RESTful Web Services. O'Reilly Media, Inc., Sebastopol (2008)
17. Rocket, T., Yin, M., Sekniqi, K., van Renesse, R., Sirer, E.G.: Scalable and probabilistic leaderless BFT consensus through metastability (2020)
18. Rodriguez, A.: Restful web services: the basics. IBM developerWorks **33**(2008), 18 (2008)

19. Russello, G., Mostarda, L., Dulay, N.: A policy-based publish/subscribe middleware for sense-and-react applications. J. Syst. Softw. **84**(4), 638–654 (2011). https://doi.org/10.1016/j.jss.2010.10.023

20. Schulte, S., Sigwart, M., Frauenthaler, P., Borkowski, M.: Towards blockchain interoperability. In: Di Ciccio, C., et al. (eds.) BPM 2019, pp. 3–10. Springer, Cham (2019). https://doi.org/10.1007/978-3-030-30429-4_1

21. Sigwart, M., Frauenthaler, P., Spanring, C., Sober, M., Schulte, S.: Decentralized cross-blockchain asset transfers. In: 2021 Third International Conference on Blockchain Computing and Applications (BCCA), pp. 34–41 (2021). https://doi.org/10.1109/BCCA53669.2021.9657007

22. Solat, S., Calvez, P., Abdesselam, F.N.: Permissioned vs. permissionless blockchain: how and why there is only one right choice. J. Softw. **16**, 95–106 (2021). https://api.semanticscholar.org/CorpusID:232264506

23. Westerkamp, M., Küpper, A.: Smartsync: cross-blockchain smart contract interaction and synchronization. In: 2022 IEEE International Conference on Blockchain and Cryptocurrency (ICBC), pp. 1–9 (2022). https://doi.org/10.1109/ICBC54727.2022.9805524

24. Wood, G., et al.: Ethereum: a secure decentralised generalised transaction ledger (2014)

25. Zamyatin, A., Al-Bassam, M., Zindros, D., Kokoris-Kogias, E., Moreno-Sanchez, P., Kiayias, A., Knottenbelt, W.J.: SoK: communication across distributed ledgers. In: Borisov, N., Diaz, C. (eds.) FC 2021. LNCS, vol. 12675, pp. 3–36. Springer, Heidelberg (2021). https://doi.org/10.1007/978-3-662-64331-0_1

26. Zhang, J., Ye, Y., Wu, W., Luo, X.: Boros: secure and efficient off-blockchain transactions via payment channel hub. IEEE Trans. Dependable Secure Comput. **20**(1), 407–421 (2023). https://doi.org/10.1109/TDSC.2021.3135076

27. Zheng, G., Gao, L., Huang, L., Guan, J.: Ethereum Smart Contract Development in Solidity, 1st edn. Springer, Singapore (2021). https://doi.org/10.1007/978-981-15-6218-1

Stand-Up Indulgent Gathering on Lines for Myopic Luminous Robots

Quentin Bramas[1], Hirotsugu Kakugawa[2], Sayaka Kamei[3(✉)], Anissa Lamani[1], Fukuhito Ooshita[4], Masahiro Shibata[5], and Sébastien Tixeuil[6]

[1] University of Strasbourg, ICube, CNRS, Strasbourg, France
{bramas,alamani}@unistra.fr
[2] Ryukoku University, Shiga, Japan
kakugawa@rins.ryukoku.ac.jp
[3] Hiroshima University, Higashihiroshima, Japan
s10kamei@hiroshima-u.ac.jp
[4] Fukui University of Technology, Fukui, Japan
f-oosita@fukui-ut.ac.jp
[5] Kyushu Institute of Technology, Kitakyushu, Japan
shibata@csn.kyutech.ac.jp
[6] Sorbonne University, CNRS, LIP6, IUF, Paris, France
sebastien.tixeuil@lip6.fr

Abstract. We consider a strong variant of the crash fault-tolerant gathering problem called stand-up indulgent gathering (SUIG), by robots endowed with limited visibility sensors and lights on line-shaped networks. In this problem, a group of mobile robots must eventually gather at a single location, not known beforehand, regardless of the occurrence of crashes. Differently from previous work that considered unlimited visibility, we assume that robots can observe nodes only within a certain fixed distance (that is, they are myopic), and emit a visible color from a fixed set (that is, they are luminous), without multiplicity detection. We consider algorithms depending on two parameters related to the initial configuration: M_{init}, which denotes the number of nodes between two border nodes, and O_{init}, which denotes the number of nodes hosting robots. Then, a border node is a node hosting one or more robots that cannot see other robots on at least one side. Our main contribution is to prove that, if M_{init} or O_{init} is odd, SUIG can be solved in the fully synchronous model.

Keywords: Crash failure · fault-tolerance · LCM robot model · limited visibility · light

1 Introduction

The Distributed Computing research community actively studies mobile robot swarms, aiming to characterize what conditions make it possible for robots that are confused (each robot has its own ego-centered coordinate system), forgetful

L. Barolli (Ed.): AINA 2024, LNDECT 200, pp. 110–121, 2024.
https://doi.org/10.1007/978-3-031-57853-3_10

(robots may not remember all their past actions) to autonomously move around and solve global problems [18]. One of these conditions is about how the robots coordinate their actions [13]: robots can either act all together (FSYNC), act whenever they want (ASYNC), or act in subsets (SSYNC).

One of the problems that researchers have explored is the *gathering problem*, which serves as a standard for comparison [18]. It is easy to state (robots must meet at the same place in a finite amount of time, without knowing where it is beforehand), but hard to solve (two robots that move according to SSYNC scheduling cannot meet in finite time [10], unless there are more assumptions).

Robot failures become more likely as the number of robots increases, or if robots are deployed in dangerous environments, but few studies address this issue [12]. A crash fault is a simple type of failure, where a robot stops following its protocol unexpectedly. For the gathering problem, the desired outcome in case of crash faults must be specified. There are two options: *weak gathering* requires all non-faulty robots to meet, ignoring the faulty ones, while *strong gathering* (or *stand-up indulgent gathering – SUIG*) requires all non-faulty robots to meet at the unique crash location. We believe that SUIG is an attractive task for difficult situations such as dangerous environments: for example, various repair parts could be transported by different robots, and if a robot crashes, the other ones may rescue and repair it after robots carrying relevant parts are gathered at the crash location.

In continuous Euclidean space, weak gathering is solvable in the SSYNC model [1,3,9,11], while SUIG (and its variant with two robots, stand up indulgent rendezvous – SUIR) is only solvable in the FSYNC model [6–8].

Some researchers have recently switched from studying robots in a continuous space to a discrete one [13]. In a discrete space, robots can only be in certain locations and move to adjacent ones. This can be modeled by a graph where the nodes are locations, hence the term "robots on graphs". A discrete space is more realistic for modeling physical constraints or discrete sensors [2]. However, it is not equivalent to a continuous space in terms of computation: a discrete space has fewer possible robot positions, but a continuous space gives more options to resolve difficult situations (e.g., by moving slightly to break a symmetry).

To our knowledge, SUIG in a discrete setting was only considered under the assumption that robots have infinite range visibility (that is, their sensors are able to obtain the position of all other robots in the system that participate to the gathering) in line-shaped networks. Such powerful sensors may seem unrealistic, paving the way for more practical solutions. With infinite visibility, Bramas et al. [5] showed that the SUIG problem is solvable in the FSYNC model only.

When infinite range visibility is no longer available, robots become unable to distinguish global configuration situations and act accordingly, in particular, robots may react differently to different local situations, yielding in possible synchronization issues [14,15]. In this paper, we consider the discrete setting, and aim to characterize the solvability of the SUIG problem when robots have limited visibility (that is, they are myopic) yet are endowed with visible lights taking colors from a finite set (that is, they are luminous). In particular, we are interested in the trade-off between the visibility range (how many hops away can we see other robots positions) and the memory and communication capacity of the robots (each robot can have a finite number of states that may be communicated to other robots in its visibility range). In more details, we study SUIG algorithms that depend on two parameters of the initial configuration: M_{init} and O_{init}. The former is the number of nodes between two border nodes, and the latter is the number of nodes with robots between two border nodes. A border node has at least one robot and no robots on one or both sides. We show that SUIG is solvable in the FSYNC model when either M_{init} or O_{init} is odd.

2   Model

We consider robots that evolve on a line shaped network. The length of the line is infinite in both directions, and consists of an infinite number of nodes $\ldots, u_{-2}, u_{-1}, u_0, u_1, u_2, \ldots$, such that a node u_i is connected to both $u_{(i-1)}$ and $u_{(i+1)}$.

Let $\mathcal{R} = \{r_1, r_2, \ldots, r_n\}$ be the set of $n \geq 2$ autonomous robots. Robots are assumed to be anonymous (i.e., they are indistinguishable), uniform (i.e., they all execute the same program, and use no localized parameter such as a particular orientation), oblivious (i.e., they cannot remember their past actions), and disoriented (i.e., they cannot distinguish left and right). Then, we assume that robots do *not* know the number of robots.

A node is considered *occupied* if it contains at least one robot; otherwise, it is *empty*. If a node contains more than one robot, it is said to have a *tower* or *multiplicity*.

The *distance* between two *nodes* u_i and u_j is the number of edges between them. The *distance* between two *robots* r_p and r_q is the distance between two nodes occupied by r_p and r_q. Two robots or two nodes are *adjacent* if the distance between them is one. Two robots are *neighboring* if there is no robot between them.

Each robot r_i maintains a variable L_i, called *light*, which spans a finite set of states called *colors*. We call such robots *luminous robots*. A light is *persistent* from one computational cycle to the next: the color is not automatically reset at the end of the cycle. Let L denote the number of available light colors. Let $L_i(t)$ be the light color of r_i at time t. We assume the *full light* model: each robot r_i can see the light color of other robots, but also its own light color. Robots are unable to communicate with each other explicitly (*e.g.*, by sending messages), however, they can observe their environment, including the positions (i.e., occupied nodes) and colors of other robots.

The ability to detect towers is called *multiplicity detection*, which can be either *global* (any robot can sense a tower on any node) or *local* (a robot can only sense a tower if it is part of it). If robots can determine the number of robots in a sensed tower, they are said to have *strong* multiplicity detection. We assume that robots do *not* have multiplicity detection capability even on their current node but still can sense the visible colors: if there are multiple robots $r_1, r_2, \ldots r_k$ in a node u, an observing robot r can detect only colors $\{L_i(t) | 1 \leq i \leq k\}$. So, r

can detect there are multiple robots at u if and only if at least two robots among $r_1, r_2, \ldots r_k$ have different colors. However, r cannot know how many robots are located in u even if it observe a single color or multiple colors at u.

We assume that robots are *myopic*. That is, they have limited visibility: an observing robot r at node u can only sense the robots that occupy nodes within a certain distance, denoted by ϕ, from u. As robots are identical, they share the same ϕ.

Let $\mathcal{X}_i(t)$ be the set of colors of robots located in node u_i at time t. If a robot r_j located at u_i takes a snapshot at t, the sensor of r_j outputs a sequence, $\mathcal{V}_j$, of $2\phi + 1$ sets of colors:

$$\mathcal{V}_j \equiv \mathcal{X}_{i-\phi}(t), \ldots, \mathcal{X}_{i-1}(t), [\mathcal{X}_i(t)], \mathcal{X}_{i+1}(t), \ldots, \mathcal{X}_{i+\phi}(t).$$

This sequence $\mathcal{V}_j$ is the *view* of r_j at u_i. To distinguish the sequence center, we use square brackets. If the sequence $\mathcal{X}_{i+1}, \ldots, \mathcal{X}_{i+\phi}$ is equal to the sequence $\mathcal{X}_{i-1}, \ldots, \mathcal{X}_{i-\phi}$, then the view $\mathcal{V}_j$ of r_j is *symmetric*. Otherwise, it is *asymmetric*. In $\mathcal{V}_j$, a node u_k is *occupied* at time t whenever $|\mathcal{X}_k(t)| > 0$. Conversely, if u_k is *empty* at t, then $\mathcal{X}_k(t) = \emptyset$ holds.

If there exists a node u_i such that $|\mathcal{X}_i(t)| = 1$ holds, u_i is *singly-colored*. Note that $|\mathcal{X}_i(t)|$ denotes the number of colors at node u_i, thus even if u_i is singly-colored, it may be occupied by multiple robots (sharing the same color). Now, if a node u_i is such that $|\mathcal{X}_i(t)| > 1$ holds, u_i is *multiply-colored*. As each robot has a single color, a multiply-colored node always hosts more than one robot.

In the case of a robot r_j located at a singly-colored node u_i, $[\mathcal{X}_i(t)]$ in r_j's view $\mathcal{V}_j$ can be written as $[L_j]$. Then, without loss of generality, if the left adjacent node of u_i contains one or more robots with color L_k, and the right adjacent node of u_i contains one or more robots with color L_l, while u_i only hosts r_j, then $\mathcal{V}_j$ can be written as $L_k[L_j]L_l$. Now, if robot r_j at node u_i occupies a multiply-colored position (with two other robots r_k and r_l having distinct colors), then

$|\mathcal{X}_i(t)| = 3$, and we can write $\mathcal{X}_i(t)$ in $\mathcal{V}_j$ as $\begin{bmatrix} L_k \\ L_l \\ [L_j] \end{bmatrix}$. When the observed node

in the view is with multiple colors, we use brackets to distinguish the current position of the observing robot in the view and the inner bracket to explicitly state the observing robot's color. Note that, because we assume that robots do not have multiplicity detection capability, at u_i, there may be two or more robots with L_k and L_l respectively, and there may be two or more robots with L_j other than r_j.

Our algorithms are driven by observations made on the current view of a robot, so we use *view predicates*: a Boolean function based on the current view of the robot. The predicate L_j matches any set of colors that includes color L_j, while predicate (L_j, L_k) matches any set of colors that contains L_j or L_k. Now the predicate $\begin{pmatrix} L_1 \\ L_2 \end{pmatrix}$ matches any set that contains *both* L_1 and L_2. Some of our algorithm rules expect that a node is singly-colored, *e.g.*, with color L_k, in that case, the corresponding predicate is denoted by $L_k!$. To express predicates in a less explicit way, we use character '?' to represent any set, including the empty set. The $\neg$ operator is used to negate a particular predicate P (so, $\neg P$ returns *false* whenever P returns *true* and vice versa). Then, the predicate $\begin{pmatrix} \neg L_1! \\ \neg L_2! \end{pmatrix}$ matches any set that is neither singly-colored L_1 nor singly-colored L_2. Also, the superscript notation P^y represents a sequence of y consecutive sets of colors, each satisfying predicate P. Observe that $y \leq \phi$. In a given configuration, if the view of a robot r_j at node u_i satisfies predicate $\emptyset^\phi[?]$ or predicate $[?]\emptyset^\phi$, then r_j is a *border robot* and u_i a *border node*.

At each time instant t, robots occupy nodes, and their positions and colors form a *configuration* $C(t)$ of the system. Then, each robot r executes Look-Compute-Move cycles infinitely many times: (i) first, r takes a snapshot of the environment and obtains an ego-centered view of the current configuration (Look phase), (ii) according to its view, r decides to move or to stay idle and possibly changes its light color (Compute phase), (iii) if r decided to move, it moves to one of its adjacent nodes depending on the choice made in the Compute phase (Move phase). We consider the *FSYNC* model in which at each *round*, each robot r executes an LCM cycle synchronously with all the other robots. We also consider the *SSYNC* model where a nonempty subset of robots chosen by an adversarial scheduler executes an LCM cycle synchronously, at each round. At time instant $t = 0$, let H_{init} be the maximum distance between neighboring occupied nodes, M_{init} be the number of nodes between two borders including border nodes, and $O_{init}(\leq M_{init})$ be the number of occupied nodes. We assume that $\phi \geq H_{init} \geq 1$, i.e., the visibility graph is connected. As previously stated, no robot is aware of H_{init}, M_{init} and O_{init}.

In this paper, each rule in the proposed algorithms is presented in the similar notation as in [17]: $< Label > : < Guard > :: < Statement >$. The guard is a predicate on the view $\mathcal{V}_j = \mathcal{X}_{i-\phi}, \ldots, \mathcal{X}_{i-1}, [\mathcal{X}_i], \mathcal{X}_{i+1}, \ldots, \mathcal{X}_{i+\phi}$ obtained by robot r_j at node u_i during the Look phase. If the predicate evaluates to *true*, r_j is *enabled*, otherwise, r_j is *disabled*. In the first case, the corresponding rule $< Label >$ is also said to be *enabled*. If a robot r_j is enabled, r_j may change its color and then move based on the corresponding statement during its subsequent Compute and Move phases. The statement is a pair of (*New color*, *Movement*). *Movement* can be $(i) \rightarrow$, meaning that r_j moves towards node u_{i+1}, $(ii) \leftarrow$, meaning that r_j moves towards node u_{i-1}, and $(iii) \perp$, meaning that r_j does not move. For simplicity, when r_j does not move (resp. r_j does not change its color), we omit *Movement* (resp. *New color*) in the statement. The label $< Label >$ is denoted as R followed by a non-negative integer (*i.e.*, R0, R1, etc.) where a smaller label indicates higher priority. If the integer in the label is followed by an alphabet (*i.e.*, R1a, R1b, etc.), the priority is determined by the lexicographic order.

Problem Definition. A robot is said to be *crashed* at time instant t if it stops executing at any time $t' \geq t$. That is, a crashed robot stops execution and remains with the same color at the same position indefinitely. We assume that robots cannot identify a crashed robot in their snapshots (i.e., they are able to see the crashed robots but remain unaware of their crashed status). A crash, if any, can occur at any phase of the execution, and break the LCM-atomic (i.e., it can occur the end of round, but also between Look phase and Compute phase or between Compute phase and Move phase). More than one crash can occur, however we assume that all crashes occur at the same node. In our model, since robots do not have multiplicity detection capability, a node with a single crashed robot and with multiple crashed robots with the same color are indistinguishable. Similarly, multiple robots with the same color at the same node have the same behavior, but some or all of them can crash.

We consider the *Stand Up Indulgent Gathering* (SUIG) problem defined in [7]. An algorithm solves the SUIG problem if, for any initial configuration C_0 (that may contain multiplicities), and for any execution $\mathcal{E} = (C_0, C_1, \dots)$, there exists a round t such that all robots (including the crashed robot, if any) gather at a single node, not known beforehand, for all $t' \geq t$. Note that, if there are multiple crashed nodes, the problem cannot be solved. Thus, we need to assume that all the crashes occur at the same node.

Because we assume that robots are anonymous and uniform, all robots have the same color in the initial configuration.

3   Impossibility Results

Several impossibility results from the literature hint at which situations are solvable for our problem. Theorem 1 and Corollaries 1–2 are for the case where robots have no lights.

Theorem 1 ([16]). *The gathering problem is unsolvable in FSYNC on line networks starting from an edge-symmetric configuration even if robots can see all the positions of the other robots with global strong multiplicity detection.*

Corollary 1 ([5]). *The SUIG problem is unsolvable in FSYNC on line networks starting from an edge-symmetric configuration even for robots with infinite visibility and global strong multiplicity detection.*

Corollary 2 ([15]). *Starting from a configuration where M_{init} is even and O_{init} is even, there exist initial configurations that a deterministic algorithm cannot gather for myopic robots.*

As above results, we suppose in the following section that either M_{init} or O_{init} is odd, that is, the initial configurations are not edge-symmetric. The following lemma is also for the case where robots have no lights.

Lemma 1 ([5]). *Even starting from a configuration that is not edge-symmetric, the SUIG problem is unsolvable in SSYNC for robots with infinite visibility and global strong multiplicity detection.*

Lemma 2. *Even starting from a configuration that is not edge-symmetric, the SUIG problem is unsolvable in SSYNC for infinite visibility, global strong multiplicity detection, infinite colors luminous robots.*

Proof. Let us suppose for the purpose of contradiction that there exists an algorithm A that solves SUIG for infinite visibility and global strong multiplicity detection luminous robots with an infinite number of colors in SSYNC. Consider the configuration C that occurs just before gathering is achieved. Now, configuration C has either three consecutive occupied nodes (let us call this configuration class C_3) or two consecutive occupied nodes (let us call this configuration class C_2). In a configuration in C_3, the border robots must be ordered to move inwards by A, otherwise gathering is not achieved in the next configuration. From a configuration in C_3, the SSYNC scheduler may select only one of the border robots for execution, then reaching a configuration in C_2. So, for any algorithm A that solves $SUIG$, an SSYNC scheduler can reach a configuration in C_2. In the sequel, we show that we may never reduce the number of occupied nodes in any execution that starts from a configuration in C_2, and hence the gathering is not solved.

Assume that we are in a configuration in C_2. Let k_1 denote the number of robots on the first occupied node u_1, and k_2 the number of robots on the second occupied node u_2. Suppose now that the particular combination of colors at both nodes yields all robots at u_1 not to move. Then, we can crash robots at u_2. As a result, gathering is never achieved, as the configuration remains in C_2 forever. The same argument holds for robots at u_2. As a result, algorithm A must command at least one robot at each node to move. Now, the scheduler executes those two robots (from the two nodes) that move. Either they both move inwards (exchanging their nodes) and the configuration remains in C_2, or at least one of them moves inwards and the resulting configuration remains in C_2, or another configuration with more occupied nodes and possibly holes. In any case, the number of occupied nodes is not reduced from two to one, so one can again construct an execution that reaches a configuration in C_2, and repeat the argument forever. Hence, algorithm A does not solve SUIG, a contradiction. □

As per Lemma 2, we assume the FSYNC model in the following section.

4   Possibility Results for Myopic Robots

4.1   The Case Where M_{init} is Odd

In this case, we show that the gathering is achieved even if robots do not have lights. For this purpose, in the following we assume that all robots have a single color W (White) which they do not change. The strategy of our algorithm is as follows: The robots on two border nodes move towards other occupied node. The formal description is shown in Algorithm 1.

Lemma 3. *Starting from a configuration C where M_{init} is odd, even if there is a crashed robot, all robots gather in $O(M_{init})$ rounds by Algorithm 1.*

Algorithm 1: Algorithm for the case where M_{init} is odd.

/* Do nothing after gathering. */
R0: $\emptyset^\phi[W!]\emptyset^\phi :: \bot$
/* Border robots move. */
R1: $\emptyset^\phi[W!](\neg(\emptyset^\phi)) :: \rightarrow$

Proof. Because we assume FSYNC model, if there is no crashed robot, it is clear that all robots gather on the central node between initial borders in $\lfloor M_{init}/2 \rfloor$ rounds. If a border robot r_b crashed at time $t < \lfloor M_{init}/2 \rfloor$ on a node u_i, it stops at u_i. If there are other (non-crashed) border robots on u_i at t, they move toward other occupied node, thus they are in u_{i+1} at $t+1$. After that, they cannot move before they become border robots. On the other hand, the other border robots move towards u_i. Thus, eventually, they arrive at u_{i+1} at $M_{init} - 2(t+1) + t$-th round, and robots on u_{i+1} become a border. After that, all robot at u_{i+1} move to u_i, and the gathering is achieved.   □

4.2   The Case Where O_{init} is Odd

In this section, we show that the gathering is achieved if robots have lights with three colors: W (White), R (Red) and B (Blue). We assume that all robots have the same color White in the initial configuration. The formal description is shown in Algorithm 2.

The transition diagram of configurations by the algorithm in the case that no crash occurs represents in Figs. 1, 2, 3, 4 and 5. In these figures, each small blue box represents a node, and each circle represents the set of robots with the color W, R and B. The robots represented by doubly lined circles are enabled. If no crash occurs during the execution, the strategy of our algorithm is as follows: Initially, all robots are White, and robots on two border nodes become Red in the first round by rule R1 (See Fig. 1(a)). The robots on two border nodes move towards other occupied node. Then, the border robots keep their lights Red or Blue, then the algorithm can recognize that they are border robots. If there exists a White robot in the adjacent node, the border robot changes its color to Blue or Red by rule R2b or R3b (See Fig. 1(b)→(c),(e)→(f)). Otherwise, it just moves without changing its color by rule R2a or R3a (See Fig. 1(a),(d)). When non-border White robots become border, they change their color to Red (resp. Blue) by rule R4a (resp. R4b) if borders that join the node have Red (resp. Blue) (See Fig. 1(f)→(a) or (b) (resp. (c)→(d) or (e))). We say that White robots are *captured* by a border if a border moves to the node occupied by the White robots. When a border node becomes singly-colored, the border robot moves toward other occupied nodes. Eventually, two borders are neighboring and they have Blue and Red respectively because O_{init} is odd. To achieve the gathering, depending on the initial occupied nodes, one of the followings occurs.

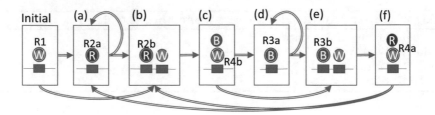

Fig. 1. Execution of border robots before the number of occupied nodes becomes three.

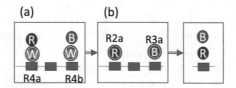

Fig. 2. Execution of Case 1.

- Case 1: If two borders are singly-colored, the distance between them is two and the central node is empty, both borders move to the central node by rules R2a and R3a (See Fig. 2(b))
- Case 2: If two borders are adjacent and singly-colored, then Blue robots join Red robots by rule R3a at the same time that Red robots become Blue by rule R4c (See Fig. 3(c)).
- Case 3: If two borders are adjacent, one border has White and Red (resp. Blue) robots and the other border is singly-colored Blue (resp. Red), then White robots become Red (resp. Blue) by rule R4a (resp. R4b) and the singly-colored Blue (resp. Red) border moves to the adjacent border by R3a (resp. R2a) (See Fig. 4(b) (resp. (c)).
- Case 4: If two borders are singly-colored Blue (resp. Red), the distance between them is two and the central node is occupied by White robots, then both borders move to the central node by rule R3b (resp. R2b) (See Fig. 5(b) (resp. (d))).

During the execution, if White robot crashes, one border eventually stops executing at the crashed node, but the other border can join the crashed border by rule R2a or R3a. For the case where Red or Blue robots crash, by special rules R5a–R5c, we are able to respond to various failure patterns.

We prove the correctness of Algorithm 2. Due to space limitation, proofs of the following lemmas are presented in the companion technical report [4].

Lemma 4. *Starting from a configuration C where O_{init} is odd, if no robot crashes, all robots gather in $O(M_{init})$ rounds by Algorithm 2.*

Lemma 5. *Starting from a configuration C where O_{init} is odd, even if a White robot crashes, all robots gather in $O(M_{init})$ rounds by Algorithm 2.*

Algorithm 2: Algorithm for the case where O_{init} is odd.

Colors

W (White), R (Red), B (Blue)

Rules

/* Do nothing after gathering. */

R0: $\emptyset^\phi[?]\emptyset^\phi :: \bot$

/* Start by the initial border robots. */

R1: $\emptyset^\phi[W!](\neg\emptyset^\phi) :: R$

/* Border robots on singly-colored nodes move inwards. */

R2a: $\emptyset^\phi[R!]\begin{pmatrix}\neg W! \\ \neg B!\end{pmatrix}(?^{\phi-1}) :: \rightarrow$

R2b: $\emptyset^\phi[R!](W!)(?^{\phi-1}) :: B, \rightarrow$

R3a: $\emptyset^\phi[B!](\neg W!)(?^{\phi-1}) :: \rightarrow$

R3b: $\emptyset^\phi[B!](W!)(?^{\phi-1}) :: R, \rightarrow$

/* When White robots become border robots, they change their color to the same color as the border robots. */

R4a: $\emptyset^\phi\begin{bmatrix}R \\ [W]\end{bmatrix}(?^\phi) :: R$

R4b: $\emptyset^\phi\begin{bmatrix}B \\ [W]\end{bmatrix}(?^\phi) :: B$

R4c: $\emptyset^\phi[R!](B!)(\emptyset^{\phi-1}) :: B$

/* Only for the case that Blue or Red robot crashes. */

R5a: $\emptyset^\phi\begin{bmatrix}R \\ B \\ [W]\end{bmatrix}(R!, B!)(\emptyset^{\phi-1}) :: R$

R5b: $\emptyset^\phi\begin{bmatrix}B \\ [R]\end{bmatrix}(R!, B!)(\emptyset^{\phi-1}) :: \rightarrow$

R5c: $\emptyset^\phi\begin{bmatrix}R \\ [B]\end{bmatrix}(R!, B!)(\emptyset^{\phi-1}) :: \rightarrow$

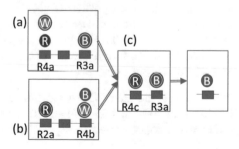

Fig. 3. Execution of Case 2.

Lemma 6. *Starting from a configuration C where O_{init} is odd, even if a Red robot crashes during the execution, all robots gather in $O(M_{init})$ rounds by Algorithm 2.*

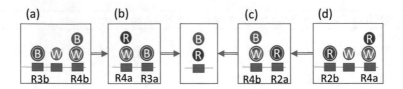

Fig. 4. Execution of Case 3.

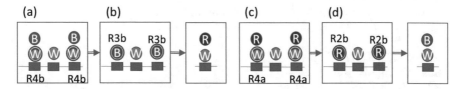

Fig. 5. Execution of Case 4.

Lemma 7. *Starting from a configuration C where O_{init} is odd, even if a Blue robot crashes during the execution, all robots gather in $O(M_{init})$ rounds by Algorithm 2.*

From Lemmas 4–7, we can deduce:

Theorem 2. *Starting from a configuration C where O_{init} is odd, Algorithm 2 solves the SUIG problem on line-shaped networks without multiplicity detection in $O(M_{init})$ rounds.*

5   Conclusion

We presented the first stand-up indulgent gathering algorithms for myopic luminous robots on line graphs. One is for the case where M_{init} is odd, while the other is for the case where O_{init} is odd. The hypotheses used for our algorithms closely follow the impossibility results found for the other cases.

Some interesting questions remain open:

1. Are there any algorithms for the case where crashed robots are located at different nodes? (In that case, one has to weaken the gathering specification, e.g., by requiring each correct robot to eventually gather at a crashed location, if any)
2. Are there any deterministic algorithms for the case where M_{init} and O_{init} are even (Such solutions would have to avoid starting or ending up in edge-view-symmetric situations)?
3. Are there any algorithms for the case where O_{init} is odd that use fewer colors than ours?
4. We present distinct solutions for the cases where M_{init} is odd and O_{init} is odd. It would be interesting to design a single algorithm that handles both cases.

References

1. Agmon, N., Peleg, D.: Fault-tolerant gathering algorithms for autonomous mobile robots. SIAM J. Comput. **36**(1), 56–82 (2006)
2. Balabonski, T., Courtieu, P., Pelle, R., Rieg, L., Tixeuil, S., Urbain, X.: Continuous *vs.* discrete asynchronous moves: a certified approach for mobile robots. In: Atig, M.F., Schwarzmann, A.A. (eds.) NETYS 2019. LNCS, vol. 11704, pp. 93–109. Springer, Cham (2019). https://doi.org/10.1007/978-3-030-31277-0_7
3. Bouzid, Z., Das, S., Tixeuil, S.: Gathering of mobile robots tolerating multiple crash faults. In: Proceedings of IEEE 33rd International Conference on Distributed Computing Systems (ICDCS), pp. 337–346 (2013)
4. Bramas, Q., et al.: Stand-up indulgent gathering on lines for myopic luminous robots (2023). https://arxiv.org/abs/2312.12698
5. Bramas, Q., Kamei, S., Lamani, A., Tixeuil, S.: Stand-up indulgent gathering on lines. In: Dolev, S., Schieber, B. (eds.) SSS 2023. LNCS, vol. 14310, pp. 451–465. Springer, Cham (2023). https://doi.org/10.1007/978-3-031-44274-2_34
6. Bramas, Q., Lamani, A., Tixeuil, S.: Stand up indulgent rendezvous. In: Devismes, S., Mittal, N. (eds.) SSS 2020. LNCS, vol. 12514, pp. 45–59. Springer, Cham (2020). https://doi.org/10.1007/978-3-030-64348-5_4
7. Bramas, Q., Lamani, A., Tixeuil, S.: Stand up indulgent gathering. In: Gąsieniec, L., Klasing, R., Radzik, T. (eds.) ALGOSENSORS 2021. LNCS, vol. 12961, pp. 17–28. Springer, Cham (2021). https://doi.org/10.1007/978-3-030-89240-1_2
8. Bramas, Q., Lamani, A., Tixeuil, S.: Stand up indulgent gathering. Theoret. Comput. Sci. **939**, 63–77 (2023)
9. Bramas, Q., Tixeuil, S.: Wait-free gathering without chirality. In: Scheideler, C. (ed.) SIROCCO 2014. LNCS, vol. 9439, pp. 313–327. Springer, Cham (2015). https://doi.org/10.1007/978-3-319-25258-2_22
10. Courtieu, P., Rieg, L., Tixeuil, S., Urbain, X.: Impossibility of gathering, a certification. Inf. Process. Lett. **115**(3), 447–452 (2015)
11. Défago, X., Potop-Butucaru, M., Raipin-Parvédy, P.: Self-stabilizing gathering of mobile robots under crash or byzantine faults. Distrib. Comput. **33**, 393–421 (2020)
12. Défago, X., Potop-Butucaru, M., Tixeuil, S.: Fault-tolerant mobile robots. In: Flocchini, P., Prencipe, G., Santoro, N. (eds.) Distributed Computing by Mobile Entities, Current Research in Moving and Computing. LNCS, vol. 11340, pp. 234–251. Springer, Cham (2019). https://doi.org/10.1007/978-3-030-11072-7_10
13. Flocchini, P., Prencipe, G., Santoro, N. (eds.): Distributed Computing by Mobile Entities, Current Researching Moving and Computing. LNCS, vol. 11340. Springer, Cham (2019). https://doi.org/10.1007/978-3-030-11072-7
14. Kamei, S., Lamani, A., Ooshita, F.: Asynchronous ring gathering by oblivious robots with limited vision. In: Proceedings of IEEE 33rd International Symposium on Reliable Distributed Systems Workshops (SRDSW), pp. 46–49 (2014)
15. Kamei, S., Lamani, A., Ooshita, F., Tixeuil, S., Wada, K.: Gathering on rings for myopic asynchronous robots with lights. In: Proceedings of 23rd International Conference on Principles of Distributed Systems (OPODIS), vol. 27 (2019)
16. Klasing, R., Markou, E., Pelc, A.: Gathering asynchronous oblivious mobile robots in a ring. Theoret. Comput. Sci. **390**(1), 27–39 (2008)
17. Ooshita, F., Tixeuil, S.: Ring exploration with myopic luminous robots. In: Izumi, T., Kuznetsov, P. (eds.) SSS 2018. LNCS, vol. 11201, pp. 301–316. Springer, Cham (2018). https://doi.org/10.1007/978-3-030-03232-6_20
18. Suzuki, I., Yamashita, M.: Distributed anonymous mobile robots: formation of geometric patterns. SIAM J. Comput. **28**(4), 1347–1363 (1999)

IoT Identity Management Systems: The State-of-the-Art, Challenges and a Novel Architecture

Samson Kahsay Gebresilassie[1]([✉]), Joseph Rafferty[1], Liming Chen[1], Zhan Cui[2], and Mamun Abu-Tair[1]

[1] School of Computing, British Telecom Ireland Innovation Centre, Ulster University, Belfast BT15 1ED, UK
{gebresilassie-s,j.rafferty,l.chen}@ulster.ac.uk
[2] British Telecom, Adastral Park, Ipswich IP5 3RE, UK
zhan.cui@bt.com

Abstract. The Internet of Things (IoT) is a technology paradigm that has transformed several domains including manufacturing, agriculture, healthcare, power grids, travel and retail. However, the growth of this interconnected world of IoT devices with their services is not without consequences, including identity-related security challenges. Security threats to identities can be vulnerabilities, misconfigurations, insecure credential storage, credential theft and social engineering. The range of different techniques that attackers use to get access to users, devices and other resources lead to serious consequences from the loss of an individual's identity to the sensitive and financial data of institutions. Thus, implementing a robust and secure identity management system (IDMS) is critical in achieving an overall secure IoT environment. Approaches for strong identity management do exist, however, they carry some deficiencies making them inadequate to address the current identity-related security challenges of IoT. These challenges include failure to provide an all-in-one decentralized IDMS inclusive of profiling (registration of entity's attributes) and identification, authentication, identity-related attack risk analysis, and trust establishment mechanisms. The purpose of this work is to investigate existing IDMS and their limitations and propose a novel architecture featuring decentralization, trust, cross-platform, and identity-related attack risk-aware mechanisms with the help of deep learning, trust, and distributed ledger technologies. The proposed IDMS architecture is also compared with existing solutions using qualitative features like availability, trust establishment, attack risk-aware capability, robustness, and cross-platform functionality.

1 Introduction

The number of IoT devices is growing massively and is predicted to increase to more than 29.2 billion in 2030 [1]. Devices and services in an IoT ecosystem can contain sensitive information such as highly personalized private data and critical infrastructure systems such as in areas like healthcare systems.

The growth of this interconnected world of devices and the wide application of the IoT ecosystem is not without consequences, particularly security challenges [2,

L. Barolli (Ed.): AINA 2024, LNDECT 200, pp. 122–135, 2024.
https://doi.org/10.1007/978-3-031-57853-3_11

3]. These challenges arise from various aspects including lack of standardization, lack of in-built security features during manufacturing, vulnerabilities of communication protocols, outdated hardware and software components [4, 5]. Attacks on IoT devices have increased in recent years as security vulnerabilities are invariably discovered [6, 7]. This makes IoT devices vulnerable to existing and new forms of attacks, resulting in serious security issues, limiting their use in critical domains. Moreover, even if IoT devices are safe on their own, they become vulnerable to a variety of attacks when connected to an insecure network including the Internet. Unauthorized access to such highly sensitive data can have serious consequences on organizations or individuals; such as stealing financial and social media accounts for economical gain, committing medical fraud, and committing crimes using the compromised identity owner's name.

An IDMS is a set of processes, policies and technologies that manage the lifecycle of digital identity, from its creation, management, and termination within a given application domain [8, 9]. It also handles identification, authentication, and authorization processes to secure resources by granting access only to authorized entities. IDMS has evolved over a series of stages from isolated to centralized, federated, user-centric, and then to decentralized. In the rapidly changing threat scenario, existing IDMSs systems are inadequate to secure critical infrastructures [10, 11]. Ensuring the privacy and security of individuals and organizations through a mature IDMS while using or exchanging this sensitive information is crucial to effectively benefit from the current and future advancements of IoT systems.

The main aim of this article is to investigate the strengths and weaknesses of existing IDMSs and recommend a novel architecture that addresses some of the current limitations. Our recommended architecture provides a more secure, decentralized, identity risk-based, and cross-platform by leveraging distributed ledger technology (DLT), trust, and DL approaches. The main contributions of the proposed architecture of this study are as follows:

- We investigate the state-of-the-art IDMS, their pros and cons with respect to IoT domain.
- We propose a novel IDMS architecture that provides more secure, decentralized, identity risk-based, and cross-platform features for IoT systems.
- We qualitatively compare our proposed architecture with existing solutions based on key features like availability, trust establishment, attack risk-aware capability, robustness, and cross-platform functionality.
- We identify current open research challenges that need to be addressed to achieve the desired IDMS solution for IoT systems.

The rest of the paper is organized as follows: Section 2 discusses some background knowledge. Section 3 presents the current state of research on IDMS for IoT systems while Sect. 4 discusses the details of the proposed architecture. Finally, the conclusion of the paper with current open research challenges and future work is presented in Sect. 5.

2 Background and Related Work

2.1 Internet of Things (IoT) Security

The Internet of Things (IoT) is an ever-expanding ecosystem of interconnected networks of billions of devices and services to deliver content-rich and contextualized services in seamless cooperation and interaction with human users [12, 13]. IoT is heterogeneous in nature consisting of different devices such as sensors, actuators, gateways, smart gadgets, edge nodes, cloud servers, etc. and software components such as APIs, middleware and digital twins. IoT is applied almost in every domain including smart homes, smart cities, smart grids, smart industries, smart agriculture, smart healthcare and smart vehicles; with IoT devices growing in billions every year.

As the IoT evolves, different architectures have been proposed and can be categorized into three main types namely layer-specific architectures, domain-specific architectures, and industry defined architectures [14–16]. The layer-specific architecture can be three layer, four layer, five layer, six layer, or seven layer architecture. In this work, we considered the five-layer IoT architecture as it is one of the most widely adopted architectures. For a variety of reasons, IoT systems have different security challenges than traditional networks [17]. To begin with, IoT systems are heterogeneous in nature, with a wide range of devices, platforms, communication methods, and protocols. Second, because devices and their owners continually change due to mobility, IoT systems have no boundaries. Third, IoT systems are made up of "things" that are not meant to be connected to the Internet. Fourth, IoT devices often have limited energy, making advanced protection procedures and tools difficult to implement. These and other security challenges lead IoT devices to be vulnerable to internal and external attacks.

In the case of external attacks, the attacker may not have direct access to the node but is capable of tampering with or replacing the genuine node with the malicious node to enable harmful operations and disrupt the network's normal performance. In internal attacks on the other hand, the attacker has access to internal resources and so promotes a variety of malicious activities. The attacks are generally explained depending on the specific layer they target considering the five-layered architecture of IoT namely: physical layer, data link layer, network layer, transport layer, and application layer as shown in Fig. 1. For example, one of the IoT's application areas is the smart home, which consists of various IoT devices, each with a specific purpose, such as smart lights,, security cameras, smart fridges, and thermostats. All of these IoT devices and control systems communicate with one another using various network protocols such as Bluetooth, Wi-Fi, ZigBee, and others. An IoT gateway is used to link these devices to the Internet. Each layer of the IoT ecosystem, which is made up of standards, services, and communication technologies, has privacy and security issues.

The main goal of cyber security is to protect individuals, data, networks, devices, organizations, and other resources from cyberattacks. Although IoT created tremendous opportunities, it has also opened new vulnerabilities and attack vectors that challenged its success. The traditional and common security goals of cyber security are Confidentiality, Integrity, and Availability, commonly called the CIA triad. However, these are not anymore sufficient goals due to advancement and emergence of new technologies including IoT besides an increased user demands. Thus, the security goals now include additional requirements including authenticity, non-repudiation, etc.

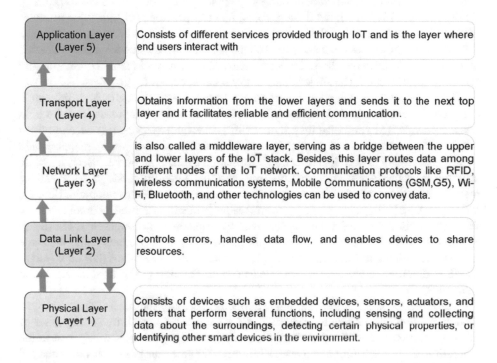

Fig. 1. Five-layer IoT architecture and main functions of each layer (adapted from [18, 19]).

2.2 Identity Management System (IDMS)

In general, cyber security is concerned with the safety and security of assets and resources in the cyber environment. Access to any resource begins with the identification of the requesting entity, verifying the identity of individuals, groups, systems, or devices is a key necessity in cyber security. An IDMS consists of many components including digital identity, entity or subject, identifier, credential, domain of application, etc. Digital identity is an electronic representation of an entity in a specific domain of application. While an entity might be a person, a group of people, an organization, a service, or a device capable of initiating or, making transactions, or performing some sort of computation the domain of application establishes the scope in which the digital identity is valid. An entity can have one or more identities within a single domain. Creating certificates, maintaining characteristics and roles, controlling accesses, and authenticating entities are all examples of tasks that may be performed by IDMS.

IDMS, also known as identity and access management (IAM), is a set of processes, policies and technologies that manage the lifecycle of digital identity from its creation and management to termination within an application domain. It also handles identification, maintaining characteristics and roles, creating certificates, authentication, and authorization processes to secure resources by granting only authorized entities [20, 21]. Traditional IDMSs have three distinct roles; identity provider (IdP), service provider (SP), and subject or entity. IdP is a service that creates and issues credentials, authenticates entities who access other services, and manages identities. The SP is an application or a service that relies on IdP for the verification of the entity's identity before granting access to its resources. The subject is an entity seeking access to resources.

IDMS has evolved over a series of stages from isolated to centralized, federated, user-centric, and then to decentralized [8, 22, 23]. The isolated approach includes both service provider (SP) and identity provider (IdP) roles both as a single service provided within a single domain. This model is unscalable as the number of SPs increases and users must remember their identification information for each SP or risk insecurity while utilizing the same identity information. In a centralized identity model, the IdP and SP roles are built to work independently yet within a single administrative domain. Here the IdP stores the identity details and performs the authentication centrally for all available services within the domain. This is achieved by implementing Single Sign-On (SSO) to provide access to users who access multiple services in the domain using the same credentials with a one-time login. Examples of this model is Kerberos [24] deployed to IDMS in different network services while Microsoft Passport [25] is used in web-based environments.

The federated model solved the single domain-based only approach of the centralized model by allowing entities to access resources in cross-domains by setting a set of rules, standards, and protocols, where different domains agree to establish trust in the federation. Shibboleth [26] and Kantara Initiative [27] are some examples of federated IDMS model. The user-centric IDMS model gives individuals control over their digital identities. When replying to an authentication or attribute requester, users have the option of selecting their credentials, giving them additional control and accountability over their identity information [28]. The decentralized (SSI) IDMS model is the most current and rising identity model is the decentralized model, usually called self-sovereign identity (SSI). This model is most typically applied by leveraging SSI principles and distributed ledger technology (DLT) or blockchain. SSI is a decentralized and owner-centric identity model that provides full control of ownership for entities over their identity. A DLT (such as blockchain) is a distributed database or ledger that is shared among the nodes of a peer to peer network with a cryptographically verified transactions in a transparent way through consensus mechanism. In identity management, blockchain enables everyone in the network to have the same measure of truth about which credentials are real and who verified the authenticity of the data stored in the credential without disclosing the authentic data. Microsoft Entra Verified ID [29] and IBM Digital Credentials [30] are among the decentralized IDMSs.

2.3   Identity-Related Vulnerabilities and Attacks

Access to any resource begins with the identification of the requesting body, hence being able to verify the identity of individuals, organizations, services, or devices is a key necessity in cyber security. Furthermore, the losses caused by cyber-attacks are closely related to identities. Attackers exploit vulnerabilities of a resource which can be an abstract or concrete private data, confidential business data, user passwords or keys, services, processes, money, reputation, etc. to gain unauthorized access. Security threats to identities can be vulnerabilities, misconfigurations, insecure credential storage, credential theft, social engineering, etc. [31]. Besides these vulnerabilities, attackers can use different techniques to get access to identity and other resources including dictionary attacks, default passwords, predictable password pattern, brute force attacks, password guessing, password reset, shared credentials, reused and recycled passwords, security questions, multi-factor authentication flaws, etc. [3, 12, 32, 33].

Dictionary attack attempts to guess passwords from a pre-compiled list of words, phrases, or commonly used passwords or encryption keys to gain unauthorized access to individual accounts or organizational resources. Using default passwords preset by manufacturers or administrators during initial setup causes a significant security risk as they are well-known and widely documented that make them easily exploitable by attackers. Using predictable password patterns such as using common words, sequences of characters, or easily guessable information, it makes password guessing or cracking simpler for attackers leading to overall accounts' security weakening. In brute force attacks, attackers systematically try all possible combinations of passwords or encryption keys until the correct one is found to gain unauthorized access to a system, account, or encrypted data. Password reset can cause security risks associated with password reset mechanisms such as email account compromise by initiating password reset process, and weak identity verification due to the use of guessable password reset questions. When multiple entities use the same (shared) credentials, the entire group is affected when any unauthorized access or malicious activity compromises any member. While multi-factor authentication (MFA) improves security by forcing users to give multiple forms of identification, it has several weaknesses and vulnerabilities that lead to different attacks. These can be manipulating the authentication process when a device used during authentication is compromised or infected with malware), biometric spoofing or replication attempts using images or molds, and insider attacks like bypassing MFA by leveraging their knowledge of the authentication process.

The consequences of identity attack can result in loss of individual identity to serious damage in sensitive and financial data. Different recommendations and approaches have been proposed to achieve this goal including end-user training, multi-factor authentication, strong password policies, password management tools, privileged access management, zero trust, etc. However, many of these security challenges still persist and new threats and vulnerabilities arises as the IoT advances. Thus, implementing a secure IDMS with features of decentralization, trust, attack risk-aware, and secure is critical in achieving an overall secure IoT environment.

2.4 Existing IDMS Solutions

Since IDMS is at the core IoT security, many experts have proposed several solutions in order to enhance the overall security posture. Different IDMS solutions have been proposed with their respective contributions although the IoT system demands more advanced solution. Deep learning-based security solutions like IDS, trust, and blockchain-based IDMS research are among the different approaches that advance the security of IoT including IDMS as discussed in Sect. 3. However, to the best knowledge of the authors, most existing solutions lack to provide full featured solution with profiling, identification, authentication, trust establishment, decentralized, and attack risk analysis-based IDMS.

Many researches show that deep learning approaches are one of the key technologies that play an important role in securing IoT systems including in building effective intrusion detection systems (IDSs) which establish a layer of security for protection, monitoring, and identifying unusual activities carried out by attackers [34, 35]. Traditional identity management systems have several difficulties with security and privacy challenges like being a single point of failure, denying full control of identity to owners, insecure, etc. Although blockchain is emerging technology, it has already been used in a variety of fields including in IDMS. Blockchain offers capabilities to overcome some of the issues of traditional IDMSs including enhanced security, decreasing dependency on a third-party authority, and some trust issues.

The study in [36] presents a novel IDMS that combines Elliptic Curve Cryptography (ECC) for authentication with anonymous credential system and traditional IDMS procedures. The proposed solution is specially tailored for the IoT designed to provide cryptographic authentication, and authorization policies. [20] mainly describes the concepts of IDMS using unified modeling language (UML) diagrams such as class, system, and sequence diagrams considering user and system requirements following the federated identity model. The study [37] proposed a solution for the identity management of devices in an IoT environment based on a central identity store where each device is associated with an account including corresponding roles. It used OAuth 2.0 tokens for device authentication and certification in network connections. The paper [38] examines the challenges of adopting centralized and federated architectures in IoT, as well as the difficulties of handling multiple identifiers and credentials. It proposes a user-centric IDMS architecture for IoT as a solution to these problems. The proposed IDMS architecture framework for IoT, which consists of three parts: the standard information model, user-centric architecture, and multi-channel authentication.

The research [39] presents a new identity and trust framework for IoT devices based on DLT, implemented on the IOTA Tangle, a direct acyclic graph-based DLT. The solution makes use of SSI, in which IoT devices assign identities to themselves that are controlled publicly and decentralized on the DLT network. The system also includes a trust technique to establish trust by automatically evaluating the trustworthiness of arbitrary identities. The authors of [40] proposes a lightweight architecture and associated protocols for consortium blockchain-based identity management for IoT to overcome the privacy, security, and scalability issues in a centralized IDMS. The technique is assessed by evaluating transaction latency and throughput using various query actions and payload sizes. In the paper [41], a decentralized identity and user-centric data management platform called OrgID is proposed that incorporates identity registration and authorization procedures. It supports self-sovereign identification architecture by leveraging blockchain technology and smart contracts. As shown in Table 2, the different IDMS models can be compared to each other using some of the key features including single point of failure, risk of attack, trust, cross-platform, availability, and security. These features are briefly described in Table 1.

Table 1. Brief description of the key features used to compare different IDMS models

Feature	Description
Single point of failure	When an IDM services fails, they result in the loss of authentication, authorization, or other critical identity-related services, affecting the overall security and functionality of the system
Risk/attack-aware	Resistant to some common risks and attacks associated with identity
Trust	Entities interacting to each other in the IoT system establish trust for secure and efficient digital interactions
Cross-platform	The ability of the IDMS to work seamlessly across different operating systems, applications, and platforms
Availability	IDMS's ability to provide reliable and uninterrupted services, ensuring that users can access the necessary resources, applications, and services whenever they need to

Table 2. Comparison of proposed architecture with some of the existing IDMSs

Ref	Model	No single point of failure	Risk/attack-aware	Trust establishment	Cross-platform	Available and secure
[20]	Federated	x	x	x	x	x
[36]	Centralized	x	x	x	x	x
[37]	Centralized	x	x	x	x	x
[38]	User-centric	x	x	x	x	x
[39]	Decentralized		x			
[40]	Decentralized		x	x		
[41]	Decentralized		x	x		
Proposed	Decentralized					

3 Proposed Architecture

To overcome the security challenges, mainly identity-related, and minimize the overall security posture of IoT, a robust IDMS solution is crucial. Current solutions are insufficient to address the current issues including being single point of failure, denies full control of identity to owners, centralized and solely controlled by IdPs, lacks proper trust establishment, and mostly proprietary solutions. They also lack to include features that can deal with attacks carried out either by external attackers from outside entities (unidentified and unauthenticated) or internal attackers from inside the domain. Moreover, simply identifying and authenticating entities like IoT devices is not enough as they may have many security vulnerabilities allowing attackers to exploit them and the network they are connected to. Thus, it is vital to ensure that IoT devices have the required trust score before they can access sensitive information or be part of critical infrastructure.

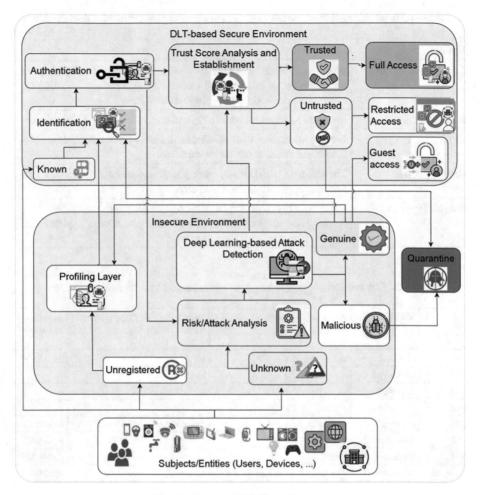

Fig. 2. Proposed IDMS architecture

This research work proposes an IDMS that is decentralized, cross-platform, attack risk-aware, trusted, and grants full control of identity to the identity owners. Deep learning methods provide improved intrusion detection approaches due to their ability to efficiently handle large-scale datasets, learn features automatically, and adapt and learn new data with less extensive retraining, as well as their flexibility and capability of capturing nonlinear relationships in data. Temper-proof transaction, decentralization, shared governance are blockchain's built-in features that play important role in advancing IDMS solutions. Entities in the IoT system not only should be identified and authenticated, they need to establish a trusted relationship for secure communication. Thus, calculating the quantifiable trust scores of the identities of IoT entities independent of a single central third party is vital in establishing a trusted IoT environment. Our proposed architecture primarily categorizes an IoT environment into secure and insecure environments. Insecure environment is the part of the IoT system where threat attack analysis, and profiling is performed. An attacker (internal or external) can use unidentified and unauthenticated devices to carry out attacks. Attacks can also be carried out from authenticated internal users which in most cases is worse than external as they usually have better knowledge of the infrastructure. This proposed solution has four main modules:

1. Risk analysis of attacks and dataset generation
2. Deep learning-based attack detection mechanism
3. Decentralized DLT-based entity profiling and authentication
4. Trust establishment through Web of Trust-based trust score approach

The overall architecture of the proposed IDMS is shown in Fig. 2.

3.1 Risk Analysis of Attacks and Dataset Generation

This module consists of full testbed that launches attacks and genuine traffic data, a parsing algorithm to structure the raw incoming network traffic, analysis and visualization, and generating a structured dataset that can be used for further attack detection process by a deep learning approach. Different hardware devices (raspberry pi, smartphone, tablet, laptop, Wi-Fi adaptor, NodeMCU, etc.), software tools (Kali Linux, Deauth, elastic stack, etc.) are in use. Thus, this module generates normal and attack traffic data, parses this raw traffic to create structured data, and prepares the dataset for further attack analysis and is uses our previous work as a foundation [42].

3.2 Deep Learning-Based Attack Detection Mechanism

With the capabilities to automatically learn complex patterns and features from raw data, capture non-linear relationships within data, adapt and learn from new data, scale with large datasets and high-dimensional input spaces, behavioural analysis, and less dependency on human expertise, deep learning models are the preferred option in developing better IDS. The capabilities of deep learning such as Convolutional Neural Networks (CNNs) applies to recognize specific patterns indicative of normal or malicious network behavior along with transfer learning (TL) to achieve combined benefits for improved attack detection, reduced training time, and adaptability to evolving attacks. Once the dataset is ready from the first module, detecting the attack is performed by this module using a deep learning approach. The dataset passes through preprocessing and the deep learning-based attack detection module detects the attacks and is an improvement to our previous work [42].

3.3 Decentralized DLT-Based Entity Profiling and Authentication

Every device capable of interacting in the IoT system can be uniquely identified. In the IDMS process, entities need to be registered by adding their attributes before they can access any resource. A unique identifier is generated for an entity from these set of attributes which makes the foundation for the next stage, authentication. Developing entity profiling and authentication processes in a decentralized environment is not well addressed by the existing solutions. Our proposed solution provides an easy to use, decentralized, and cross-platform solution to profile and authenticate entities in the IoT system.

3.4 Trust Establishment Through Web of Trust

Although there is no standard definition for trust, it can be briefly stated as a general expectancy of good behaviour on the part of others over whom we have no control which depends on the participating entities, situation, environment and other factors with measurable and non-measurable variables [43, 44]. IoT devices are frequently open to the public and mostly communicate through wireless mediums leading to the exploitation of vulnerabilities. Establishing trust among the devices and other entities in the IoT framework is critical in ensuring secure communication. Thus, the main goal of trust is to ensure that entities in the IoT system operate and stay trusted by others while being safeguarded from being manipulated by unauthorized entities. Most current trust models function in their respective domains with their own single root of trust operating independent and without cooperation with other domains while other trust solutions involve third-party-based key exchanges or certificates. To prevent trust manipulation, trust establishment should be used in conjunction with authentication methods.

The proposed trust solution in this work provides a decentralized Web of Trust (WOT)-based trust score approach that is suitable for IoT systems. In this study, WOT establishes trust on IoT by relying on the recommendations and opinions of the nodes within the DLT(blockchain) network. WOT calculates indirect trust for each entity that participate in the IoT network from the other entities' recommendations and opinions when a trustor does not know or have no initial trust. As a result, each entity is assigned a trust score calculated from both direct trust which determines a consensus on the reliability, safety, and credibility of the entity among the participants of the IoT network. WOT trusts entities with higher trust score while those with lower trust score considers them as risky and may issue warnings or alerts to inform the other entities about potential risks, scams, or security issues, or completely deny them any access. In the secure environment part of our solution, the trusted device category indicates that the device has been authenticated and pass through the trust score analysis module to be authorized as a trusted and fixed relationship, and so has full access to all services. The untrusted device category, on the other hand, refers to the fact that the device has been properly authenticated but lacks a permanent established trusted relationship, limiting it to specified services or totally denied access to any resource if the trust level is very low.

4 Use Cases

Our proposed architecture can be applied to any IoT use cases that require decentralized, cross-platform, and secure IDMS including smart homes, smart city, smart healthcare, industrial IoT, etc. The flow in our proposed architecture for the identification, authentication, trust score calculation, and attack risk analysis can be applied in quite the same approach to all IoT use cases. A given IoT entity is registered with its attributes in the profiling stage if it is a legally owned new entity or passed through the attack risk analysis if the device is unknown which could be malicious or a guest entity. Once a device has a unique identification, it's regarded as a known entity. New entity passing through the profiling and identification stages becomes ready for authentication through a DLT system. Unknown entity is categorized as either malicious or genuine entity following the result of the attack risk analysis. An authenticated and genuine entity passes through the trust score calculation phase to determine whether it is trusted or untrusted. Trusted entity is granted full access while untrusted entity is granted either restricted access or denied and quarantined based on the trust score level. With this approach, any domain application of IoT can apply the proposed architecture to benefit decentralized, trusted, attack risk-aware, and cross-platform IDMS to achieve an overall security goals of confidentiality, integrity, availability, authenticity, and non-repudiation.

5 Conclusion

This paper examines key security challenges faced be current IDMSs for IoT systems. A robust IDMS architecture is also proposed to minimize some of the key security challenges that the existing IDMSs are facing. A comparison is performed between the proposed architecture and its counterpart IDMS using key features including availability, trust establishment, attack risk-aware capability, robustness, and cross-platform functionality. The implementation of the proposed architecture is the future work consisting of the four modules.

Acknowledgments. This research has been supported by the BT Ireland Innovation Centre (BTIIC) project, funded by BT, and Invest Northern Ireland.

References

1. IoT connected devices worldwide 2019–2030 | Statista. https://www.statista.com/statistics/1183457/iot-connected-devices-worldwide/. Accessed 06 Dec 2023
2. Fateh, M., Sial, K.: Security issues in internet of things: a comprehensive review, 207–214 (2019)
3. Siboni, S., Sachidananda, V., Meidan, Y., Bohadana, M., Mathov, Y., Bhairav, S.: Security testbed for internet-of-things devices. IEEE Trans. Reliab. 1–22 (2018). https://doi.org/10.1109/TR.2018.2864536
4. Dorri, A.: Towards an optimized blockchain for IoT. In: 2017 IEEE/ACM Second International Conference on Internet-of-Things Design and Implementation, pp. 173–178 (2017)
5. Lake, D., Milito, R., Morrow, M., Vargheese, R.: Internet of things: architectural framework for ehealth security. J. ICT Stand. **1**, 301–328 (2014). https://doi.org/10.13052/jicts2245-800x.133
6. Khanam, S., Ahmedy, I.B., Idna Idris, M.Y., Jaward, M.H., Bin Md Sabri, A.Q.: A survey of security challenges, attacks taxonomy and advanced countermeasures in the internet of things. IEEE Access **8**, 219709–219743 (2020). https://doi.org/10.1109/ACCESS.2020.3037359
7. Asharf, J., Moustafa, N., Khurshid, H., Debie, E., Haider, W., Wahab, A.: A review of intrusion detection systems using machine and deep learning in internet of things: challenges, solutions and future directions. Electron **9** (2020). https://doi.org/10.3390/electronics9071177
8. Gebresilassie, S.K., Rafferty, J., Morrow, P.J., Chen, L.L., Cui, Z., Abu-Tair, M.: Distributed, secure, self-sovereign identity for IoT devices, 1–6 (2020)

9. Chen, J., Liu, Y., Chai, Y.: An identity management framework for internet of things. In: 2015 IEEE 12th International Conference on E-Business Engineering, pp. 360–364 (2015). https://doi.org/10.1109/ICEBE.2015.67
10. Bergmann, G.: User centric identity management. Dtsch. Drucker Stuttgart. **42**, 16–17 (2006)
11. Gebresilassie, S.K., Rafferty, J., Morrow, P., Chen, L., Abu-Tair, M., Cui, Z.: Distributed, secure, self-sovereign identity for IoT devices (2020)
12. Doctoral, P., Bengal, W.: Internet of things: A to Z. I, 50–63 (2015)
13. Chibelushi, C., Eardley, A., Arabo, A.: Identity management in the internet of things: the role of MANETs for healthcare applications. Comput. Sci. Inf. Technol. **1**, 73–81 (2013). https://doi.org/10.13189/csit.2013.010201
14. Hang, L., Kim, D.H.: Design and implementation of an integrated IoT blockchain platform for sensing data integrity. Sensors (Switzerland). **19**, (2019). https://doi.org/10.3390/s19102228
15. Alansari, Z.: Internet of things: infrastructure , architecture, security and privacy. **2018**, 16–17 (2018)
16. Pratap Singh, S., Kumar, V., Kumar Singh, A., Singh, S.: A survey on internet of things (IoT): layer specific vs. domain specific architecture BT. In: Presented at the Second International Conference on Computer Networks and Communication Technologies (2020)
17. Ahmed, A.I.A., Ab Hamid, S.H., Gani, A., khan, S., Khan, M.K.: Trust and reputation for internet of things: fundamentals, taxonomy, and open research challenges. J. Netw. Comput. Appl. **145** (2019). https://doi.org/10.1016/j.jnca.2019.102409
18. Wu, M., Lu, T.J., Ling, F.Y., Sun, J., Du, H.Y.: Research on the architecture of internet of things. In: ICACTE 2010 - 2010 3rd International Conference on Advanced Computer Theory and Engineering Proceedings, vol. 5, pp. 484–487 (2010). https://doi.org/10.1109/ICACTE.2010.5579493
19. Aydos, M., Vural, Y., Tekerek, A.: Assessing risks and threats with layered approach to internet of things security. Meas. Control (UK) **52**, 338–353 (2019). https://doi.org/10.1177/0020294019837991
20. Athamena, B., Houhamdi, Z.: Identity management system model in the internet of things. TEM J. **9**, 1338–1347 (2020). https://doi.org/10.18421/TEM94-04
21. Alkhalifah, A., D'Ambra, J.: The role of identity management systems in enhancing protection of user privacy. In: Proceedings of 2012 International Conference on Cyber Security, Cyber Warfare and Digital Forensic, CyberSec 2012, pp. 144–149 (2012). https://doi.org/10.1109/CyberSec.2012.6246091
22. Jiang, J., Duan, H.: A federated identity management system with centralized trust and unified single sign-on (2011). https://doi.org/10.1109/ChinaCom.2011.6158260
23. Characterization, a P., Schneider, F.B.: Federated identity management systems: IEEE Secur. Priv. **11**, 36–48 (2013)
24. Neuman, C.B., Ts'o, T.: Kerberos: an authentication service for computer networks. IEEE Commun. Mag. **32**, 33–38 (1994). https://doi.org/10.1109/35.312841
25. Oppliger, R.: Microsoft .NET passport and identity management. Inf. Secur. Tech. Rep. **9**, 26–34 (2004). https://doi.org/10.1016/S1363-4127(04)00013-5
26. Shibboleth. https://www.shibboleth.net/. Accessed 07 Dec 2023
27. Kantara Initiative: Trust through ID Assurance. https://kantarainitiative.org/. Accessed 07 Dec 2023
28. Ahn, G.-J., Ko, M., Shehab, M.: Privacy-enhanced user-centric identity management. In: 2009 IEEE International Conference on Communication, pp. 1–5 (2009). https://doi.org/10.1109/ICC.2009.5199363
29. Microsoft Entra Decentralized ID Whitepaper
30. IBM Digital Credentials. https://wiki.digitalcredentials.ibm.com/#/. Accessed 07 Dec 2023
31. Haber, M.J., Rolls, D.: Identity Attack Vectors (2020). https://doi.org/10.1007/978-1-4842-5165-2

32. Tabassum, A., Lebda, W.: Security framework for IoT devices against cyber - attacks
33. Farha, F., Ning, H., Liu, H., Yang, L.T., Chen, L.: Physical unclonable functions based secret keys scheme for securing big data infrastructure communication. Inf. Sci. (NY) **503**, 307–318 (2019). https://doi.org/10.1016/j.ins.2019.06.066
34. Tsimenidis, S., Lagkas, T., Rantos, K.: Deep Learning in IoT Intrusion Detection. Springer, New York (2022). https://doi.org/10.1007/s10922-021-09621-9
35. Chen, Z., et al.: Machine learning-enabled iot security: open issues and challenges under advanced persistent threats. ACM Comput. Surv. **55**, 1–35 (2022). https://doi.org/10.1145/3530812
36. Bernabe, J.B., Hernandez-Ramos, J.L., Gomez, A.F.S.: Holistic privacy-preserving identity management system for the internet of things. Mob. Inf. Syst. **2017** (2017). https://doi.org/10.1155/2017/6384186
37. Trnka, M., Cerny, T.: Identity management of devices in internet of things environment. In: 2016 6th International Conference on IT Convergence and Security, ICITCS 2016 (2016). https://doi.org/10.1109/ICITCS.2016.7740343
38. Chen, J., Liu, Y., Chai, Y.: An identity management framework for internet of things. An identity management framework. Internet Things, 360–364 (2015). https://doi.org/10.1109/ICEBE.2015.67
39. Luecking, M., Fries, C., Lamberti, R., Stork, W.: Decentralized identity and trust management framework for internet of things. IEEE International Conference on Blockchain Cryptocurrency, ICBC 2020 (2020). https://doi.org/10.1109/ICBC48266.2020.9169411
40. Bouras, M.A., Lu, Q., Dhelim, S., Ning, H.: A lightweight blockchain-based iot identity management approach. Future Internet **13**, 1–14 (2021). https://doi.org/10.3390/fi13020024
41. Gilani, K., Ghaffari, F., Bertin, E., Crespi, N.: Self-sovereign identity management framework using smart contracts. Presented at the June 9 (2022). https://doi.org/10.1109/noms54207.2022.9789831
42. Gebresilassie, S.K., Rafferty, J., Chen, L., Cui, Z., Abu-Tair, M.: Transfer and CNN-based de-authentication (disassociation) DoS attack detection in IoT Wi-Fi networks. Electron. **12** (2023). https://doi.org/10.3390/electronics12173731
43. Yan, Z., Zhang, P., Vasilakos, A.V.: A survey on trust management for internet of things. J. Netw. Comput. Appl. **42**, 120–134 (2014). https://doi.org/10.1016/j.jnca.2014.01.014
44. Truong, N.B., Lee, H., Askwith, B., Lee, G.M.: Toward a trust evaluation mechanism in the social internet of things. Sensors (Switzerland) **17**, 1–24 (2017). https://doi.org/10.3390/s17061346

STARM: STreaming Association Rules Mining in High-Dimensional Data

Rania Mkhinini Gahar[1]([⊠]), Olfa Arfaoui[2], Adel Hidri[3], Suleiman Ali Alsaif[3], and Minyar Sassi Hidri[3]

[1] National Engineering School of Tunis, OASIS Research Laboratory, University of Tunis El Manar, Tunis, Tunisia
rania.mkhininigahar@enit.rnu.tn
[2] National Engineering School of Tunis, RISC Research Laboratory, University of Tunis El Manar, Tunis, Tunisia
olfa.arfaoui@isln.u-carthage.tn
[3] Computer Department, Deanship of Preparatory Year and Supporting Studies, Imam Abdulrahman Bin Faisal University, Dammam, Saudi Arabia
{abhidri,saalsaif,mmsassi}@iau.edu.sa

Abstract. Predictive analytics involves using Data Mining algorithms to discover knowledge from large databases. The Association Rules (ARs) mining technique is considered to be one of the most prevalent data mining techniques in this context. When it comes to Big Data, we talk about data stream mining which is the process of extracting knowledge from continuous data streams. In this paper, STARM (STreaming Association Rules Mining) is proposed as an efficient and distributed algorithm for mining ARs. Based on the transaction-sensitive sliding-window model, the Apriori algorithm is applied to data streams to extract frequent itemsets (FI) that are then generated into ARs via Spark streaming framework. A Dimensionality Reduction (DR) step takes place as a data preprocessing step that may reduce the search space. The conducted experiments show that the proposed streaming model achieves state-of-the-art performance.

Keywords: Association Rules · Dimensionality Reduction · Spark Streaming · Apriori · Sliding Window

1 Introduction

We live today in a global dematerialized economy, where the main source of wealth creation is information, converted into knowledge or know-how. The environment in which companies operate is becoming increasingly complex. With the success of digitization, and the advent of the Internet, the web represents a colossal pool of data that continues to grow every day, and with the emergence of connected objects, companies have entered the Big Data era. They are flooded with information of all kinds. And the volume of stored data is enough to turn heads. Not to mention their ever-increasing variety (texts, images, sounds, etc.), always in correlation with the data circulating online

© The Author(s), under exclusive license to Springer Nature Switzerland AG 2024
L. Barolli (Ed.): AINA 2024, LNDECT 200, pp. 136–146, 2024.
https://doi.org/10.1007/978-3-031-57853-3_12

[21]. Big Data consists of collecting as much information as possible within the company, by analogy to *Open Data*, which concerns information available to the general public on open sources. Big Data mining can benefit from frequent itemsets (FI) mining; the FI can be used to generate Association Rules (ARs) [12, 13]. The last few years have been distinguished by the appearance of numerous applications that process data generated continuously at high speeds. This data is known as a *data stream*. Indeed, we find this kind of dynamic database in many areas (banking transactions, use of the Web, network monitoring, etc.). These data flows are characterized by the velocity, the continuity in the arrival of the data, and the large size of the data generated. The extraction of frequent patterns from data streams poses many problems, among these problems let us quote for example the impossibility of blocking the data stream, in order to produce results in real-time.

FIs are used under the Spark streaming framework for generating ARs, taking into account the reduction of the search space by introducing a phase of dimensionality reduction. To overcome these problems and others, we propose a new approach that implements FI under the Spark streaming framework.

The rest of this paper is organized as follows: Sect. 2 introduces the problem statement. Section 3 briefly presents works that are based on streaming ARs mining. Section 4 details the proposed STreaming ARs mining (STARM) in high-dimensional data. Section 5 discusses the computational results of the proposed approach. Section 6 summarizes the paper and highlights future work.

2   Problem Statement

Data streams with their characteristics raise implicit challenges during the ARs mining process through the data flow. Indeed, these data are continuous, unlimited, and of high velocity. Not only, but with the big data deluge, several dimensions have been presented at once in order to further describe the data. In this sense, we will expose two challenges: streaming FI mining and high dimensionality.

2.1   Streaming Frequent Itemsets Mining

Different from the data of traditional static databases, data flows are distinguished from other types of flows by their continuity and their unlimitedness. They usually arrive at high velocity and are characterized by a data distribution that often changes over time. As the number of applications for data stream mining increases rapidly, there is a growing need to perform ARs mining on stream data. Two classes can play the role of classifying data streams and are defined in properties 1 and 2 respectively:
Offline data streams identify a discontinuity or regular influxes [15].
Online data streams specify a real-time updated data that come one by one over time. A transaction data stream is a sequence of incoming transactions, and a stream snippet is called *a window*. A window can be:

- Time-based, or
- Either a sliding window.

Let's consider the data Stream defined as follows:

$$DS = Bb_{iai}, Bb_i + 1_{ai}+, ..., B_{bnan} \qquad (1)$$

where DS is an infinite batch sequence. Each batch is associated with a period $[a_k, b_k]$. We consider that B_{bnan} is the most recent batch. Any B_{bkak} is a data sequence stream such that $B_{bkak} = [S_1, S_2, S_3, ..., S_j]$. Each data sequence S has a list of itemsets. We will also consider $L = |Bb_{iai}| + |Bb_i + 1_{ai} + 1|... + |Bb_{nan}|$ the length of the data Stream such that $|Bb_{kak}|$ Constitutes the cardinality of the Bb_{kak} set. To illustrate this, we will consider the batch set of Table 1. The support of a sequence S at a time interval $[a_i, b_j]$ is defined by the ratio between the number of (customers, patients, etc.) having a sequence S belonging to the current time interval, and the total number of (customers, patients, etc.). However, having defined *minsup*, the problem of extracting sequential patterns from data streams would be to find all frequent patterns S_k in an arbitrary time interval such as:

Table 1. Batches (B_0^1, B_1^2) B_2^3 overview

Batch	Sequence	Itemsets
B_0^1	S_a	(1) (2) (3) (4) (5)
	S_b	(8) (9)
B_1^2	S_c	(1) (2)
B_2^3	S_d	(1) (2) (3)
	S_e	(1) (2) (8) (9)
	S_f	(2)(1)

$$\sum_{t=a_i}^{b_i} support_t(S_k) \geq minsup \times |B_{a_i}^{b_i}| \qquad (2)$$

Table 2 represents the frequent sequential patterns according to a time interval $[a_i, b_i]$ and a min support of 50%. From this example, it can be noticed that the support of sequential patterns can vary considerably depending on the time interval. Hence the need for approaches that allow us to manage this fact while respecting all the performance criteria.

Table 2. FI according to a time interval

Intervals	Batches	Frequent Itemsets
[0,1]	B_0^1	<(1) (2) (3) (4) (5)>
		<(8) (9)>
[0,2]	B_0^1 B_1^2	<(1) (2)>
[0,3]	B_0^1 B_1^2 B_2^3	<(1) (2)>, <(1)>, <(2)>

2.2 Dimensionality Reduction

We are witnessing a deluge of digital data, in the form of images, sounds, texts, physical measurements as well as all the available information on the Internet. The complexity of this problem comes from the very large number of variables each year. For example, an image typically has more than a million pixels, and therefore, more than a million variables that must be taken into account when answering a question. The interaction of these variables produces a gigantic number of possibilities. It is the curse of dimensionality [24]. This expression designates a major problem that often appears in data science, more precisely when we are left with a huge features number relative to the observations number. In reality, the problem does not concern exactly the dimension, even if it is intuitively easier to understand expressed like that, but rather the complexity of the phenomenon. Indeed, if there are many input features that can take many different values but in reality, the behavior is easily generalizable, then we need fewer training examples since we can deduce that it happens between the examples by extrapolation. Though, massive data is a generous power source and a major challenge for existing data analysis tools. An important choke point from this aspect is an enormous number of characteristics or dimensions associated with a measured quantity. This embarrassment is frequently referred to as *the curse of dimensionality* [2].

Besides that, the algorithms developed throughout the literature find it difficult to evolve with respect to the amplification of the data dimensions number. Among the solutions suggested in this context, is one which articulates the fact of mapping the data residing in the research space of high dimensionality towards another one having a minimal dimensionality, and this by keeping almost the initial data structure. The Dimensionality reduction (DR) phenomenon is considered one of the optimization problems in machine learning. Its role lies in reducing attributes number, and eliminating data that appears irrelevant, noisy, and redundant, in order to achieve accuracy and as well as save time needed for calculation, and simplify the result. Certainly, talking about Big Data, analyzing it with high dimensionality presents one of the greatest difficulties because its complexity is very high [11,14]. Several definitions have been given to the DR. In order to simplify it, we consider a dataset represented as an $n \times m$ matrix. These data have an intrinsic dimensionality k (where $k < m$, and often $k \ll m$). The DR techniques mission [10] is to transform the initial dataset having a dimensionality m into another having dimensionality k while keeping the data structure as much as possible.

3 Related Work

The introduction of Data Mining in information systems has become very exciting. ARs mining techniques are widely used to extract useful knowledge. It is based on two steps: the frequent items search, and the ARs generation. These latter are one of the most popular data mining methods in the field of marketing and distribution. The majority of challenging applications focus on learning algorithms operating in dynamic environments, where data is collected over time. The task of extracting FI from data streams may face many new challenges. Besides the one-scan constraint and the limited memory requirement, the exponential itemsets explosion compounds the concerns. The most difficult problem in extracting FI from data streams is that previous FI can become frequent, and in reverse, those that were frequent before may become frequent. Three main approaches can take place (see Fig. 1):

- Landmark windows: Approaches that do not distinguish recent elements from older ones.
- Sliding windows: Approaches that give more importance to recent transactions (using sliding windows or decay factors).
- Damped windows: Approaches for mining at different temporal granularities.

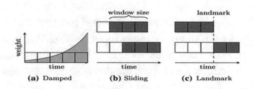

Fig. 1. Overview of time window models [22]

Apriori [1] and FP-Growth [3] are the bases of each ARs mining algorithms.

3.1   Landmark Windows

This model takes into consideration all data that has arrived from a specific point in time called the benchmark (usually the system boot time). In this model, knowledge discovery is performed on all the data mass collected between the benchmark and the moment of launching the extraction process [15,20], the authors proposed the *Stream-Mining* algorithm. It computes the k-itemsets after the first pass and ensures a provable link on the correctness of the results after the first pass over the dataset. In addition, the *Lossy Counting* algorithm presented in [7] produced a relative set of FI from a data stream. The latter is divided into a series of segments and each segment is made up of transactions.

3.2   Sliding Windows

Talking about the sliding window model, we can say that it only processes window elements keeping only FI. As for the size of the sliding window, we can say that it can vary depending on the applications it can use [8]. In [6], the authors shed light on the problem of exploring closed FI and this on a sliding window of dataflow through limited memory space. Developed in C++ language, their *Moment* algorithm uses a structure based on a prefixes tree in memory, called Closed Enumeration Tree (CET), in order to maintain selected elements set, having a dynamic aspect and this on a sliding window. In [23], the authors developed a Compact Pattern Stream tree (*CPS-tree*) to capture the recent stream data content and efficiently remove the obsolete, old stream data content. In [17], the authors proposed a new approach whereby the data stream is monitored continuously to detect any occurrence of a conceptual shift. To prevent the mining of

FI at every time point, the proposed data structures are maintained incrementally to monitor the changing of FI. This approach like the ones above is not applicable in a distributed environment. Several techniques for parallelizing sliding window processing over data streams on a shared-nothing cluster of commodity hardware have been also proposed [9].

3.3 Damped Window

The Damped model also called the Time-fading model is considered a variation of the landmark model. This model decreases the effect of old information on the result of extracting frequent patterns from data streams. The *estDec* algorithm proposed by Chang and Lee [4] adopted Hibder's mechanism [5] to estimate the frequency of itemsets. It used a parameter d $(0 < d < 1)$, to decrease the effect of old transactions on the extraction result. When a new transaction arrives, the frequency of an old itemset is reduced by the factor d. Another algorithm called *TUF-Streaming* [18] used the time-fading model in an uncertain data environment to mine *frequent* patterns from streaming data. The algorithms already discussed before use several methods and structures to discover sets of frequent elements on the data stream. Their drawbacks are that they do not support distributed parallelism which involves newer frameworks with more memory capacity and a faster distribution process. It is then time to develop an algorithm that meets all the advantages of the previous algorithms while supporting their massive character.

4   Streaming Association Rules Mining in High-Dimensional Data

In this section, we suggest a new parallel approach based on the Apriori algorithm, using Spark Streaming, called the STARM algorithm. The latter has the advantage of exploiting ARs in streaming data. The Apriori algorithm resides in the fact of processing only static aspect data, and not dynamic growth data. To overcome this problem, we improve the Apriori algorithm to perform the ARs extraction in data streams. We use the Spark Streaming framework to split the streaming data into batches in the first place, update the FI list and then the frequent pattern tree in memory in the second place, and finally extract the ARs based on the dynamically generated frequent pattern tree.

4.1   Model Overview

The proposed model is based essentially on 2 RDDs. The first one will be devoted to the DR process, which in turn introduces a statistic descriptive method Principal component analysis (PCA) [16], while the second at the searching of the FI and then ARs generation. Figure 2 shows the general overview of the proposed approach. As we see multiple data transactions are the feed source of our approach. Because of their high-dimensional nature, these later need to be reduced. After being ingested in the Spark streaming framework, they will be split in order to be processed by the different maps in the RDD1. Thereafter, the resulting PCA matrix will be the input of the Reduce task to generate as a final output the cleaned and reduced data. Once reduced, they feed the RDD2 to generate firstly the FI through the Map phase and then ARs in the Reduce one.

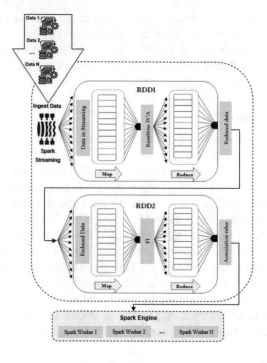

Fig. 2. Model overview.

4.2 STARM Algorithm

The main idea of our approach is given by the Algorithm 1. It is clear that in this later high-dimensional data represents the power source of our approach. This is their ingestion into Spark streaming framework. Once acquired, the data takes the route to distributed parallel processing scattered over two RDDs. The first one deals with DR via the CPA descriptive method. Once reduced, the data passes to another process to generate the FIs in the first place from the RDD2 map function and the ARs in the second place from the Reduce function of the same RDD.

Algorithm 1: STARM

Input: Data with high-dimensional nature
Output: ARs
1: Configure Sliding Windows (Number of windows, Window Size)
2: RDD1:

- Reducing data ingested in Spark Streaming via PCA descriptive method

3: RDD2:

- Searching the FI: Map function.
- ARs generation: Reduce function.

4: Send result to Spark engine.

---

5   Computational Results

To check the efficiency and performance of the proposed approach, we compare our algorithm, called STARM, with both algorithms MR-Apriori and SSPFP [19] according to the runtime.

5.1   Experimental Protocol and Computing Environment

The setup is a local Spark cluster on a workstation having Intel(R) Core(TM) i5-5200U CPU @2.20 GHz with 16 cores, 16 GB RAM, and 1 TB of disk storage. Spark-2.1.1, Hadoop-2.6.0 are installed over Centos Ubuntu distribution 14.04 64-bit running on the workstation. Java7 is the programming language used. Algorithms are scalable to the multi-node cluster.

5.1.1   Spark Streaming

Apache Spark is an open-source distributed computing framework that allows complex analyzes to be carried out on a large scale and whose operation is detailed in the next chapter. Spark is used here for its streaming engine which is an extension of the core Spark API. It enables fast, scalable, and fault-tolerant processing of data streams extracted from *Kafka*. Thus, when Spark streaming receives data from Kafka, it divides it into mini batch RDDs which are in turn processed by the *Spark Engine* to output results in the form of batches.

5.1.2   Datasets

All staked datasets can be found via the FIMI Frequent Itemset Exploration Implementations Repository http://fimi.cs.helsinki.fi/). Their varied characteristics residing respectively in the number of transactions, the elements varieties, and the rarity-density level can be useful for the contribution to our experiments. They allow us to measure our study's effectiveness. Table 3 includes a brief summary of information on these datasets.

Table 3. Characteristics of the Datasets.

Datasets	Dataset size	Total items	Average	Min	Max
KOSARAK	990002	41270	8.09	1	2498
T10I4D100 K	99935	870	10.1	1	29
Retail	88125	16469	10.3	1	76
PUMSB	49030	2113	74.0	74	74

5.2   Results

It is essential to launch some configurations concerning the sliding windows before starting. The main parameters to define are respectively the number of windows and the size of each one. All these parameters are determined in return to the characteristics of the machine operating system, the hard disk type, the capacity of the latter, as well as the capacities of the processor and the memory. Table 4 deduces the execution time resulting from the configuration already described. STARM algorithm is applied to massive datasets and compared with the two parallel algorithms MR-Apriori and SSPFP. By referring to the results obtained with STARM (see Fig. 3), we distinguish that they are much better in what concerns the execution time, and this is with respect to the parallel algorithms MR-Apriori and SSPFP. Thus, the results obtained can be considered encouraging and therefore prove the performance of our approach.

Table 4. Running time due to different sliding window configurations

Number of windows	Window size	Total Runtime (sec)
10	16700	38.2
20	10400	27
30	7200	20.3
40	5600	16.4
50	3910	11.1
60	1450	8.7
70	1040	3.8

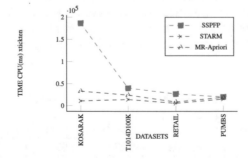

Fig. 3. The variation of the execution time vs MinSup for the parallel algorithms.

6   Conclusion and Future Work

In this paper, we study the problem of distributed ARs mining over streaming data and propose a distributed algorithm named STARM based on Spark Streaming is a popular streaming data processing platform. The STARM algorithm uses sliding window technology to access partial data in the streaming data and involves a distributed DR as a preprocessing step. The experiment shows that using the proposed algorithm not only attains highly accurate mining results but also runs significantly faster and consumes less memory for mining recent FI over data streams. As perspectives, we consider the application of our proposed approach to real-world cases such as content-based recommendation systems.

References

1. Agrawal, R., Srikant, R., et al.: Fast algorithms for mining association rules. In: Proceedings of the 20th International Conference on Very Large Databases (VLDB), vol. 1215, pp. 487–499. Santiago, Chile (1994)
2. Bellman, R., Kalaba, R.: Dynamic programming and statistical communication theory. Proc. Natl. Acad. Sci. U.S.A. **43**(8), 749 (1957)
3. Borgelt, C.: An implementation of the FP-growth algorithm. In: Proceedings of the 1st International Workshop on Open Source Data Mining: Frequent Pattern Mining Implementations, pp. 1–5 (2005)
4. Chang, J.H., Lee, W.S.: Finding recent frequent itemsets adaptively over online data streams. In: Proceedings of the Ninth ACM SIGKDD International Conference on Knowledge Discovery and Data Mining, pp. 487–492 (2003)
5. Cheng, J., Ke, Y., Ng, W.: \delta-tolerance closed frequent itemsets. In: Proceedings of the Sixth International Conference on Data Mining (ICDM), pp. 139–148. IEEE (2006)
6. Chi, Y., Wang, H., Yu, P.S., Muntz, R.R.: Moment: maintaining closed frequent itemsets over a stream sliding window. In: Proceedings of the Fourth IEEE International Conference on Data Mining (ICDM), pp. 59–66 (2004)
7. Cormode, G.: Fundamentals of analyzing and mining data streams. Monde des Util. Anal. Données **36**, 1–5 (2007)
8. Datar, M., Gionis, A., Indyk, P., Motwani, R.: Maintaining stream statistics over sliding windows. SIAM J. Comput. **31**(6), 1794–1813 (2002)
9. De Matteis, T., Mencagli, G., De Sensi, D., Torquati, M., Danelutto, M.: Gasser: an auto-tunable system for general sliding-window streaming operators on GPUs. IEEE Access **7**, 48753–48769 (2019)
10. El Moudden, I., ElBernoussi, S., Benyacoub, B.: Modeling human activity recognition by dimensionality reduction approach. In: Proceedings of the 27th International Business Information Management Association Conference-Innovation Management and Education Excellence Vision, vol. 2020 (2016)
11. Gahar, R.M., Arfaoui, O., Hidri, M.S., Alouane, N.B.-H.: Dimensionality reduction with missing values imputation. arXiv preprint arXiv:1707.00351 (2017)
12. Gahar, R.M., Arfaoui, O., Hidri, M.S., Hadj-Alouane, N.B.: Parallelcharmax: an effective maximal frequent itemset mining algorithm based on mapreduce framework. In: Proceedings of the IEEE/ACS 14th International Conference on Computer Systems and Applications (AICCSA), pp. 571–578. IEEE (2017)
13. Gahar, R.M., Arfaoui, O., Hidri, M.S., Hadj-Alouane, N.B.: An ontology-driven mapreduce framework for association rules mining in massive data. Procedia Comput. Sci. **126**, 224–233 (2018)
14. Gahar, R.M., Arfaoui, O., Hidri, M.S., Hadj-Alouane, N.B.: A distributed approach for high-dimensionality heterogeneous data reduction. IEEE Access **7**, 151006–151022 (2019)
15. Jin, R., Agrawal, G.: An algorithm for in-core frequent itemset mining on streaming data. In: Proceedings of the Fifth IEEE International Conference on Data Mining (ICDM) (2005)

16. Karamizadeh, S., Abdullah, S.M., Manaf, A.A., Zamani, M., Hooman, A.: An overview of principal component analysis. J. Sig. Inf. Process. **4**(3B), 173 (2013)
17. Koh, J.-L., Lin, C.-Y.: Concept shift detection for frequent itemsets from sliding windows over data streams. In: Proceedings of the International Workshops on Database Systems for Advanced Applications, pp. 334–348 (2009)
18. Leung, C.K.-S., Jiang, F.: Frequent pattern mining from time-fading streams of uncertain data. In: Cuzzocrea, A., Dayal, U. (eds.) DaWaK 2011. LNCS, vol. 6862, pp. 252–264. Springer, Heidelberg (2011). https://doi.org/10.1007/978-3-642-23544-3_19
19. Liu, L., Wen, J., Zheng, Z., Hansong, S.: An improved approach for mining association rules in parallel using spark streaming. Int. J. Circuit Theory Appl. **49**(4), 1028–1039 (2021)
20. Liu, X., Guan, J., Ping, H.: Mining frequent closed itemsets from a landmark window over online data streams. Comput. Math. Appl. **57**(6), 927–936 (2009)
21. Sahlberg, P.: Education policies for raising student learning: the Finnish approach. J. Educ. Policy **22**(2), 147–171 (2007)
22. Silva, J.A., Faria, E.R., Barros, R.C., Hruschka, E.R., de Carvalho, A.C.P.L.F., Gama, J.: Data stream clustering: a survey. ACM Comput. Surv. (CSUR) **46**(1), 1–31 (2013)
23. Tanbeer, S.K., Ahmed, C.F., Jeong, B.-S., Lee, Y.-K.: Sliding window-based frequent pattern mining over data streams. Inf. Sci. **179**(22), 3843–3865 (2009)
24. Verleysen, M., François, D.: The curse of dimensionality in data mining and time series prediction. In: Proceedings of the International Work-Conference on Artificial Neural Networks, pp. 758–770 (2005)

Regulation Compliance System for IoT Environments: GDPR Compliance as a Use-Case

Mamun Abu-Tair[1]([✉]), Aftab Ali[1], Samson Kahsay Gebresilassie[1], Joseph Rafferty[1], and Zhan Cui[2]

[1] School of Computing, Ulster University, Belfast, UK
m.abu-tair@ulster.ac.uk
[2] British Telecom, London, UK

Abstract. Due to the increasing number of IoT devices and their different manufacturers, it is necessary to ensure that IoT devices are compliant with laws, regulations and standards before they are placed into operation within their application area, such has within organisations, companies or homes. This paper describes how to ensure regulatory/standards compliance of the IoT devices operating within a home, organisation or company; in particular when those devices attach to homes or organisations network and starts working under that network's authority. Through application of the proposed solution, IoT devices will be able to demonstrate their compliance with regulations and standards without exchanging extraneous information about themselves or related devices. Conversely, a governance framework will be able to integrate this information to verify the IoT devices compliance to the standards and regulations without any third party involvement. The proposed solution can be applied in home environments where a centralised source of truth, such as a BT HomeHub Internet access point, can ensure that all IoT devices trying to connect through it are compliant with legal obligations and regulations such as the General Data Protection Regulation (GDPR).

1 Introduction

Worldwide, the number of Internet of Things (IoT) connected devices will be around 26.22 billion at the end of 2019. This figure will almost triple and reach 75.44 billion devices by the end of 2025 [1]. IoT devices play many positive roles in life and this involvement is expected to increase in the near future due to IoT increased functionality and automation. With this increasing popularity, many vendors have developed IoT devices different functionalities to improve the quality of life in homes/organisations. However, this is good side of the story, in the real life the IoT devices can be targeted by hackers and breach the users' privacy and security. [2].

Additionally, by deploying more IoT devices in environments (whether a smart home, company, ..etc.), we increase the security threat and our privacy may be in danger. Although, IoT security and privacy features are among some of the characteristics that consumers consider, the device cost and functionality could be more important aspects in some consumers' views.

© The Author(s), under exclusive license to Springer Nature Switzerland AG 2024
L. Barolli (Ed.): AINA 2024, LNDECT 200, pp. 147–160, 2024.
https://doi.org/10.1007/978-3-031-57853-3_13

Cesar Cerrudo, Chief Technology Officer (CTO) at IOActive Labs, showed how vulnerable our IoT environments are by hacking the United State (US) traffic control systems [3]. Cerrudo built a system using $100 worth of equipment to cause traffic disturbance for some major US cities.[1]

This above example shows the importance of implementing a regulation checking mechanism for all the IoT devices operating in a smart environment. More specifically, the smart environments should be provided with a mechanism to decide whether they allow or disallow any IoT device operating within their operational space. In the near future the smart environments should have the following characteristics:

- Every smart environment has it is own digital identity which uniquely identifies the environment and describes it's specification.
- Every IoT device has it is own digital identity which describes its functionality.
- Governmental and standards bodies should have the ability to check if the IoT/smart environment complies with laws and standards. The Governmental and standards bodies should be able to issue certificates for the IoT devices/smart environments if they comply to specific laws/standards.
- The smart environment should be provided with a mechanism to validate any certificate/claim presented by any IoT device.

1.1   Contributions

In this paper we propose a standards and regulations compliance system for smart environments. The proposed system deploys the concept of Self-Sovereign Identity to provide the device with the ability to control their own identity. By developing a system on the top of the Self-Sovereign concepts the IoT devices and the smart environments should have full control to decide what information they are going to share. The following are three Self-Sovereign Identity concepts adopted by the proposed system:

- It will allow smart environments to obtain the identity assurances that it needs.
- It will prevent data breaches as no data about the IoT devices is stored at any smart environment's components.
- It will allow the IoT devices/smart environments to decide on how to store and share their data.

Although the self sovereign identity addresses the issue of storing/sharing the IoT devices'/smart environments' data, it cannot address the problem of verifying the identity of the device/ smart environment. To achieve this goal, the proposed system uses the Hyperledger Indy blockchain distributed ledger to manage the verification process. Hyperledger Indy is a distributed ledger built for decentralized identity management. It provides tools, libraries, and reusable components for maintaining and using independent digital identities.

---

[1] https://www.youtube.com/watch?v=_j9lELCSZQw&t=2337s.

1.2 Structure of This Paper

In Sect. 2 we provide background information and a rationale for our proposed system. In Sect. 3 we explain our proposed system and provide a detailed description of its components. We present and discuss our use-case in Sects. 3 followed by an operational example in Sect. 5. Section 6 provide a detailed discussion for our proposed solution. We conclude with a short summary in Sect. 7.

2 Related Work and Background

The context of this paper is smart environments and more specifically smart home environments. We present our background work in five parts: IoT security and privacy, self-sovereign identity, blockchain and distributed ledgers, existing compliance solutions and the General Data Protection Regulation (GDPR).

2.1 IoT Security and Privacy

Security and privacy are integral parts of any IoT platform. This subsection will cover the latest research in these two research areas.

Abeshu and Chilamkurti, in [4], provided an overview of cybers-attack in "fog-to-thing" computing environments, incorporating a fog computing ecosystem and IoT applications. The authors discussed the limitations of the existing cryptographic solutions which were initially developed to provide security within traditional distributed database computer systems. The authors found that existing machine-learning-based attack detection mechanisms are not applicable to IoT applications for several reasons:

- System development and implementation flaws.
- The increase is the quantity and variety of attacks.
- Hacking skills have proven to inevitability defeat detection mechanisms.

To address this problem, the authors proposed a distributed deep learning driven, fog-to-things, computing attack detection scheme. The authors concluded that by using deep learning techniques, the accuracy and the efficiency of cyber-attack detection could be improved in IoT/Fog environments.

Privacy is considered one of the key challenges nowadays for IoT systems as any IoT system should maintain the privacy of users' data throughout its storage, processing, and in communication transit.

In [5] the authors propose a lightweight secure and privacy-preserving communication protocol for Smart Home Systems (SHSs). The proposed protocol incorporates a high entropy, chaos-based, cryptographic scheme to generate keys to be applied to symmetric encryption of communications. Integrity of communications is assured through the use of Message Authentication Codes (MACs) in data transmissions. The study provides a comprehensive security analysis and performance evaluation of the proposed protocol in terms of computational complexity, memory cost, and communication overhead.

As shown in the section, the IoT environments are suffering from multiple security threats and privacy breaches. Although many of the researchers try to address the issue in different ways, none of them used the blockchain solution to build a trusted smart environments. In such environments, only a certified IoT devices can be attached to the environments

2.2 Self-Sovereign Identity

The basic concept of self-sovereign identity which was introduced by Christopher Allen [6] is to allow people and businesses to store their own identity data locally, and provide them with the ability to validate and verify their identity without sharing their sensitive information.

To consider an identity a self-sovereign the following conditions must be met according to Allen:

- Existence: The identity can be only for devices/users who/m already exist.
- Control: The users/devices should have a full control of their identity.
- Access: The users/devices should have full access to their own data at any time.
- Transparency: Any system dealing with the users/devices data should be transparent in terms of functionality and management.
- Persistence: the identity should last as long as the users/devices wish.
- Portability: the identity should be portable. The device/user/organisation should have the ability to carry their identity with them.
- Interoperability: the identity should be widely used across different platforms and systems
- Consent: the users/devices must agree on how to use their identities.
- Minimalization: the identification process should be based on minimal necessary data to accomplish the task and avoid sharing unnecessary data.
- Protection: The identity management system should protect the rights of the users/devices/organisations.

2.3 Existing Compliance Solution

Regulations and standards compliance systems attracted many researchers in the past few years. This is due to the increasing awareness of data privacy and increasing attacks on personal information.

Stuurman and Kamara in [7] discuss the importance of the standards and regulations compliance systems for an IoT environment. They propose a data protection model to ensure the compliance with data protection regulations in IoT environments. The proposed data model is consistent with GDPR.

In [8] the authors provide a blockchain based compliance system for Everything IoT (EIoT) environments. More specifically, the proposed solution ensures the compliance of Bulk Electric System (BES) operations to the North American Electric Reliability Corporation (NERC) Critical Infrastructure Protection (CIP) requirements. The proposed solution provide an extra layer of security for the whole BES supply chain.

Bourgeois *et al.* in [9] propose a GDPR compliant architecture for IoT environments. The proposed architecture identifies three GDPR principles: private-by-default, analytics transparency and accountable analytics to protect the privacy of the IoT data in an academic environment.

In [10] the authors provide a regulation compliance framework for healthcare systems. The proposed framework ensures that the healthcare systems are compliance to the Health Insurance Portability and Accountability Act (HIPAA).

The authors in [11] discuss compliance to security standards no always assure security for all components of the IT systems. The authors explain many interesting aspects need to be addressed before adhering to standards and regulations such as equal security for all IT components of a system.

Different than the above research, the proposed system will use the blockchain technology to ensure GDPR compliance for all devices connected to a smart environment. Although we are using GDPR compliance as a use case; however; our systems can be easily adapted to cover any other regulation or standard.

3   Proposed Regulation Compliance System

Identity management in information and online systems is considered as one of the key issues of IoT environments. This is because digital identities are open to abuse from operators/organisations and the online identity claims verification is very difficult for individuals [8]. To solve this problem, the self-sovereign and blockchain concepts can be used to realise a novel solution addressing these needs. Figure 1 presents our proposed architecture which we incorporate in our use-case in Sect. 3. The proposed architecture is based on the Government for the Province of British Columbia ("BC-Gov") implementation of the Hyperledger Indy identity management system [7].

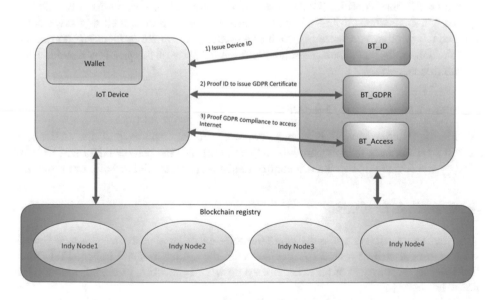

Fig. 1. Proposed Architecture.

The elements of this solution are:

- The blockchain ledgers network
- The entities which can issue and verify assertions
- The identity management modules

These are explained below.

3.1   The Blockchain Ledgers Network

The blockchain ledgers network is a network of Hyperledger Indy nodes, each of which contains a blockchain ledger. The four nodes are used as distributed ledgers in order to build a decentralised identity management system. It is worth noting that there is no personal data written to the ledgers. The ledgers are used to anchor encrypted data without publishing them. Hyperledger Indy uses the ledgers to allow communication peers to establish encrypted connections to exchange the personal data.

3.2   The Issuers/Verifiers Entities

The issuers are responsible for issuing the verifiable claims for the IoT devices in an efficient way. These claims can be cryptographically verified by any another part asserting proof of some aspect such as capabilities or compliance. For example, an IoT device requests proof of regulation compliance from the regulator body. The regulator bodies will provide the credentials to the IoT devices which include a proof of regulation compliance and any other necessary information. This information will be kept on the IoT device itself and not shared with any other third party. The issuer will update the ledger with a Distributed Identifier (DID) anchor and the credentials definitions only. The IoT device can share some attributes from their credentials as proof of their regulation compliance to any other verifier entity, such as a home environment or industrial consumer. The verifier entity can check the source and the integrity of the proof by checking the distributed ledgers.

3.3   The Identity Holder Modules

An Identity Holder Module is a repository/wallet to save and keep all the credentials issued by the issuers for this particular device/entity. All the personal data will be handled by the device and has the full control to share the credentials wholly or partially.

4   Use-Case Scenario

This section describes a use-case where the self-sovereign identity management framework can be used to provide IoT devices with the ability to share information about their identity securely in order to establish a network connection through for example a home hub/gateway.

This use-case was enabled and explored though use of a developed prototypical system which implements the architecture presented in Sect. 3.

Figure 2 illustrates the different components of the prototype application and illustrates the relationship between each of them. Describing a scenario of a smart home environment. In this scenario, all devices which operate within the home should comply with legal obligations and the regulations of the home owner or a network services provider. Figure 2 presents a blockchain based solution which enables this.

In the figure, any IoT device which request access to the home network should be registered with regulations and standards bodies such as a GDPR compliance underwriter. The GDPR body is responsible for device compliance checking, issuing certificates and registering them to the blockchain. The HomeHub in Fig. 3 has the responsibility to verify any claim provided by IoT devices by checking the blockchain registry.

With the current enforcement of many laws and regulations (such as GDPR) it becomes necessary to ensure all users/devices connected within your premises comply with laws/regulations/standards. In this use-case we consider the case of IoT compliance to GDPR while restricting Internet access through the BT Homehub only to IoT devices with GDPR compliance proof.

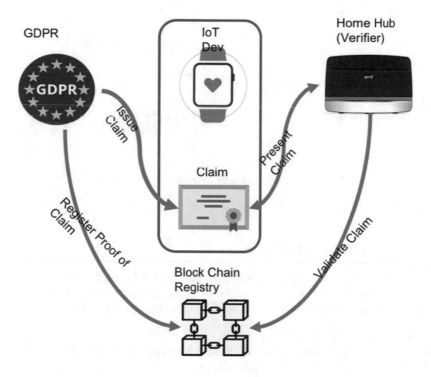

Fig. 2. Developed use-case for smart home environments.

In the proposed prototype the following components are introduced:

- BT device registry: To assign a unique ID for IoT devices.
- BT GDPR: BT GDPR is acting as an issuer in the IoT scenario which produces a certificate to ensure the IoT devices follow GRPR regulations.
- IoT Devices: an IoT device which is equipped with a blockchain wallet to store the credentials.
- HomeHub: the homeHub is acting as verifier and controls network access for IoT devices as informed by verification of the IoT identity and GDPR checking process.

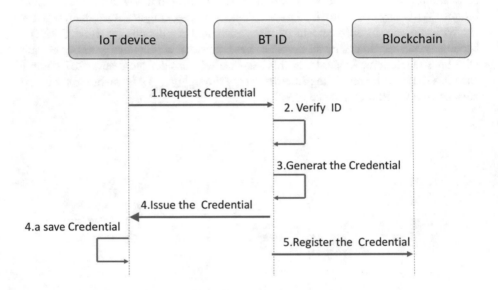

Fig. 3. The process of issuing an identity to a device.

4.1   Issuing an ID for an IoT Device

To register the initial identity of a device, the device should register with an organisational/governmental body (BT ID in our prototype) to recognise it as a legally operating device. In the case of the request being approved, the BT ID will do the following:

- Generate a credential for the IoT device.
- Send the credential to the IoT device to save it and keep it on its wallet.
- Register this transaction to the blockchain which will allow any other third party to verify it in the future.

Figure 3 presents this process.

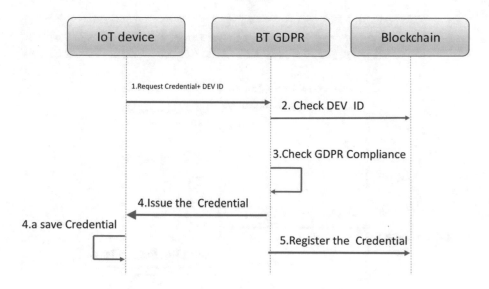

Fig. 4. The process to assert GDPR compliance to a device.

4.2   GDPR Compliance Module

In our use case, all IoT devices should prove two main things: (i) the IoT device is registered with the BT identity body and (ii) the IoT device complies with GDPR. However,
in our use-case the IoT device cannot request proof of GDPR compliance until it has
a valid identity (i.e. registered with BT ID). If the IoT device can provide the proof of
its identity, then it can request/apply for a GDPR compliance check. Figure 4 shows the
steps of the GDPR request procedure as follows:

- The IoT device asks for a GDPR Check.
- The BT GDPR (BT_GDPR) module asks the IoT device to presents its claim.
- The IoT device presents the credential (the IoT ID) to BT_GDPR.
- BT_GDPR will verify the issuer (BT ID) and the integrity of the data by checking
  the blockchain ledger.
- BT_GDPR provides the IoT device with GDPR credentials and registers it into the
  distributed block chain ledger.

4.3   BT Internet Access Module

This module is responsible for ensuring the IoT device has GDPR compliance before
granting it access to the internet through the BT HomeHub. As a prerequisite, any IoT
devices which want to connect through the HomeHub need to provide proof of their
compliance to the GDPR regulation. The BT Internet access module will issue an internet access certificate, send it to IoT device to keep in its wallet, register it to the block
chain distributed ledger and grant access through the HomeHub. This process is presented in Fig. 5 as follow:

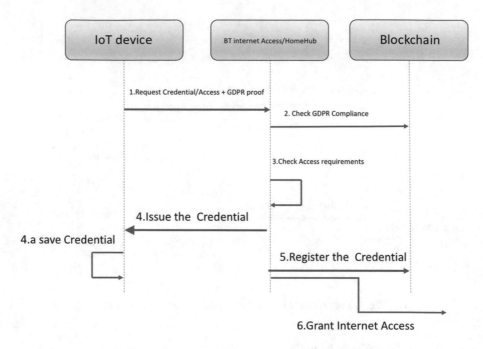

Fig. 5. The process of enabling internet access to an IoT device.

- The IoT device requests internet access from the HomeHub asserting its identity and GDPR compliance proof.
- The HomeHub verifies the IoT compliance with GDPR by checking the data integrity and issuer data with the blockchain.
- The HomeHub generates the internet access credentials.
- The HomeHub sends the internet access credentials to the IoT device to save it in its wallet.
- The HomeHub registers the internet access credentials into the blockchain.
- The HomeHub grants internet access to the IoT device.

5   Operational GDPR Compliance Example

In this section, we show an operational example of our proposed GDPR compliance system. The example illiterates how a sensor obtains the GDPR credentials and presents it to the Home Hub to operate within the smart environment.

5.1   ID Credentials

The first stage of the system is to issue an identity credentials for each of its components: sensor, GDPR and BT Home Hub. This will allow trust to be built between them and initiate the secure communication channels to exchange the other credentials. The credentials include the name of the device/organisation, the email address of the administrator and type.

GDPRCertificate 1.3

certificate_n: GDPR123456789

expiry_date: 2022

issue_date: 2018

name: Sensor_1

Fig. 6. Issuing a GDPR certificate for Sensor_1

5.2   Issuing and Creating Credential

The main purpose of the proposed system is to ensure GDPR compliance for all IoT devices operating within the working spaces of a smart environment. The BT GDPR module is responsible for creating the GDPR scheme where the GDPR regulation body defines the GDPR certificate and the associated attribute with it. The attributes of GDPR certificates includes the name of the device/organisation, certificate issuing date, certificate expiry date and certificate number. Upon request, the BT GDPR will check if a device or organisation is GDPR compliant and issue a certificate accordingly. Figure 6 shows the GDPR certificate generated by BT_GDPR upon the request from Sensor_1.

5.3   Access Request and Proof Request

In our use-case the any device requests to operate within authorities of a smart environments must comply with GDPR regulation. Upon a request to connect generated by Sensor_1 the Home hub will request a proof of GDPR compliance. Figure 7 shows the requested attribute needed for the home hub to verify the device/organisation GDPR compliance. As shown in the figure the Home hub requests only two attributes: expiry date and certificate number. Figure 8 shows Sensor_1 receiving the request where it can accept or reject sharing their own information. If Sensor_1 provides the Home hub with the necessary information, the home hub can verify the integrity of the data by using the blockchain registry network as shown in Fig. 9.

```
{
 "name": "Certificate-Data",
 "version": "0.1",
 "requested_attributes": {
  "attr1_referent": {
    "name": "expiry_date",
    "restrictions": [
      {"cred_def_id": "TJVMc9WJmRjqNUy8yRZrNm:3:CL:20:MyGDPRCertificate"}
    ]
  },
  "attr2_referent": {
    "name": "certificate_n",
    "restrictions": [
      {"cred_def_id": "TJVMc9WJmRjqNUy8yRZrNm:3:CL:20:MyGDPRCertificate"}
    ]
  }
 },
 "requested_predicates": {}
}
```

Fig. 7. Home Hub proof request: GDPR attributes

Fig. 8. Sensor_1 receiving proof request from Home Hub

Certificate-Data

expiry_date: 2022

certificate_n: GDPR123456789

Fig. 9. Home Hub verifies Sensor_1's data

6 Discussion

In this paper, we proposed a GDPR compliance system based on the Self-Sovereign Identity and Hyperledger Indy framework. The proposed system address all Self-Sovereign Identity's characteristics as follows:

- Existence: The proposed solution introduces a BT_ID identity to check the existence of a device/organisation and issue and ID credentials accordingly.
- Control: The devices/organisations on the proposed solution have full control of their identity and how they want to interact with other components on the system.
- Access: All the devices/organisations can access their own credentials (i.e. ID or GDPR) at any time.
- Transparency: The proposed system has transparent and clear management procedures in how to issue the credentials, proof request and verifying the provided data.
- Persistence: the ID and GDPR credentials can last as long as the device/organisation exist.
- Portability: the ID and GDPR credentials are portable as they are stored on the device/organisation wallet and can be used if requested.
- Interoperability: the ID and GDPR credentials can be used in multiple other organisations/ smart home environments if requested.
- Consent: As shown in Fig. 8 the home Hub should ask the device/organisation for identity information and the device/organisation can approve or reject such a request.
- Minimalization: The GDPR compliance identification process requested as minimal information as possible to complete the process. Figure 7 shows the only two necessary attributes are needed to complete the GDPR compliance check.
- Protection: The rights of any device/organisation is protected on the proposed solution including the rights of any device/organisation opt-out from the smart environments or revoke their credentials.

7 Conclusion

We have proposed a GDPR compliance system based on the Self-Sovereign Identity concepts and Hyperledger Indy framework. The proposed solution provides an assurance for all users in smart home environments that all devices connected to the environment comply with GDPR. By complying with GDPR regulations the users'/organisations' privacy will be protected.

The proposed solution satisfies the 10 Self-Sovereign Identity concepts as discussed in Sect. 6.

In our use-case the IoT scenario should adhere to different regulations. Although the proposed solution only provides a solution for the GDPR compliance problem, it could be adapted to address any other standards or regulations.

Acknowledgements. This research work was conducted under the BT Ireland Innovation Centre (BTIIC) project and was funded by Invest Northern Ireland and BT.

References

1. Statista: The Statistics Portal, Internet of Things (IoT) connected devices installed base worldwide from 2015 to 2025 (in billions), March 2019. https://www.statista.com/statistics/471264/iot-number-of-connected-devices-worldwide/
2. Zou, J., Ye, B., Qu, L., Wang, Y., Orgun, M.A., Li, L.: A proof-of-trust consensus protocol for enhancing accountability in crowdsourcing services. IEEE Trans. Serv. Comput. 1 (2018). https://doi.org/10.1109/TSC.2018.2823705
3. Cerrudo, C.: Hacking US Traffic Control Systems, March 2019. https://defcon.org/images/defcon-22/dc-22-presentations/Cerrudo/DEFCON-22-Cesar-Cerrudo-Hacking-Traffic-Control-Systems-UPDATED.pdf
4. Abeshu, A., Chilamkurti, N.: Deep learning: the frontier for distributed attack detection in fog-to-things computing. IEEE Commun. Mag. **56**(2), 169–175 (2018)
5. Song, T., Li, R., Mei, B., Yu, J., Xing, X., Cheng, X.: A privacy preserving communication protocol for IoT applications in smart homes. IEEE Internet Things J. **4**(6), 1844–1852 (2017)
6. Allen, C.: The Path to Self-Sovereign Identity, April 2016. http://www.lifewithalacrity.com/2016/04/the-path-to-self-soverereign-identity.html
7. Stuurman, K., Kamara, I.: Iot standardization - the approach in the field of data protection as a model for ensuring compliance of IoT applications?. In: 2016 IEEE 4th International Conference on Future Internet of Things and Cloud Workshops (FiCloudW), pp. 336–341 (2016). https://doi.org/10.1109/W-FiCloud.2016.74
8. Mylrea, M., Gourisetti, S.N.G.: Blockchain for supply chain cybersecurity, optimization and compliance. In: Resilience Week (RWS) 2018, pp. 70–76 (2018). https://doi.org/10.1109/RWEEK.2018.8473517
9. Bourgeois, J., Kortuem, G., Kawsar, F.: Trusted and GDPR-compliant research with the internet of things. In: Proceedings of the 8th International Conference on the Internet of Things, IOT '18, pp. 13:1–13:8. ACM, New York, NY, USA (2018). https://doi.org/10.1145/3277593.3277604, http://doi.acm.org/10.1145/3277593.3277604
10. Wu, R., Ahn, G.-J., Hu, H.: Towards Hipaa-compliant healthcare systems. In: Proceedings of the 2nd ACM SIGHIT International Health Informatics Symposium, IHI '12, pp. 593–602. ACM, New York, NY, USA (2012). https://doi.org/10.1145/2110363.2110429, http://doi.acm.org/10.1145/2110363.2110429
11. Duncan, B., Whittington, M.: Compliance with standards, assurance and audit: does this equal security?. In: Proceedings of the 7th International Conference on Security of Information and Networks, SIN '14, pp. 77:77–77:84. ACM, New York, NY, USA (2014). https://doi.org/10.1145/2659651.2659711, http://doi.acm.org/10.1145/2659651.2659711

Two-Dimensional Models of Markov Processes for System Availability

Orhan Gemikonakli[1], Eser Gemikonakli[2], and Enver Ever[3(✉)]

[1] Final International University, Beşparmaklar Caddesi, No: 6, Çatalköy, Girne, Mersin-10, Turkey
`orhan.gemikonakli@final.edu.tr`
[2] Department of Computer Engineering, Faculty of Engineering, University of Kyrenia, Kyrenia, Mersin 10, Turkey
`eser.gemikonakli@kyrenia.edu.tr`
[3] Computer Engineering, Middle East Technical University Northern Cyprus Campus, 99738 Guzelyurt, Mersin 10, Turkey
`eever@metu.edu.tr`

Abstract. System availability is a mission-critical issue. Availability is mainly due to system or component failures and recovery/repair. Both of these processes are stochastic. They are associated with two random variables: mean time between failures and mean time to repair. Models have been developed to estimate the availability of certain systems, especially computer-based processing systems. Mostly, one-dimensional mathematical models are used for this purpose. However, certain applications, such as the availability of consensus protocols, may require two-dimensional models. This is because of security threats that bring another dimension to availability models. Hence, two dimensions are needed for such analysis. An example of this case is Blockchain, which is prone to failures and may have security vulnerabilities. In this paper, following a brief review of known approaches, two-dimensional approaches to availability estimation are introduced. These approaches find applications in the availability of distributed systems using consensus protocols such as PAXOS and IBFT/IBFT2.

1 Introduction

Most of the work reported in the literature is concerned with the one-dimensional availability models of queuing systems. Mathematical modelling addresses pure performance, availability, and performability problems for one- and two-dimensional Markov Processes [7,9,11,13]. In the past decades, solutions have also been developed for three- and four-dimensional processes. Alongside hybrid [4], iterative [6], approximate [5], and numerical solutions, the best-known solution techniques are Seelen's method [18], Matrix-geometric Solution [16], the Gauss-Seidel iterative method [8], and the Spectral Expansion method [1,15]. Some two-dimensional models lend themselves to product-form solutions [22]. Another useful technique is to use a reward model [14,17,19]. However, most of the work reported is on pure performance or performability studies. The need for availability estimation may require multi-dimensional models and solutions. Some progress has been made in this. One of the methods proposes

L. Barolli (Ed.): AINA 2024, LNDECT 200, pp. 161–172, 2024.
https://doi.org/10.1007/978-3-031-57853-3_14

a solution for four-dimensional processes using Spectral Expansion together with a Markov reward model (MRM) [2] for performability evaluation. Distributed systems employing consistency protocols such as PAXOS and IBFT2 require two- and three-dimensional solution techniques for availability and performability analysis, respectively. Work is also in progress for multi-dimensional (beyond four dimensions) modelling. This paper focuses on solving system availability problems. Two-dimensional models and their solutions are presented with ideas for carrying existing work forward. The work can be extended to consider the performability of queueing systems.

The rest of the paper is structured as follows. Section 2 provides a review of existing research on availability modelling and solution techniques. Section 3 describes the methodology employed in this study. Section 4 presents the results obtained using the proposed model and solution approaches. Finally, Sect. 5 summarizes the work and presents a conclusion.

2 Related Work

Solving one-dimensional queuing models involving Markovian processes is well established. Performance evaluation of such systems has a long history. In solving such systems for performance evaluation, it is assumed that there are two stochastic processes: an *input process* and a *service process*. The inter-arrival and service times are exponentially distributed with mean values of $1/\lambda$ and $1/\mu$, respectively. A two-dimensional model (e.g. a fault-tolerant multi-server system) can also be solved for state probabilities in the presence of break-downs and repairs. To obtain the state probabilities of a fault-tolerant, $c - server$ system, it is possible to employ one of the two-dimensional solution approaches, such as the ones discussed in [5, 19, 22]. The proposed solutions offer performability measures for such systems. For performability evaluation, in addition to the aforementioned processes, two more stochastic processes are considered: failures and recovery/repair, which are associated with the mean time to failure (MTTR), and the mean time to repair (MTTR) denoted by $1/\xi$ and $1/\eta$ respectively.

When the systems are considered from an availability point of view, well-known approaches are presented in, [7, 20, 21]. One of the most commonly used approaches given for an N-server system is in Eq. 1

$$A_i = \left[i! \sum_{k=0}^{N} \frac{\left(\frac{\eta}{\xi}\right)^{k-i}}{k!} \right]^{-1} , i = 0, 1, \ldots, N \tag{1}$$

This approach addresses many availability problems by considering only two stochastic processes, which are failures and repairs. Today, security breaches impose further threats to system availability. Hence, further work is needed to accommodate these processes for the availability evaluation of computer networks, Blockchain, IoT systems, etc. In this work, the repair process following a breakdown has been used for system recovery from a security breach. It is possible to separate the two processes and consider them separately.

2.1 Solution Techniques for Two-Dimensional Processes

As introduced in the introduction section, there are various approaches to solving two-dimensional processes. In this context, hybrid [4], iterative [6], approximate [5], Markov reward model [10, 17], and numerical solutions show themselves as flexible techniques. Amongst these approaches, well-known ones are Spectral Expansion, Matrix-geometric Solution, Seelen's method, and Block Gauss-Seidel Method. While the Spectral Expansion approach gives an exact solution, it suffers from the state explosion problem. On the other hand, Seelen's method and Matrix-geometric Solution are less favourable compared to the Spectral Expansion technique. This is well documented in the literature.

The **Block Gauss-Seidel** method which was developed in 1989, is an iterative method used for solving all kinds of simultaneous equations iteratively [8]. This method reduces the finite-state problem to a linear equation involving vector-generating functions and some unknown probabilities. The computations appear to be serial. Since each component of the new iterate depends upon all previously computed components, the updates cannot be done simultaneously.

As the other two-dimensional modelling techniques do, **Seelen's Method** gives an approximate solution for Quasi-Birth-and-Death (QBD) and the multi-band QBD (QBD-M) systems [18]. The Markov Chain is first truncated to a finite state, approximating the original process. Then, it is used together with a dynamically adjusted relaxation factor in an efficient iterative solution algorithm. In each iteration, the algorithm computes one of the state probabilities. The use of an appropriate value for the relaxation parameter is important to obtain the most accurate results. However, this method does not define any solutions to determine the value of the relaxation parameter [1].

In the **Matrix-geometric** solution method, first, a non-linear matrix equation is formed from the system parameters. Then, the minimal non-negative solution R is computed by using an iterative algorithm. This method has probabilistic interpretation for each step of the computations. The main drawback of this method is that the number of iterations for computing R cannot be predetermined, and there is a significant computational requirement to obtain R. Another observation is that in the matrix-geometric method, for some values of certain parameters, the computational requirements are uncertain and relatively large [15].

The **Spectral Expansion** is a solution technique which is useful in the performance and dependability modelling of discrete event systems. It solves Markov models of certain kinds that arise in several practical system models. The reported applications and results include performability modelling of several types of multiprocessors, multi-task execution models, and networks of queues with unreliable servers [1, 3]. In the spectral expansion method, accurate eigenvalues and eigenvectors are necessary since the performance measures can be quite sensitive to these. Gauss-Siedel and Seelen's Methods are not preferred methods compared to Matrix-geometric and Spectral Expansion techniques. The latter two methods have been critically analyzed, and the performances of the two methods have been compared [15]. The authors stated that the spectral expansion method is a better solution, especially when more heavily loaded systems are studied, and batch arrivals (or departures) are included in the model.

In [19], performability models based on Markov reward models (MRM) are presented. The behaviour of multiprocessor systems has been described as a continuous time Markov chain and associated reward rates for performance measures of each state. A systematic study has been performed on complex multiprocessor systems for computing the distribution of accumulated reward of Markov reward models. A Markov chain is presented for a heterogeneous multiprocessor system. Results mainly focus on the distribution of cumulative bandwidth for each processor. Queuing issues and performability measures related to queuing theory were not investigated. Results show that instantaneous measures do not show the dynamic behaviour of the system. The importance of repair facilities on performability distribution is underlined. Also, results indicate that the change in failure and repair behaviour affect the complementary distribution of accumulated reward used for performability. The potential use of MRM goes beyond two-dimensional systems. MRM is a promising approach that can be used when the rest of the techniques fail. Two product-form solutions can be used together to solve a two-dimensional system. One obvious example is reported in [2].

These methodologies are applicable to the solution of two-dimensional availability models. However, they suffer some problems reported in the literature, and in some cases (e.g. large state space, extreme numerical values) they fail to perform.

2.2 Beyond Two-Dimensional Processes

Models for three-dimensional Markov Processes have many application areas. One such area is a two-stage tandem network. The network may have single or multiple servers at each stage. These servers are prone to failures. In this case, each stage can be modelled as a two-dimensional process and solved using one of the approaches stated above (e.g. Spectral Expansion). However, transitions between stages do not allow the system to have an MRM solution. An approximate solution is given in [6]. However, this is a CPU-intensive solution.

It is possible to use two-dimensional approaches for availability modelling. As an example, consider the availability of an IoT system with Byzantine fault tolerance for mission-critical applications. One form of unavailability will be due to breakdowns. This is well exploited in the literature. However, if a system has $N = H + F$ servers where H servers are trustworthy, but the remaining servers are Byzantine (e.g. hacked and used maliciously), this will be another form of unavailability. Now we have four processes for availability analysis: break-down at a rate ξ per server, repair at a rate η, a state change from H to F due to a security breach at a rate ζ, and recovery from F to H at a rate φ. In this work, η and φ are combined, and the former is used. In [13], this problem is partially addressed, ignoring malicious attacks and recovery and considering the case with constant H and F. A more effective model and solution approaches are presented in this study, together with a roadmap for a complete solution to the problem using a three-dimensional model.

A four-dimensional approach can be used to solve a number of multi-process problems. One such case may be extending the work presented in [4]. In that paper, a two-stage system was considered, assuming that the first stage had a highly reliable single server. Initial work shows that it may be possible to consider two fault-tolerant networks in tandem, each having several parallel servers.

It is worth exploring more than four dimensions. Fault-tolerant tandem networks with more than two stages may require multiple dimensions for a solution. Initial work suggests that circular formations with multiple interacting planes (two-dimensional lattices) can be used for that purpose.

3   Work Done

In this section, firstly, a mathematical solution to the problem of the availability of the Istanbul Byzantine Fault-Tolerant (IBFT2) protocol will be presented. In [13], an implementation of a distributed system using a Byzantine Fault-Tolerant (BFT) consensus protocol is considered. A numerical solution was given for pure availability. The work is further extended [12] incorporating Byzantine failures (i.e. malicious hacking of nodes in a Blockchain). In IBFT, if consensus is not reached (i.e. receiving confirmation from more than 2/3 of the N servers), then messages will not be committed. For ease of presentation, here we will assume $N = 4$. Figure 1 shows the state diagram of such a system. Here, it is assumed that in $N = H + F$, H and F are constant at the start. Changes are due to breakdowns and repairs only. Two variables can be introduced to explain this state; $h = 0, 1, \ldots, H$ and $f = 0, 1, \ldots, F$. As is the case in [13], here, we assume that F nodes are unresponsive.

3.1   Two-Dimensional Model for Availability

Figure 1 shows a two-dimensional state diagram to model the availability of Blockchain networks. Here, the limitation is that the problem can only be solved for predefined values of F and H.

It is important to calculate state probabilities q_f and r_h. The former of these probabilities represent the probability of the system state being at a block of states, $(f, j), f = 0, 1, \ldots, F$ (the probability of having f operative Byzantine servers). r_h represents the probability of having h operative, non-Byzantine (trustworthy or uncompromised) servers. Here, two variables are introduced; $h = 0, 1, \ldots, H$ and $f = 0, 1, \ldots, F$. Changes in the h and f values are due to the failure and repair of server nodes. It can be noted that, in fact $q_f = \Sigma_{h=0}^{H} r_h, f = 0, 1, \ldots, F$. Since both r_h and q_f are associated with one-dimensional processes, a one-dimensional Markov chain approach can be used to obtain both sets of probabilities using the well-known availability formula given in [20].

$$q_f = \left[f! \sum_{k=0}^{F} \left(\frac{(\eta/\xi)^{k-f}}{k!} \right) \right]^{-1}, f = 0, 1, \ldots, F$$

$$r_h = \left[h! \sum_{k=0}^{H} \left(\frac{(\eta/\xi)^{k-h}}{k!} \right) \right]^{-1}, h = 0, 1, \ldots, H \qquad (2)$$

However, this solution means that $\Sigma_{h=0}^{H} = 1$ for each f-column. Hence, defining the availability as in Eqn. 3 solves the problem.

$$a_{f,h} = q_f r_h, f = 0, 1, 2, \ldots, F \text{ and } h = 0, 1, 2, \ldots, H \qquad (3)$$

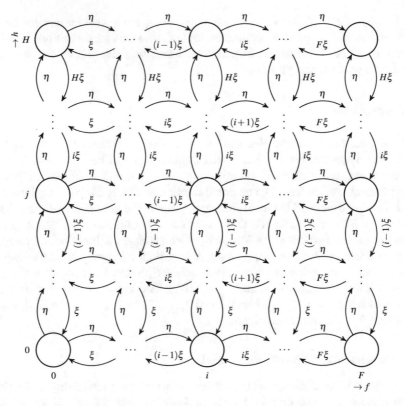

Fig. 1. Availability model for IBFT2 Blockchain Consensus Protocol, $f = 0, 1, 2, \ldots, F, h = 0, 1, 2, \ldots, H$

This is a much simpler and faster solution than the numerical solution presented in [13]. The availability is computed and presented in turn.

3.2   Extended Availability Model

Consider a multi-server distributed system where server nodes may fail due to software or hardware failures or may have been compromised. The compromised servers may be unresponsive or acting maliciously. Firstly, let's assume that the compromised servers become unresponsive. It can be assumed that the system starts with N healthy servers, e.g., $h = N$ where $h = 0, 1, 2, \ldots, N$ where h represents the number of operative and uncompromised servers. Then, through breakdowns and server compromises, the system ends up in $h < N$.

When specific features of such a system are considered:

1. The breakdowns mean that the number of healthy servers can take any value $0 \leq h \leq N$. The total number of operative servers at time t can be denoted as $0 \leq N(t) \leq N$ where $N(t) = h + f$. $N(t)$ changes with break-downs and repairs.
2. Compromising servers causes a drop in the number of healthy servers, however, this does not change $N(t)$. A successful security breach changes the state from (f,h) to $(f+1,h-1))$ and a recovery changes the state from (f,h) to $(f-1,h+1$ keepinh $N(t) = h + f$ constant.

This is illustrated in Fig. 2. The stages are described by (f,h), for $N = 4$. The h nodes are compromised at a rate $h\zeta$ and reinstated at a rate η.

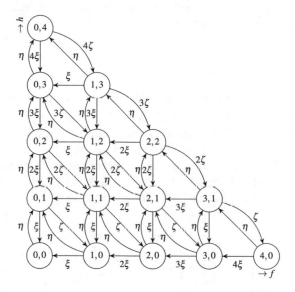

Fig. 2. Availability model for a BFT consensus protocol for $N = 4$ and variable F and H.

3.2.1 Balance Equations

The balance equations can be obtained from Fig. 2 as follows:

$$\eta p_{00} - \xi p_{10} - \xi p_{01} = 0$$
$$(2\eta + \xi)p_{10} - \xi p_{11} - \zeta p_{01} - 2\xi p_{20} = 0$$
$$2(\eta + \xi)p_{20} - \xi p_{21} - \zeta p_{11} - 3\xi p_{30} = 0$$
$$(2\eta + 3\xi)p_{30} - \xi p_{31} - \zeta p_{21} - 4\xi p_{40} = 0$$
$$(\eta + 4\xi)p_{40} - \zeta p_{31} = 0$$
$$(\eta + \xi + \zeta)p_{01} - \eta p_{00} - 2\xi p_{02} - \xi p_{11} - \eta p_{10} = 0$$
$$2(\eta + \xi + \zeta)p_{11} - 2\zeta p_{02} - 2\xi p_{21} - 2\xi p_{12} - \xi p_{10} = 0$$

$$(2\eta + 3\xi + \zeta)p_{21} - 2\zeta p_{12} - 2\xi p_{22} - 3\xi p_{31} - \eta p_{20} = 0$$
$$(\eta + 4\xi + \zeta)p_{31} - 2\zeta p_{22} - \eta p_{40} - \eta p_{30} = 0$$
$$(\eta + 2\xi + 2\zeta)p_{02} - \eta p_{01} - 3\xi p_{03} - \xi p_{12} - \eta p_{11} = 0$$
$$(2\eta + 3\xi + 2\zeta)p_{12} - 3\zeta p_{03} - 2\xi p_{22} - 3\xi p_{13} + \eta p_{21} = 0$$
$$(\eta + 4\xi + 2\zeta)p_{22} - 3\zeta p_{13} - \eta p_{31} - \eta p_{21} = 0$$
$$(\eta + 3\xi + 3\zeta)p_{03} - \eta p_{02} - 4\xi p_{04} - \xi p_{13} - \eta p_{12} = 0$$
$$(\eta + 4\xi + 3\zeta)p_{13} - 4\zeta p_{04} - \eta p_{22} - \eta p_{12} = 0$$
$$4(\xi + \zeta)p_{04} - \eta p_{03} - \eta p_{13} = 0$$

$$(4)$$

In turn, these balance equations can be solved for state space probabilities using the solution of the system of equations $AX = B$.

4   Results

Figure 3 shows the average availability as a function of the total number of nodes and $\frac{\eta}{\xi}$. The system has been assumed available for $h > \frac{2N}{3}$. As the number of nodes increases, the availability deteriorates. This is because the failure rate is proportional to the number of nodes when a single repair facility is assumed. For $\eta/\xi = 100$, the availability changes between 96–98% depending on the number of nodes. This shows that, for high availability, $\eta/\xi > 100$ is desirable. This can be achieved by reducing repair time. One way of reducing the repair time can be to increase the number of repair facilities.

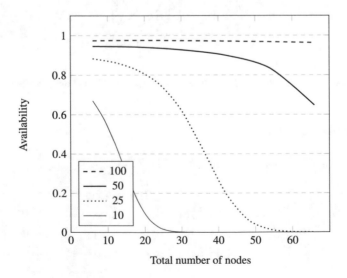

Fig. 3. System availability as a function of the total number of nodes and η/ξ

So far, it has been assumed that any combination of (h, f) for $N = h + f$ is equally likely. Next, a new model has been introduced considering security breaches and recovery as additional Markov processes. The state diagram of this model is shown in Fig. 2. Here, once a node fails, repair work re-instates the server as a h-server regardles of its status prior to failure. The model is more difficult to solve. The solution approaches summarised earlier (such as the Spectral Expansion method) fail to give results. Hence, numerical techniques were used to solve $N = 4$. In addition to the two random variables introduced earlier, ζ has been used to denote the rate of successful attacks on servers (Fig. 4).

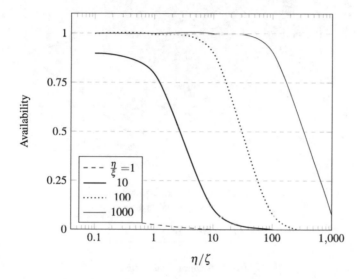

Fig. 4. System availability as a function of the total number of nodes and η/ξ and η/ζ

The results clearly show the impact of a serious level of security breaches. Even with $N = 4$, the availability can rapidly diminish as security is compromised. However, speedy recovery from failure has a significant impact on system availability. Increasing the number of repairmen will certainly improve the performance considerably (Fig. 5).

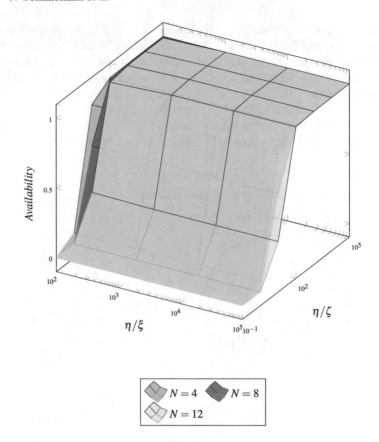

Fig. 5. System Availability

5   Conclusion

Following a brief review of some widely used solution techniques for two-dimensional stochastic processes, this study discusses new techniques for solving multi-dimensional Markov Processes for the availability of multi-server systems, emphasising Blockchain mechanisms. An approach similar to MRM is used for two-dimensional models. While MRM is used in performability analysis of multi-server systems poses some difficulty due to the presence of zero-transition states, availability models do not suffer such complications. The models are then extended to cover three-dimensional availability models focusing on Blockchains. Here, unavailability is not only due to server breakdowns but also due to security breaches. Finally, a new concept, *fake availability*, has been introduced. In this case, there seems to be a sufficient number of servers for a consensus; however, a number of these servers may well be acting maliciously. Numerical results have been presented for all cases considered. The models can further be extended to include tandem systems prone to failures and/or security breaches. In such cases, multi-dimensional models would be required.

References

1. Chakka, R.: Spectral expansion solution for some finite capacity queues. Ann. Oper. Res. **79**, 27–44 (1998)
2. Ever, E.: Fault-tolerant two-stage open queuing systems with server failures at both stages. IEEE Commun. Lett. **18**(9), 1523–1526 (2014)
3. Ever, E., Gemikonakli, E., Nguyen, H.X., Al-Turjman, F., Yazici, A.: Performance evaluation of hybrid disaster recovery framework with d2d communications. Comput. Commun. **152**, 81–92 (2020)
4. Ever, E., Gemikonakli, O., Kocyigit, A., Gemikonakli, E.: A hybrid approach to minimize state space explosion problem for the solution of two stage tandem queues. J. Netw. Comput. Appl. **36**(2), 908–926 (2013)
5. Gemikonakli, O., Ever, E., Kocyigit, A.: Approximate solution for two stage open networks with Markov-modulated queues minimizing the state space explosion problem. J. Comput. Appl. Math. **223**(1), 519–533 (2009)
6. Gemikonakli, O., Sanei, H., Ever, E.: Approximate solution for the performability of markovian queuing networks with a large number of servers. In: 5th International Workshop on Signal Processing for Wireless Communication SPWC 2007 (2007)
7. Goyal, A., Lavenberg, S.S., Trivedi, K.S.: Probabilistic modeling of computer system availability. Ann. Oper. Res. **8**, 285–306 (1987)
8. Grassmann, W.K.: Computational methods in probability theory. Handbooks Oper. Res. Management Sci. **2**, 199–254 (1990)
9. Haverkort, B.R., Ost, A.: Steady-state analysis of infinite stochastic petri nets: comparing the spectral expansion and the matrix-geometric method. In: Proceedings of the Seventh International Workshop on Petri Nets and Performance Models, pp. 36–45. IEEE (1997)
10. Howard, R.: Dynamic probabilistic systems, vol. ii: Semi-Markov and decision processes. john wiley and sons, inc. (1971)
11. Kirsal, Y., Gemikonakli, O.: Performability modelling of handoff in wireless cellular networks with channel failures and recovery. In: 2009 11th International Conference on Computer Modelling and Simulation, pp. 544–547 (2009)
12. Marcozzi, M., Gemikonakli, O., Gemikonakli, E., Ever, E., Mostarda, L.: Availability evaluation of IoT systems with byzantine fault-tolerance for mission-critical applications. Internet Things **23**, 100889 (2023)
13. Marcozzi, M., Gemikonakli, O., Gemikonakli, E., Ever, E., Mostarda, L.: Availability model for byzantine fault-tolerant systems. In: International Conference on Advanced Information Networking and Applications, pp. 31–43. Springer (2023)
14. Meyer, J.: On evaluating the performability of degradable computing systems. IEEE Trans. Comput. **100**(8), 720–731 (1980)
15. Mitrani, I., Chakka, R.: Spectral expansion solution for a class of Markov models: application and comparison with the matrix-geometric method. Perform. Eval. **23**(3), 241–260 (1995)
16. Neuts, M.F.: Matrix-geometric solutions in stochastic models, volume 2 of johns hopkins series in the mathematical sciences (1981)
17. Reibman, A., Smith, R., Trivedi, K.: Markov and Markov reward model transient analysis: an overview of numerical approaches. Eur. J. Oper. Res. **40**(2), 257–267 (1989)
18. Seelen, L.: An algorithm for ph/ph/c queues. Eur. J. Oper. Res. **23**(1), 118–127 (1986)
19. Smith, R., Trivedi, K.S., Ramesh, A.: Performability analysis: measures, an algorithm, and a case study. IEEE Trans. Comput. **37**(4), 406–417 (1988)
20. Trivedi, K.S.: Probability & statistics with reliability, queuing and computer science applications. John Wiley & Sons (2008)

21. Trivedi, K.S., Sathaye, A.S., Ibe, O.C., Howe, R.C.: Should i add a processor? (performance evaluation). In: Twenty-Third Annual Hawaii International Conference on System Sciences, vol. 1, pp. 214–221. IEEE (1990)
22. Yaqoob, M., Gemikonakli, O., Ever, E.: Modelling heterogeneous future wireless cellular networks: an analytical study for interaction of 5g femtocells and macro-cells. Futur. Gener. Comput. Syst. **114**, 82–95 (2021)

Enabling AI in Agriculture 4.0: A Blockchain-Based Mobile CrowdSensing Architecture

Ankit Agrawal$^{(\boxtimes)}$, Bhaskar Mangal, Ashutosh Bhatia, and Kamlesh Tiwari

Birla Institute of Technology and Science, Pilani, Rajasthan, India
{p20190021,p20210473,ashutosh.bhatia,
kamlesh.tiwari}@pilani.bits-pilani.ac.in

Abstract. Agriculture 4.0 relies on extensive data for predictive services, necessitating effective data collection. Mobile CrowdSensing (MCS), with its cost-effectiveness and scalability, addresses this need but faces centralization limitations. Blockchain-based frameworks have been proposed to mitigate these issues but often focus solely on data collection, lacking a comprehensive end-to-end architecture for smart agriculture. Recent literature has explored the integration of the Internet of Things (IoT), edge computing, fog computing, and cloud computing capabilities to establish centralized end-to-end architectures. Nonetheless, these architectures come with their own set of centralized limitations. In the context of contemporary technologies, the integration of blockchain and digital twin (DT) holds the potential to revolutionize the field of smart agriculture. This paper introduces a holistic end-to-end, layered, and service-oriented architecture for Agriculture 4.0, integrating mobile crowdsensing, blockchain, and DT. Unlike existing architectures, this approach aims to overcome centralization limitations, leveraging the strengths of emerging technologies. The proposed architecture extends current capabilities for more efficient and secure Agriculture 4.0 practices. We deploy the suggested architecture onto the Ethereum blockchain, demonstrating its practicality through the obtained results.

1 Introduction

The continuous evolution of technologies transformed the industrial revolutions from Industry 1.0 to Industry 4.0, which is expected to transform the agricultural industry revolutions towards Agriculture 4.0. The technologies characterizing Industry 4.0 include IoT, big data, cloud computing, AI, 5G, and Blockchain, enabling the agricultural ecosystem to be more intelligent, secure, advancing towards high automation and decision making capabilities [1]. Agriculture 4.0 utilizes IoT technology to collect a large amount of Spatio-temporal data in real-time. The realm of big data science often integrates various technologies such as machine learning (ML), artificial intelligence (AI), cloud computing, and others to perform analytics on the collected data, enabling decision-making capabilities. On the other hand, Blockchain technology maintains data integrity and enhances trust among the system entities. In addition, the DT concept plays a crucial role in Agriculture 4.0 by offering a virtual representation of physical

L. Barolli (Ed.): AINA 2024, LNDECT 200, pp. 173–186, 2024.
https://doi.org/10.1007/978-3-031-57853-3_15

agricultural assets, processes, and systems [2]. This innovative approach brings numerous benefits, including enhanced precision farming, optimized resource efficiency, and informed decision-making through real-time monitoring and predictive analytics. DTs empower farmers with a comprehensive toolset to drive sustainability, productivity, and resilience in modern agriculture practices.

Besides the benefits of using AI in Agriculture 4.0, there are certain challenges of using AI technology. To harvest the full potential of AI in Agriculture 4.0, the AI models should be fed with real-time data for making timely predictions and to rewire themselves as per the "concept drift [3]." Another critical issue is the scalability requirement for data collection. The volume and diversity of input data measure the strength of an AI system. Artificial neural networks, fuzzy control systems, and other forms of AI engines require massive amounts of data for proper training before they can be successfully applied in the agricultural sector to guide farmers toward precision agriculture. However, data collection becomes a hurdle for such a large amount of data.

Deploying sensing infrastructure to collect the data across many farm fields in the country and providing its maintenance is neither feasible nor economically viable for any third-party service provider, including the government. Agricultural Mobile crowdsensing (AMCS) is considered a cost-effective and scalable solution to solve this issue [4]. The existing AMCS systems are centralized and have their own limitations, such as being less secure due to a single point of failure, the threat of data modification, being vulnerable to attacks, and having trust deficit issues due to lack of transparency. Most importantly, such systems do not allow data sharing among multiple stakeholders, especially farmers. Moreover, two critical issues in the data collection process through mobile crowdsensing demand attention: data ownership and transfer and lack of motivation among farmers to actively participate in the data collection process.

Numerous blockchain-based MCS frameworks, [5,6], have been proposed to address centralization issues. However, these frameworks primarily focus on raw data collection based on requester specifications, neglecting the crucial aspect of providing services based on the acquired data. An essential service of an MCS system involves delivering validated data to the requester. Existing frameworks often rely on common methods like finding data similarity and truth discovery to validate data over the blockchain. Nevertheless, these approaches face challenges when applied to agricultural data due to the dynamic nature of field properties. Addressing this issue requires an additional implementation of ML/AI methods, which is currently lacking in established blockchain-based MCS systems due to the impracticality of integrating such methods directly into the blockchain. Consequently, an opportunity exists to create an environment that facilitates the seamless integration of ML/AI into existing blockchain-based MCS systems. Beyond these frameworks, the literature also presents layered IoT architectures discussed in Sect. 2 that explore integrating advanced technologies, including IoT, edge computing, fog computing, Cloud computing, DT, and blockchain, to enhance service capabilities. However, there is a notable absence of architectures that comprehensively integrate all these technologies to harness their collective advantages.

To remove these obstacles in enabling Agriculture 4.0, we propose an end-to-end, layered, and service-oriented architecture, integrating advanced technologies like IoT, DT, mobile crowdsensing, blockchain, and AI. The proposed architecture enables real-time, scalable, cost-effective agricultural data collection using blockchain-based mobile crowdsensing that caters to centralization problems, data ownership, and incentivization. Also, it enables AI to provide services from the collected data.

2 Related Work

This section discusses a succinct overview of smart agriculture architectures, primarily centered on sensor/things, edge, fog, and cloud computing paradigms. A recent survey [7] on smart agriculture has discussed the role of these computing paradigms in detail. The IoT ecosystem comprises three key components: Perception, Communication, and Intelligence/Control Layers [8]. Numerous proposed architectures address challenges related to data traffic, communication reliability, and efficient resource utilization. Examples include layered architectures [7, 9–11].

Authors in [12] proposed an extended layered centralized architecture encompassing the security and privacy requirements, such as authentication, access control, authorization, integrity, availability, entity, data, and location privacy. However, the application of traditional centralized security measures faces constraints related to a singular point of failure, traceability, verifiability, and scalability. The authors in [13] proposed a blockchain-IoT layered architecture. The additional blockchain layer is responsible for the storage and security of data. However, storing data in the blockchain may create storage scalability issues. The authors in [14] integrate blockchain technology into the existing layered architecture to identify the anomalies through the smart contract. However, the data collection is solely done in a centralized environment. The authors in [15] proposed a blockchain-assisted architecture for smart agriculture to maintain communication among the stakeholders securely. The proposed architecture utilized a private blockchain explicitly designed to secure supply chain processes and forecast commodity prices.

A few articles incorporate the concept of DTs in the existing centralized layered architecture. For instance, [16] proposed a novel architecture by integrating the concept of DTs into an existing layered architecture to address resource scheduling, resource allocation, and task scheduling challenges in smart agriculture. Note that the discussed architectures are not designed explicitly for mobile crowdsensing. On the contrary, our proposed architecture presents a core layer that seamlessly integrates DT and Blockchain, amplifying the capabilities of the existing architecture. This integration establishes an environment capable of providing secure and privacy-preserving data collection and AI-enabled agricultural services and opens a new window to generate frameworks on top of the proposed architecture.

3 Proposed Architecture

This section discusses the proposed end-to-end layered and service-oriented architecture for mobile crowdsensing, enabling AI to perform data analysis and predictions. The layers include the physical sensing layer, virtual sensing layer, DT layer, and application layer. The DT layer is a core layer in the architecture and comprises the storage, blockchain, and AI/ML/Cloud layers. Each layer provides different services and has different responsibilities and purposes, as shown in Fig. 1. The primary services of the proposed architecture include data services and agricultural services. These services span data collection, local and persistent data storage, data delivery, and data analysis and/or prediction, all carefully designed to prioritize the security and privacy of both the data and the users.

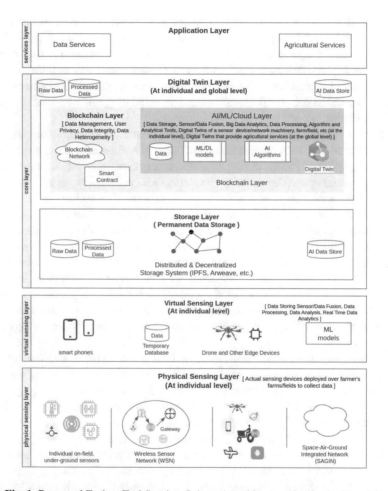

Fig. 1. Proposed End-to-End Service-Oriented Architecture for Agriculture 4.0

3.1 Physical Sensing Layer

This layer outlines the methodology for collecting data from farmers' fields. The physical sensing layer is responsible for interacting with the physical environment and employs IoT sensing devices for data collection. Various sensing technologies play a pivotal role in this process, encompassing individual or networked on-ground and underground sensors, as well as remote sensing technologies like drones, aircraft, and satellites. Farmers or service providers can deploy sensors in fields, on agricultural machinery, vehicles, or robots, forming wireless sensor networks, or utilizing remote sensing technologies for comprehensive data collection. Remote sensing technologies are particularly effective for gathering imagery data. The collected data from these physical IoT devices can be transmitted to the owner's smartphone using wireless technologies such as Wi-Fi, ZigBee, 5G, etc.

3.2 Virtual Sensing Layer

This layer is termed the virtual sensing layer because physical sensing devices do not directly engage with remote devices or third-party applications, mitigating the impact of cybersecurity attacks. High-end devices like smartphones and drones play a crucial role in collecting data from physical sensing devices. Subsequently, these devices validate and analyze the collected data using ML models through mobile or standalone applications. Such advanced devices possess the capability to process raw discrete and imagery data, extracting valuable information. Moreover, they facilitate edge computing capabilities in agriculture. In the contemporary landscape, smartphones are considered IoT sensing devices, leveraging their embedded sensors to sense the environment and collect data. Another reason for labeling this layer as the virtual sensing layer is that high-end devices conceal the actual sensing devices, responsibly interacting with and sending data in a privacy-preserving manner using blockchain identities to the upper layer.

3.3 Digital Twin (DT) Layer

The DT layer assumes a pivotal role in various domains, particularly in Agriculture 4.0, owing to the versatile services it offers. Functioning at individual and global levels, the DT layer entails the creation of virtual counterparts corresponding to physical entities within a farmer's domain, be they living, non-living, process, or system. Living entities encompass animals, plants, and more, while non-living entities include agricultural vehicles, machinery, robots, fields, and farms. Farmers or service providers can fashion the DT for living and non-living entities, delivering services such as real-time monitoring, system failure analysis, optimization and updates, technology integration, and energy consumption analysis at an individual level. The article [17] describes the utilization of DTs in different applications of smart agriculture.

The service providers can also harness DTs of processes or systems to deliver agricultural services globally to the application layer. For such global services, a DT needs robust data storage, the capability to handle big data, and an infrastructure for implementing ML, DL, or AI over big data to analyze data and offer predictions, all while ensuring security, privacy, and trust. The DT layer is structured into three components to fulfill these requisites: storage, blockchain, and AI/ML/Cloud. It's important to highlight that the development of DTs in the context of smart agriculture is predominantly in its conceptual stages. Creating a DT is an optional feature in the proposed architecture, allowing service providers to leverage the DT layer for services without necessarily creating a DT.

Storage Layer

The gathered data is initially stored locally for individual-level data cleaning, processing, and analysis. However, opting for local storage on personal devices or local servers introduces potential drawbacks, including data loss, security vulnerabilities, scalability issues, and maintenance overhead. Although local storage may appear cost-effective in the short term, it is crucial to consider the long-term expenses associated with hardware maintenance, upgrades, and the potential costs linked to data loss. Therefore, there is a need for persistent data storage for extended usability. In contemporary times, numerous distributed and decentralized storage systems like InterPlanetary File System (IPFS), Arweave, etc., are accessible, providing enduring data storage solutions. These external storage systems offer various advantages, including decentralization, censorship resistance, data immutability, content addressing, and enhanced security. Often, these systems incorporate built-in security features, such as encryption, ensuring data protection during storage and transmission. Users at an individual level can trust the integrity and confidentiality of their stored information, while additional security measures can be implemented to safeguard data security and privacy when utilizing external storage systems.

Blockchain Layer

The blockchain layer assumes a pivotal role in the proposed architecture, facilitating the implementation of a blockchain-based mobile crowdsensing system for data acquisition. It serves as an intermediary between data producers and consumers, with the specifics of the data acquisition process discussed in Sect. 4.1. Leveraging blockchain for data acquisition provides inherent advantages in managing data and handling heterogeneity. The technology ensures data immutability and user privacy, facilitates the use of smart contracts, and significantly establishes trust. Users interacting with the blockchain employ a pseudonymous identity, preserving privacy as personal information is not linked to this pseudonym. It's worth noting that while blockchain can be used for data storage, the potential scalability issues associated with massive data sizes led us to consider an external storage system in the proposed architecture. Additionally, it is crucial to highlight that implementing ML, DL models, or AI algorithms directly over the blockchain is not deemed feasible.

AI/ML/Cloud Layer

The Cloud Layer, a pivotal component in our architecture, is dedicated to leveraging AI/ML technologies to extract valuable insights from the data collected in the proposed agriculture. This layer serves as the powerhouse for performing big data analytics, ML/DL models, and AI algorithms over the collected agricultural data. It leverages big data analytics tools to process and analyze large datasets efficiently. This includes trend analysis, predictive modeling, recommendations, and identifying patterns to support data-driven decision-making, empowering stakeholders with actionable insights from historical and real-time data.

The Cloud Layer caters to individual and global needs within the agriculture domain. It provides scalable and flexible AI-as-a-services tailored to stakeholders' unique needs. These services encompass infrastructure for creating DTs, big data storage, data validation, and a suite of other services that can be seamlessly integrated into the overall agriculture ecosystem. Stakeholders, from individual farmers to agricultural service providers, can harness the cloud infrastructure for various purposes. For instance, in Empowering individual farmers, the cloud infrastructure allows them to create and manage DTs for their specific agricultural plots. Through the analysis of DTs, farmers can gain insights into crop health, soil conditions, and optimal resource utilization, enhancing their decision-making capabilities. Agricultural service providers can leverage cloud services to deliver innovative solutions to consumers globally. This includes precision farming services, crop monitoring, disease prediction, and customized recommendations based on data-driven insights.

3.4   Application Layer

This layer serves as the user interface, where stakeholders can interact with the agricultural data, analytics, and AI-driven functionalities. This layer describes various software applications and services directly interacting with users tailored to their specific needs within the agriculture domain. The software applications may include desktop or mobile, primarily web3 applications.

This layer's key features and capabilities may include: 1) Tailored for farmers and landowners, precision agriculture applications empower users to monitor and manage their agricultural operations with precision. Features may include real-time crop health monitoring, weather forecasting, and automated recommendations for optimal resource utilization. 2) Users can visualize and interact with DTs created in the cloud layer. The application layer enables a graphical representation of agricultural plots, presenting data on soil conditions, crop growth, and other relevant parameters. 3) Decision support systems within this layer provide stakeholders with actionable insights derived from big data analytics, ML models, and AI algorithms.

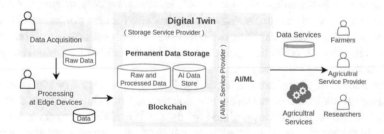

Fig. 2. Data flow for the proposed architecture

4   Working Flow of the Proposed Architecture: An Implementation Perspective

The proposed architecture aims to achieve key objectives centered around providing data and agricultural services. The data services may include the raw data or processed data. Agricultural services can encompass a wide range of offerings that contribute to various aspects of farming, land management, and agricultural productivity. Here are some possible agricultural services: Crop Monitoring Services, Weather Forecasting Services, Precision Farming Services, Soil Health Assessment Services, Pest and Disease Management Services, Smart Irrigation Services, and others. This architecture has two distinct user entities: producers and consumers. Producers are the entities responsible for offering services, while consumers are those who utilize these services. The specific identity of producers and consumers varies based on the nature of services the proposed architecture facilitates. To illustrate, for data services, producers may include farmers, agricultural industry farms, and other relevant entities, whereas consumers could comprise researchers, agricultural service providers, government agencies, and others. Similarly, producers might be agricultural service providers and government bodies in the context of agricultural services, while consumers could be farmers, agricultural industry farms, and additional stakeholders. This dynamic interplay between producers and consumers forms a pivotal aspect of the architecture's functionality, ensuring a tailored and effective service delivery system.

The working flow of the proposed architecture is visually represented in Fig. 2, outlining the processes related to data and agricultural services. Focusing on data services, the architecture's physical and virtual sensing layers are crucial for data collection. The physical sensing layer initiates the data acquisition process by utilizing actual physical sensors owned by data producers. Subsequently, the acquired raw data is transmitted to the virtual sensing layer, which incorporates edge computing capabilities to process the raw data from the physical layer, if necessary. Following data collection based on the specific requirements of data consumers, the data producers leverage the DT layer to securely deliver the data to the consumers while upholding privacy standards. The DT layer encompasses various functionalities within the proposed architecture for data services, including storage services offered by external storage service providers, facilitating interaction or communication between data producers and consumers through blockchain, enabling AI/ML service providers to generate synthetic data for data producers, and providing an infrastructure for data producers to create the DTs. The Sect. 4.1 further explains the collaborative process between producers and consumers in the data collection phase through the utilization of a blockchain-based Mobile Crowdsensing (MCS) system.

In the context of agricultural services, the DT layer assumes a pivotal role in facilitating interactions between producers and consumers through blockchain technology. In addition to the functionalities outlined for the DT layer in the preceding description, it holds the responsibility of delivering global services derived from the aggregated agricultural data to the consumers. This layer serves as a valuable resource for service providers to generate DTs corresponding to various processes, services, and systems. It is noteworthy that the creation of DTs is an optional step for service providers, and they retain the capability to offer services without necessarily generating DTs. This flexibility caters to diverse preferences within the architecture's framework.

4.1 Data Service: Data Collection Using Blockchain-Based Mobile Crowdsensing System

The data acquisition involves collecting data from the farmers utilizing blockchain technology to ensure transparency, security, and trustworthiness. Since agricultural needs can be region-specific, they are often influenced by various factors that vary from one region to another. Agriculture is deeply connected to the local climate, soil conditions, water availability, pests and diseases, and other environmental factors. Thus, we assume that the world map is divided into multiple zones, and each zone is uniquely identified using a zone ID. Moreover, diverse agricultural applications may necessitate distinct types of data. The data requirements depend on each application's goals, objectives, and processes. However, there are common types of data that can be relevant across various agricultural applications. In response, a smart contract has been devised to oversee sensing tasks, with each task aligned to a specific region and a singular sensor type, effectively managing data heterogeneity [5]. Dividing a complex task into subtasks corresponding to each sensor type and zone allows sharing among multiple subscribers. The data shared by multiple subtasks is transmitted only once, leading to optimizing the use of available sensing and blockchain resources. Each instance of the smart contract handles a particular sensing task that can be defined using the following parameters: zone ID, sensor type, sampling interval (SI) list of a day, frequency in each SI, task duration in number of days, number of required data collectors, and the data submission frequency, and others.

The system comprises three entities: the data producer, the consumer, and the blockchain-assisted MCS system, as shown in Fig. 3. The data producer gathers data according to the consumer's specifications. The blockchain acts as a decentralized and trusted third-party intermediary between the data producer and consumer. Smart contracts are employed to implement and manage the system's various phases, including user registration, task creation, task subscription, reservations, selection of data producers, data submission, and reward distribution. While data validation is a crucial aspect of the MCS system, employing similarity matching techniques becomes challenging due to variations in the values of similar or different crop parameters managed by individual data producers. Implementing ML-based validation techniques directly over the blockchain is impractical. Therefore, data validation can be conducted individually on the virtual sensing layer by utilizing edge computing capabilities or globally on the cloud layer of the proposed architecture.

Two smart contracts are used to execute these phases and to maintain the data ownership. The first smart contract encompasses user registration and task creation phases, while the second smart contract handles the functions for the remaining phases. In the system's initial steps, users are required to register. Each user interacts with the system using their blockchain address as an identity. Upon calling the *"registration"* function specified in the smart contract, the blockchain initially validates the identity's authenticity by verifying the signature attached to the transaction. Following the confirmation of authenticity, the smart contract stores the blockchain address and other relevant parameters, including reputation, among others. The data consumer initiates another function, *"datarequest"*, wherein they provide updated requirements for the previously discussed parameters. This function leads to creating an instance of the second smart contract.

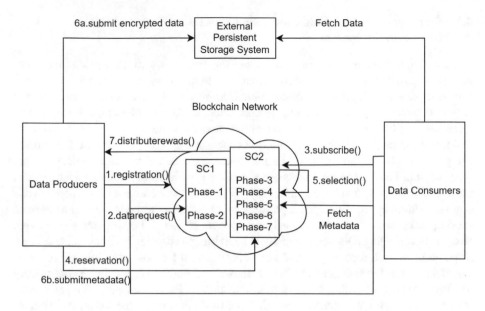

Fig. 3. Data Acquisition Process using blockchain-based MCS system

Alternatively, the data consumer can invoke the "subscribe" function to subscribe to another ongoing sensing task that aligns with their sensing requirements. This design permits multiple data consumers to subscribe to the same sensing task, addressing the issue of data redundancy and data sharing.

After the creation of a new sensing task, data producers can express their interest in providing the required data by invoking the *"reservation"* function of the second smart contract. This smart contract incorporates a *"selection"* function that considers factors such as reputation to determine the necessary number of data producers. Subsequently, the selected producers utilize their sensing technologies to collect and validate the data on the second layer of the proposed architecture using the application provided by the data consumer. Upon successful validation, each producer submits the encrypted data to the external storage system, such as IPFS, and the respective data fetching addresses are submitted to the smart contract by calling *"submitmetadata"* function. The data submission is done after considering the data submission frequency parameter set by the data consumer based on the need, enabling real-time data collection. To ensure authorized access, each producer encrypts the key used for data encryption with the consumer's public key and submits the encrypted key(s) to the smart contract. Following the submission of keys, the smart contract has a *"distributerewards"* function used to distribute the rewards to the producers, which can be utilized to obtain agricultural services from service providers. Now, the data consumer can retrieve metadata (data fetching address and key) from the smart contract, use the data fetching address to obtain encrypted data from an external storage system, and use the key to decrypt the data.

4.2   Agricultural Services

The envisioned architecture expands upon the depicted data acquisition process outlined in Fig. 3 to encompass the provision of agricultural services. Within this framework, service consumers, primarily farmers, have the capability to request various agriculture services through the blockchain-assisted MCS system. To facilitate this, service providers can leverage the cloud sublayer embedded in the core architecture, employing the collected data to generate the requested services. Subsequently, these generated services are delivered to the service consumers, thus completing the service cycle within the proposed system. This integration offers a seamless and efficient exchange, where farmers utilize the rewards acquired from the initial data collection process to compensate service providers through the blockchain-assisted MCS system. This comprehensive approach not only enriches the system's functionalitics but also establishes a sustainable ecosystem that encourages continued data participation and service utilization.

Table 1. Smart contracts deployment cost (in gas)

Smart Contracts	Deployment Cost
BRSCAgri	1579620
TSSCAgri	1246499

5   Result

We instantiate the proposed architecture on the Ethereum blockchain, employing Solidity as the programming language. Leveraging the brownie framework, we develop, test, and deploy smart contracts on Ganache. Utilizing Python scripts, we deploy the smart contracts on Ganache, an invaluable tool for establishing a decentralized Ethereum network locally featuring multiple accounts. The implementation encapsulates the MCS system phases (1. Registration, 2. Task Creation, 3. Task Subscription, 4. Reservation, 5. Selection, 6. Data Submission, 7. Reward Distribution) within two smart contracts. These contracts interact seamlessly to ensure the system's coherent functionality. To assess the smart contracts' performance, we scrutinize two Ethereum blockchain-related metrics: gas consumption and latency. "gas" denotes a unit measuring resource consumption, encompassing computation and storage necessities for specific operations. Conversely, latency refers to the block generation time on Ganache, encompassing a transaction's time to execute and successfully validate.

Table 1 provides insights into gas consumption during the deployment of smart contracts. Each MCS phase is represented by a dedicated function in the smart contracts, complemented by two additional functions (8. adding service providers with services and 9. service subscriptions by service consumers) related to the services offered by the proposed architecture. In the assessment of these phases, the considered metrics hinge on various factors, such as the number of reservations & selections and the number of sensor values submitted. In our experimental setup, five data producers initiate reservations, three are subsequently selected, and each chosen data producer submits 12 sensing values. Figure 4 depicts each function's total gas and time consumption.

The factors influencing gas and time consumption in the selection phase encompass the number of reservations (NR) and the required number of data producers required (NP). To comprehend the impact of these parameters, we explore three scenarios. In the first scenario, we maintain a constant number of reservations and manipulate the number of selections. Figure 6a illustrates the effects of this scenario on gas and time consumption. Similarly, in the second scenario, we fix NP (with 70% selections) and vary NR, with Fig. 6b depicting the consequences on Gas and Time consumption. In the third scenario, we vary both NR and NP, and Figs. 6c and 6d showcase the effects of this scenario on gas and time consumption, respectively. Notably, the reward distribution function is contingent on the number of selections. Thus, Fig. 5 elucidates the impact of the number of selections on gas and time consumption during the reward distribution process.

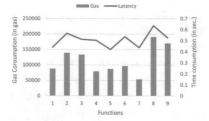

Fig. 4. Total gas and time consumption of each function

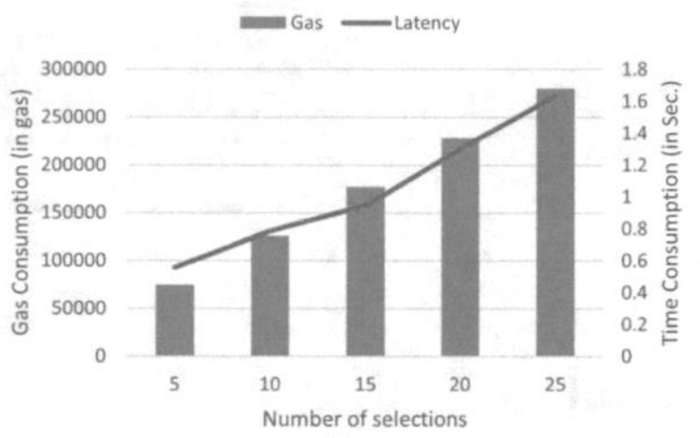

Fig. 5. Effect of number of selections on reward distribution

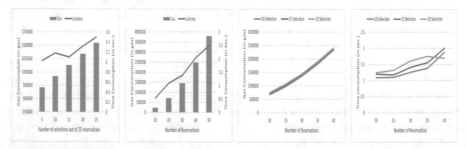

(a) Effects of scenario 1 on gas and time consumption   (b) Effect of scenario 2 on gas and time consumption   (c) Effects of scenario 3 on gas consumption   (d) Effects of scenario 3 on time consumption

Fig. 6. Effect of parameters on the selection phase

6 Conclusion

This paper introduces a comprehensive and service-oriented architecture tailored for Agriculture 4.0. By integrating cutting-edge technologies, the architecture enhances the capabilities of existing architectures. Our proposal incorporates a blockchain-enabled mobile crowdsensing system, presenting a cost-effective and scalable solution for real-time data collection while enabling data ownership and sharing. The blockchain also enables the integration of incentivization mechanisms to motivate the data producers to participate in the data collection. Integrating DT technology within our architecture transforms individual-level data collection and global agricultural services. In conjunction with blockchain technology, the DT layer establishes a robust foundation for secure interactions between providers and consumers, facilitating the seamless exchange of data and agricultural services. Blockchain's inherent features, including transparency, traceability, and trust, strengthen the integrity of these exchanges within the agricultural ecosystem.

References

1. Karunathilake, E.M.B.M., Le, A.T., Heo, S., Chung, Y.S., Mansoor, S.: The path to smart farming: innovations and opportunities in precision agriculture. Agriculture **13**(8), 1593 (2023)
2. Purcell, W., Neubauer, T.: Digital twins in agriculture: a state-of-the-art review. Smart Agricultural Technol. **3**, 100094 (2023)
3. Bhat, S.A., Huang, N.F.: Big data and AI revolution in precision agriculture: survey and challenges. IEEE Access **9**, 110209–110222 (2021)
4. Sun, Y., et al.: On enabling mobile crowd sensing for data collection in smart agriculture: a vision. IEEE Syst. J. **16**(1), 132–143 (2021)
5. Agrawal, A., Choudhary, S., Bhatia, A., Tiwari, K.: Pub-SubMCS: a privacy-preserving publish-subscribe and blockchain-based mobile crowdsensing framework. Futur. Gener. Comput. Syst. **146**, 234–249 (2023)
6. Shen, X., Xu, C., Zhu, L., Lu, R., Guan, Y., Zhang, X.: Blockchain-based lightweight and privacy-preserving quality assurance framework in crowdsensing systems. IEEE Internet Things J. **11**(1), 974–986 (2023)
7. Kalyani, Y., Collier, R.: A systematic survey on the role of cloud, fog, and edge computing combination in smart agriculture. Sensors **21**(17), 5922 (2021)
8. Sinha, A., Shrivastava, G., Kumar, P.: Architecting user-centric internet of things for smart agriculture. Sustainable Comput. Inform. Syst. **23**, 88–102 (2019)
9. Alharbi, H.A., Aldossary, M.: Energy-efficient edge-fog-cloud architecture for IoT-based smart agriculture environment. IEEE Access **9**, 110480–110492 (2021)
10. Tsipis, A., Papamichail, A., Koufoudakis, G., Tsoumanis, G., Polykalas, S.E., Oikonomou, K.: Latency-adjustable cloud/fog computing architecture for time-sensitive environmental monitoring in olive groves. AgriEngineering **2**(1), 175–205 (2020)
11. Montoya-Munoz, A.I., Rendon, O.M.C.: An approach based on fog computing for providing reliability in IoT data collection: a case study in a Colombian coffee smart farm. Appl. Sci. **10**(24), 8904 (2020)
12. Vangala, A., Das, A.K., Chamola, V., Korotaev, V., Rodrigues, J.J.: Security in IoT-enabled smart agriculture: architecture, security solutions and challenges. Clust. Comput. **26**(2), 879–902 (2023)

13. Dey, K., Shekhawat, U.: Blockchain for sustainable e-agriculture: literature review, architecture for data management, and implications. J. Clean. Prod. **316**, 128254 (2021)
14. Chaganti, R., Varadarajan, V., Gorantla, V.S., Gadekallu, T.R., Ravi, V.: Blockchain-based cloud-enabled security monitoring using internet of things in smart agriculture. Future Internet **14**(9), 250 (2022)
15. Khan, A.A., et al.: A blockchain and metaheuristic-enabled distributed architecture for smart agricultural analysis and ledger preservation solution: a collaborative approach. Appl. Sci. **12**(3) (1487) (2022)
16. Kalyani, Y., Bermeo, N.V., Collier, R.: Digital twin deployment for smart agriculture in Cloud-Fog-Edge infrastructure. Int. J. Parallel Emergent Distrib. Syst. **38**(6), 461–476 (2023)
17. Pylianidis, C., Osinga, S., Athanasiadis, I.N.: Introducing digital twins to agriculture. Comput. Electron. Agric. **184**, 2021, 105942 (2021). ISSN 0168-1699

HPC-SBC: An Experimental Effort to Evaluate Storage in High-Performance Computing Configurations Using a Context Approach

Fernando de Almeida Silva[✉], João Pedro de Souza Jardim da Costa,
José Maria N. David, and Mario A. R. Dantas

Computer Science (PGCC), Federal University of Juiz de Fora (UFJF), Juiz de Fora, Brazil
{fernandoalmeida.silva,jose.david,mario.dantas}@ufjf.br,
joao.costa@estudante.ufjf.br

Abstract. In this article we present the challenges and efforts to evaluate the secondary memory performance of nodes from cluster architectures, for utilization in cluster-like environments in HPC, or data centers, such as the Grid'5000 computing environment. The proposal is a tool for general analysis of these experimental environments and including a preliminary study of the tools available to evaluate the performance of the nodes storage. There were some limitations in evaluating tools for secondary memory performance analysis. As a result, the proposal tool, called HPC-SBC (HPC Storage Benchmark Context), is presented. The proposal is an experimental effort to evaluate storage in high-performance computing configurations adopting a context approach. The initial results indicate the success of the proposal, as it is shown the nodes with the best storage performance.

1 Introduction

The increasing production and consumption of data has changed the way we deal with information. However, studies and research not only require high-power processing, but also generate a huge volume of data. Thus, data acquisition, storage, analysis and visualization play an important role today. This scenario highlights the importance of storage devices, as the type and quality of secondary memory can affect the performance of experiments.

This difference between processing elements and I/O performance is referred to as I/O-gap. Since further advances in reducing the latency seem to be very limited, the typical solution to increase the performance is to exploit some kind of parallelism. For example, starting multiple I/O-operations in parallel, rather than executing them sequentially. For this to happen, choosing the storage devices with the best performance could be more advantageous than choosing randomly. However, how evaluate previously the available storage devices in high performance computing context, our experimental effort focus it.

The contribution of this work is the development of the HPC-SBC (HPC Storage Benchmark Context) tool, which is a tool to evaluate the performance of storage in secondary memory, analyzing transfer rate, for use in HPC cluster environments or data

L. Barolli (Ed.): AINA 2024, LNDECT 200, pp. 187–198, 2024.
https://doi.org/10.1007/978-3-031-57853-3_16

centers. This research also aims to present an overview of the Grid'5000 computing environment, including a preliminary study of the tools available to evaluate the performance of nodes' secondary memory storage devices. In addition to the gaps found that motivated the development of the HPC-SBC tool, for an experimental storage evaluation effort adopting a context approach, in high performance configurations.

This work is structured as described below. Section 2 presents related work on HPC environments such as Grid'5000. Its main characteristics, tools and challenges for measuring node storage performance. Section 3 presents the development of the HPC-SBC tool to fill the gap found in the environment, as explained in the previous section. The experiments and results obtained are presented in Sect. 4. Finally, Sect. 5 provides discussions, limitations and potential of this research, as well as possibilities for future work and addition to final considerations.

2   Related Works

HPC systems consist of a very specialized and complex combination of hardware and software elements, focusing mainly on providing high processing power for large-scale parallel and distributed applications. Thus, low communication latency and data access are considered increasingly relevant requirements [1]. Figure 1 presents an overview of a general infrastructure model common in many modern HPC environments. Even in the simplified illustration, the complexity of these environments can be seen.

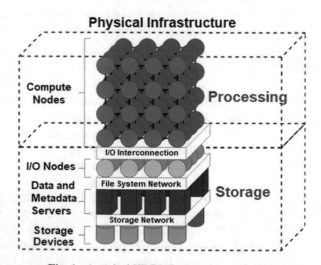

Fig. 1. A typical HPC infrastructure model [1].

Therefore, high-performance environments such as HPC systems often face I/O bottlenecks. One factor that increases this problem is the gap between processing power and storage performance. Since storage technologies evolve at a slower speed than processing, this also contributes to the increase in the I/O performance problem. Furthermore, high-performance applications often move Exabytes of computed or generated data between nodes, and their performance depends on the I/O process. Typically, a clustered architecture such as the Grid'5000 computing environment is used to run these applications. Therefore, I/O performance bottlenecks are an important area of study to find solutions that improve I/O performance [2].

Grid'5000 is a flexible, large-scale testbed for experiment-based research across all areas of computer science, with a focus on parallel and distributed computing, including cloud, HPC, and Big Data and AI. It has hundreds of servers, or nodes, spread across nine locations in France and Luxembourg. They are currently grouped into nine sites: Grenoble, Lille, Luxembourg, Lyon, Nancy, Nantes, Rennes, Sophia, and Toulouse. Figure 2 illustrates the architecture of servers in the computing environment.

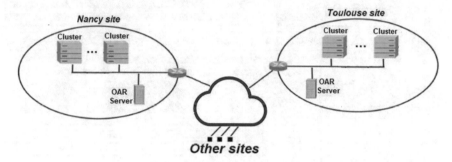

Fig. 2. Overview of the architecture of the Grid'5000 computing environment (modified from [3]).

In Grid'5000, each site has one or more clusters, composed of one or more nodes that are considered homogeneous among themselves in terms of memory and processing elements. However, they are heterogeneous when compared with nodes from different clusters, whether from the same site or others. This difference is even more evident in storage devices, as the same node can have different types of storage [4].

Conducting experiments on distributed systems often requires complex instrumentation to automate experiments involving a large number of machines. On Grid'5000, users generally connect via SSH to the sites' front ends and use the available commands to manage resources. The control infrastructure is based on OAR Server, which is an HPC resource manager and related tools. Each site runs an independent instance of OAR, allowing you to reserve resources that can range from an entire cluster to a single processing element. Users looking for specific resources can consult through the Grid'5000 web page the clusters (sets of homogeneous nodes) available at different sites (sets of clusters) [5].

Figure 3, illustrates examples of Gantt diagrams, which are provided from some tools available from the environment, which can be found in [ttps://intranet.grid 5000.fr/oar/[**Site**]/drawgantt-svg/].

The tool meets the demands for displaying the nodes, available or not, of each cluster on the analyzed website in a quick and visual way. For example, a cutout of Fig. 3 represented by Fig. 4, it shows that at the current moment, indicated by the line red vertical, all nodes in the "troll" cluster are currently available for experiments. This

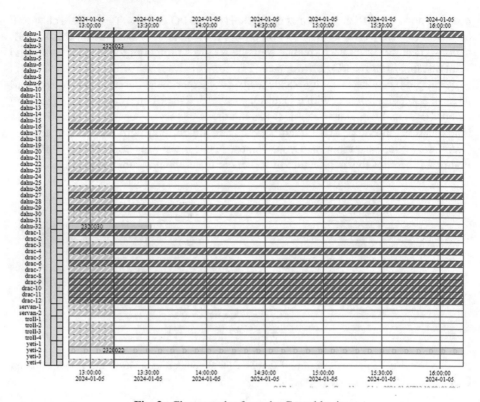

Fig. 3. Cluster nodes from the Grenoble site.

is indicated by the horizontal white bands to the right of the red vertical line. However, in the "yeti" cluster only three of the four nodes are available: yeti-1, yeti-3, yeti-4.

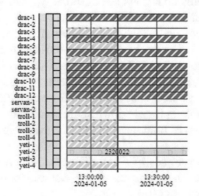

Fig. 4. Tool indicates nodes availability.

Another tool similar to the previous one also displays, through Gantt diagrams, the nodes and their respective storage devices, which may or may not be reserved for experiments through the OAR Server. The Tool is available via the link [ttps://intranet.grid5000.fr/oar/[**Site**]/drawgantt-svg-disks/]. The Fig. 5 is a cutout of the tool, showing the nodes of the chiclete and chifflot clusters on the Lille site and their respective storage devices.

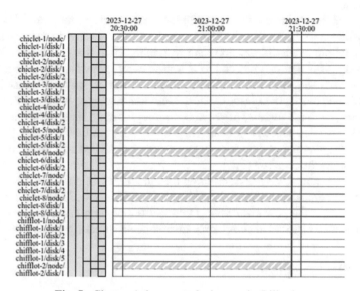

Fig. 5. Cluster node storage devices at the Lille site.

It should be noted that, unlike the previous one, this tool displays, for some sites, only some nodes. For example, Fig. 6 shows that the tool displays 7.5% of the total nodes on the Grenoble website, displaying nodes from one of the five clusters, as shown in Fig. 6.

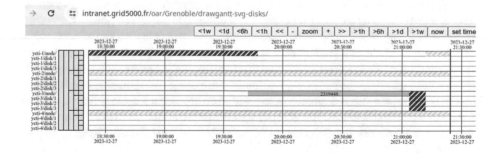

Fig. 6. Partial display of nodes and their storage devices from the Grenoble website.

Furthermore, when querying some sites, no nodes from any cluster are displayed. For example, it shows the message that returns when trying to obtain information about nodes and storage devices of clusters from the Luxembourg site.

Another aspect to mention is that the tool does not provide the throughput of each node's storage devices. Therefore, when planning the implementation of experiments, it is important to consider this information. Especially in experiments that heavily use storage devices. This could avoid I/O congestion, which could impact the performance of these tests.

But one of the characteristics in designing the structure of the Grid'5000 environment, in addition to the tools themselves, was the attention given to the possibility that external collaborators could develop their own tools. Both the Reference API and the OAR provide APIs that can be used to automate complex tasks, thus allowing you to overcome gaps found. For example, a shortcoming found in Grid'5000 is its distributed nature: each site has its own instance of the resource manager, which can make it difficult to reserve a large number of resources across multiple sites at the same time. So a Grid'5000 user developed a tool called Funk that uses the Benchmark API and OAR and reservation planning to find resources that match a desired specification and a time slot that matches the required [6].

Therefore, a similar way to developing the Funk tool was necessary, developing an application that quickly obtains and provides this information. The next section shows the main points in the effort to develop this tool for probing and benchmarking node storage devices in distributed computing environments similar to Grid'5000.

3   HPC-SBC Proposal

At the beginning of the development of HPC-SBC, it was taken into account that the performance of storage devices can be measured in MB/s or IOPS (input/output operations per second). Among the tools and commands for performance evaluation there are: Fio, dd, hdparm, gnome-disks, rsync and others. Although comparing them all to each other is beyond the scope of this work.

In addition, for development in the Grid'5000 environment, the Execo and EnOSlib libraries are available for creating tools for automating experiments [7]. Being Execute is a Python library for experiments, which allows asynchronous control of local or remote, autonomous or parallel processes [8]. Similarly, EnosLib is a library for Python programming language dedicated to research and experiments in distributed computing.

The choice of the EnosLib library is technically justified, as it has relevant characteristics such as abstractions to help researchers implement extensible and replicable artifacts through providers for other distributed computing environments in addition to Grid'5000. The provider is a function that takes as an argument a declarative description of the resources expected in that computing environment, and in the current version, EnosLib has providers for: Vagrant, Grid'5000, Vmon, OpenStack, Distem and FIT IoT-Lab [9].

Resource abstraction allows developers to write code that is virtually independent of concrete resources, while providers do the heavy lifting of acquiring resources from the environment, thus allowing tools to run in different computing environments.

Also highlights [9] that Enos Lib takes into account experimentation best practices and leverages modern toolkits in automatic deployment and configuration systems. Integrating state-of-the-art services that are useful in most experiment-oriented activities, such as network traffic modeling, container technologies, and stack monitoring, for example.

For the development of HPC-SBC, a powerful feature of the EnOSlib library was used, which is to support network emulation for quick use, allowing a virtual network to be created in a few seconds from selected nodes. Then, this virtual network can replicate the parallel execution of the same commands for all machines, but returning the results in a unified way. Figure 7 shows the workflow of the HPC-SBC tool.

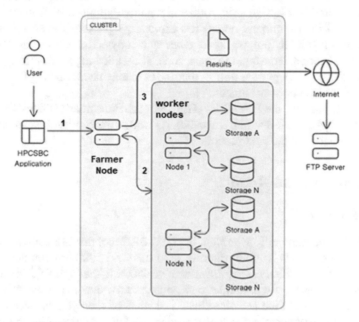

Fig. 7. Systematic view of HPC-SBC software components.

Among the commands to be executed by all machines, hdparm is justified, as on Linux machines it is an integrated I/O testing tool, which can be activated via the command line. In addition to displaying storage device statistics, it can also use hardware-related parameters, which proves useful when analyzing different types of storage devices [10].

As the environment has heterogeneous storage devices between nodes in different clusters, the "df-h" command was included to detect the devices. This command checks the availability of connected devices in data centers, being among the main native tools and commands for storage devices [11]. Additionally, this command is very useful in identifying the types of storage devices on the node, identifying which devices are mounted and the amount of space used and available on each [12].

Upon receiving the results, the HPC-SBC organizes and structures this data, creating a CSV file with the name of the evaluated cluster and the date and time of the evaluation. Next, the tool uploads the file with the results obtained, automating the sending to the FTP server. As in [13], the process of uploading files to an FTP server uses the ftplib Python library which is used to connect to an FTP server and transfer data. This library introduces some concepts, such as content movement, paired documents and error handling, for example.

Similar to [14] the HPC-SBC at the end of the process automatically sends the file with the results obtained to the DLPTest FTP server. This server is public and open to everyone, in addition, the server was chosen due to its existing infrastructure, with easy file management and periodic automated deletion of files, but if you wish to change the FTP server, its settings are easily identified in the source code.

To run the HPC-SBC, Grid'5000 offers the Jupyter lab tool, which provides a web interface for submitting jobs to Grid'5000 and running Python code. Allowing to follow the evolution of the experiment during the exploratory phase while evolving the code development. Jupyter lab tool, works together with JupyterHub which has two kernels, one for python3 code and the other for bash code. Thus, allowing users, through the "pip" command, to install any python library that the user may need, making them available in the virtual laboratory environment [7].

The source code for the HPC-SBC (HPC Storage Benchmark Context) is available on GitHub (github.com/fernandojf1/hpcsbc). As well as the results obtained in the tests, which are described in the next section.

4 Experiments and Results

4.1 Experiments

In Grid'5000 in addition to JupyterHub, the HPC-SBC tool that is a command line tool, can be run remotely via SSH. Furthermore, in the Grid'5000 environment by a web interface too that is available at [ttps://intranet.grid5000.fr/shell/[**site**]/]. The tool can be run without parameters, for example: python hpcsbc.py, thus using the default values and testing two nodes of the neowise cluster, which is on the Lyon website.

When using parameters in the tool it reaches its full potential, allowing the user to evaluate another cluster and the number of nodes evaluated, up to the limit of available nodes. The available number of nodes can be easily obtained by the tool shown in Fig. 3. The HPC-SBC allows you to evaluate clusters from any site, even if you are remotely accessing a node from another site.

In the battery of tests using the HPC-SBC tool, we were able to simultaneously evaluate 31 nodes of the paravance cluster at the Rennes site. Thus, evaluating 43% of the cluster's storage devices in a few minutes. This indicates the robustness of the tool in different analysis volumes.

At another point in the tests, two tests were run simultaneously to evaluate the uvb cluster nodes, through two different machines accessed remotely. Each test evaluated 10 nodes available, totaling 20 nodes from the same cluster, indicating the ability to use the tool simultaneously.

At the end of the evaluations. Similar to [15] the data files obtained from each experiment were combined into a CSV file by the Linux command "cat". The combined CSV file was viewed in Microsoft EXCEL for initial analysis.

4.2 Results

Initial analysis of the data obtained by the HPC-SBC tool found that 717 devices from 137 nodes were evaluated. Among the storage devices evaluated, one had an upper limit (MAX) that was 36 times greater than the device with a lower limit (MIN), according to Table 1.

Table 1. Lower limit (MIN) an Upper limit (MAX) values of devices.

Node	Site	Storage Device	Transfer rate (MB/sec)	Performance
uvb-5	Sophia	sda3	78,49	Lower
servan-1	Grenoble	nvme1n1p1	2836,82	Upper

After initial analysis, the combined file of all experiments was imported into a database for deeper analysis. For example, identify the node with the largest difference between the upper and lower limit between storage devices. Of the 139 nodes evaluated, 6 nodes were found to have a relevant difference between the storage devices of the same node. For example, the dahu-15 node obtained a value close to 3 times higher than the other on one device, according to Table 2.

Table 2. Values Lower limit (MIN) an Upper limit (MAX) of the same node (dahu-15).

Node	Site	Storage Device	Transfer rate (MB/sec)	Performance
dahu-15	Grenoble	sdb2	189,51	Lower
dahu-15	Grenoble	sda5	531,29	Upper

Another analysis carried out was the percentage of test coverage in relation to the current Grid'5000 hardware, as shown in Table 3.

Table 3. Test coverage of Grid'5000 environment.

Hardware Grid'5000 [16]		Analyzed	Coverage %
sites	9	8	89%
clusters	69	17	25%
nodes	770	137	18%
SSDs and HDDs on nodes	1468	717	49%

Among the analyzes of the data obtained by the HPC-SBC tool, a survey was carried out to obtain the upper limit (MAX) and lower limit (MIN) per site, as shown in Fig. 8.

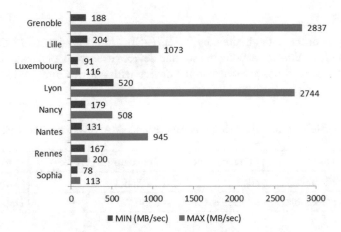

Fig. 8. Upper limit (MAX) and lower limit (MIN) of storage devices per site.

The graph from Fig. 8, the Grenoble and Lyon sites stand out at the upper limit (MAX). However, the lower limit (MIN) is close to the other sites, indicating that it is necessary to look for which clusters have the best performance. Therefore, an analysis was carried out to obtain the upper limit (MAX) and lower limit (MIN) of the storage devices per cluster, according to Fig. 9.

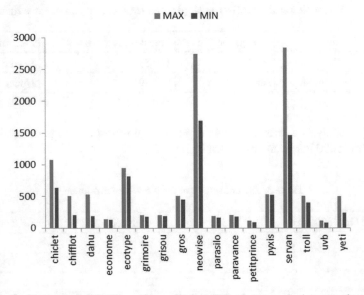

Fig. 9. Upper limit (MAX) and lower limit (MIN) of storage devices per cluster.

Figure 9 shows that the neowise cluster from the Lyon site and the servan cluster from the Grenoble site stand out in terms of the upper limit (MAX) and lower limit (MIN) values of the storage devices, when compared to the other clusters evaluated.

Furthermore, the lower limit (MIN) of both is greater than the upper limit (MAX) of the other clusters evaluated.

5 Conclusions and Future Works

The large amount of data created, coupled with the growing gap between processing power and storage latency, poses a challenge in terms of data transfer and storage. Thus, in this article we present the challenges and efforts to evaluate the performance of the secondary memory of nodes in the HPC computing environment, such as the Grid'5000, including a preliminary study of the available tools. As a result, the HPC-SBC (HPC Storage Benchmark Context) tool was developed, which is an experimental effort to evaluate storage in high-performance configurations using a context approach.

In the first battery of tests, the HPC-SBC tool achieved to evaluate 18% of the nodes and 49% of the storage devices in Grid'5000 environment. Furthermore, analysis of the results of the evaluated nodes, provided by HPC-SBC, showed which nodes, clusters and sites had superior and inferior storage performance.

Therefore, the initial results of the HPC-SBC tool indicate the success of the proposal, pointing HPC-SBC users to the nodes with the highest storage performance. This is relevant for experiments with high storage demands, such as I/O for example, and it also could be considered as an possible extension to the EnosLib.

As future work, other environments should also be evaluated, considering possible adjustments to the HPC-SBC tool to diversify the application. For example, when evaluating other types of storage devices and other forms of parallelism, allowing the evaluation of multiple storage devices from the same node, which are virtually joined together. In addition, carry out research aimed at developing software solutions to suggest devices in their best storage scenarios.

Acknowledgments. This work was partially funded by CAPES/Brazil, CNPq/Brazil, FAPEMIG/Brazil, INESC P&D Brasil/Brazil, PROQUALI Program from UFJF. We like to thanks the people from Grid'5000 environment, where experiments were carried out. The Grid'5000 is supported by a scientific interest group hosted by INRIA and including CNRS, RENATER and several Universities as well as other organizations.

References

1. Pioli, L., et al.: Characterization research on I/O improvements targeting DISC and HPC applications. In: IECON 2020 the 46th Annual Conference of the IEEE Industrial Electronics Society, pp. 2095–2100 (2020)
2. Pioli, L., Ströele, V., Dantas, M.A.R.: An effort to characterize enhancements I/O of storage environments. Int. J. Grid Utility Comput. **14**(1), 51–61 (2023)
3. Khalil, F., Miegemolle, B., Monteil, T., Aubert, H., Coccetti, F., Plana, R.: Simulation of micro Electro-Mechanical systems (MEMS) on grid. In: VECPAR'08–8th International Meeting High Performance Computing for Computational Science (2008)

4. Vasconcelos, M.F.S., Cordeiro, D., Dufossé, F.: Indirect network impact on the energy consumption in multi-clouds for follow-the-renewables approaches. In: 11th International Conference on Smart Cities and Green ICT Systems. SCITEPRESS-Science and Technology Publications, pp. 44–55 (2022)
5. Bertot, L., Nussbaum, L., Margery, D.: Implementing SFA support on an established HPC-flavored testbed: lessons learned. In: IEEE INFOCOM 2020-IEEE Conference on Computer Communications Workshops (INFOCOM WKSHPS), pp. 830–835. IEEE (2020)
6. Margery, D., Morel, E., Nussbaum, L., Richard, O., Rohr, C.: Resources description, selection, reservation and verification on a large-scale testbed. In: Leung, V.C.M., Chen, M., Wan, J., Zhang, Y. (eds.) TridentCom. LNICSSITE, vol. 137, pp. 239–247. Springer, Cham (2014). https://doi.org/10.1007/978-3-319-13326-3_23
7. Bertot, L., Nussbaum, L.: Leveraging notebooks on testbeds: the Grid'5000 case. In: IEEE INFOCOM 2021-IEEE Conference on Computer Communications Workshops (INFOCOM WKSHPS), pp. 1–6. IEEE (2021)
8. Chardet, M., Coullon, H., Pérez, C., Pertin, D., Servantie, C., Robillard, S.: Enhancing separation of concerns, parallelism, and formalism in distributed software deployment with Madeus (2020). (hal-02737859). https://inria.hal.science/hal-02737859. Accessed Jan 2024
9. Cherrueau, R.A., et al.: EnosLib: a library for experiment-driven research in distributed computing. IEEE Trans. Parallel Distrib. Syst. 33(6), 1464–1477 (2022)
10. Godinho, A., Rosado, J., Sá, F., Cardoso, F.: IoT single board computer to replace a home server. In: 2023 18th Iberian Conference on Information Systems and Technologies (CISTI), pp. 1–6. IEEE (2023)
11. Huawei Technologies Co., Ltd.: OpenStack. Cloud Computing Technology, pp. 251–294. Springer, Singapore (2022)
12. Findlay, B.: Techniques and methods for obtaining access to data protected by Linux-based encryption–a reference guide for practitioners. Forensic Sci. Int. Digit. Invest. 48 (2024)
13. DuraiPandian, M., Abhishek, N., Dipsijo, M.P.: Security auding system in cloud computing. J. Xi'an Shiyou Univ. Nat. Sci. Ed. 19(4), 646–650 (2023)
14. Rifai, A.D., Ramadhani, E.: Implementation of logging feature in android payment SDK using scrum method. IJISTECH (Int. J. Inf. Syst. Technol.) 5(4), 346–353 (2021)
15. Nakayama, H., et al.: Inactivation of axon guidance molecule netrin-1 in human colorectal cancer by an epigenetic mechanism. Biochem. Biophys. Res. Commun. 611, 146–150 (2022)
16. Grid'5000 Hardware (2023). https://www.grid5000.fr/w/Hardware#Summary. Accessed Jan 2024

Effective Parallel Formal Verification of Reconfigurable Discrete-Event Systems Formalizing with Isabelle/HOL

Sohaib Soualah[1,2]([✉]), Mohamed Khalgui[3], and Allaoua Chaoui[4]

[1] Faculty of Sciences of Tunis, University of Tunis El Manar, 2092 Tunis, Tunisia
soualahsohaib94@gmail.com
[2] LISI Laboratory, National Institute of Applied Sciences and Technology,
University of Carthage, 1080 Tunis, Tunisia
[3] National Institute of Applied Sciences and Technology, University of Carthage,
1080 Tunis, Tunisia
[4] MISC Laboratory, Faculty of NTIC, University Constantine 2 Abdelhamid Mehri,
Constantine, Algeria
allaoua.chaoui@univ-constantine2.dz

Abstract. This paper addresses the formal verification of reconfigurable discrete event systems (RDESs) using Isabelle/HOL proof assistant. A reconfigurable system transitions from one mode to another during its operation to adapt its behavior to the relevant environment. By including such a feature, RDESs become complex and are often costly in terms of computation time and memory. The verification of RDESs consists of two main steps: state space generation and state space analysis. In order to improve these two steps, we propose in this paper, a new approach for verifying system properties that is performed on a developed distributed architecture. The proposed approach allows to avoid redundant computation and reduce execution time by considering the relationships between properties and creating a parallel algorithm that ensures a suitable execution order for each property that is not costly and efficient. The proposed approach is evaluated by exploiting a case study that illustrates the impact of using this approach. The results demonstrate the significance of this paper.

Keywords: Discrete-event Systems · Reconfiguration · Theorem prover · Isabelle/HOL · Parallel Computing · Formal Verification

1 Introduction

Reconfigurable Discrete Event Systems (RDESs), as introduced by [13], represent a new class of complex systems encompassing manufacturing systems [5], real-time systems, and intelligent control systems [6]. These systems adhere to various conditions such as synchrony, monitoring, and connectivity, making them a pivotal aspect of future system trends. Despite their promise, the inherent complexity of RDESs poses a significant challenge to formal verification. Model checking, a widely adopted formal technique for automatic verification, involves validating the satisfaction of functional properties to ascertain the behavior of the formal model. Notably, Isabelle/HOL proof assistant has been proposed as an alternative verification solution [11], forming the basis for our approach in this paper.

The growing interest from industry and academia in the verification of RDESs [4] has spurred research efforts aimed at developing new formalisms and improving verification methods. Addressing the challenge of validation time, [8] proposed a method that guides Computation Tree Logic (CTL) formulas verification, reducing the number of properties to be verified by identifying relationships between them. However, limitations arise when the relationship rate among properties is low, impacting performance. Another notable approach involves the use of Petri nets, specifically reconfigurable timed net condition/event systems (R-TNCESs) [4], extending the work of [12]. While this method enhances the efficiency of accessibility graph generation, its application with large-scale systems on a centralized machine proves resource-intensive [7].

In the context of formal method problems, [1] leveraged cloud-based solutions for efficient CTL properties verification using a distributed fixedpoint algorithm. However, the system's complexity poses challenges, requiring substantial data resources [3]. Furthermore, while cloud-based architectures [2,3], have been proposed to address computational power and data availability limitations, coordination issues between workers have been overlooked, potentially hindering timely verification in the presence of faulty properties.

While each of the aforementioned works contributes significantly to the verification task, they often neglect certain parameters. Notably, there is a gap in related work addressing the verification of reconfigurable systems using Isabelle/HOL while considering the interplay between system behavior and component communication. In this paper, our focus is on reconfigurable discrete event systems, with their behavior and communication modeled using Isabelle/HOL [11]. Our contribution involves a novel verification method aimed at enhancing the verification process, with a specific emphasis on improving state space analysis.

Our proposed approach introduces a distributed architecture comprising local, organizational, and storage workstations to achieve parallel verification of system properties. To reduce execution time and the number of properties for verification, we build upon the work presented in [10,11], distinguishing between properties, identifying relationships, and decomposing complex properties into simpler ones. We employ two algorithms, one for property simplification and another for distributed verification, with results aggregated at the storage workstation. The subsequent verification is performed on local workstations using the Isabelle/HOL tool.

In summary, our work focuses on:

1. Developing a distributed working environment for parallel verification.
2. Simplifying system properties to leverage existing relationships for efficient verification.

3. Distributing properties to local workstations for the verification process.
4. Receiving and storing verification results.

To substantiate the feasibility of our approach, we present a case study, detailing each step of the verification process. Additionally, we conduct comparisons with previous works to highlight the advantages of our proposed methodology. The results demonstrate a significant reduction in execution time and the number of properties for verifying reconfigurable systems.

The remainder of this paper is organized as follows: Section 2 presents essential concepts. Section 3 outlines our parallel formal verification approach. Section 4 delves into the application of our proposed methodology and presents performance evaluations. Finally, Sect. 5 concludes our work.

2   Background

2.1   System Formalization

In our previous work [11], we defined a meta-model to model RDESs using model-driven engineering (MDE). MDE is a software development methodology used by engineers. We propose a new type of Isabelle/HOL equivalence to this meta-model. We establish the link between MDE and Isabelle/HOL by defining a set of reconfiguration rules to allow for the automatic generation of the system in IsabelleHOL. Based on our previous work [11], an RDES consists of:

- A set of reconfiguration rules that it uses to transition from one configuration to another.
- A configuration is a stable state with a specific duration in which a system performs an activity (i.e., the system's components are in specific communication with each other).

Definition 1. An *RDES* is a structure defined as follows:
    $RDES = (B, RR)$ where:

- B is the system's behavior.
- RR is the set of reconfiguration rules.

Definition 2. *RDES* Behavior
    The behavior of a system B is a set of m configurations, represented as follows:
    $B = Conf_0, Conf_1, Conf_2, ..., Conf_i, ..., Conf_m$ where:

- $Conf\_0$ is the initial configuration.
- Each configuration $Conf\_i$ is represented by the following tuple:
- $Conf\_i = (U, L)$ where:

a. U is the set of units. b. L is the set of links between units.

Definition 3. *RDES* Reconfiguration Rules

RDES reconfiguration rules (*RR*) are a set of transformations between configurations. That is, $RR = r\_1, ..., r\_m$ allows for automatic transformations between configurations. A reconfiguration rule of an *RDES* r_i (*Conf*, *Conf'*) is a structure that changes the system from one configuration *Conf* to another *Conf'*. It is defined as follows:

$$r\_i\,(Conf, Conf') = (Condition, Operation, S-Conf, D-Conf)\ \text{where:}$$

- *Condition* True, False is the precondition of r_i.
- *Operation* includes the addition or removal of units and links from a source configuration $Conf\_i$ to obtain a target configuration $Conf\_j$.
- $S-Conf$ denotes the configuration $Conf\_i$ before the application of r_i.
- $D-Conf$ denotes the target configuration $Conf\_j$ after the reconfiguration rule r_i is applied.

When a reconfiguration scenario is applied, the reconfiguration rule r_i is used for the transformation from one configuration Conf_i to another configuration $Conf\_j$. If *Condition* = *True*, r_i is executable; otherwise, it is non-executable.

2.2 Properties Formalization

Isabelle/HOL provides concepts and types to formalize different kinds of properties. In Isabelle/HOL, properties are referred to as lemmas or theorems. The general schema for proofs in Isabelle/HOL is presented as a lemma. A lemma consists of the following elements:

- Name: The name of the proof.
- Formula: The property to be verified.
- Proof goal: The desired result.

Formulas are written in Isabelle/HOL syntax, which includes the following operators: → (implies), ∀ (for all), ∃ (exists), ∨ (or), and ∧ (and). Let *P* denote a proof written in Isabelle/HOL, *f* denote a formula in *P*, and *Ci* denote a configuration of the system where $i \in \{1, ..., n\}$. Figure 1 presents the general schema for proofs in Isabelle/HOL as a lemma.

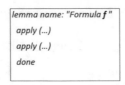

Fig. 1. The Global schema of proof.

2.3   Classification of Properties

In our previous [10], we proposed an optimal strategy for guiding Isabelle/HOL proofs that minimizes the validation time by reducing the number of properties to be verified. This work is based on the classification of properties, and the reduction is obtained by extracting possible relationships. Then, we expressed these relationships using a set of newly proposed operators: *dominance* and *equivalence*. Let P denote a proof written in Isabelle/HOL, f denote a formula in P, and Ci denote a configuration of the system where $i \in \{1, \dots, n\}$. We define the following operators to express the relationship between two proofs:

Internal dominance: ($PhDomCiPk$) means that formula fh of Ph on configuration Ci dominates formula fk of Pk on the same configuration because of the existence of a relationship between formulas of the proofs (*i.e., fh is a subformula of fk or the goal of Ph is a subgoal of Pk*).

External dominance: ($PhDomCi, CjPk$) means that formula fh of Ph on configuration Ci dominates formula fk of Pk on configuration Cj because of the existence of a relationship between formulas of the proofs (*i.e., fh is a subformula of fk or the goal of Ph is a subgoal of Pk*).

Equivalence: ($PhEquiCi, CjPk$) means that proof Ph on configuration Ci is equivalent to proof Pk on configuration Cj (*i.e., fh is equivalent to fk or the goal of Ph is the same as the goal of Pk*).

Relationships between physical processes are reflected in the properties. Therefore, any two properties that share a goal or subgoal can be in a relationship, as described in [10]. This relationship can be: **(i) Dominance relationship:** When the final result of the entire formula depends only on the dominant part of the formula due to the dominance relationship between the physical processes. **(ii) Equivalence relationship:** When two properties are equivalent. We denote two types of properties: atomic goal properties and composed goal properties.

3   The Proposed Approach

3.1   Motivation

The emergence of new complex systems makes the task of verifying RDECSs complex. Real-world RDECSs can include millions of transitions, which significantly increases the number of properties that need to be verified, and therefore the time it takes to complete the verification process. In this paper, we propose a new approach to improve the verification process. Therefore, we aim to make the verification process more efficient by reducing the time it takes and the number of properties that will be verified. The improvement is obtained through the quality of the proposed environment during the verification process, which would allow for high efficiency in less time, as we can have many computing resources working in parallel. Also, considering the possible relationships between the properties allows to reduce their number, which allows us to gain more time. Figure 2 shows the different stages that we go through to reach the proposed verification process in this paper.

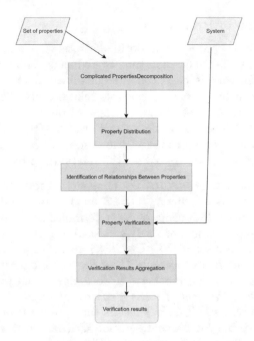

Fig. 2. Steps of our proposed verification.

3.2 Formalization

This subsection presents our proposed approach to achieving an efficient verification process. The approach can be summarized in the following steps:

- **Step 1:** *Complicated Properties Decomposition.* In this step, we formalize properties in Isabelle/HOL. We utilize this formalization to decompose complex proofs into simpler ones. Two possible types of complex proofs exist:
  - Decomposable: The operators ∨ (or) and ∧ (and) are not linked to parentheses. These properties can be directly split into a set of sub-proofs. (e.g., $= P1 \land P2$ gives $1 = P1$ and $2 = P2$).
  - Non-decomposable: The operators ∨ (or) and ∧ (and) are linked to parentheses. For this type of property, we first apply distribution and then re-check whether they are decomposable or not.
- **Step 2:** *Property Distribution.* Using Algorithm 1 and taking advantage of the quality of the proposed verification environment, we distribute and divide the properties among a set of local stations. The organizational station works to coordinate in the implementation of this, which ensures a balanced division between the local stations in terms of the assigned verification task.
- **Step 3:** *Identification of Relationships Between Properties.* Using Algorithm 2 and the set of properties assigned to each local station, we identify the type of relationships between the properties to be verified. The result obtained is exploited to avoid duplication. Compared to classical verification, we have to

verify all the properties, while in this proposed approach some of them can be ignored.

- **Step 4:** *Property Verification.*Each local station can execute the assigned verification process by calling the Isabelle tool. After the completion of each verification process, the local station sends the result to the organizational and aggregating station.
- **Step 5:** *Verification Results Aggregation.* The organizational and aggregating station receives the results received from the local stations.

The results are stored in the storage station.

We denote the local station as **LS**, the organizational station as **OS**, and the storage station as **SS**.

3.3   Effective Parallel Verification Environment

In this subsection, we present the working environment to be developed. Figure 3 shows the components of this environment, which are distinguished into three components as follows:

- Local workstation: It is a high-performance computer that can perform computational tasks (property relationships identification and verification process. It also sends and receives data).
- Organizational workstation: It is responsible for distributing, aggregating verification results, and overall supervision of the process in terms of coordinating with a set of local stations.
- Storage station: This station plays the role of the database that stores the results. It is accessed to save a part of data or read a saved result.

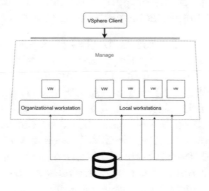

Fig. 3. Our working environment.

3.4 Implementation

The implementation of the proposed environment is divided into two main steps:

- **Development of the software:** This step involves the development of the software components of the environment, including the algorithms for property decomposition, property distribution, property relationship identification, property verification, and verification results aggregation.
- **Deployment of the environment:** This step involves deploying the software components on the three stations of the environment. Using virtualization technology, we create the working environment using VMware VSphere. Figure 3 shows that the core of the environment consists of three parts as follows:
  - *Local workstations:* A unique computing machine for performing computing tasks
  - *Organizational workstation:* The supervisor of storing results and dividing properties
  - *Storage station:* A memory that can be accessed directly from each station to store the results.

The group of workstations is a set of virtual machines, each of which takes on one of the three roles mentioned above. It can be said that the working environment, which we call *EW*, is a group of *VWs* virtual machines that share the network and computing resources such as memory, network, and storage.

In the above section, we use the following notation:

- OS: Represents the organizational workstation.
- LS: Represents the local workstation.
- SS: Represents the storage station.
- Decomposable: Represents a property that can be decomposed.
- Non-decomposable: Represents a property that cannot be decomposed.

To implement the verification process, the **OS** performs a classification and simplification of the properties using the property simplification algorithm, which works as follows, as shown in Fig. 4. The communication between the **OS** and **LSs** is based on **Algorithms** 1 and 2. The **OS** uses **Algorithm** 1 to distribute the properties to the **LSs**. **Algorithm** 2 provides the main function for the **LS**. We use a property distribution block based on the mathematical operation **Mod()** for balanced computation. Each **LS** receives the message from the **OS** using the function **receive()** from **Algorithm** 2. At each LS level, the function **generateRelations()** is called, which allows ignoring some properties as they are redundant, resulting in a smaller number of properties. After verifying each property, the LS sends the result to the SS using the function **send()** from *Algorithm* 2. Upon completion of the verification process, the **OS** receives the results sent from the **LSs** using the function *receive()* from *Algorithm* 1.

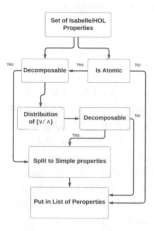

Fig. 4. Simplification Algorithm working.

Algorithm 1: Algorithm 1.

Input: *MListe* list local workstations,;
PListe list Proofs,;
i Int;;
for *property P,P = 1..(PListes.size())* **do**
  | Send(*P*,*MListe*.get(*i Mod (MListes.size())*));
end
Receive();

Algorithm 2: Algorithm 2

Input : *P* property,;
Isasys system in Isabelle,;
Output: *Res* bool;;
Receive(*P*);
/*check database if result the verification of P exist*/;
if *(not P verified)* **then**
  | *Res* =Verify(*P*, *Isasys*);
  | /*Put result the verification of P in the database */;
  | PutInDB(*P*, *Res*);
end
Send(*Res*);

---

4   Experimentation

The experimentation of the proposed environment is performed using a case study of an RDECS model. The case study is used to evaluate the effectiveness of the environment in terms of reducing the verification time and the number of properties to be verified.

4.1   Case Study

In this section, we use the FESTO production system as a case study. The important properties that need to be verified in the system are safety, liveness, and deadlock freedom. In this example, the validity of the safety property is verified when at most one unit is active in a state (i.e., it is impossible to activate two units in different configurations at the same time). The liveness and deadlock freedom properties are verified when any request to change the configuration is satisfied after a limited time. To verify the validity of the system, we verify that every configuration respects the activation conditions. Some details of the formal verification (i.e., proofs in Isabelle/HOL) are given in Table 1. As shown in Fig. 5, 6, 7, and 8 the system is modeled using Isabelle/HOL. the system component (UNIT) shown in Fig. 5, all possible links between units (Arc) shown in Table Fig. 6, different configurations (SC), reconfiguration rules (RR) shown in Fig. 7, and all systems (Isa_System) are shown in Fig. 8.

```
definition Piece Injection::UNIT where " Piece Injection UName
=("Piece Injection"),state=False "
definition converter::UNIT where " converter UName
=("converter"),state=False "
definition Tester1::UNIT where " Tester1UName
=("Tester1"),state=False "
definition Evacuate::UNIT where " Evacuate UName
=("Evacuate"),state=False "
definition Elevator::UNIT where " Elevator UName
=("Elevator"),state=False "
definition Disc::UNIT where " Disc UName
=("Disc"),state=False "
definition Driller1::UNIT where " Driller1UName =("Driller1")
,state=False "
definition Tester2::UNIT where " Tester2UName
=("Tester2"),state=False "
definition Elevator2::UNIT where" Elevator2UName =("
Elevator2"),state=False "
```

Fig. 5. FESTO MPS Units Isabelle formalization.

```
definition Piece Injection_to_converter::Arc where " Piece
Injection_to_converter AName =(" Piece
Injection_to_converter"),S_unit =( Piece Injection) ,D_unit
=( converter) "
definition converter_to_Tester1::Arc where "
converter_to_Tester1AName =("
converter_to_Tester1"),S_unit = (converter),D_unit =
(Tester1)"
definition Tester1_to_Evacuate::Arc where "
Tester1_to_Evacuate AName =("
Tester1_to_Evacuate"),S_unit = (Tester1),D_unit =
(Evacuate)"
definition Arc4 ::Arc where " Evacuate_to_Disc AName =("
Evacuate_to_Disc"),S_unit = (Evacuate),D_unit = (Disc)"
definition Disc_to_Driller1::Arc where " Disc_to_Driller1
AName =(" Disc_to_Driller1"),S_unit = (Disc),D_unit =
(Driller1)"
definition Driller1_to_Tester2::Arc where " Driller1_to_
Tester2 AName =(" Driller1_to_Tester2"),S_unit =
(Driller1),D_unit = (Tester2)"
definition Tester2_to_Elevator2:: Arc where " Tester2_to_
Elevator2 AName =(" Tester2_to_Elevator2"),S_unit =
(Tester2),D_unit = (Elevator2)"
```

Fig. 6. FESTO MPS Arcs Isabelle formalization.

```
definition Light1 ::SC where " Light1 CName=("
Light1"),allUnits= [Piece Injection, converter, Tester1,
Evacuate, Elevator1, Disc, Driller1, Tester2, Elevator2],
allArcs=[ Piece Injection_to_ converter,
converter_to_Tester1, Tester1_to_Evacuate,
Evacuate_to_Disc,Disc_to_Driller1, Driller1_to_ Tester2,
Tester2_to_ Elevator2] "
definition M_to L1 ::RR where " M_to L1 RName =(" M_to
L1"),Prec M=("H"),Next M=("M") ,c=("two Driller active") "
definition M_to L2::RR where " M_to L2 RName =(" M_to
L2"),Prec M=("H"),Next M=("L") ,c=("One Driller active") "
definition H_to L1::RR where " H_to L1RName =(" H_to
L1"),Prec M=("M"),Next M=("L") ,c=("One Driller active") "
definition H_to L2::RR where " H_to L2 RName =(" H_to
L2"),Prec M=("M"),Next M=("H") ,c=("Two Driller active") "
definition H_to M::RR where " H_to L2 RName =(" H_to
M"),Prec M=("M"),Next M=("H") ,c=("Two Driller active") "
```

```
definition Festo ::Isa System where "Festo SName =("Festo"),
RRs=[M_to_L1, M_to_L2, H_to_L1, H_to_L2, M_to_H,
H_to_M], SCs=[ Light1, Light2, medium, high]"
```

Fig. 7. FESTO MPS configurations (SC), reconfiguration rules (RR) Isabelle formalization.

Fig. 8. FESTO MPS Isabelle formalization.

4.2 Application

To ensure the validity of the system modeling, we need to verify the properties of the system. First, we apply the property decomposition task, which is performed on the organizational workstation. For this, we adopt the algorithm proposed in [9]. Table 1 shows the properties and the results of the property decomposition task. After completing this step, we proceed to the second step, which is the distribution of the properties to the local workstations. Table 2 shows the properties assigned to each local workstation. Each local workstation generates the potential relationships between the properties assigned to it. Table 3 shows the property relationship identification process on each local workstation. Each local workstation can ignore some properties based on the result of the property relationship identification process. Each local workstation becomes ready for the verification process with the minimum number of properties to be verified. The verification results are stored on the storage station.

4.3 Evaluation

Figure 10 shows two curves corresponding to the verification process using our proposed approach and without it. The plot shows the time taken with respect to the number of properties when the system is run twice. The blue curve corresponds to the verification without the proposed method. The green curve corresponds to the distributed verification using the proposed method. It is important to note that the verification time decreases gradually when we use the proposed method. Figure 9 compares the results of the proposed approach with the approach used in the methods that do not consider the analysis of the properties and the verification environment. It shows that the proposed approach allows to reduce the execution time thanks to the consideration of the efficient parallel verification and the analysis of the relationships between the properties. The

Table 1. Isabelle/HOL proofs and their decomposing

Isabelle/HOL proofs	Decomposition
*lemmal*1	Non-decomposable
*lemmal*2	**lemma** *ll*2: "((getUnit (Unit7))))" by (simp add: Unit7_def) *lemmalll*2: "((getSUnit (Unit8)))" by (simp add: Unit8_def)
*lemmal*3	Non-decomposable
*lemmal*4	Non-decomposable
*lemmal*5	Non-decomposable
*lemmal*6	Non-decomposable
*lemmal*7	**lemma** **ll7:** "((getUnit (Unit9)) (getUnit(Unit8)))" by (simp add: Unit9_def Unit8_def) *lemmalll*7: "((getUnit (Unit9)) (getUnit (Unit7)))" by (simp add: Unit9_def Unit7_def)
*lemmal*8	*lemmall*8: "((getUnit (Unit8)) (getUnit (Unit9))))" by (simp add: Unit8_def Unit9_def) *lemmalll*8: "((getUnit (Unit7)) (getUnit (Unit9))))" by (simp add: Unit7_def Unit9_def)
*lemmal*9	Non-decomposable
*lemmal*10	**lemma** **ll10:** "((getUnit (Unit8))) (getUnit (Unit9))" by (simp add:Unit8_def Unit9_def) *lemmalll*10: "((getUnit (Unit7))) (getUnit (Unit9))" by (simp add:Unit7_def Unit9_def)
*lemmal*11	Non-decomposable

Table 2. The Assignment proofs to LSs.

LSs	Properties
LS 1	*lemmal*1, *lemmall*2, *lemmalll*7, *lemmall*11
LS 2	*lemmalll*2, *lemmall*7, *lemmal*9
LS 3	*lemmal*3, *lemmal*5, *lemmal*4, *lemmal*6
LS 4	*lemmall*8, *lemmall*10, *lemmalll*8, *lemmalll*10

Table 3. The Relationships between proofs.

LS 1	$l1$ DOMC_1,C_3 $l11$ $ll2$ DOMC_1 $lll7$
LS 2	$lll2$ DOMC_1 $ll7$
LS 3	$l3$ EQUIC_1,C_2 $l5$ $l4$ EQUIC_1,C_2,C_3 $l6$
LS 4	$ll8$ EQUI_$2,C_3$ $ll10$ $lll8$ EQUI_$2,C_3$ $lll10$

relationships between the properties can be large or small, or medium. Therefore, the effectiveness of the proposed approach is directly proportional to the ratio of the relationships between the properties. In other words, the better the

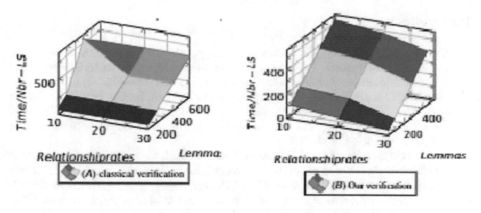

Fig. 9. A figure with two subfigures

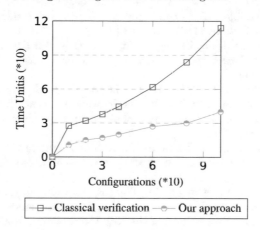

Fig. 10. More number of configurations: Classical verification Vs Our approach.

ratio, the better the approach. As mentioned earlier, before verifying any property, there is a step of connecting to the storage station, which may contain the result. This step takes some extra time when the ratio of the relationships between the properties is weak.

5 Conclusions

The proposed approach and environment have the potential to improve the efficiency of the verification process of RDECS models. The approach reduces the number of properties to be verified by decomposing complex proofs into simpler ones and by identifying the relationships between properties. The environment provides a high-performance platform for executing the verification process in parallel.

References

1. Camilli, M., Bellettini, C., Capra, L., Monga, M.: CTL model checking in the cloud using mapreduce. In: 2014 16th International Symposium on Symbolic and Numeric Algorithms for Scientific Computing, pp. 333–340. IEEE (2014)
2. Choucha, C.E., Ben Salem, M.O., Khalgui, M., Kahloul, L., Ougouti, N.S.: On the improvement of r-tncess verification using distributed cloud-based architecture. In: Proceedings of the 15th International Conference on Software Technologies: ICSOFT, pp. 339–349 (2020)
3. Choucha, C.E., Ramdani, M., Khalgui, M., Kahloul, L.: On decomposing formal verification of ctl-based properties on iaas cloud environment, pp. 544–551 (2020)
4. Hafidi, Y., Kahloul, L., Khalgui, M., Li, Z., Alnowibet, K., Qu, T.: On methodology for the verification of reconfigurable timed net condition/event systems. IEEE Trans. Syst. Man Cybern. Syst. 50, 3577–3591 (2018)
5. Khalgui, M., Hanisch, H.-M.: Automatic nces-based specification and sesa-based verification of feasible control components in benchmark production systems. Int. J. Model. Ident. Control 12(3), 223–243 (2011)
6. Khalgui, M., Mosbahi, O., Zhang, J., Li, Z., Gharbi, A.: Feasible dynamic reconfigurations of petri nets-application to a production system. In: ICSOFT, vol. 2, pp. 105–110 (2011)
7. Koubâa, A., Qureshi, B., Sriti, M.-F., Javed, Y., Tovar, E.: A service-oriented cloud-based management system for the internet-of-drones. In: 2017 IEEE International Conference on Autonomous Robot Systems and Competitions (ICARSC), pp. 329–335. IEEE (2017)
8. Ramdani, M., Kahloul, L., Khalgui, M.: Automatic properties classification approach for guiding the verification of complex reconfigurable systems. In: ICSOFT, pp. 625–632 (2018)
9. Soualah, S., Hafidi, Y., Khalgui, M., Chaoui, A., Kahloul, L.: Formalization and verification of reconfigurable discrete-event system using model driven engineering and isabelle/hol. In: Proceedings of the 15th International Conference on Software Technologies: ICSOFT, pp. 250–259 (2020)
10. Soualah, S., Hafidi, Y., Mosbah, O., Khalgui, M., Chaoui, A., Laid, K.: Formal verification of reconfigurable discrete-event systems using isabelle/hol theorem prover on cloud environment. In: Proceedings of the 35rd Annual European Simulation and Modelling Conference, ETI, EUROSIS (2021)
11. Soualah, S., Khalgui, M., Chaoui, A., Laid, K., Hafidi, Y.: Efficient verification of reconfigurable discrete-event system using isabelle/hol theorem prover. In: Proceedings of the the 34rd Annual European Simulation and Modelling Conference, ETI,EUROSIS, pp. 139–146 (2020)
12. Zhang, J., Khalgui, M., Li, Z., Mosbahi, O., Al-Ahmari, A.M.: R-tnces: a novel formalism for reconfigurable discrete event control systems. IEEE Trans. Syst. Man Cybern. Syst. 43(4), 757–772 (2013)
13. Zhang, J., Li, H., Frey, G., Li, Z.: Shortest legal firing sequence of net condition/event systems using integer linear programming. In: 2018 IEEE 14th International Conference on Automation Science and Engineering (CASE), pp. 1556–1561. IEEE (2018)

Data Poisoning Attacks in Gossip Learning

Alexandre Pham[1(✉)], Maria Potop-Butucaru[1], Sébastien Tixeuil[1,2], and Serge Fdida[1]

[1] Sorbonne Université, CNRS, LIP6, 75005 Paris, France
{alexandre.pham,maria.potop-butucaru,sebastien.tixeuil,
serge.fdida}@lip6.fr
[2] Institut Universitaire de France, Paris, France

Abstract. Traditional machine learning systems were designed in a centralized manner. In such designs, the central entity maintains both the machine learning model and the data used to adjust the model's parameters. As data centralization yields privacy issues, Federated Learning was introduced to reduce data sharing and have a central server coordinate the learning of multiple devices. While Federated Learning is more decentralized, it still relies on a central entity that may fail or be subject to attacks, provoking the failure of the whole system. Then, Decentralized Federated Learning removes the need for a central server entirely, letting participating processes handle the coordination of the model construction. This distributed control urges studying the possibility of malicious attacks by the participants themselves. While poisoning attacks on Federated Learning have been extensively studied, their effects in Decentralized Federated Learning did not get the same level of attention. Our work is the first to propose a methodology to assess poisoning attacks in Decentralized Federated Learning in both churn free and churn prone scenarios. Furthermore, in orde r to evaluate our methodology on a case study representative for gossip learning we extended the gossipy simulator with an attack injector module.

Keywords: Gossip learning · Decentralized federated learning · Poisoning attacks · Methodology

1 Introduction

Machine learning algorithms use statistical methods to make predictions or classifications, and to discover patterns and insights in data. Typically, a ML approach consists in collecting data, selecting a model, and tuning model parameters based on collected data (in the model training phase) before actually using the trained model.

In traditional systems relying on ML, a central entity manages both the ML model and the data collected for training the model. Such data centralization is problematic with regard to risk and responsibilities [13] when the data used to train the model is sensitive and should be kept private, or simply to obey regulations.

In 2016, Google introduced Federated Learning (FL) [13] as a solution to privacy-wise Machine Learning. In this framework, the central entity only manages a global model and coordinates the training across local devices (e.g. smartphone, laptop etc.), which use their own data (that they do not share). Then, devices send their newly

L. Barolli (Ed.): AINA 2024, LNDECT 200, pp. 213–224, 2024.
https://doi.org/10.1007/978-3-031-57853-3_18

adjusted local model's parameters to the central entity that aggregates it with other local models to create a new global model, and this process repeats. This method enables private collaborative learning. However, despite its reduced role, the existence of a central server yields a single point of failure, and an obvious attack target, as summarized by Lui et al. [11], and by Xia et al. [19].

Decentralized Federated Learning (DFL) aims to do FL without relying on a central server [2], using *P2P* or *Gossip Communications*. The former relies on an existing architecture to operate, while the latter assumes direct communications in a neighborhood. In DFL, FL based attacks and defenses have also been studied. For example, Bernstein et al. [3,4] with *SignSGD* in the FL context, where they firstly present and study their work as a compression mechanism, and then study it as a defense mechanism. Later, Qu et al. adapted *SignSGD* in the DFL context [16]. In the following, we focus on Gossip Learning (GL), which recently was instrumental in many applications such as building a recommendation system [1], or improving Channel State Information feedback performance [8].

Ormándi et al. [14] is one of the eldest traces of GL. Each node periodically sends their model to a neighbor. When a node receives a model, the node merges it with its current model and considers the result as its new model. Later, Hegedűs et al. [9] improved the work by Ormandi et al [14] by introducing two useful mechanisms in communication restrained networks. After, Danner et al. [5] improved the merging process used by Hegedűs et al. [9]. It should be noted that none of the aforementioned papers address attack or defense in this context.

In this work, we assess the impact of attackers (also called *Byzantine* or *malicious*) nodes that try to poison the global model build through a GL algorithm. Our benchmark GL algorithm was proposed by Hegedűs et al. [9] as best-in-class in this line of research. Close to our work, Giaretta and Girdzijauskas [7] studied the applicability of GL, highlighting problems and providing fixes when data distribution is correlated to either degree distribution or communication speed. However, they do not consider Byzantine nodes in their work.

Our Contributions. Our contribution is threefold. First, we propose a methodology to assess the effects of a poisoning attack in a system that executes a gossip learning algorithm. Second, in order to evaluate our methodology, we implemented an extension of the popular *gossipy*[1] simulator. Our extensions are publicly available (with execution details)[2] and implement a poison injector in *gossipy*, which allows to assess the performances of both clean and corrupted dataset simultaneously. Finally, we apply our methodology on the state of the art gossip learning algorithm by Hegedűs et al [9]. Our findings show that the resilience of this algorithm to poisoning attacks depends on several factors such as topology, Byzantine nodes distribution, and churn. Our analysis pave the way to efficiently design countermeasures against poisoning attacks. We organize this work as follows. Section 2 briefly describes our case study. In Sect. 3 we describe our methodology. We present our results based on this methodology in Sect. 4. We discuss and conclude with Sect. 5.

---

[1] https://github.com/makgyver/gossipy/.

[2] https://gitlab.lip6.fr/apham/data-poisoning-attacks-in-gossip-learning.

2   Case Study: State-of-the-Art Gossip Learning

Gossip Learning is a way to do Federated Learning in decentralized systems without a central coordinator via gossip exchanges of parameters or updates, directly between nodes. Recently, Hegedűs et al [9] proposed a GL algorithm that outperforms FL [13] with respect to the global performances (in the ML perspective) when taking into account communication resources. Their algorithm, *Partitioned Token Gossip Learning Algorithm (PTGLA)* uses two interesting mechanisms for communication-restrained networks such as IoTs. The first one is a compression mechanism to decrease message size, by not sending all parameters during each exchange, but part of it called Partition and, we denote S the total number of partitions that the system is using, a high S implies smaller messages. The second mechanism enables us to control the flow of communications, thanks to *Tokens*, by achieving a balance between sending messages *proactively*, i.e. periodically, which may lead to communication flooding and sending messages *reactively*, i.e. after an event e.g. a local update or a message received, which may lead to starvation (no message circulates in the network).

In the sequel, we study the resilience of this algorithm to various poisoning attacks.

3   Methodology

In this section, we present the various elements of our methodology: the choice of the topologies, Dataset and ML algorithm, the attack, the churn settings, and the metrics used to assess the impact of the attacks.

Topologies for GL infrastructure. We investigate the following topologies with n nodes:

- 20 fan-out network, as originally used by Hegedűs et al [9];
- random 20-regular with bidirectional links and same number of links for every node;
- Watts-Strogatz ($k = 20, p = 0.5$) for its small-world property [17];
- Erdős-Rényi $\left(\text{with } p = \frac{2\log(n)}{n} \right)$ for its balanced property [6];
- Zipf law graphs, where we require that each node has at least one neighbor.

For the last two topologies, we restrict our simulations to instances where the network is connected.

Dataset for ML Model Training. In this work, we use the MNIST handwritten digit database [10] as our dataset. It is made of 2 sets: a training set and a test set that we denote $\mathscr{D}_{\text{training}}$ and $\mathscr{D}_{\text{test}}$, of cardinal 60000 and 10000 respectively[3]. An instance of this dataset (either from the training or test) is a couple $(x, y) \in \mathbb{R}^{784} \times [\![0,9]\!]$, x represents a 28×28 pixel images with the number y written on it. Each node k will have as a local training set $\mathscr{D}^k_{\text{training}}$ of cardinal 250, which is drawn from $\mathscr{D}_{\text{training}}$ in an i.i.d. fashion[4]

---

[3] This is done to evaluate the model against unseen data, but close to data that were used for adjusting model's parameters. This allows us to see whether the model generalize well.

[4] This means that data is equally distributed among nodes, every node has approximately 25 images of each number.

as it is the best case scenario for GL as pointed out in Giaretta and Girdzijauskas [7] and such as $\bigcap_{i=1}^{n} \mathscr{D}^{i}_{\text{training}} = \emptyset$.

ML Model. Similarly to Hegedűs et al. [9], each node k trains a (Multinomial) Logistic Regression model to correctly classify images from the dataset described previously. Hegedűs et al. [9] study thoroughly the choice of hyperparameters η (learning rate) and λ (L2 regularization coefficient). In our work, we fix $\eta = 1$ and $\lambda = 0$. We choose those values as they yield best performance on the churn-free Erdős-Rényi case, but, the choice of those 2 values is not trivial and should be studied for all topologies as done by Hegedűs et al [9].

Byzantine Attacks. As done by Wu et al. [18] in the FL setting, Byzantine nodes insert a pattern on samples that they also mislabel. In this study, Byzantine nodes insert a 9-pixel trigger pattern to 20% of the image of their dataset, which they relabel to 0. Their goal is to make honest nodes classify marked images as 0, without decreasing the classification performances on untampered data (i.e. without the trigger pattern). As Byzantine nodes attack the GL system by tampering with their dataset to disturb the network, this is called a *data poisoning attack*, and as their goal is to introduce a hidden objective, this kind of attack is also called *backdoor attack*. To assess the performances and the impact of the attack on benign nodes, we draw 2000 images from the test set that we divide into two sets of equal size. The first half will be used to assess the performances on the usual classification task. For the remaining half, we remove all (x,y) where $y = 0$, and apply the same modifications on x that Byzantine nodes do on the remaining samples. We call those two sets *test* and *backdoor* set respectively from now and show samples from both in Fig. 1.

In order to choose the Byzantine nodes[5], we will use two strategies: *classical* and *random*. In the *classical* strategy, we select nodes with the highest degree first. In the *random* strategy, we select nodes by randomly sampling without replacement.

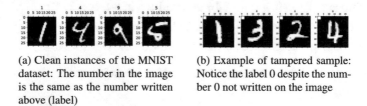

(a) Clean instances of the MNIST dataset: The number in the image is the same as the number written above (label)

(b) Example of tampered sample: Notice the label 0 despite the number 0 not written on the image

Fig. 1. Examples of clean and tampered data from the MNIST dataset.

Churn. Churn refers to the fact that in a network, devices can join and leave freely. Hegedűs et al [9] used smartphone traces to get a realistic churn scenario (where nodes can disconnect and reconnect). Based on their results, to get close to their churn scenario, nodes have a 20% chance of being online at each round.

---

[5] We borrow the idea behind these strategies from Magnien et al. [12], where they use these strategies in order to select nodes to be removed from a graph to study the graph connectivity.

Metrics. We are interested here in the average *accuracy* of honest nodes on the test and backdoor sets. *Accuracy* is defined as the number of correct predictions over the total number of predictions, using the dataset as a reference. For the *test* set, the higher, the better: (honest) nodes should perform well on data that has not been marked with the pattern and for the *backdoor* set, it is the opposite: nodes should perform badly on (maliciously) marked images.

4   Simulation Results

To implement the study based on the methodology defined in the previous section, we develop an extension of the *gossipy* simulator that makes use of the PyTorch [15] library for its ML part. The *gossipy* simulator enables us to test multiple GL algorithm (PTGLA is already implemented[6]) in a unified manner. However, *gossipy* has two main limitations: it does not take into account the possible existence of Byzantine alongside honest nodes, and it does not allow assessing the performances of both a clean and a corrupted dataset simultaneously. By contrast, our extensions, available (with execution details) on Gitlab[7] address those shortcomings. Simulations were done using Python 3.9.13 or 3.9.15, using CPU only, on Intel Xeon E5-2650v3, and Intel Xeon Gold 6330 machines respectively. We define $n \in \{100, 150\}$ as the number of nodes and f as the number of Byzantine nodes in a simulation (in the worst case, $f = 0.3n$). Each curve shown here is an average over at least 10 random runs. In the following, we present two scenarios, churn and churn-free, starting with the churn-free scenario, i.e. when nodes are always online. Here, Byzantine nodes want honest nodes to classify clean inputs (shown in Sect. 3) with high accuracy, while also classifying tampered inputs (shown in Sect. 3) the way Byzantine nodes want, i.e. tampered inputs are classified as 0.

Churn-Free Scenario. In Fig. 2, we represent performances across the 2 sets defined in Sect. 3 (*test* and *backdoor* sets) when $n = 100$ (resp. $n = 150$) with $f = 30$ (resp. $f = 45$) Byzantine nodes placed with the *random* strategy. We use Byzantine-free simulations as baselines, selecting the number of partitions S that yields the best performances on the test set. On this set (left column of Fig. 2), we can see that $S = 1$ and sometimes $S = 4$ (which are the smallest values studied here for S) are special cases across all 4 topologies shown in Fig. 2, as honest nodes perform poorly on the usual classification task (compared to the baseline, which is not what Byzantine nodes want), and is very sensitive to the attack as seen in the right column of Fig. 2. This can be explained by the fact that if a Byzantine node directly sends (proactively or reactively) a partition to an honest node, as honest nodes are always online, they always receive and process the partition, and may send it reactively to another node, speeding up the spread of malicious partitions. When Byzantine nodes are placed randomly, if the topology is 20 fan-out, honest nodes are the most sensitive against the attack, while they are more resilient when the topology is Watts-Strogatz.

Besides, on the test set, we observe that when $S \neq 1$ and 4, there are no noticeable differences between the choice of S, as honest nodes perform very closely to the

---

[6] https://github.com/makgyver/gossipy/tree/3d655829805fc0dc2f01f5b0862240fca08ffe1c.

[7] https://gitlab.lip6.fr/apham/data-poisoning-attacks-in-gossip-learning.

baseline. This can be explained by the i.i.d. data distribution: honest nodes, having the same 'learning material', learn quickly and correctly to classify unaltered inputs. On the backdoor set (right column of Fig. 2), still excluding $S = 1$ and 4, we see that the accuracy of honest nodes on the backdoor set is constrained between 0.1 and 0.2, and that, increasing S also increase the resiliency of honest nodes against this particular attack. This can be explained by the fact that parameters that interact with the trigger pattern are too diluted among all partitions as the number of partitions S increase (which is the opposite when S is small), and that Byzantine nodes follow the algorithm (except when training). Hence, corrupted parameters (that interacts with the trigger pattern) are less likely to be updated by honest nodes. However, Byzantine nodes successfully introduce (albeit not as high as they want) the unwanted behavior, without affecting drastically the normal classification task, as we compare the results of the different result of S studied and the baseline on the right column of Fig. 2: Byzantine nodes induce from 10 to 20 times more misclassification compared to the baseline. Overall, for the 4 topologies studied, namely Erdős-Rényi, 20 fan-out, 20 random-regular and Watts-Strogatz, when Byzantine nodes are placed randomly, against this particular attack, honest nodes are more resilient when the system is using a (very) high number of partitions S.

In Fig. 3, we compare the *classical* and *random* strategy for Byzantine nodes, in systems that have hubs or where the degree distribution is skewed. Usually, for S fixed, the *classical* strategy is clearly more detrimental for honest nodes than the *random strategy*. On the left column of Fig. 3, we can see that the accuracy of honest nodes on the test set drop drastically when Byzantines nodes are placed *classically* compared to the *random* strategy. In average, we observe a difference of 40% and 38% on the accuracy for the test set, for Watts-Strogatz and Zipf-based topologies respectively, which is not what Byzantine nodes want for honest nodes, that do classify the way Byzantine nodes want in the backdoor set (right column of Fig. 3): in average, we observe a difference of 73% and 48% on the accuracy for the backdoor set between the 2 strategies for Watts-Strogatz and Zipf-based topologies respectively.

Considering the case where Byzantine nodes are placed randomly, Watts-Strogatz, and the other three topologies studied previously, are more suitable for honest nodes compared to the Zipf-based topology as they are more resilient against this attack.

Interestingly, for Zipf-based topology, when Byzantine nodes are placed randomly, excluding $S = 1$, on the right column, it sometimes seems detrimental for honest nodes that the system use a high number of partitions S compared to the other 4 topologies already studied previously.

Churn Scenario. Here, we study the case where nodes can disconnect and reconnect as described in Sect. 3 and have a 20% chance to be online.

In Fig. 4, we fix the number of partitions to $S = 8$ and number of nodes to $n = 150$, and study the effect of Byzantine placement strategy and numbers up to $f = 45$ when nodes degree follow a Zipf law distribution. We can see, on the right column of Fig. 4, that Byzantine nodes, when placed with the *classical* and *random* with $f = 5$, for the random and classical strategy, it is almost as if there is no attack (case $f = 0$). While the *classical* strategy is more harmful to the network for f fixed, considering the backdoor set, we note the fact that $f = 40$ Byzantine nodes selected with the *random* strategy

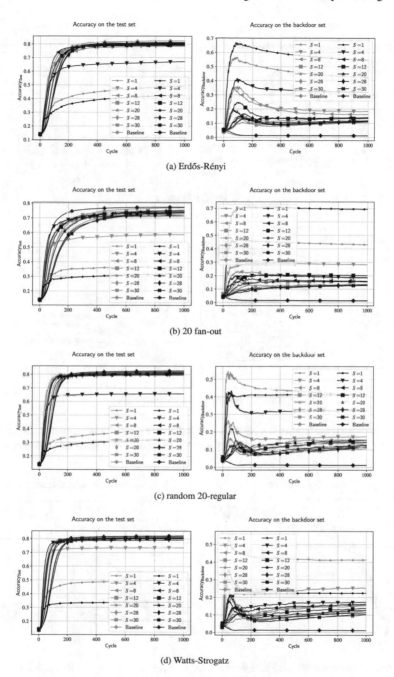

Fig. 2. Accuracy on the test set (left, higher is better) and backdoor set (right, lower is better) for different topologies with $n = 100$ (gray) and $n = 150$ (black) with random Byzantine placement strategy in a churn-free system with $f = 30$ and 45 respectively. 'Baseline' curves represent the results in Byzantine-free simulations, with the best choice of S among values studied here with Byzantine.

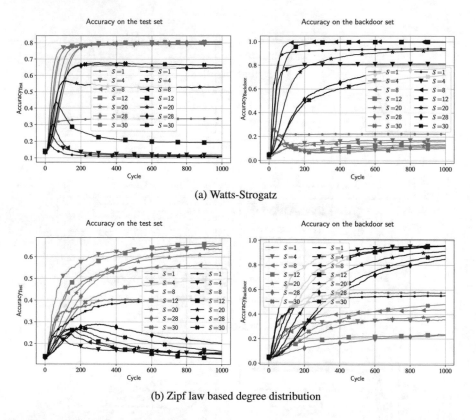

(a) Watts-Strogatz

(b) Zipf law based degree distribution

Fig. 3. Accuracy on the test set and backdoor set for different topologies with $n = 150$ and $f = 45$ with random Byzantine placement strategy (gray) and classical Byzantine placement strategy (black).

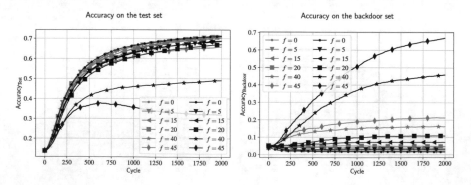

Fig. 4. Accuracy on the test and backdoor set for $n = 150$, $S = 8$ for $f \in \{0, 5, 15, 20, 25, 40, 45\}$ with the random (gray) and classical (black) placement strategy when nodes degree follow a Zipf law.

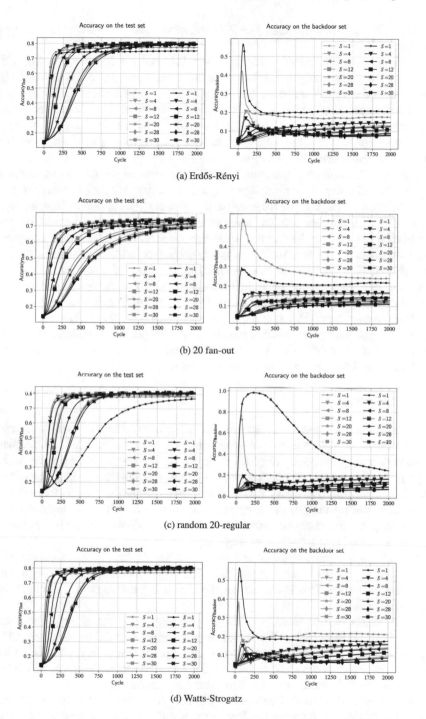

Fig. 5. Accuracy on the test set and backdoor set for different topologies with $n = 100$ (gray) and 150 (black) with $f = 30$ and 45 respecively.

is about 5% more harmful than $f = 20$ Byzantine nodes selected with the *classical* strategy.

In Fig. 5, we again study $n = 100$ (resp. $n = 150$) and $f = 30$ (resp. $f = 45$), with the *random strategy*, similarly to Fig. 2. We see that honest nodes perform as well as in the *churn-free* scenario on the test set (left column of Fig. 5), still due to the fact that data is distributed in an i.i.d. fashion. In this scenario, honest nodes are also slower compared to the churn-free scenario, which is also expected as they are offline most of the time, hence, they are less learning exchanges between nodes for a given number of rounds compared to the churn-free scenario.

Interestingly, while $S = 1$ is again a special case here, unlike the churn-free scenario, honest nodes classify badly clean inputs at the beginning, but they converge to a much better solution at the end, as they successfully manage to classify clean inputs with a better success rate and are less affected by the attack compared to the churn-free scenario. This is due to the fact that all nodes are most of the time offline, Byzantine nodes included, hence, the attack is not as powerful as in the churn-free scenario, but we notice, that the results are close with the other value of S studied. In the journal version of this work, we are planning to study the churn scenario with the assumption that Byzantines nodes are always online. Overall, we notice that values presented in the churn-free scenario (Fig. 2) and the churn scenario (Fig. 5) are relatively close. This might be explained by the fact that the churn is probabilistic, combining with the fact that the choice of the ith to send is random, hence, over time, the churn-free scenario, looks very similar to the churn-free case (except when $S = 1$).

5 Conclusion

In this paper, we proposed a methodology to study the resilience of gossip learning in the presence of poisoning attacks. As case study we target the compression mechanism of GL algorithm proposed by Hegedűs et al. [9]. We investigate its resilience in a broad range of topologies (e.g. Erdős-Rényi, 20 fan-out, random 20-regular, Watts-Strogatz and Zipf-based graph) in both churn-free and churn scenarios. Our findings show that the communication optimizations and the choice of the underlying topology are not always favorable to the honest nodes. Moreover, the distribution of the Byzantine nodes has also a strong impact on the accuracy metric.

Usually, in the churn-free and churn cases, when Byzantine nodes are placed randomly, the use of a low number of partitions (i.e. bigger messages), is usually detrimental for honest nodes. This is not necessarily true if the topology is Zipf-based, in the churn-free scenario.

In the churn-free scenario, when Byzantine nodes are placed using the *classical* strategy in Watts-Strogatz and Zipf topologies, the attacks exhibit a completely different impact on the network compared to when they are placed randomly. We observe that in average, Byzantine nodes cause a drop of 40% and 38% on the usual classification task and a 73% and 48% increase on the backdoor task respectively.

For honest nodes, when Byzantine nodes are placed randomly, we do not observe noticeable differences between the churn-free and churn scenario (except when nodes use bigger messages, as they are more resilient against the attack in the latter scenario), this might be attributed to the probabilistic churn.

In the future we plan to extend this work to more complex datasets and data distribution, and study other proposals for GL algorithm including some that take into account the possibility of poisoning data and models.

Acknowledgements. The work presented in this document has received funding from the EU Horizon Europe research and innovation Programme under Grant Agreement No. 101070118.

References

1. Belal, Y., Bellet, A., Mokhtar, S.B., Nitu, V.: PEPPER: empowering user-centric recommender systems over gossip learning. In: Proceedings of the ACM on Interactive, Mobile, Wearable and Ubiquitous Technologies, 6(3), 101:1–101:27, September 2022
2. Beltrán, E.T.M., Pérez, M.Q., Sánchez, P.M.S., Bernal, S.L., Bovet, G., Pérez, M.G., Pérez, G.M., Celdrán, A.H.: Decentralized federated learning: fundamentals, state of the art, frameworks, trends, and challenges. IEEE Commun. Surv. Tutor., 1 (2023)
3. Bernstein, J., Wang, Y.-X., Azizzadenesheli, K., Anandkumar, A.: signSGD: compressed optimisation for non-convex problems. In: Proceedings of the 35th International Conference on Machine Learning, pp. 560–569. PMLR, July 2018
4. Bernstein, J., Zhao, J., Azizzadenesheli, K., Anandkumar, A.: signSGD with majority vote is communication efficient and fault tolerant. In: International Conference on Learning Representations, September 2018
5. Danner, G., Hegedűs, I., Jelasity, M.: Improving gossip learning via limited model merging. In: Advances in Computational Collective Intelligence, Communications in Computer and Information Science, pp. 351–363. Springer, Cham (2023). https://doi.org/10.1007/978-3-031 41774 0_28
6. Erdős, P., Rényi, A.: On random graphs I. Publicationes Mathematicae Debrecen **6**, 290–297 (1959)
7. Giaretta, L., Girdzijauskas, Š.: Gossip Learning: Off the Beaten Path. In: 2019 IEEE International Conference on Big Data (Big Data), pp. 1117–1124, December 2019
8. Guo, J., Zuo, Y., Wen, C.-K., Jin, S.: User-centric online gossip training for autoencoder-based CSI feedback. IEEE J. Sel. Top. Signal Process. **16**(3), 559–572 (2022)
9. Hegedűs, I., Danner, G., Jelasity, M.: Decentralized learning works: an empirical comparison of gossip learning and federated learning. J. Parallel Distributed Comput. **148**, 109–124 (2021)
10. LeCun, Y., Cortes, C., Burges, C.J.: MNIST handwritten digit database. ATT Labs [Online], 2 (2010)
11. Liu, P., Xiangrui, X., Wang, W.: Threats, attacks and defenses to federated learning: issues, taxonomy and perspectives. Cybersecurity **5**(1), 1–19 (2022)
12. Magnien, C., Latapy, M., Guillaume, J.-L.: Impact of random failures and attacks on Poisson and power-law random networks. ACM Comput. Surv. **43**(3), 13:1–13:31 (2011)
13. Brendan McMahan, H., Moore, E., Ramage, D., Hampson, S., Arcas, B.J.: Communication-Efficient Learning of Deep Networks from Decentralized Data, February 2016
14. Ormándi, R., Hegedűs, I., Jelasity, M.: Gossip learning with linear models on fully distributed data. Concurrency Comput. Pract. Exp. **25**(4), 556–571 (2013)
15. Paszke, A., et al.: PyTorch: an imperative style, high-performance deep learning library. In: Advances in Neural Information Processing Systems, vol. 32. Curran Associates, Inc. (2019)
16. Youyang, Q., Chenhao, X., Gao, L., Xiang, Y., Shui, Yu.: FL-SEC: privacy-preserving decentralized federated learning using SignSGD for the internet of artificially intelligent things. IEEE Internet Things Mag. **5**(1), 85–90 (2022)

17. Watts, D.J., Strogatz, S.H.: Collective dynamics of 'small-world' networks. Nature **393**(6684), 440–442 (1998)
18. Wu, C., Yang, X., Zhu, S., Mitra, P.: Mitigating Backdoor Attacks in Federated Learning, January 2021
19. Xia, G., Chen, J., Chaodong, Yu., Ma, J.: Poisoning attacks in federated learning: a survey. IEEE Access **11**, 10708–10722 (2023)

The Home-Delivery Analysis of Prescription Drugs for Chronic Diseases in the Post-Pandemic Era–An Example of Local Community Pharmacy

Wei-Yuan Ho[1]([⊠]) and Hsing-Chung Chen[2,3]([⊠]) [iD]

[1] Ph.D. Program in Artificial Intelligence, College of Information and Electrical Engineering, Asia University, Taichung, Taiwan
topo22079309@gmail.com.tw
[2] Department of Computer Science and Information Engineering, Asia University, Taichung 413305, Taiwan
cdma2000@asia.edu.tw
[3] Department of Medical Research, China Medical University Hospital, China Medical University, Taichung 404327, Taiwan
shin8409@ms6.hinet.net

Abstract. In 2023, the Taiwanese pharmacy industry faces novel challenges post-epidemic. This study examines the role of door-to-door delivery of chronic disease prescriptions in boosting sales for community pharmacies during this period. Utilizing AI analysis and questionnaires, the research assesses how prescription delivery influences sales of pharmaceutical products and customer satisfaction at Pharmacy C in Taichung City. The objectives include evaluating the impact of prescription delivery on sales of drugs and related medical products at Pharmacy C, analyzing customer satisfaction and loyalty towards delivery services using AI, and correlating these services with sales performance. The study also aims to share best practices and suggest improvements for community pharmacies by examining the case of Pharmacy C. In conclusion, the research offers strategies for community pharmacies to enhance sales, customer satisfaction, and loyalty through AI. It presents successful cases with potential benefits for the entire medical pharmacy industry and helps pharmacies understand the business potential of AI and home delivery services, aligning with current trends.

1 Introduction

Taiwan has entered aged society since 2018 [1], and chronic diseases have become common health problems among middle-aged and elderly people [2]. Therefore, the demand for medical long-term care services is increasing rapidly. Moreover, patients' medication receiving behavior of with chronic diseases has changed largely in the post-epidemic era. The objective of this research was to study how the home delivery service of chronic disease medication influent the sale performance of Pharmacy C in Taichung City. In addition, we use AI technology to analyze patient satisfaction of home delivery

© The Author(s), under exclusive license to Springer Nature Switzerland AG 2024
L. Barolli (Ed.): AINA 2024, LNDECT 200, pp. 225–234, 2024.
https://doi.org/10.1007/978-3-031-57853-3_19

service as well as the correlation of these factors and sales performance. This study can provide some novel strategies to enhance customer satisfaction and increase pharmacy sales performance.

As Taiwan progresses into an aged society, chronic diseases like cardiovascular diseases, diabetes, and hypertension are becoming increasingly prevalent among the middle-aged and elderly. This trend poses significant challenges to the public health system and underscores the need for a deeper understanding of these diseases' distribution and their strain on medical resources. The shift towards an aged society also amplifies the demand for long-term medical care, necessitating innovative healthcare models.

The COVID-19 pandemic has notably altered patient behavior, especially among those with chronic conditions. There's been a surge in the demand for telemedicine and medication delivery services, leading to a transformation in how patients access their medications. This behavioral shift presents both challenges and opportunities for pharmacy operations, calling for adaptations in service delivery.

Moreover, the pandemic has catalyzed the digital transformation of healthcare services. The expansion of online consultations and medication delivery is reshaping the operational and service models of pharmacies, making the adoption of digital solutions imperative.

In this context, AI technology emerges as a vital tool for pharmacies. It can enhance service efficiency and quality by analyzing customer behavior, offering personalized medication recommendations, and optimizing inventory management. AI's analytical and predictive capabilities empower pharmacies to make data-driven decisions, improving medication delivery, customer satisfaction, and sales performance.

This study proposes innovative AI-driven strategies to boost customer satisfaction and enhance pharmacy sales. By conducting an empirical analysis of Pharmacy C in Taichung City, the research aims to offer actionable insights and strategies. These findings are expected to help pharmacies navigate new market demands and trends, showcasing the practical application and value of AI in modern healthcare challenges.

2   Research Background and Motivation

The outbreak of the COVID-19 epidemic in 2020 has had a huge impact on the global medical system [3]. In addition to changes in people's medical treatment behavior, the medication habits of patients with chronic diseases have also been influenced. Before the pandemic, patients with chronic diseases or those with consecutive prescriptions usually went to the hospital to see a doctor and receive medicines. However, after the outbreak, many patients with chronic diseases were unwilling to go to the hospital to receive medicines because they were worried about the increased risk of infection.

This resulted in the instability of many patients with chronic diseases, which had a negative impact on their health.

Taiwan has entered an aging society, and the accompanying chronic diseases have become common health problems. In order to ensure that patients with chronic diseases can use drugs safely [4] and stably, relevant government units and medical groups must deal with this issue. According to the definition of the World Health Organization (WHO), when the proportion of the elderly population over 65 years old reaches 7% of the total population, it is called an "aging society"; when the proportion of the elderly population reaches 14% of the total population, it is called an "aged society". "; When the proportion of the elderly population reaches 20% of the total population, it is called a "super-aged society"[1]. Due to the continuous advancement of medical technology, the average life span in Taiwan continues to increase and the birth rate decreases. Therefore, the number and proportion of the elderly population are gradually increasing, making the problem of population aging increasingly serious. In 1993, the proportion of the elderly population over 65 years old accounted for more than 7% of the total population for the first time; by 1997, this proportion had exceeded 10% [1]. According to statistics, in 2018, 14% of Taiwan's population was aged 65 and over, marking the transition to an 'aged society' [5]. Projections indicate that by 2028, this demographic will comprise 22.5% of the population [6] (see Table 1 for details). Furthermore, domestic surveys from 2002 highlight a significant prevalence of hypertension, diabetes, and hyperlipidemia among the elderly, particularly within the 60–69 age group [7]. This trend underscores the growing need for focused medication and healthcare strategies for elderly patients with chronic diseases.

Table 1. Estimation of the structure of Taiwan's elderly population

Year	65 years of age or older	
	Population	% of the total population
1997	2.397 million people	10.4%
2018	3.48 million people	14.7%
2028	5.361 million people	22.5%

Source: Executive Yuan Economic Development Council, Taiwan, R.O.C.[5–7]

Since patients with chronic diseases often require long-term use of medicines to control their conditions, middle-aged and elderly people, the majority of whom take many medicines, have reduced body organ functions and have limited knowledge about medicines, often suffer from poor efficacy or the emergence of adverse reactions. If the domestic medical system cannot ensure the accuracy, effectiveness, and safety of medication for the elderly and chronically ill patients, society as a whole will inevitably pay more for medical costs. According to the analysis of the elderly population structure in townships, there are 23 townships across the country that belong to "super-elderly communities", with the elderly population accounting for more than 20%, and most of them are located in rural areas with inconvenient transportation. With the advancement of medical and health technology, the average life span of Chinese people continues to extend [8], and the population structure also tends to become older. Moreover, the prevalence of chronic diseases and functional disorders is gradually increasing, and the number of disabled people will also increase significantly. In that case, the demand for long-term care will increase. According to AI statistics, there are currently more than 85,000 people in Taichung City need long-term care services, of which more than 80% are elders and people with disabilities. Judging from the administrative division and distribution of Taichung City, Beitun District in Taichung City has the largest number of people in need of long-term care services, reaching 8,680 people, followed by Xitun District with 6,490 people, and Dali District with 6,424 people [9]. In administrative districts with larger populations, there are relatively more elderly people, so the demand for long-term care services is also higher.

Faced with the needs of an aging society, our government and private organizations have taken action to solve the problems. Many non-governmental organizations and companies that serve the elderly and senior citizens have been slowly established. These non-governmental organizations have innovative strategies with flexibility and diversified development space [10]. It can make up for the shortcomings of national policies and even come up with some great policy to solve social problems. There are some social issues appeared due to the demographic changes and socio-economic problems in Taiwan. It is facing on that Taiwanese society's difficulties due to demographic changes and socio-economic problems; social problems abound. To solve the current tricky problems, research and discussion on social services have become increasingly urgent. Recently, due to the serious problem of population aging in Taiwan, Taiwanese society is facing numerous social problems due to demographic changes and socio-economic difficulties. In addition, to solve the current thorny problems [11], it is increasingly urgent to use AI to conduct research and discussion on innovative medical and drug services. The government is actively using AI to evaluate and promote long-term care insurance and long-term care medical care, so many medical service industries targeting the elderly have emerged. Since the launch of National Health Insurance, patients have been forced to change their medicine-receiving habits and start receiving medicines from pharmacies authorized by the National Health Insurance. Subsequently, through government promotion [12], hospitals were encouraged to release prescriptions and sign in to community health insurance special pharmacies to receive medicines, resulting in an increase in the number of people going to special health insurance pharmacies to receive medicines year by year. The current statistical table of the number of authorized medical service agencies of the Central Health Insurance Agency of the Ministry of Health and Welfare evaluated by AI (report number: MHAH6552R01) [13]: On April 12, 2016, it was announced that there was a total of 5,949 authorized health insurance pharmacies. Pharmacists in community health insurance authorized pharmacies, he is independent and professional. When accepting prescriptions, he could evaluate the appropriateness of people's medication and give good advice. This study uses AI technology trend research to analyze the innovative pharmaceutical services of pharmacists at community C prescription home dispensing pharmacies in Taichung City, Taiwan, who deliver medicines to the community. It uses AI evaluation to understand whether there is an improvement in the business performance of the overall pharmacy [14] and whether the development of innovative pharmaceutical service models can solve the problem of chronic disease medication for aging elderly patients in the post-epidemic era. Through providing better service, we can set an excellent example among the few home dispensing pharmacies in Taiwan. This is the main motivation for this paper to study AI as a new trend.

3 Research Purpose

In view of the above, this study is a retrospective study. The case is the prescription exploration of C Community Home Dispensing Pharmacy in Taichung City, Taiwan. It is hoped to use AI technology to explore the number and chronic prescription signatures of community home delivery through the C Community Health Insurance Home Dispensing Bureau database. Information, as well as an analysis of the application costs. The purpose of this study is to explore the AI analysis and AI service model of community home delivery pharmacy in case C from the perspective of home delivery of drugs, and to analyze how community home delivery pharmacy in C implements this new model in pharmaceutical services. Further explore whether the pharmaceutical services provided by AI technology can improve business performance. Based on the above research background and motivation, this study aims to achieve the following three purposes:

1. Understand the service model of C Home Pharmacy.
2. Understand how C House Pharmacy implements this new AI technology service model.
3. C Home Pharmacy uses AI technology to evaluate whether business performance has grown relatively.

By exploring the service model and performance data of C Prescription Home Dispensing Pharmacy, and comparing the three-year performance trend analysis of C Prescription Home Dispensing Pharmacy during the epidemic, we find out whether this AI technology service operation model can effectively improve the business performance of C Prescription Home Dispensing Pharmacy. AI technology is used as an assessment. The ultimate purpose of this study is to learn a way to make some positive development of community home dispensing pharmacies in the future. We also hope to provide community pharmacies with relevant AI management models and practical suggestions in the new era of AI medical care.

4 Research Process

This research follows the following process: first, determine the research purpose and theme based on the research background and motivation, then collect data and organize relevant literature, and establish the conceptual structure of the research with reference to relevant theories. Next, perform AI technical data processing and trend analysis, and finally draw conclusions and write a report.

5 Research Methods

The research method of this study adopts the data and trend analysis method in the quantitative research type. Also, we use AI technology to analyze and compare the number of patients who received prescriptions for chronic diseases and their expenses in the three years during the epidemic period at the C Community Home Dispensing Pharmacy Bureau. The analysis time is January 1, 2019. The results of this study can provide content and directions for improving and increasing services for community health insurance pharmacies that want to engage in home delivery of drugs in the future. It can be provided as a reference for government and industry authorities, and is expected to solve the problems faced of long-term care and drug use.

6 Basic Data Analysis of Research Results

In this study, the prescription statistics from C's home delivery pharmacy are categorized into four variables: gender, age, area of residence, and annual health insurance declaration fee. The data were then organized and analyzed to summarize the demographic characteristics of the patients. It was found that the majority of patients receiving home delivery prescriptions reside in the Western District of Beizhong, constituting approximately 45.89% of the sample, as detailed in Table 2 and Fig. 1 (Tables 3, 4 and 5).

Table 2. Table of the number of people receiving prescriptions by year in each district (Source: Ho, Wei-Yuan (2023) [16])

Number	In 2019	In 2020	In 2021	total	percentage
Central	4227	4376	5365	13968	29.35%
Western	6537	7520	7783	21840	45.89%
North	2491	3113	4710	10314	21.67%
Other districts	742	401	327	1470	3.09%
sum	13997	15410	18185	47592	100%

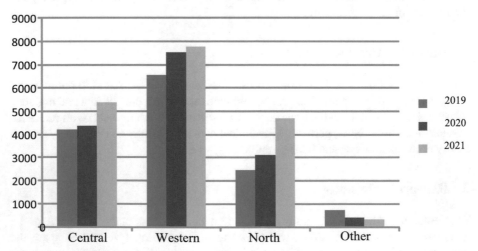

Fig. 1. Comparison of the number of people receiving prescriptions by year in each district (Source: Ho, Wei-Yuan (2023) [16])

Table 3. The statistics of prescription receipts by sex, age and month in 2019 (Source: Ho, Wei-Yuan (2023) [16]) collected from C's home delivery pharmacy in Taichung City, Taiwan.

gender	years	2019											
	age	1	2	3	4	5	6	7	8	9	10	11	12
woman	0–9	0	0	0	0	0	0	0	2	3	3	1	5
	10–19	3	1	1	2	3	3	4	3	5	5	3	4
	20–29	7	16	19	15	11	9	25	23	13	9	6	12
	30–39	29	45	50	38	41	40	54	41	47	41	42	37
	40–49	95	84	94	81	75	84	102	124	82	99	78	82
	50–59	38	32	44	41	34	46	52	44	43	77	47	48
	60–69	76	52	70	77	60	72	89	80	67	93	90	69
	70–79	59	46	66	52	62	55	80	74	77	86	68	78
	80–89	105	107	104	99	111	113	150	141	128	121	133	134
	90–99	125	97	104	115	110	113	165	152	146	161	137	137
	100 +	33	30	35	36	31	32	50	36	42	37	26	27
man	0–9	0	0	0	0	0	0	0	0	0	0	0	0
	10–19	7	12	9	5	6	15	8	5	7	11	10	11
	20–29	11	15	24	15	18	18	27	16	15	18	11	8
	30–39	26	18	18	25	22	20	30	29	27	35	29	38
	40–49	59	41	47	46	49	66	78	62	63	53	51	70
	50–59	77	64	64	84	69	61	91	91	64	94	82	83
	60–69	97	90	78	97	95	82	108	104	97	93	101	96
	70–79	99	89	103	109	88	86	87	102	85	100	101	112
	80–89	66	56	44	80	57	76	115	99	75	100	87	90
	90–99	71	47	65	75	69	73	89	80	77	99	75	82
	100 +	20	14	19	20	14	8	9	10	12	7	13	9
The total number of transactions in 2019		1103	956	1058	1112	1025	1072	1413	1318	1175	1342	1191	1232
Total number of transactions in 2019		13997											

7 Technology Options to Provide Instant Consultation

Online face-to-face and AI medication guidance and online face-to-face medication consultation. You can also use AI, line, e-mail and telephone to contact with patients. The C prescription home dispensing pharmacy provides instant drug consultation. AI technology can immediately provide medication consultation services when patients have medical problem. (For any key medication issues, as long as you call, AI technology

Table 4. 2019 Prescription Declaration Fee Table (Unit: NTD) collected by Pharmacy C in Taichung City, Taiwan. Source: Ho, WeiYuan (2023) [16].

Year	2019					
Month	1	2	3	4	5	6
Total Expenses for Reporting	$432,529	$355,326	$385,598	$416,084	$383,265	$443,396
month	7	8	9	10	11	12
Total Expenses for Reporting	$605,162	$561,621	$497,700	$515,659	$518,388	$527,404
Total Total Filing Fees	$5,642,132					

Table 5. 2019 Gender Statistics from C's home delivery pharmacy in Taichung City, Taiwan. (Source: Ho, Wei-Yuan (2023) [16]).

2019		
age	man	woman
0–9	0	14
10–19	106	37
20–29	196	165
30–39	317	505
40–49	685	1080
50–59	924	546
60–69	1138	895
70–79	1161	803
80–89	945	1446
90–99	902	1562
100 +	155	415
sum	6529	7468

can immediately provide medication consultation services.) This kind of instant service can also attract new customers to the pharmacy. AI big data leads service innovation is one of the focuses today. Accurate analysis of big data can understand customer preferences, explore customized needs, and provide personalized services. People use their mobile phones to make appointments for medication services and conduct medication consultations. They not only use data but also create data. Starting from data analysis, we understand consumer behavior and preferences, use technological tools as intermediaries. We also use AI artificial intelligence technology to predict potential customer flows and service needs, and master rich data to provide personalized services, directly communicating with drug dealers., nursing institutions, and hospitals negotiate, use databases to control service costs, and establish a new supply chain. In this digital era, consumers create digital footprints all the time, and big data has become the source of service innovation. From product innovation to value creation is the key to transformation in the era of smart economy. For C prescription home dispensing pharmacies, they need to increase the channels to contact with customers, obtain consumer behavior data, understand customer needs, promote the diversified integration of data through technology, and master the data through various AI technology analysis tools or technologies. Connotation, develop new types of services, and develop diversified emerging service models. Mastering data innovation is equivalent to mastering the competitiveness of business opportunities. C Prescription Home Pharmacy adopts new concepts of service innovation to face the wave of data economy. It starts with data to understand consumer behavior and preferences, provides the information or services consumers need, and improves service experience. Facing the future digital economy with data value, data services will play an important role in business model and service innovation.

8 Conclusion and Suggestions

The most important service innovations of C Prescription Home Dispensing Pharmacy are the service of AI mobile pharmacists and the service of home delivery drugs. C Prescription Home Dispensing Pharmacy uses the AI mobile pharmacist model to build channels for home dispensing pharmacies and deliver medicines to homes to achieve the purpose of serving the public. C Prescription Home Dispensing Pharmacy focuses on the pharmacist industry, which is different from its pharmacy peers. Diversified operations can also be based on the business model and operation of C Prescription Home Dispensing Pharmacy [15], which can also be set as business model innovation. C Prescription Home Dispensing Pharmacy breaks the geographical restrictions and challenges the business model of traditional pharmacies. Taiwanese society has entered an aging society, and the issue of elderly care has become increasingly important. Especially in the post-epidemic period, it is safer and more convenient to wait for medicine at home than go to community pharmacies to get medicine. Elderly people over 65 years old are also prone to problems with improper medicine, so the elderly use medicine. We need to be more cautious about this issue. The C prescription home dispensing pharmacy hopes to provide public health services and use AI technology to improve the problems of pharmaceutical care services. It can also reduce the side effects of multiple medications taken by the patient. Therefore, pharmacists should reduce the risk of inappropriate drug use, and reduce the occurrence of another disease in the elderly through pharmaceutical care. In addition, C Prescription Home Dispensing Pharmacy conducted data analysis through home delivery of prescription drugs and made a chart. It was found that the number of home delivery prescriptions and prescription health insurance declaration fees have a year-by-year growth trend. The conclusion can be learned that the pharmacy's home delivery of prescription drugs has positive effect on the pharmacy. The growth of business performance served by AI technology is helpful and worth promoting.

Acknowledgements. This work was supported by the Chelpis Quantum Tech Co., Ltd., Taiwan, under the Grant number of Asia University: I112IB120. This work was supported by the National

Science and Technology Council (NSTC), Taiwan, under NSTC Grant numbers: 111–2218-E-468–001-MBK, 110–2218-E-468–001-MBK, 110–2221-E-468–007, 111–2218-E-002–037 and 110–2218-E-002–044. This work was also supported in part by Asia University, Taiwan, and China Medical University Hospital, China Medical University, Taiwan, under grant numbers below. ASIA-111-CMUH-16, ASIA-110-CMUH-22, ASIA108-CMUH-05, ASIA-106-CMUH-04, and ASIA-105-CMUH-04.

References

1. Lin, Y.-Y., Huang, C.-S.: Aging in Taiwan: building a society for active aging and aging in place. Gerontologist **56**(2), 176–183 (2016). https://doi.org/10.1093/geront/gnv107
2. Fu, S., Huang, N., Chou, Y.J.: Trends in the prevalence of multiple chronic conditions in Taiwan from 2000 to 2010: a population-based study. Prev. Chronic Dis. **23**(11), E187 (2014). https://doi.org/10.5888/pcd11.140205.PMID:25340359;PMCID:PMC4208993
3. Canadian Pharmacists Association. Blueprint for Pharmacy: The Vision for Pharmacy. Ottawa, Canada: Canadian Pharmacists Association (2008)
4. Chen, H.S., Guo, F.R., Chen, C.Y., Chen, J.H., Kuo, T.S.: Review of telemedicine projects in Taiwan. Int. J. Med. Inform. **61**(2–3), 117–129 (2001)
5. Population Aging [Internet]. National Development Council. https://www.ndc.gov.tw/en/Content_List.aspx?n=85E9B2CDF4406753
6. Chen, C.-H.: Taiwan's Rapidly Aging Population: A Crisis in the Making? (2023)
7. 2002 Survey on the Prevalence of Hypertension, Hyperglycemia and Hyperlipidemia in Taiwan [Internet]. Health Promotion Administration, Ministry of Health and Welfare (2015). https://www.hpa.gov.tw/EngPages/Detail.aspx?nodeid=1062&pid=6072
8. Bai, R., Liu, Y., Zhang, L., Dong, W., Bai, Z., Zhou, M.: Projections of future life expectancy in China up to 2035: a modelling study. The Lancet Public Health. **8**(12), E915–E922 (2023)
9. Taichung City Government Accounting Office. Brief Analysis of Municipal Statistics of the Accounting and Statistics Department No. 111–007 (2022)
10. Suhui, H., Chang Cheng, L.: Analyze the functions and social value of non-profit organizations from the perspective of public-private collaboration - taking long-term care policy as an example. J. Chin. Adm. **23**, 65–83 (2018)
11. Hepler, C.D., Strand, L.M.: Opportunities and responsibilities in pharmaceutical care. Am. J. Hosp. Pharm. **47**(3), 533–543 (1990)
12. Lai, X.: Promote continuous prescription release strategy for chronic diseases. J. Taiwan Pharm. **28**(2) (2012)
13. Kucukarslan, S., Schommer, J.C.: Patients' expectations and their satisfaction with pharmacy services. J. Am. Pharm. Assoc. **42**(3), 489–495 (2002)
14. Larson, L.N., Rovers, J.P., Mackeigan, L.D.: Patient satisfaction with pharmaceutical care: update of a validated instrument. J. Am. Pharm. Assoc. **42**(1), 44–50 (2002)
15. Saini, L.: Long-term care pharmacy operations (2023)
16. Ho, W.-Y.: Performance Analysis of Community Pharmacy Home Delivery of Chronic Disease Prescription Drugs in the Post-Pandemic Era: A Case Study of Pharmacy C in Taichung City (2023)

A Language-Agnostic Compression Framework for the Bitcoin Blockchain

Orestes Papanastassiou[✉] and Alex Thomo

University of Victoria, Victoria, BC, Canada
{orestespap,thomo}@uvic.ca

Abstract. This research addresses the growing interdisciplinary interest in Bitcoin by proposing a versatile compression framework for transforming raw blockchain data into a streamlined compact format suitable for high-performance analysis. Our approach focuses on developing a language-agnostic API, ensuring accessibility across programming languages. Beyond data extraction, our framework outputs the Bitcoin user transaction graph, facilitating network analysis, forensics, and pattern detection. Processed data are exported to the HDF5 file format for compatibility with mainstream analysis tools. A proof-of-concept CPython implementation demonstrates the framework's feasibility, showcasing its real-world applicability for data-driven investigations in Bitcoin research.

Keywords: bitcoin · blockchain · cryptocurrencies · big data

1 Introduction

The Bitcoin blockchain[1] is a decentralized, distributed ledger that records Bitcoin transactions. This ledger is public and immutable, making it a valuable source of information for various research and analytical purposes. However, its sheer size presents significant challenges to researchers and analysts. The ability to extract, store, and analyze blockchain data efficiently is desired by academics and industry.

This research addresses a critical gap in the current landscape by proposing an abstract algorithmic framework for the extraction and compression of Bitcoin's transaction ledger and user transaction graph. The data are converted into a form that we call *normal form*, which adheres to a data-oriented design [3].

Data-oriented design is a design paradigm applied to algorithms and data structures to maximize the use of the CPU and GPU cache. This involves minimizing the size of the working set (program input, i.e. a matrix) to fit as much data as possible into the cache, using a contiguous memory layout whenever possible, and designing algorithms so that data reside in the cache until they are no longer needed by the program [8]. A cache line is the currency between the CPU and the main memory [5]. Reading from memory is a significant bottleneck for

---

[1] https://bitcoin.org/bitcoin.pdf.

© The Author(s), under exclusive license to Springer Nature Switzerland AG 2024
L. Barolli (Ed.): AINA 2024, LNDECT 200, pp. 235–249, 2024.
https://doi.org/10.1007/978-3-031-57853-3_20

modern CPUs, thus high-performance data analysis necessitates fitting as much data as possible into consecutive cache lines so that the size of the subset of the working set retrieved every time the CPU reads data from main memory is maximized.

Our normal form makes it possible for the entire transaction ledger or user transaction graph to fit into the main memory of a 64GB commodity research machine, with $O(1)$ indexing of the underlying data for fast lookups. The framework facilitates the extraction of the raw blockchain data and transforms them into our normal form with a minimal memory footprint, such that the entire blockchain can be extracted on a commodity machine with just 32GB of memory.

To maintain the framework's language-agnostic theme end-to-end, output data are exported to the HDF5 file format [6], which can be processed by any mainstream programming language and data analysis software. Furthermore, the output can be seamlessly stored in a conventional SQL/No-SQL DBMS. We also present a proof-of-concept implementation in Python to demonstrate the feasibility and efficiency of our approach. We chose Python for our proof-of-concept implementation because of its interdisciplinary popularity and the challenges associated with implementing the framework efficiently using this language. The main challenge is CPython's inefficient memory consumption; the increased productivity and simplicity offered by its object-oriented and dynamically typed nature come with trade-offs.

Our contributions can be summarized as follows.

- **Algorithmic Framework/API:** We introduce a comprehensive language-agnostic algorithmic framework/API designed to efficiently extract and compress Bitcoin's transaction ledger and user transaction graph. This framework is crucial for handling the massive scale of blockchain data, which contains over 800 million transactions as of January 1st, 2023[2].
- **Cache Optimized Data Layout with O(1) Indexing:** Our *normal form* data structure facilitates $O(1)$ indexing for high-performance data analysis. The entire Bitcoin transaction ledger or user transaction graph can be accommodated in the main memory of a commodity research machine.
- **HDF5 Format for Cross-Platform Compatibility:** Output data are saved in the HDF5 file format, facilitating the use of the processed data with various mainstream programming languages and data analysis tools, and can be seamlessly stored in both SQL and No-SQL databases.
- **CPython Proof-of-Concept Implementation:** To validate our work, we present a CPython-based proof-of-concept that overcomes the language's memory hurdles.

2   Related Works

BlockSci [7] is a comprehensive state-of-the-art blockchain exploration tool implemented in C++ with a CPython interface for seamless high-level querying

---

[2] https://blockchain.com/explorer/charts/n-transactions-total.

of the data. Similar to our work, BlockSci's data layout is designed to maximize the utility of the CPU cache. The project is no longer maintained, which makes it challenging to install and use because of compatibility issues with newer versions of Linux and CPython. Because it is a packaged querying tool, there is little flexibility in changing the set of transaction attributes and migrating the data to a different DBMS.

BTCSpark [11] is built on top of Apache Spark and Spark SQL, a robust and distributed data processing framework, to address the challenges associated with large-scale Bitcoin data analysis. This approach is particularly valuable in scenarios where the volume and complexity of Bitcoin data require efficient and seamless distributed processing.

BitSQL [9] is an SQLite and MariaDB-based blockchain querying tool. The data layout is similar to the one proposed by this work, but redundantly stores both raw transaction IDs/addresses and their integer hashes, leading to an output that is double in size compared to ours. Furthermore, our research dedicated a significant amount of time to integrating various SQL/NoSQL DBMSs into the pipeline and failed to replicate the claimed performance figures.

[2] contributes a general extraction framework for the Ethereum and Blockchain blockchains. The framework extracts raw blockchain data, as well as off-chain data (i.e. exchange rate, user IP addresses), and stores them in an SQL or No-SQL database. The purpose of this framework is to facilitate individual as well as cross-chain data analytics.

3   Background

3.1   Bitcoin Transactions

The Bitcoin blockchain records transactions in a sequence of blocks, created by miners in a peer-to-peer network. Miners compile transactions into blocks, earning new Bitcoins and fees as rewards.

A Bitcoin transaction consists of an arbitrary number of inputs and outputs. The currency of the Bitcoin blockchain is the set of unspent transaction outputs. Each transaction output specifies a value of Bitcoins to be spent and contains a script (*ScriptPubKey*) dictating the conditions under which these Bitcoins can be spent. An unspent output is spent when referenced by a transaction input alongside the appropriate unlocking script (*scriptSig*). Bitcoin users/entities are an abstraction of output script addresses.

In a valid transaction, the sum of the Bitcoin value of the inputs must be greater than or equal to the sum of the value of the outputs - users cannot spend more money than they have. When the input value is greater than the output value, miners can claim the remainder as a fee. If the total input value is greater than the debt that needs to be settled, the remaining Bitcoins can be reclaimed by appending an output(s) to the transaction's output set referencing the desired amount and address. This is known as a *change* output.

The first transaction in a block is called *Coinbase* and can be used by miners to claim the newly minted Bitcoins and any aggregated transaction fees. *Coinbase* outputs create new Bitcoins, and thus increase the money supply.

3.2 User Transaction Graph and Address Clustering

A directed graph $G = (V, E)$ consists of vertices V and directed edges E, where each edge points from one vertex to another. The blockchain can be represented as a directed multigraph, where directed edges represent transaction outputs and vertices/nodes represent users. A multigraph is a graph that can have multiple edges between a pair of endpoints.

Bitcoin users are inferred via address clustering. The *common-input-ownership*[3] heuristic is the most widely used address clustering method (CIO henceforth) The heuristic explicitly assumes that a multi-input transaction is signed (via the *ScriptSig*) by the same user, even though it is technically possible for each input to be signed by a separate user [4]. CIO is a computationally cheap algorithm for address cluster inference and is supported by our framework.

3.3 Notations

We list some recurring notations throughout the paper. Whenever we use *list* the order matters, and whenever we use *set* the order does not matter.

1. $S(.)$ is the size of a data structure/type in bytes.
2. h: Block height. This is a block's ordered position (index) in the blockchain.
3. B_h: List of blocks at height h. A block contains a list of transactions.
4. T_h: List of transactions at height h, all the transactions in blocks in B_h.
5. A_h: Address set at height h, all the addresses in transactions in T_h.
6. U_h: Sub-list of T_h containing all the transactions of T_h with at least one unspent output.
7. b: Block : $b \in B_h$.
    (a) $b.h$: Height of block b.
    (b) $b.t$: Transaction list of block b; $b.t \subset T_h$.
        i $b.t[i]$: i^{th} transaction in $b.t$.
8. t: Transaction.
    (a) $t.id$: Unique ID of the transaction.
    (b) $t.b$: index of t in block b; $0 \le t.b < |b.t|$.
    (c) $t.h$: Transaction's block height; $0 \le t.h < |B_h|$.
    (d) $t.norm$: Transaction's *normal form*.
    (e) $t.in$: List of transaction inputs (see below for the definition of inputs).
    (f) $t.out$: List of transaction outputs (see below for the definition of outputs).
    (g) $t.uout$: List of transaction unspent outputs.
    (h) $t.ts$: Transaction timestamp.
    (i) $t.insum$: Sum of inputs' referenced unspent output values. Each input references an unspent output from an older transaction.

---

[3] https://en.bitcoin.it/wiki/Common-input-ownership_heuristic.

      i $t.insum.f$: Denomination flag.

      ii $t.insum.v$: Monetary value (Bitcoin or Satoshi, depending on f).

  (j) $t.outsum$: Sum of a transaction's output values. Same attributes as $t.insum$.

  (k) $t.fee$: Fee, if $t.insum > t.outsum$.

  (l) $t.norm$: normal form, list of integers (see Methodology).

9. in: Input.

  (a) $in.i$: i^{th} input in $t.in$.

  (b) $in.id$: Referenced output's transaction ID ($t^i.id$, $t^i \neq t$).

  (c) $in.outi$: Referenced output's position in its own transaction's (t^i) output list ($t^i.out[in.outi]$).

  (d) $in.addr$: Input address.

10. out: Output, $uout$: Unspent output.

  (a) $out.addr$: Output address.

  (b) $out.f$: Denomination flag.

  (c) $out.v$: Output Bitcoin value.

11. G_h: User transaction graph at height h.

  (a) V_h: Node set.

  (b) E_h: Edge set.

4 Methodology

4.1 Normal Form, Denomination Flags, and MurmurHash

We transform raw Bitcoin transaction data into a list of unsigned 4-byte integers that we call *normal form*. This list contains input/output addresses, output values, sum of input values, and timestamp. The pipeline can be easily modified to include other attributes, such as output script type.

Let $|t.in| = n$ and $|t.out| = m$. The normal form structure is as follows:

1. The first two elements are n and m.
2. The following n elements are the *MurmurHash*-ed input addresses in the order they appear in the transaction.
3. The following m elements are the *MurmurHash*-ed output addresses ".
4. The following $2 \cdot m$ elements are ($flag, value$) pairs, one pair per output address. *flag* encodes *value*'s denomination (Satoshi or Bitcoin).
5. The last three elements are a ($flag, value$) pair encoding the total monetary value of the inputs ($t.insum$), and transaction timestamp.

Transaction outputs are denominated in Satoshis ($1\ BTC = 100e6\ SAT$) on the blockchain. Outputs greater than $2^{32} - 1\ SAT$ exceed the limit of a 4-byte unsigned integer ($uint32$). For such large Satoshi values, we convert them into Bitcoin, round them to the fourth decimal, and scale them back to an integer. If still too large, the offset from $2^{32} - 1$ is stored. Negative fees due to rounding errors are set to zero.

Denomination flags indicate the transaction value's format and spent status. This facilitates the construction of the unspent transaction output set ($UTXO$).

Flags $(0, 3)$ represent SAT, $(1, 4)$ BTC, and $(2, 5)$ BTC as an offset from $2^{32} - 1$. Unspent output flags are < 3; when an unspent output is spent, it is marked as spent by incrementing its flag by three. To convert back into SAT, for $flag \in (1, 4)$ multiply the output value by $1e4$, and for $flag \in (2, 5)$ add $2^{32} - 1$ and multiply by $10e4$.

Raw Bitcoin transaction IDs ($txid$) are 64-character long HEX strings. Addresses are $base58Check$ encoded alphanumeric strings of 58 characters [1]. $txid$ ($t.id$) can be represented as a HEX string of 64 characters (32 bytes) and an address can be represented as a 58-byte array of ASCII characters. A raw Bitcoin address and transaction ID occupy an entire cache line and half a cache line respectively (a cache line is usually 64 bytes). Our solution uses the $MurmurHash^4$ non-cryptographic hash function to convert addresses and transaction IDs into 4-byte unsigned integers.

4.2   Transaction Indexing

Normal-form transactions are stored in a one-dimensional array of unsigned integers (tArr). To facilitate efficient traversal of the transaction array and $O(1)$ lookups we introduce two auxiliary offset arrays; toff and boff. toff maps an ordered transaction index (i) to the slice of tArr that contains ordered transaction i.

$$\forall i \in [1, |T_h|) : \text{tArr}[\text{toff}[i]:\text{toff}[i+1]) = T_h[i].norm \tag{1}$$

The i-th blockchain transaction's normal form is stored in the range [toff[i],toff[i+1]) of tArr. Similarly, boff enables transaction indexing by block height. Given a block height h, we can retrieve the block's transaction set from tArr as follows,

$$\forall h \in [1, |B_h|) : \text{tArr}[\ \text{toff}[\text{boff}[h]] : \text{toff}[\text{boff}[h+1]]) = B_h[h].t \tag{2}$$

We can also index transactions by block height and transaction block position $(t.h, t.b)$. For $a = \text{toff}[\text{boff}[t.h]+t.b]$, $b = \text{toff}[\text{boff}[t.h]+t.b+1]$,

$$tArr[a : b) = B_h[t.h].t[t.b].norm \tag{3}$$

4.3   Retrieving Input Addresses

The first stage parses the raw data and maps transactions to normal form. Let $|B_h| = N$ and $\forall b \in B_h : |b.t| = M$, we have the following intermediate representation,

$$\text{tArr} = [B_h[0].t[0].norm, \ldots, B_h[N-1].t[M-1].norm] \tag{4}$$

$$= [T_h[0].norm, \ldots, T_h[(N \cdot M) - 1].norm] \tag{5}$$

In our CPython implementation, we store tArr across multiple arrays/files on $5e6$ transaction intervals because of memory limitations. In the intermediate

---
[4] https://github.com/aappleby/smhasher.

representation, inputs are represented as a triplet of integers compressed into one integer between stages one and two. For now, assume that each input is represented as an integer triplet (a set of three integers). For every $t \in T_h$ we have,

$$\forall i \in [0, |t.inp|) : (t^i.h, t^i.b, t.in[i].outi) \land (t^i \neq t) \tag{6}$$

The items in the triplet refer to the referenced output's transaction block height, position in the block, and the output's position in t^i respectively. This is used in the next stage to extract $in.addr, in.f$ and $in.v$. Given an arbitrary transaction id as input, the corresponding transaction can be retrieved in $O(1)$ via,

$$UMap_h(id) \rightarrow [t.h, t.b, |t.uout|], \ t = B_h[t.h].t[t.b] \tag{7}$$

$UMap_h$ is a key-value data structure that maps a transaction ID to a triplet containing the transaction's block height, position in the block, and number of unspent outputs. When a transaction is added to $UMap$, $|t.uout|=|t.out|$. When a future $in \in t'.in$ references $t.uout$, $|t.uout|$ is decremented by one. Every time block data are moved to secondary storage, $UMap$ is updated such that every t with $t.uout \in \varnothing$ is removed. The blockchain's $UTXO$ set at some height h can be reconstructed via $UMap_h$.

4.4 UMap as a CPython *dict*

As of v3.12, CPython's *dict* is based on a dynamic indexing array and a compact hash table[5]. Each entry in the hash table is represented by a hash (h) ($\leq$ 8-byte signed integer,) and two pointers (p) for the key-value (k, v) pair (8 bytes/pointer). An entry's position in the hash table is stored in the dynamic array as an $\leq$ 8-byte unsigned integer (i).

In general, a map data structure's memory footprint is lower bounded by the size of its payload and upper bounded by the payload times the resizing factor[6]. For large payloads, the resizing factor for a Python dictionary is two, which we can use to approximate an upper bound. The following formula calculates a dictionary's payload ($P(k, v, i, h)$) size (n is the number of (k, v) pairs),

$$S(P) = n \cdot (S(k) + S(v) + 2 \cdot S(ptr) + S(h) + S(i)) \tag{8}$$

$$S(P(k, v, uint32, int64)) = n \cdot (S(k) + S(v) + 28) \tag{9}$$

We estimate a *dict*'s memory footprint to fall within this range,

$$S(P) \leq S(dict) \leq 2 \cdot S(P) \tag{10}$$

Considering the current size restrictions of the Bitcoin protocol $|b.t|$ and $|t.out|$ rarely exceed $10e4$, and is virtually impossible to exceed $1e6$. The average $|b.t|$ for the past six years has hovered around $2e3$, briefly reaching a peak of

---

[5] https://mail.python.org/pipermail/python-dev/2012-December/123028.html.
[6] https://courses.csail.mit.edu/6.006/spring11/rec/rec07.pdf.

$\approx 3.4e3$ before reverting to the mean[7]. The record $|t.in|$ and $|t.out|$ stands at $20e3$ and $13e3$ respectively[8]. As of January 1st, 2023, $|B_h| \approx 770e3$, so $b.h$ can be represented by a seven-digit integer to accommodate block heights $< 1e7$.

$EncodeT$ transforms a $(t.h, t.b, j)$ triplet tr into a 22-digit unsigned integer. $DecodeT$ is the inverse map,

$$EncodeT(tr) \rightarrow ((1e7 + t.h) \cdot 1e7 + t.b) \cdot 1e7 + j \qquad (11)$$

$$DecodeT(int) \rightarrow (t.h, t.b, j), \ int \in N \qquad (12)$$

where, $t.h = (int) \ div \ (1e14) - 1e7$, $t.b = ((int) \ mod \ (1e14)) \ div \ (1e7)$, and $j = (int) \ mod \ (1e6)$. Depending on the context, j can represent one of three attributes. If tr describes an in, then $j \rightarrow in.i$ or $in.outi$, if tr describes an out, then $j \rightarrow out.i$, and if t is meant to keep track of $|t.uout|$, $j = |t.uout|$. The maximum value in $EncodeT$'s range is interpreted as $t.h = 9999999$, and $t.b = j = 999999$. Any unsigned integer in this range is represented as a 36-byte int instance in CPython, whereas an implementation representing the triplet as a $tuple$ of three int objects requires 148 bytes of memory. As a result, our encoding consumes $\approx 50\%$ less memory resources than a $tuple$-based representation.

4.5   Handling Hash Collisions

Unambiguous resolution of hash collisions in $UMap$ leads to garbage output data. To avoid ambiguity, we leverage that the chronological traversal of transactions guarantees that previous transaction inputs do not reference the currently processed transaction's ID.

To add $t.id$ to $UMap$, we $MurmurHash$ the ID, check the hash's membership in $UMap$'s keyset, and if the result is positive there is a hash collision. This means that an older transaction ID is mapped to the same hash. In this case, we add the raw transaction ID to $UMap$'s keyset, represented as a string of 64 HEX characters. A low hash collision rate means that the memory footprint of string keys is virtually zero. For any arbitrary transaction t' referenced by some input in another transaction t, we check the membership of $t'.id$ in $UMap.keys$. If it exists, t' is represented by $t'.id$ in $UMap$, and if it doesn't, it is represented by the $MurmurHash$ of $t'.id$.

5   Algorithms

5.1   First Stage: Transforming Raw Transaction Data into Normal Form

The pipeline begins by traversing the raw transaction data to convert them into normal form and create the (toff, boff, tArr) arrays and $UMap$ in the process. Each transaction t is temporarily stored in an array ($tempt$) that is unpacked

---

[7] https://blockchain.com/explorer/charts/n-transactions-per-block.
[8] https://coinmetrics.io/batching/.

into tArr before moving on to the next one. For each transaction input (in), $in.id$ (the equivalent of $t'.id$) is looked up in $UMap.keys$ according to the procedure described above.

The desired $(t'.h, t'.b, |t'.uout|)$ triplet is stored as an integer in $UMap[key]$, and is unpacked via $outT = DecodeT(UMap[key])$. The input is now represented by $EncodeT((outT[0], outT[1], in.outi))$, which is used in the second stage to retrieve the input's address and referenced output's Bitcoin value, both of which are stored in the referenced output. As per our pruning optimization, $UMap[key]$ ($|t'.uout|$) is decremented by one.

Next, each output address in $t.out$ is hashed and appended to $tempt$, followed by each output's $(out.f, out.v)$ pairs. The last three elements in the array are $t.insum.f = 0, t.insum.v = 0, t.ts$. t is added to the unspent transaction set, represented by $UMap$. The algorithm checks $t.id$'s hash membership in $UMap.keys$ to determine whether there is a hash collision or not. It then encodes the desired triplet $inInt = Encode(t.h, t.b, |t.uout|)$ and adds a new entry to $UMap$, $UMap[key] = inInt$. Depending on whether there is a hash collision or not, key is either $t.id$ or $t.id$'s hash.

5.2 Second Stage: Resolving *uout* References

The list of files produced by the first stage is traversed to retrieve $in.addr$ and $t.insum$. The latter is necessary for the calculation of $t.fee$. If an input references an unspent output stored in another file (file'), the input is stored in a map ($lookupMap$) alongside other inputs with out-of-file references, so that their data are retrieved after tArr is traversed. The map is then grouped by file ID so that $uout$ references located in the same file are grouped, and thus each file is loaded once. Map keys are file IDs, pointing to (in, out) sequences. Both in and out are encoded as $EncodeT((t.h, t.b, j))$.

Each transaction (t) input is decoded to extract the corresponding $uout$ reference, represented by the compressed $(t'.h, t'.b, out.i)$ triplet. If $t'.h$ is in the file's block height range, then t' is in the same file. The referenced output (out) is retrieved from tArr using the toff and boff offset arrays, the corresponding $uout$ reference in $t.in$ is replaced by $out.addr$, $t.insum.v$ is incremented by $out.v$, $t.insum.f$ is assigned the appropriate flag, and $out.f$ is incremented by three to mark it as spent. If $t'.h$ is not in the file's block height range, then the input and the referenced output are appended to $lookupMap[t.h]$, and the corresponding $uout$ reference in $t.in$ is set to NULL. Once all transactions in the file are traversed, the missing data are retrieved by traversing $lookupMap$ and loading the appropriate files.

5.3 Third Stage: Address Clustering

This stage applies the CIO heuristic to the address set (A_h), implemented based on the Weighted Quick Union Find algorithm [12,13]. [10] implements CIO using a variation of *Union-Find*.

WQUF consists of two core operations; *Find* and *Union*. If an arbitrary element belongs to a set, *Find* returns the set's root, and if it doesn't, it returns the empty set operator. *Union* merges two distinct sets by attaching the root of the smaller set to the root of the larger set. It then increments the size of the larger set by the size of the smaller set to reflect the size of the expanded set and finally returns the root of the larger set.

5.4   Fourth Stage: User Transaction Graph

For every $t \in T_h$, the cluster address (set root) of $t.in$ is retrieved via $parentMap[t.in[0].addr]$, and an outgoing edge is connected to each output cluster ($\forall out \in t.out : parentMap[out.addr]$). This process leads to the creation of $|t.out|$ number of edges. $parentMap$ maps a MurmurHash-ed Bitcoin address to its cluster address (set root).

A dummy *Coinbase* node (user) is added to the graph to represent transfers of newly minted Bitcoins and transaction fees. For each t, if $t.fee > 0$, an edge is created between the input cluster and *Coinbase*, weighted by the transaction fee. The graph is implemented by a key-value data structure, where each key is a cluster ID that points to a list of outgoing edges,

$$edge = (parentMap[out.addr], out.\text{f}, out.v, t.ts) \tag{13}$$

6   Results

(See Table 1)

Table 1. The first column lists experiment parameters, the second column represents Experiment Small, and the third Experiment Large.

Experiments	Small	Large
Height range $h \in [0, H]$	$[0, 3e5]$	$[0, 769842]$
$|A_H|$	3.733$e7$	1.084$e9$
$|TX_H|$	40$e6$	792$e6$
CPU	i7-3770	Xeon E5-2620
DRAM	12GB	128GB
DIMMs	2x4 GB, 2x2 GB	8x16 GB
L1	256 KiB	384 KiB
L2	1 MiB	1.5 MiB
L3	-	15 MiB

6.1   Compression

We define compression rate as the percentage data size reduction by the compressed data (a) relative to the uncompressed data (b),

$$CR(a,b) = (1 - (S(a)/S(b))) \cdot 1e2 \tag{14}$$

The unclustered blockchain's (u) HDF5 output consists of tArr, toff, and $UMap$ (key-value pairs stored in a 1D array) and the clustered blockchain's (c) HDF5 output consists of the same arrays plus $parentMap$. In Experiment Large, the observed $S(u)$ (minus $UMap$), $S(c)$ (minus $UMap$, $parentMap$), $S(UMap)$, and $S(parentMap)$ are 53.8, 44.5, 1.2, and 6.3 GB respectively. With $S(Bitcoin)$ ≈ 446 GB on January 1st 2023, our normal form yields $CR(u, Bictoin) \approx 88\%$.

The graph $(\hat{G}_H)$ is stored in three arrays; $nodes$ contains cluster addresses, $edges$ contains the outgoing edges for each cluster, and eoff contains the $edges$ offsets for each cluster (identical to tArr and toff). The observed $S(\hat{G}_H)$ (minus $UMap$, $parentMap$) in B is ≈ 43 GB, with 4 GB evenly divided between $nodes$ and eoff arrays, and 39 GB for $edges$. The total size (with the two maps) is $\approx$ 47.3 GB.

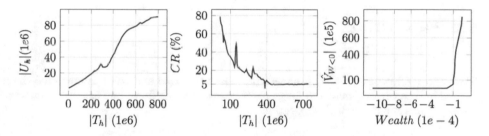

Fig. 1. [Left]: X and Y-axis represent the sizes of the transaction set $(|T_h|)$ and unspent transaction set $(|U_h|)$ respectively, as h grows to H. [Middle]: The graph shows the observed compression rate from pruning $UMap$ $(CR(UMap', UMap))$ as a function of the size of the transaction set $|T_h|$, as h grows to H. [Right]: Cumulative Distribution Function of negative wealth. The X-axis represents negative wealth values and the y-axis node count.

6.2   Memory

$UMap$ and $parentMap$ are the data structures with the largest memory footprint in the API. An alternative solution to $dict$ would be a key-value pair DB, in which case a sample of the key-value pairs reside in the cache (DRAM in this case) and the rest in storage. We tried two state-of-the-art solutions, $RocksDB$ and $Redis$, and in both cases, elapsed time increased at least three-fold. Approximately half of the queries are cache misses, and retrieving data from storage is orders of magnitude slower than main memory. The decision to store intermediate data ([toff, boff, tArr]) in *pickle* files has a similar reasoning. We tried two RDBMs, *MySQL* and *SQLite*, and elapsed time increased at least ≈ three-fold.

Figure 1 [left] shows that $|U_h| \approx |T_h|/10$ throughout the experiment. For $n = |U_H| = |T_H|/10 \approx 90e6$, $S(P)$ yields 8.7GB, thus our estimate is $S(UMap_H) \in [8.7, 17.4]$ GB. The observed $S(UMap_H)$ is 13.1 GB, ≈50% greater than our lower bound, and ≈ to the median value of our estimated size range. As of October 2023, $|T_{h=H}| \approx 900e6$, so our pruning optimization makes it possible to extract 90% of the unclustered transaction data with just 32GB of DRAM.

$UMap$ is pruned every time the extracted block data are stored in secondary storage and removed from the cache. Data are dumped in storage on $5e6$ transaction intervals, thus Fig. 1 [middle] contains 160 such measurements. The pruning optimization's returns are diminishing as h, and therefore $|T_h|$, increases. For $|T_h| \geq 400e6$, the achieved compression rate stays at 5%. At $p = (400e6, 40e6)$, $UMap$ is pruned an additional 80 times. However, had the pruning stopped at p, $|UMap.keys_{h=H}|$ would have grown to $40e6 \cdot (1.05)^{80} \approx 200e6$.

$parentMap.keys$ is the equivalent of A_H. To estimate its size at the end of B, we set $n = |A_H| \approx 1e9$ and get $S(parent) \in [92, 184]$ GB. The observed $S(parent)$ is 115GB, approximately 25% larger than our lower bound, and 17% less than the median value of our estimated size range.

6.3   Elapsed Time

Table 2. First column lists experiment parameters, second and third columns list the results in Experiment Small and Large respectively, and fourth column lists Experiment Large/ Experiment Small timing ratios.

Experiment timings	Small	Large	t_L/t_S
Normal Form	2h 45 min	67h 50 m	25
UOUT References	12 min	68 h	340
Address clustering	2.5 min	2h 30 m	60
Graph creation	7 min	3 h 30 m	30
Data export	3 min	1 h 15 m	25
Total	3h 10 min	5d 23 h	

Table 2 lists the timings for each stage. Experiment Large's working set is ≈ 20 times larger than Experiment Small's, and since the framework's complexity is $O(n)$, we should expect timings in Large to be at least 20 times greater than Small. The empirical average $t_L/t_S \approx 45$ reflects the difference in processing power between the two machines and the growing number of CPU cache misses in Experiment Large due to the much larger working set (and an unexpected issue with stage two). We suspect that a critical factor in stage one's poor performance could be the third-party library used to read the raw blockchain data[9]. A main priority for our future work is further investigating the potential bottlenecks caused by this API.

Stage two's poor performance can be attributed to its worst complexity case turning out to be its average/expected complexity due to the inherently random nature of unspent output references. The worst case is that each file has at least one *uout* reference for each previous file. That proved to be the case for most of our block files. As a result, most files had to access all previous files to retrieve input data, equivalent to performing a file scan each time.

6.4   Information Loss from Floating Point Arithmetic

Floating point arithmetic approximation error inevitably leads to nodes with negative wealth, spending more Bitcoins than they receive. We classify a user transaction graph $\hat{G}_h$ as a *valid economy* if it satisfies the following constraints:

1. $\forall n \in (\hat{V}_h - \{CB\}) : \widehat{Wealth}_h(n) \gtrsim 0$
   (a) $\widehat{Wealth}_h(n) = inVal_h(n) - outVal_h(n)$
2. $Supply_h \approx \hat{Supply}_h$
   (a) $\hat{Supply}_h = outVal_h(CB) - inVal_h(CB)$
3. $Supply_h \approx Sum(\widehat{Wealth}_h(n) \ \forall n \in \hat{V}_h)$

Where $inVal_h$ and $outVal_h$ represent the total value of a node's incoming and outgoing edges at h respectively. $Supply_h$ represents the Bitcoin supply in the blockchain at h, and $\hat{Supply}_h$ is the supply in $\hat{G}_h$. $\widehat{Wealth}_h(n)$ represents individual node wealth, where the real precise value is unknowable. CB represents the Coinbase node. The first constraint states that a node cannot spend more Bitcoins than it owns. The following constraints can be deduced from the first one and all constraints are relaxed, therefore a node's wealth can be *slightly* negative.

Table 3. First column lists the parameters of the experiment, second column lists the ground truth data (the blockchain), and third column lists the observed values.

Experiment Large	$G_{h=H}$	$\hat{G}_{h=H}$
$Supply_H$	19.249e6	19.237e6
$|V_{H_{W<0}}|/|V_H|$	0	0.1416
$min(Wealth_H)$	0	$-129e-3$
$Supply_H - Sum(Wealth_H)$	0	17.28

Table 3 lists the necessary data to prove that our $\hat{G}_H$ satisfies all three constraints. All measurements are made at the end of Experiment Large, at height

---

[9] https://github.com/alecalve/python-bitcoin-blockchain-parser.

$h = H = 769,842$. The observed $\hat{Supply}_H / Supply_H = 99.94\%$ means that the Bitcoin supply in $\hat{G}_H$, based on $outVal_H(CB) - inVal_H(CB)$, reflects the real supply at height H. $\hat{G}_H$ then satisfies the second constraint.

The second statistic measures the proportion of $|\hat{V}_H|$ with negative wealth $(v \in \hat{V}_H : Wealth_H(v) < 0)$. Given that $|\hat{V}_H| \approx 490e6$, and $|\hat{V}_{H_{W<0}}|/|\hat{V}_H| \approx 14\%$, $\approx 69e6$ nodes in the graph have a negative net worth. This does not invalidate our user graph; Fig. 1 [right] shows the cumulative distribution function of negative user wealth at $h = H$, where 99% of negative $Wealth$ values fall in the $[1e-4, 0)$ range. The most negative observed $Wealth$ value is ≈ -0.13 Bitcoins. $\hat{G}_H$ then satisfies the first constraint.

The last statistic measures the difference between the supply in $\hat{G}_H$ and total user wealth. The value of the real figure is zero because user wealth cannot exceed $Supply_H$. In $\hat{G}_H$, this figure is ≈ 17 Bitcoins. In relative terms $(17/Supply_H) \approx 8.83e - 57$, so the observed figure is very close to zero. $\hat{G}_H$ satisfies the last constraint and thus qualifies as a *valid economy*.

References

1. Antonopoulos, A.M.: Mastering Bitcoin, , 2 edn., pp. 55–88. O'Reilly Media, Inc. (2017)
2. Bartoletti, M., Lande, S., Pompianu, L., Bracciali, A.: A general framework for blockchain analytics. In: Proceedings of the 1st Workshop on Scalable and Resilient Infrastructures for Distributed Ledgers, pp. 1–6. Association for Computing Machinery (2017)
3. Bayliss, J.D.: The Data-Oriented Design Process for Game Development, vol. 55, pp. 31–38. IEEE Computer Society (2022)
4. Di Francesco, D., Maesa, A.M., Ricci, L.: Data-driven analysis of bitcoin properties: exploiting the users graph. Int. J. Data Sci. Anal. **6**, 63–80 (2018)
5. Drepper, U.: What every programmer should know about memory (2007). https://people.freebsd.org/~lstewart/articles/cpumemory.pdf. Accessed 09 Nov 2023
6. Folk, M., Heber, G., Koziol, Q., Pourmal, E., Robinson, D.: An overview of the hdf5 technology suite and its applications. In: Proceedings of the EDBT/ICDT 2011 Workshop on Array Databases, pp. 36–47. Association for Computing Machinery (2011)
7. Kalodner, H., et al.: {BlockSci}: Design and applications of a blockchain analysis platform. In: 29th USENIX Security Symposium (USENIX Security 20), pp. 2721–2738. USENIX Association (2020)
8. Kowarschik, M., Weiß, C.: An overview of cache optimization techniques and cache-aware numerical algorithms, pp. 213–232. Springer (2003)
9. Mun, H., Lee, Y.: Bitsql: a sql-based bitcoin analysis system. In: 2022 IEEE International Conference on Blockchain and Cryptocurrency (ICBC), pp. 1–8. IEEE (2022)
10. Ron, D., Shamir, A.: Quantitative analysis of the full bitcoin transaction graph. In: Sadeghi, A.-R. (ed.) FC 2013. LNCS, vol. 7859, pp. 6–24. Springer, Heidelberg (2013). https://doi.org/10.1007/978-3-642-39884-1_2
11. Rubin, J.: Btcspark: scalable analysis of the bitcoin blockchain using spark (2015). https://rubin.io/public/pdfs/s897report.pdf. Accessed 19 Nov 2023

12. Sedgewick, R., Wayne, K.: Algorithms, pp. 216–233. Addison-Wesley Professional, 4 edition (2011)
13. Robert Endre Tarjan: Efficiency of a good but not linear set union algorithm. J. ACM (JACM) **22**(2), 215–225 (1975)

Analyzing Topic Models: A Tourism Recommender System Perspective

Maryam Kamal[1], Gianfranco Romani[1], Giuseppe Ricciuti[2],
Aris Anagnostopoulos[1] , and Ioannis Chatzigiannakis[1(✉)]

[1] Department of Computer, Control and Management Engineering (DIAG),
Sapienza University of Rome, 00185 Rome, Italy
ichatz@diag.uniroma1.it
[2] Amarena Company Srl, Via Casilina 3T, 00182 Rome, Italy

Abstract. Topic Modeling is a well-known text-mining strategy that detects potential underlying topics for documents. It plays a pivotal role in recommender systems for processing proliferated user-generated content (UGC) for personalized recommendations. Its application presents unique challenges in tourism sector due to the diversity, dynamicity, and multimodality of tourism data. This study presents a comprehensive analysis of selected promising topic models specifically in context of tourism recommender systems. The study conducts experimental evaluation of models' performance on five datasets, and highlights their advantages and unique characteristics based on multiple evaluation parameters. Results reveal no best approach in general, rather optimality of models depend on data characteristics, as thoroughly discussed in this paper. It further discusses open issues for the tourism context-related application of topic models, and future research directions.

Keywords: Topic Modeling · Text Mining · Comparative Analysis · Touristic Experiences

1 Introduction

Over past few years, the technological evolution and increased adoption of web-based platforms have caused sheer expansion in the volumes of user-generated content (UGC) [10,12]. Particularly, for the tourism industry, UCG has become an integral part of all tourism activities. However, the explosion of UGC on web along with broad diversity of content makes it imperative to acquire the interpretation and profiling of content and users. This has necessitated the development of advanced tourism recommender systems primarily relying on tourists' UGC. Processing textual UGC such as tourists' experiences and reviews is crucial for recommender systems, here topic modeling (TM) serves as a pivotal strategy. Topic Modeling links a vast volume of unstructured UGC and the diverse needs for personalized tourism recommendations.

This research is supported by Amarena Company srl.

© The Author(s), under exclusive license to Springer Nature Switzerland AG 2024
L. Barolli (Ed.): AINA 2024, LNDECT 200, pp. 250–262, 2024.
https://doi.org/10.1007/978-3-031-57853-3_21

Topic modeling (TM) is a well-known data mining technique that detects potential latent topics for documents based on semantic relevance of words and documents [11]. It plays multifaceted role in context of recommender systems, such as enhancing the personalization aspect of an RS by identifying users' preferences as topics from their shared experiences. At the same time, it extracts the potential interest topics for users from the sheer volume of UGC, streamlining the recommendations. This has made topic modeling one of the most in-demand techniques in the domain of tourism, where topics and labels are required to associate diverse preferences of tourists to related offerings by the travel business, considering the travelers' reviews and user-generated content.

The application of topic models for tourism-related data is particularly unique and challenging. The aim is to acquire topics considering the underlying sentiments, preferences, experiences, and expectations of tourists. Simultaneously, the diversity in tourists' generated content and the strong co-occurrence of emotion-oriented vocabulary are different from the characteristics of blog posts expressing other opinions. Moreover, in comparison to other types of data such as, in microblog services like Twitter, documents reporting touristic experiences are longer, while on the contrary, they are much shorter when compared to articles found in journals or encyclopedias [15]. Such structural differences in the corpus and the vocabulary have a significant impact on the performance of topic models. It is therefore important to evaluate and analyze how topic models perform on tourism-related data and understand the reasons for such performances. To address this need, our study provides a comprehensive analysis of promising topic models in the context of tourism recommender systems. In particular, the Latent Dirichlet Allocation (LDA) [5], Non-Negative Matrix Factorization (NMF) [13], Top2Vec [2], Bidirectional Encoder Representations from Transformers (BERTopic) [7], RoBERTa [14], Contextualized Topic Model (CTM) [4], and Embedded Topic Model (ETM) [6] are analyzed comprehensively.

2 Background

2.1 Preliminaries

This section briefly provides an understanding of the base principles and mechanisms for each selected topic model.

Latent Dirichlet Allocation (LDA) is a generative probabilistic model for text, using Dirichlet hyperparameters α and β. The goal is to maximize the probability of document corpus D given these hyperparameters, as in Eq. 1.

$$\text{Maximize } P(D|\alpha, \beta) \tag{1}$$

Top2Vec uses word and document embeddings to discover latent semantic structures in text. It automatically determines the number of topics and does not require preprocessing like stopword removal.

Non-Negative Matrix Factorization (NMF decomposes a term-document matrix A into two non-negative matrices W and H, as shown in Eq. 2.

$$A = W \times H \tag{2}$$

It iteratively updates these matrices to extract topics from data.

BERTopic employs BERT embeddings and transformer embeddings, using class-based TF-IDF ($cTF - IDF scoring$) to evaluate term significance in clusters. *RoBERTa* is an optimized version of BERT, focusing on word context for topic prediction.

Contextualized Topic Model (CTM) includes CombinedTM, which combines contextual embeddings with bag-of-words, and ZeroShortTM, which supports multilingual topic modeling.

Embedded Topic Model (ETM) integrates LDA with a variational autoencoder and word embeddings, generating topic proportions for each document.

2.2   Recent Studies

In recent years, multiple studies and researchers have found topic models significantly helpful to tourism-related concerns. Prominently, topic modeling is used to discover preferences in travel itineraries, to study customers' opinions, and to

Table 1. Application of topic models by recent studies in tourism

Studies using TM in Tourism field			
Study	Objectives	Model(s) Used	Evaluation Metrics
Y. Guo et al. (2017) [8]	Tourist satisfaction analysis	LDA	Jaccard coefficient, human analysis and Standford Topic Modelling Toolbox
J. Bao et al. (2017) [3]	Bikesharing	LDA	Perplexity
H. Quab Vu et al. (2019) [16]	Analysis of travel itineraries	LDA	Perplexity, topic concentration
N. Hu et al. (2019) [9]	Customers' complaints	STM	Several analysis on the topics obtained. No specific metric score
Q. Yan et al. (2022) [17]	interaction actors and experience detection	LDA	Content Analysis
N. Zhao. et al. (2023) [18]	Cultural tourism promotion	LDA	Perplexity and Classification

make recommendations. Since our study involves the application of topic models in the context of touristic experiences, we have summarized some recent relevant studies for topic modeling in tourism, in Table 1.

3   Materials and Methods

3.1   Datasets

For experimental evaluations and analysis of topic models, we have used a total of five unique datasets. Three datasets are exclusively designed for this study, namely AirBnB Touristic Experiences (ATE), TripAdvisor Tourist Activities (TAT), and KuriU (KU). While two datasets are publically available, 20News-Group (20NG) and TourPedia (TP). We collected exclusive datasets by web-scraping online posted touristic experiences on leading web-based tourism platforms; AirBnB for ATE and TripAdvisor for TAT. KU is a sampled dataset of the touristic experiences recommender system research project, KuriU, whose module is this study. Note that KU is an Italian language dataset, while we filtered English documents from other datasets. This is to analyze multi-lingual behavior of models. The datasets include touristic experience data for the city of Rome and their statistical summary is mentioned in Table 2.

Table 2. Statistics of the datasets

Dataset	# of Docs	# of Words	Vocabulary Size	Avg. Words Per Doc
ATE	611	111,169	12,661	182
TAT	1860	192,087	13,723	103
KU	5,724	1,556,416	138,095	272
TP	8,000	191,996	27,012	24
20NG	18,846	3,423,145	29,548	182

3.2   Evaluation Parameters

The study evaluates topic models on the following parameters:

Topic Diversity (TD): It measures the distinctiveness of the document clusters produced by the models, using Eq. 3. The value of topic diversity usually ranges between 0 and 1, where a value close to 1 means higher topic diversity while a value closer to 0 means a lower topic diversity. A model is appreciated if it produces higher topic diversity for a given dataset.

$$TD = \frac{n(U)}{K * n(T)} \qquad (3)$$

Here, in Eq. (3), n(U) represents the cardinality of the set of unique words U. K represents the top K words for all topics. T represents the set of topics generated by the model where n(T) is the cardinality of the set T.

Inverted RBO (IRBO): It illustrates to what extent topics differ from each other. It ranges from 0 to 1, where 0 means fully identical and 1 means fully diverse topics. It penalizes topics with common words at different rankings less than topics sharing the same words at the highest ranks.

Topic Coherence: It measures the interpretability and coherence of the topics produced by a model and its association with the considered data. A higher value of topic coherence represents better results of a topic model in terms of producing coherent topics. Let N top words of a topic, $P(w_i, w_j)$ refers to the probability of occurrence of words w_i and w_j together, while $P(w_i)$ and $P(w_j)$ is the probability of occurrence of these words individually, we used the following four types of topic coherences:

– C_{uci} that is calculated by using Eq. 4,

$$C_{uci} = \frac{2}{N(N-1)} \sum_{i=1}^{N-1} \sum_{j=i+1}^{N} \log \frac{P(w_i, w_j) + \epsilon}{P(w_i)P(w_j)} \tag{4}$$

– C_v that is estimated by using by Eq. 5 and Eq. 6,

$$\vec{v}(W') = \left\{ \sum_{w_i \epsilon W'} \left(\frac{\log \frac{P(w_i, w_j) + \epsilon}{P(w_i)P(w_j)}}{-\log(P(w_i, w_j) + \epsilon)} \right)^{\gamma} \right\}_{j=1,\ldots,|W|} \tag{5}$$

$$\Phi_{s_i}(\vec{u}, \vec{w}) = \frac{\sum_{i=1}^{|W|} u_i . w_i}{\|\vec{u}\|_2 . \|\vec{w}\|_2} \tag{6}$$

In Eq. (5) the context vector is $\vec{v}(W')$ and Φ is the confirmation measure.
– C_{umass} calculated by using Eq. 7,

$$C_{umass} = \frac{2}{N(N-1)} \sum_{i=2}^{N} \sum_{j=1}^{i-1} \log \frac{P(w_i, wj) + \epsilon}{P(wj)} \tag{7}$$

– C_{npmi} is an improvisation of the C_{uci} coherence that uses normalized point-wise mutual information (NPMI).

3.3   Experiment and Results

This subsection presents the results and analysis from our experimental evaluations. The implementations are conducted using Python version 3.9.7 on Jupyter Notebook and Google Colab. The coherence evaluation parameters are estimated using Gensim toolkit, while topic diversity measures are estimated using Octis toolkit. Each model is tested with ten iterative runs and the results mentioned in this section are average recorded for each experiment. For the experimentation, we used the default text embedding models for each strategy, which are: *Doc2Vec* for Top2Vec, *roberta-base-nli-stsb-mean-tokens* for RoBERTa, and *all-MiniLM-L6-v2* for BERTopic (English datasets) while *paraphrase-multilingual-MiniLM-L12-v2* for Italian language dataset. Note that we pre-defined the number of topics for LDA, NMF, CTM, and ETM using the elbow method, while Top2Vec, BERTopic, and RoBERTa are modeled to decide the best suitable number of topics by themselves. Note that as per the requirement of some models, LDA, NMF, and ETM were provided pre-processed data, including removal of stop-words and special characters, and lemmatization.

Topic Diversity: An interesting quality determinant explored in this study is topic diversity. A model is well-appreciated if it estimates higher topic diversity with a suitable number of topics. Figure 1 shows the results obtained in this regard, where Fig. 1a illustrates a comparison of the models with respect to average topic diversity (TD) and Fig. 1b shows average Inverted RBO (IRBO) achieved for each dataset. Here Top2Vec shows higher topic diversity on average, for both cases, considering all datasets. An interesting finding is for TP dataset from Fig. 1a, which illustrates a reduced variation of topic diversity among models and BERTopic as the best method. Similarly, it is interesting to observe from Fig. 1b that BERTopic and RoBERTa show much less IRBO when applied to a small-sized dataset with shorter document lengths like ATE. Note that although Top2Vec provides higher topic diversity on average, the number of clusters (topics) it has produced is also considerably less for almost every dataset (Fig. 1a). This might also indicate a high diversity within a topic cluster which is expected to be less for a good topic model.

Topic Coherence: Remark that the higher the coherence score, the better coherent the topics, except for C_{umass}, where a lower value represents better coherence, according to Gensim implementation [1].

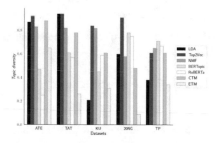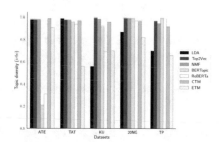
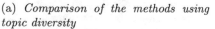

(a) *Comparison of the methods using topic diversity*

(b) *Comparison of the methods using IRBO score*

Fig. 1. Topic modeling evaluation based on Diversity metrics

From Fig. 2a, NMF shows better C_{uci} for relatively smaller-sized datasets as ATE and TAT, but as the size of the datasets grows, ETM starts depicting better results. On average ETM concludes to deliver maximum coherence as compared to the others, in terms of C_{uci}. For C_v from Fig. 2b, while NMF shows better coherence on average for 3 out of 5 datasets, its performance degrades when applied to the largest dataset, 20NG, here Top2Vec exhibits better C_v

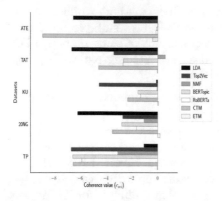

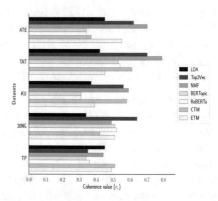

(a) *Comparison of the methods using C_{uci} score*

(b) *Comparison of the methods using C_v score*

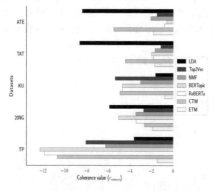

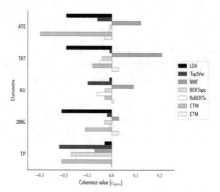

(c) *Comparison of the methods using C_{umass} score*

(d) *Comparison of the methods using C_{npmi} score*

Fig. 2. Topic modeling evaluation based on Coherence metrics

than others. This may imply a sensibility of NMF to the sizes of the datasets, where this model seems suitable for small to medium-sized datasets when considering C_v. From Fig. 2c for C_{umass}, LDA outperforms others on average, while Top2Vec shows better performance for the Italian language dataset KU. Note that although LDA shows better C_{umass} on average, BERTopic outperforms all in the case of the medium-sized English dataset TP. This implies the adoption of LDA for small and large-sized English datasets when considering C_{umass} coherence. Top2Vec might be applied if dealing with multi-lingual medium-sized datasets while BERTopic is suggested for medium-sized English datasets when C_{umass} is concerned. Another interesting shortcoming is from Fig. 2d, where NMF illustrates better C_{npmi} in almost every dataset (except for TP) and on average as a whole. Notice that for TP, ETM outperforms all in terms of C_{npmi}.

Table 3. Comparisons of the results on ATE dataset.

Models	C_{uci}	C_v	C_{umass}	C_{npmi}	Topic Diversity (TD)	IRBO	Number of topics
LDA	−6.56	0.45	**-8.45**	−0.19	0.87	0.98	14
Top2Vec	−3.42	0.62	−1.52	−0.06	**0.92**	0.98	6
NMF	**0.01**	**0.70**	−2.09	**0.12**	0.83	0.98	14
BERTopic	−0.10	0.34	−0.63	−0.01	0.47	0.21	3
RoBERTa	−0.14	0.34	−0.83	−0.01	0.25	0.31	10
CTM	−8.93	0.37	−5.51	−0.30	0.88	**0.99**	14
ETM	−0.40	0.55	−1.85	−0.03	0.65	0.91	14

Table 4. Comparisons of the results on TAT dataset.

Models	C_{uci}	C_v	C_{umass}	C_{npmi}	Topic Diversity (TD)	IRBO	Number of topics
LDA	−6.68	0.42	**-8.68**	−0.19	**0.94**	**0.99**	16
Top2Vec	−3.42	0.70	−1.17	−0.01	**0.94**	0.98	6
NMF	**0.59**	**0.79**	−1.70	**0.21**	0.82	0.98	16
BERTopic	−2.66	0.53	−2.02	−0.04	0.61	0.96	45
RoBERTa	−2.69	0.54	−1.86	−0.05	0.57	0.94	44
CTM	−4.61	0.61	−4.44	−0.08	0.78	0.97	16
ETM	−0.03	0.45	−1.72	0.03	0.26	0.56	16

Table 5. Comparisons of the results on KU dataset.

Models	C_{uci}	C_v	C_{umass}	C_{npmi}	Topic Diversity (TD)	IRBO	Number of topics
LDA	−0.11	0.37	−1.65	−0.01	0.21	0.56	22
Top2Vec	−4.57	0.56	**-5.38**	−0.10	**0.84**	**0.99**	50
NMF	−0.05	**0.59**	−3.03	**0.09**	0.82	0.98	22
BERTopic	−1.53	0.31	−4.72	−0.03	0.45	0.93	75
RoBERTa	−1.32	0.31	−4.72	−0.06	0.59	0.69	14
CTM	−2.34	0.58	−4.94	−0.03	0.61	0.96	22
ETM	**0.06**	0.39	−0.79	0.01	0.31	0.70	22

Also for 20NG, ETM and NMF deliver the same readings. Hence we can state that NMF performs better for small to medium-sized datasets, while ETM performs better for medium to large-sized datasets if C_{npmi} is concerned.

Table 6. Comparisons of the results on TP dataset.

Models	C_{uci}	C_v	C_{umass}	C_{npmi}	Topic Diversity (TD)	IRBO	Number of topics
LDA	−1.06	0.45	−3.62	−0.03	0.38	0.70	14
Top2Vec	−6.72	0.35	−8.10	−0.22	0.61	0.97	41
NMF	−3.10	0.44	−6.27	−0.07	0.65	0.95	14
BERTopic	−6.59	0.34	**-12.35**	−0.17	**0.71**	**0.99**	142
RoBERTa	−5.91	0.36	−11.91	−0.15	0.67	**0.99**	106
CTM	−6.57	**0.51**	−10.70	−0.21	0.61	0.92	14
ETM	**−0.03**	0.49	−1.49	**−0.01**	0.34	0.66	14

Table 7. Comparisons of the results on 20NG dataset.

Models	C_{uci}	C_v	C_{umass}	C_{npmi}	Topic Diversity (TD)	IRBO	Number of topics
LDA	−6.23	0.34	**−5.92**	−0.21	0.60	0.87	111
Top2Vec	−2.72	**0.64**	−2.74	−0.02	**0.91**	**0.99**	83
NMF	−1.05	0.49	−3.46	**0.03**	0.58	**0.99**	111
BERTopic	−2.80	0.51	−5.06	−0.03	0.78	**0.99**	216
RoBERTa	−1.64	0.52	−3.43	−0.01	0.75	0.97	90
CTM	−3.53	0.42	−2.67	−0.11	0.48	0.97	111
ETM	**0.19**	0.51	−1.91	**0.03**	0.09	0.82	111

Considering the c_v as the closest coherence measure to human judgment, we can state that NMF produces more human interpretable topics as compared to others. However, the diverse shortcoming points to insightful implicit findings of the study that the coherence of topic models is significantly influenced by the type and size of the datasets along with the number of topics the model uses. This behavior can be observed in Tables 3, 4, 5, 6 and 7, where results are mentioned in detail. A overall view of evaluations are illustrated in Fig. 3.

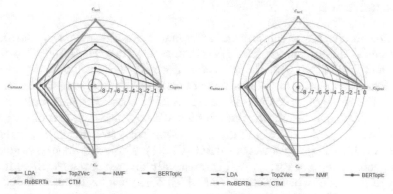

(a) *Comparison of the topic models for ATE dataset*

(b) *Comparison of the topic models for TAT dataset*

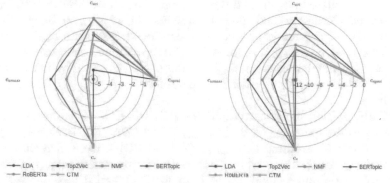

(c) *Comparison of the topic models for KU dataset*

(d) *Comparison of the topic models for TP dataset*

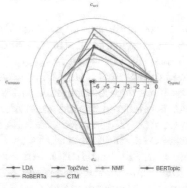

(e) *Comparison of the topic models for 20NG dataset*

Fig. 3. Evaluation of the topic models based on the results on each dataset

4   Conclusion

Our study delineates a comprehensive review of promising novel and devised topic models. These include LDA, NMF, and Top2Vec, BERTopic, RoBERTa, CTM, and ETM. Further, our study presents an in-detail experimental evaluation-based comparative analysis of these models in a touristic experiences context. The analysis is conducted based on topic coherence and topic diversity in terms of multiple significant parameters. We considered four topic coherence parameters: C_{uci}, C_v, C_{umass} and C_{npmi} along with two diversity parameters: Topic Diversity (TD) and Inverted RBO (IRBO). The experimental evaluations are conducted over five variant and contextually diverse datasets where four are related to touristic experiences, out of which three are exclusively designed for the purpose of this study. The study contributes significant conclusive quantitative results and reveals many valuable implicit deductions. The diverse quantitative findings of the study implicitly reveal that there is no conclusive winner among the considered models and the performance and suitability of the models are correlated to the size and type of data. For this reason, we have concluded the suitability of the models as per the mentioned attributes of the datasets. From Table 3, we observed that for ATE, NMF performs better as compared to others for 3 out of 6 parameters, C_{uci}, C_v and C_{npmi}, followed by LDA, Top2Vec and CTM which performed better for 1 parameter each, C_{umass}, TD and IRBO respectively.

Similarly, Table 4 illustrates results for TAT where NMF performs better on 3 out of 6 parameters, C_{uci}, C_v and C_{npmi}. While LDA also shows better performance for 3 out 6 parameters, C_{umass}, TD, and IRBO. Here LDA outperforms others majorly for diversity while NMF outperforms others majorly for coherence. Top2Vec produces equal TD as LDA for TAT and it also delivers better TD for ATE. Hence we conclude that the use of NMF is preferred for small to medium-sized datasets where document length is moderately shorter on average for better coherence, while Top2Vec or LDA delivers better diversity in such cases.

Further, from Table 5, we conclude that on average Top2Vec outperforms others for medium-sized datasets having multi-lingual documents. Since Top2Vec outperforms others for 3 out of 6 parameters, C_{umass}, TD, and IRBO, followed by NMF that outperformed others for C_v and C_{npmi}, we suggest the suitability of Top2Vec for such cases if moderate coherence is preferred along with high diversity. Conversely, NMF is preferred if good coherence is required irrespective of high diversity. Moreover, from Table 7 we conclude that Top2Vec performs better on average for large-sized English datasets as it delivers better results for 3 out of 6 parameters C_v, TD, and IRBO. Although Table 6 reveals that BERTopic outperforms others quantitatively for majority parameters (C_{umass}, TD and IRBO) for medium-sized datasets, however, RoBERTa exhibits considerably better qualitative aspects than BERTopic for such datasets with marginal difference in readings in terms of C_{umass}, TD and IRBO. Hence, we suggest the use of Top2Vec for large-sized English datasets and RoBERTa for medium-sized English datasets. In both cases, ETM may also be used if only the coherence parameter is of concern since it delivers better coherence for both cases in terms of C_{uci} and C_{npmi}.

5   Open Issues and Future Research Directions

The diverse domain of touristic experiences causes heterogeneous issues such as the unavailability of versatile and diverse public datasets, changing tourist preferences, data multimodality, and more. Such current limitations can be interesting possible future directions of this study.

References

1. Alenezi, T., Hirtle, S.: Normalized attraction travel personality representation for improving travel recommender systems. IEEE Access (2022)
2. Angelov, D.: Top2vec: distributed representations of topics. arXiv preprint arXiv:2008.09470 (2020)
3. Bao, J., Xu, C., Liu, P., Wang, W.: Exploring bikesharing travel patterns and trip purposes using smart card data and online point of interests. Netw. Spat. Econ. **17**, 1231–1253 (2017)
4. Bianchi, F., Terragni, S., Hovy, D.: Pre-training is a hot topic: Contextualized document embeddings improve topic coherence (2020)
5. Blei, D.M., Ng, A.Y., Jordan, M.I.: Latent Dirichlet allocation. J. Mach. Learn. Res. **3**(Jan), 993–1022 (2001)
6. Dieng, A.B., Ruiz, F.J., Blei, D.M.: Topic modeling in embedding spaces. Trans. Assoc. Comput. Ling. **8**, 439–453 (2020)
7. Grootendorst, M.: Bertopic: Neural topic modeling with a class-based tf-idf procedure. arXiv preprint arXiv:2203.05794 (2022)
8. Guo, Y., Barnes, S.J., Jia, Q.: Mining meaning from online ratings and reviews: tourist satisfaction analysis using latent Dirichlet allocation. Tour. Manage. **59**, 467–483 (2017)
9. Hu, N., Zhang, T., Gao, B., Bose, I.: What do hotel customers complain about? text analysis using structural topic model. Tour. Manage. **72**, 417–426 (2019)
10. Kamal, M., Chatzigiannakis, I.: Influential factors for tourist profiling for personalized tourism recommendation systems–a compact survey. In: 2021 International Conference on Innovative Computing (ICIC), pp. 1–6. IEEE (2021)
11. Korenčić, D., Ristov, S., Repar, J., Šnajder, J.: A topic coverage approach to evaluation of topic models. IEEE Access **9**, 123280–123312 (2021)
12. Krumm, J., Davies, N., Narayanaswami, C.: User-generated content. IEEE Pervasive Comput. **7**(4), 10–11 (2008)
13. Lee, D., Seung, H.S.: Algorithms for non-negative matrix factorization. In: Advances in neural information processing systems 13 (2000)
14. Liu, Y., et al.: Roberta: a robustly optimized bert pretraining approach. arXiv preprint arXiv:1907.11692 (2019)
15. Lui, M., Lau, J.H., Baldwin, T.: Automatic detection and language identification of multilingual documents. Trans. Assoc. Comput. Ling. **2**, 27–40 (2014)
16. Vu, H.Q., Li, G., Law, R.: Discovering implicit activity preferences in travel itineraries by topic modeling. Tour. Manage. **75**, 435–446 (2019)

17. Yan, Q., Jiang, T., Zhou, S., Zhang, X.: Exploring tourist interaction from user-generated content: topic analysis and content analysis. J. Vacation Mark., 13567667221135196 (2022)
18. Zhao, N., Fan, G., Qi, Z., Shi, J.: Exploring the current situation of cultural tourism scenic spots based on lda model-take Nanjing, Jiangsu province, China as an example. Procedia Comput. Sci. **221**, 826–832 (2023)

Dissecting the Hype: A Study of WallStreetBets' Sentiment and Network Correlation on Financial Markets

Kevin Wang[1], Bill Wong[1], Mohammad Ali Khoshkholghi[2], Purav Shah[2],

Ranesh Naha[3(✉)], Aniket Mahanti[1], and Jong-Kyou Kim[4]

[1] University of Auckland, Auckland, New Zealand
{kwan772,bwon783}@aucklanduni.ac.nz, a.mahanti@auckland.ac.nz
[2] Middlesex University, London, UK
{a.khoshkholghi,p.shah}@mdx.ac.uk
[3] Federation University, Melbourne, Australia
n.naha@federation.edu.au
[4] University of New Brunswick, New Brunswick, Canada
jongkyou.kim@unb.ca

Abstract. The emergence of online investment communities like WallStreetBets (WSB) on Reddit has revolutionised retail trading, characterised by collaborative, meme-infused market influence. This study analyses WSB's ecosystem, examining how social sentiment, network structure, and user interactions impact stock volatility. Our analysis leverages sentiment analysis and network theory on millions of posts to understand community dynamics and their effectiveness in predicting stock prices. This research highlights the rising influence of social media in financial markets, especially with the recent surge in retail investors and their market impact.

1 Introduction

In recent years, the convergence of social media and financial markets has birthed a formidable force, one capable of catalysing seismic shifts within the economic landscape. The subreddit WSB, a community within the Reddit platform, exemplifies the potency of this convergence. In January 2021, a short squeeze of GameStop's stock was driven by the subreddit WSB. This collective of retail investors demonstrated the significant influence social media can exert on financial markets. GameStop's stock, heavily shorted at approximately 140% of its public float, experienced an unprecedented rise from $17.25 to over $500 pre-market value. The stock's volatility persisted post-peak, with remarkable surges and declines, exemplifying the impact of digitally coordinated investment strategies on market dynamics.

Our study delves into the narrative and data-driven aspect of the WSB, particularly highlighting events such as the GameStop short squeeze, providing insights to media outlets, financial institutions, and regulatory bodies alike. This phenomenon catalysed a broader discourse about the power of social platforms in shaping market

© The Author(s), under exclusive license to Springer Nature Switzerland AG 2024
L. Barolli (Ed.): AINA 2024, LNDECT 200, pp. 263–273, 2024.
https://doi.org/10.1007/978-3-031-57853-3_22

behaviour, leading to concern and calls for regulation. The events also sparked rigorous academic interest in understanding the mechanisms behind such collective actions and their influence on finance systems. The emergence of this 'meme stock' movement, where traders/investors rally around certain stocks promoted on social media, has raised profound questions about conventional financial theories, market infrastructure, and the role of public discourse in the valuation of securities.

Our study presents a multi-faceted analysis encompassing sentiment and network analyses derived from a vast collection of Reddit posts and comments. Our research narrows its focus on the WSB subreddit, considering its influence on the stock prices of companies like GameStop (GME) and Tesla (TSLA). By scrutinising the temporal evolution of posting behaviour, user influence, and sentiment trends, we draw correlations that contribute to a nuanced understanding of how social media discourse may reflect or even drive investor sentiment and stock market dynamics. Additionally, we explore the user interactions manifested in the WSB community, revealing the digital Pareto Principle at play and its potential implications for market manipulation.

2   Related Work

2.1   Social Media Characterisation

Wealth and Influence Dynamics in Social Media: Disparities, Algorithmic Biases, and Influence Cascades: Guidi et al. [8] examined Blockchain Online Social Media (BOSM), uncovering a highly polarised environment where wealth and influence are concentrated, highlighting the critical need to address socio-economic disparities in such platforms. Glenski and Weninger's study [7] on Reddit revealed how algorithmic biases influence user behaviour and content propagation, indicating the potential of these insights in driving strategic user engagement during important financial events. Bakshy et al. [2] investigated Twitter's role in information cascades, showing the complexity of influencer dynamics in spreading information and suggesting the need for a diversified approach in leveraging online influence, especially in influencer marketing and during periods of market volatility. These studies collectively shed light on the impact of social media on socio-economic aspects and user engagement behaviours.

In a particularly relevant exploration, Thukral et al. [19] studied the user behaviours and interaction patterns on Reddit, unveiling a series of behavioural phenomena and interaction dynamics. The researchers delved deep into the underlying behavioural patterns and dynamics that shape these interactions. The concept of "Limelight Score" is introduced to quantify the depth of discussion around a single comment, revealing that a significant portion of discussions can be centred around one pivotal comment.

2.2   Social Media and the Financial Market

Social Media's Influence in Financial Markets: Sentiment Analysis and Predictive Insights: Studies reveal the profound impact of social media on financial markets. Analysis of Reddit's r/personalfinance subreddit [20] shows how financial discussions shape user behaviour, with network analysis and topic extraction methods providing insights into collective financial decision-making. Twitter's influence on stock price predictions is illuminated in studies by Jessica and Raymond Sunardi Oetama [13, 18], suggesting the integration of social media sentiment with traditional financial forecasting. Further, research [1] on multi-modal sentiment analysis, including financial video news, indicates the potential for a more comprehensive approach to predicting market fluctuations. The integration of sentiment analysis with algorithmic predictions in market forecasting [9, 16] demonstrates enhanced accuracy in trend prediction. A notable study [17] focuses on Twitter sentiment as a predictive factor for stock market movements, using sentiment analysis and machine learning to correlate public sentiment with stock price changes.

Utilising Reddit Insights for Precise Stock Price Prediction: A highly relevant study [10] ventures into the potential of utilising discussions from WSB to predict stock prices, concentrating on the enhancement of prediction accuracy using techniques of trust filter and a sliding window mechanism. The researchers not only collected sentiment of WSB posts through an individualised approach but also introduced a trust score per record and user, to identify and harness reliable stock-related discussions from the seemingly chaotic pool of user-generated content. In addition to exploring the effects of applying a sliding window approach to aggregate Reddit data, the research acknowledges the prevailing impact of sentiment within financial forums. Employing deep learning research within Natural Language Processing, it adeptly evaluates trends and establishes connections between historical sentiment and financial data.

2.3 Comparison to Prior Studies

In contextual comparison with prior works, our study offers several distinct contributions. While previous research has provided valuable insights into the relationship between social media sentiment and financial market predictions, our investigation digs deeper into the network dynamics of WSB. Unlike prior studies that focus predominantly on sentiment analysis, our research integrates network information, such as user influence and interaction patterns, to fortify the predictive models. This comparative analysis not only positions our study within the existing literature landscape but also accentuates our innovative approach.

Our study reveals novel metrics for understanding the gravitas of network influence within WSB, which, to our knowledge, has not been extensively explored before. It is this synthesis of sentiment analysis and network dynamics that delineates our contribution. We are bridging the gap by correlating the quantifiable measures of user influence with stock price movements, which provides a more granular understanding of the market sentiments as reflected in social media interactions. This work, therefore, not only into the unexplored domain of WSB but also proposes a methodologically robust framework that enhances the accuracy and reliability of stock market predictions. The comprehensive insights derived from our research underscore the role WSB plays in relation to the financial markets.

3   Methodology

This section explains the structured approach employed to examine the correlation between social media sentiment, particularly emanating from the WSB forum, and the price movements of stocks such as GME, TSLA, and S&P 500 (SPY). It explicates the phases of data collection, sentiment analysis, network analysis, and predictive modelling, along with their respective evaluation metrics.

Data Collection: The workflow of data collection involved using the Python Reddit API Wrapper (PRAW), Pushshift API, and The-eye Reddit archive to collect historical posts and comments from the subreddit WSB (Fig. 1a). From this endeavour, around two million posts were collected from WSB. Another two million comments pertinent to TSLA and GME were extracted from The-eye archive, which hosts an exhaustive repository of WSB comments preceding March 2023. The extraction of the two million comments is based on the inclusion of words "$TSLA", "$GME", "Tesla" or "Gamestop" in the body of each comment. Hence any comments mentioning these words will be extracted from the historical WSB comment dataset. This extraction enables a more targeted analysis of stock-specific discussions. To further refine the dataset for subsequent network analysis, we leveraged the relational structure of Reddit's data to fetch the authors of parent posts for each comment. By correlating each comment with its parent post ID, we could accurately attribute comments to their original discussions and extract the associated parent post author, a critical step for the construction of the communication graph detailed later in the methodology.

Financial datasets encompassing daily trading metrics for the stocks TSLA, GME, and SPY were obtained from the Yahoo Finance API, complementing the social media data. Data cleansing and validation comprised the elimination of duplicate records and the dropping of Not-a-number (Nan) records, which together ensured the robustness of the data foundation for the study.

Characterisation Metrics: The sentiment of each post and comment was quantified using FinBERT[6], which exhibited superior accuracy (74%) over VADER [3] (50%) in our preliminary tests on a financial dataset [14]. The Louvain algorithm was applied for community detection within the WSB network [5], and the Google PageRank algorithm assessed the relative importance of each user within the WSB community [15]. Temporal network changes were tracked via metrics such as the number of edges, nodes, and average degrees, providing a dynamic view of community interaction over time.

The study also involves building ML models for stock price predictions leveraging financial attributes, sentiment data and network data. The financial attributes include open, high, low, and close prices and volume of a given stock. The sentiment data is the daily net sentiment score of the given stock in WSB. And the network data is the PageRank importance score of authors in WSB. The models used are Random Forest (RF) [4], Multilayer Perceptron (MLP) [12] and Long-short Term Memory (LSTM) [11]. These models are evaluated and compared using the Root-mean-square Error (RMSE) metric.

Data Overview: The data compiled presented a robust foundation for analysis, summarised in the following table:

Table 1. WSB data overview

Dataset	Number of Posts	Number of Comments	Number of Authors	Average comments per post
WSB (TSLA)	42528	72766	30554	2.4
WSB (GME)	226330	2000847	73974	27.0
WSB (all)	2275310	96317	634201	29.0

The collected WSB dataset is uploaded to Kaggle[1]. Table 1 summarizes data from the WallStreetBets (WSB), covering three distinct datasets: WSB (TSLA), WSB (GME), and WSB (all). The WSB (TSLA) dataset includes 42,528 posts, 72,766 comments from 30,554 authors, averaging 2.4 comments per post. In contrast, the WSB (GME) dataset is larger, with 226,330 posts, 2,000,847 comments from 73,974 authors, and an average of 27.0 comments per post. The most comprehensive dataset, WSB (all), encompasses 2,275,310 posts, 96,317 comments from 634,201 authors, averaging 29.0 comments per post.

4 Results

4.1 Sentiment and Network Insights Mostly Enhance Model Performance

The outstanding pattern recognition ability of ML models led to their domination in stock price/trend prediction. It certainly is a big interest for many researchers and traders to investigate methods and techniques that can further enhance ML model performances due to the financial benefits it could potentially bring. We conducted model evaluations on RF, FNN, and LSTM to assess the impact of sentiment data and network data on improving model performance, we started with a base model that solely utilises financial attributes. Subsequently, we integrated sentiment scores and PageRank importance scores into the model input for further evaluation. The specific configurations of these models are detailed as follows:

- RF exclusively with Financial Attributes (RF-FA)
- RF integrating Financial Attributes and Sentiment Score (RF-FA-SS)
- RF amalgamating Financial Attributes with a PageRank-weighted Sentiment Score (RF-FA-SS-PW)
- FNN focusing solely on Financial Attributes (FNN-FA)
- FNN combining Financial Attributes and Sentiment Score (FNN-FA-SS)
- FNN merging Financial Attributes with a PageRank-weighted Sentiment Score (FNN-FA-SS-PW)
- LSTM centred on Financial Attributes (LSTM-FA)
- LSTM integrating Financial Attributes and Sentiment Score (LSTM-FA-SS)
- LSTM synthesising Financial Attributes with a PageRank-weighted Sentiment Score (LSTM-FA-SS-PW)

---

[1] Kaggle WSB Dataset Link: https://www.kaggle.com/datasets/kevinwang313/wallstreetbets-dataset.

Table 2. RSME of all models for predicting TSLA and GME stock price.

	TSLA	GME		TSLA	GME
RF-FA	13.90	2.22	FNN-FA-SS-PW	10.41	2.37
RF-FA-SS	13.53	2.08	LSTM-FA	11.28	1.98
RF-FA-SS-PW	13.01	2.07	LSTM-FA-SS	10.51	1.67
FNN-FA	11.46	1.96	LSTM-FA-SS-PW	10.39	1.69
FNN-FA-SS	13.33	1.65			

A clear observation from our experiments was the enhancement in model performance upon the inclusion of sentiment data. This is supported by the uniform decline in RMSE values when comparing models equipped solely with Financial Attributes against those equipped with both Financial Attributes and Sentiment Scores. Such consistent improvements imply a correlation between WSB sentiment and stock price dynamics. While our sentiment analysis indicated a lagging sentiment effect on stock price, it showcased minimal evidence of sentiment data leading stock price movements. This prompts us to postulate the existence of a non-linear association between sentiment data and stock price movement, a relationship potentially not captured by Pearson correlation detection. A fascinating discovery from the model evaluation pertains to the distinct impacts of PageRank Weighting (PW) on TSLA and GME predictive modelling. The RSME of all models is shown in Table 2 for predicting TSLA and GME stock price. For TSLA, the incorporation of PW persistently mitigated the RSME across all models with a 9% decrease in RMSE on average. In contrast, all GME models had no statistically significant enhancements by incorporating PW. This discrepancy shows the possible distinguished user behaviours between meme stocks and traditional stocks, inferring a notable variance in network interactions within WSB. The chaotic dynamics of meme stock networks on WSB may lead to more volatile and potentially deceptive sentiment data, introducing additional noise and thereby obstructing model generalisation. Conversely, the sentiment distribution for traditional stocks like TSLA appears more stable and resilient to manipulation, rendering the insights from influential voices invaluable. Our results resonate with multiple studies affirming the efficacy of sentiment data in enhancing model performance [1,9,10,16,17]. While WSB sentiment may not emerge as a singular influential factor upon which retail traders should depend, its incorporation increased the robustness of price-prediction models. Our study delineates itself by unravelling the utility of PW in refining sentiment data for predicting conventional stock prices. Employing PW amplifies the significance of individuals possessing heightened influence and followership. In typical stock discussions, these influential individuals often demonstrate a profound comprehension of stocks and predominantly disseminate quality insights, thereby enhancing the calibre of the sentiment data. However, in the context of meme stock discussions, these influencers occasionally resort to exaggeration to instigate excitement, thereby infusing noise into the sentiment data. Huynh et al. utilised trust filters on WSB data, revealing notable improvements in their findings [10]. Trust filters were employed to exclude data originating from bots and less active users by scrutinising their posts and comments. While this approach shares some similarities with our approach, our methodology transcends the trust filter technique by integrating network information. Rather than discarding data, we retain all senti- ment data but modulate its significance based on a network analysis of all individuals discussing a specific stock.

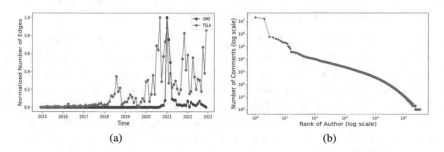

(a)   (b)

Fig. 1. (a) Normalised number of edges over time for TSLA and GME (b) Zipf's distribution showing authors and number of comments

4.2 User Interactions Exhibit Notable Disparities Between

The temporal analysis reveals interesting insights into the WSB networks. It shows how certain metrics change over time, therefore showing the growth and progress of the network which could reveal interesting patterns of the network. Figure 1a shows a comparison between the edge number of GME and TSLA network graphs over time. The number of edges for both the GME network and TSLA network are normalised to highlight the difference in its patterns. An edge in the network graph represents a com- ment which is the primary method of interaction in Reddit. For stocks like GME, these graphs illuminate a significant surge in activity during events like the short squeeze, which then quickly returns to pre short squeeze level. In contrast, stocks such as Tesla demonstrate a steadier and more consistent growth in interactions. Our data shows that meme stocks like GME, are predominantly initiated by a limited number of authors. In comparison, interactions around more conventional stocks are broader and more evenly distributed. There is a noticeable difference between user behaviours of meme stocks and traditional stocks in WSB. Often, individuals or groups will artificially inflate the value of these stocks in what is known as "pump and dump" tactics. They will generate hype around a stock, driving up its price, only to sell their shares once the price peaks, leaving other investors at a loss. When we see an unusually high level of activity or discussion around a particular stock, it could be a warning sign. It might indicate that the stock is not fairly evaluated, and there is a risk that its value could plummet soon after. This is captured by Fig. 1a, where the GME interactions had an unusual surge, and it dropped back to normal levels soon after. Retail investors must be aware of these pat- terns and exercise caution when considering investments in such stocks. This analysis shows the distinct user behaviour of meme stocks and non-meme stocks. Highlighting the potential market manipulations. However, on the other hand, the GME short squeeze does not represent the characteristics of all meme stocks. A more comprehensive study on a wide range of meme stock behaviours would be beneficial in crafting regulations and defence mechanisms for preventing market manipulations.

4.3 The WSB Community Demonstrates Pronounced Susceptibility to Manipulation Through Influential Dynamics

To verify the constant Pareto Principal findings in various studies [2] and to identify dominant actors within the WSB community, network graphical analyses were employed. In this context, the Pareto Principle or the 80–20 rule, oftentimes observed within various systems, is explored alongside the PageRank algorithm to navigate through the intricate web of user interactions and influence within WSB. Our results showed that WSB not only adhered to the Pareto Principle but exhibited an even more skewed interaction distribution, with a mere 20% of user posts captivating a staggering 97.70% of comments, far surpassing the conventional 80–20 distribution. This phenomenon further aligns with Zipf's distribution (Fig. 1b), characterised by a slope of -1.72, signifying an extreme imbalance in the distribution of user interactions and influence. Such an extreme manifestation of the Pareto Principle in WSB implies the susceptibility to market manipulation by a small set of top authors wielding considerable influence over the majority of its user base. The execution of PageRank analysis highlighted that a subset of authors, specifically "bigbear0083" and "OPINION_IS_UNPOPULAR", exert significant influence within the WSB community. These highly influential figures introduce a nuanced complexity and potential for bias. The capacity for influential entities, especially moderators, to amplify or perhaps suppress particular narratives holds the potential to significantly skew the WSB network's dialogues, thereby introducing a heavy bias within the forum. Our results indicate a high potential for manipulation; however, it's crucial to recognise that numerous influential individuals may offer unbiased, truthful information, including valuable financial advice and insightful market analysis. These influencers might possess a strong sense of integrity that deters them from engaging in manipulation.

4.4 Sentiment Pattern Shows Moderate Correlation and Lagging Effect Compared to Price Movements

Our sentiment analysis focused on potential correlations between WSB sentiment data and stock price movements, particularly for enhancing stock price predictions. We analysed daily net sentiment data, summing up sentiment scores from comments related to specific stocks, against daily stock prices. We found a peak Pearson correlation of 0.40 between GME stock price changes and 7-day Exponential Moving Average (EMA) of GME net sentiment (Fig. 2a), during the January 2021 short squeeze. For TSLA, the peak correlation was 0.29, observed between the 7-day EMA of net sentiment and the subsequent day's closing price (Fig. 2b). Both GME and TSLA sentiment indicators showed a lagging trend behind stock price movements, indicating stock prices influenced WSB sentiment rather than vice versa. While recognising the inherently lagging nature of the EMA indicator, even the most up-to-date net sentiment score manifested this lagging tendency. This suggests that shifts in GME's stock price were precursors to the sentiment alterations within WSB posts related to GME, thereby influencing community reactions based on stock performance. The study from [1] also unveiled a lagging correlation between Reddit activity and stock price changes in a one-to-four-day period, lacking any linear relationship between sentiment and stock price changes. The similarity between our results and the study provides more reliability in the validity of the sentiment analysis findings and implications. While the intention was to harness the results of sentiment analysis to enhance stock price prediction, our results show the opposite relationship, where price has better predictive power. Although using WSB sentiment data for stock price prediction may not improve performance, the sentiment analysis predominantly concentrated on linear correlations. Hence, there is no evidence to reject the possibility for the ML models to discover non-linear predictive patterns, which enhances prediction performance.

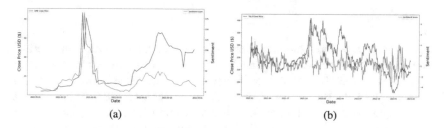

(a)   (b)

Fig. 2. (a) GME Close Price and 7-day EMA of Sentiment (b) TSLA Close Price and 7-day EMA of Sentiment

5 Discussion

The integration of sentiment and network data into machine learning models for stock price prediction, as demonstrated in our study, has clear implications for various stakeholders in the financial markets. The reduced RMSE in models incorporating sentiment data points to the value of trader sentiment, particularly from forums like WSB, in understanding stock dynamics. This is crucial for investors and traders who seek to harness alternative data to gain an edge in the market. For meme stocks like GME, our results indicate that sentiment data may add noise rather than value, potentially due to the stock's volatile nature and the influence of a few key actors. In contrast, the predictive accuracy for more traditional stocks like TSLA improves with the inclusion of sentiment and network insights, highlighting the stability and informational value of the sentiment in these markets. Our study contributes to the literature by elucidating the variable impact of sentiment data on different types of stocks. The use of PageRank-weighted sentiment scores is a novel approach that refines the influence of sentiment based on the credibility and network position of the individuals within the WSB community. This could be particularly beneficial for financial analysts and researchers who require a more granular understanding of market sentiment. Our findings could be of significant interest to regulatory authorities. The notable variations in user engagement, particularly with meme stocks, highlight the need for regulatory measures to deter market manipulation. Although much of our analysis points to a centralised character within the WSB community, it is possible that key influencers are offering impartial and honest financial perspectives. However, incidents like the GME short squeeze and the creation of various meme stocks emphasise the urgency for improved regulation to reduce the susceptibility of WSB to market manipulation. In terms of practical applications, our findings can enhance algorithmic trading strategies by integrating sentiment analysis, especially when using advanced ML models like LSTMs. For the research community, our approach can be replicated or expanded upon with other datasets or within different market conditions to further validate and build on our conclusions.

6 Conclusion

In conclusion, this research has shed light on the network structures, user interactions, and sentiment within the WSB community on Reddit. The network analysis has shown a heavily skewed distribution of user interactions within WSB, where a small subset of influential individuals exert substantial influence over most users. This phenomenon is a clear representation of the Pareto principle within social media platforms, which was also reported in prior studies [8], thus not only supporting our findings but also contributing to a holistic understanding of user interactions across online social platforms. The sentiment analysis indicates a moderate correlation between WSB sentiment data and stock price movements, with a lagging effect observed. This suggests that shifts in stock prices are precursors to changes in sentiment within the WSB community, degrading the predictive power provided by the sentiment data. The integration of insights from both network and sentiment analyses into machine learning models shows enhancement in the predictability of stock price movement, particularly with traditional stocks like Tesla, thereby validating the efficacy of this integration. However, it also brought to light the nuanced and distinct behaviours and volatile sentiment distribution associated with meme stocks like GME, providing a fresh perspective into WSB user behaviour and their influence on market dynamics. The discovery that the proposed techniques amplify the performance of price prediction is significant as it offers tangible advantages to future researchers by enabling more accurate and effective financial forecasting. Overall, these findings have addressed all research objectives by concluding the above results regarding the WSB network, sentiment, and its implications on stock price prediction. The exploration of this study is limited to the sentiment and networks of GME and TSLA within the WSB community due to time and resource constraints. Future research could broaden the scope of this study and conduct a comprehensive network and sentiment analysis across a wider spectrum of stocks within WSB.

References

1. Alzazah, F., Cheng, X., Gao, X.: Predict market fluctuations based on the TSI and the sentiment of financial video news sites via machine learning. In: 2023 15th International Conference on Computer and Automation Engineering (ICCAE), pp. 302–308. IEEE (2023)
2. Bakshy, E., Hofman, J.M., Mason, W.A., Watts, D.J.: Everyone's an influencer: quantifying influence on twitter. In: Proceedings of the Fourth ACM International Conference on Web Search and Data Mining, pp. 65–74 (2011)
3. Berri, A.: Sentimental analysis using vader (2020). https://towardsdatascience.com/sentimental-analysis-using-vader-a3415fef7664
4. Biau, G., Scornet, E.: A random forest guided tour. TEST **25**, 197–227 (2016)
5. De Meo, P., Ferrara, E., Fiumara, G., Provetti, A.: Generalized Louvain method for community detection in large networks. In: 2011 11th International Conference on Intelligent Systems Design and Applications, pp. 88–93. IEEE (2011)
6. Genc, Z.: FinBERT: financial sentiment analysis with BERT (2020). https://medium.com/prosus-ai-tech-blog/finbert-financial-sentiment-analysis-with-bert-b277a3607101
7. Glenski, M., Weninger, T.: Predicting user-interactions on reddit. In: Proceedings of the 2017 IEEE/ACM International Conference on Advances in Social Networks Analysis and Mining 2017, pp. 609–612 (2017)
8. Guidi, B., Michienzi, A., Ricci, L.: Assessment of wealth distribution in blockchain online social media. IEEE Trans. Comput. Soc. Syst. (2022)
9. Guo, X., Li, J.: A novel twitter sentiment analysis model with baseline correlation for financial market prediction with improved efficiency. In: 2019 Sixth International Conference on Social Networks Analysis, Management and Security (SNAMS), pp. 472–477. IEEE (2019)
10. Huynh, D., Audet, G., Alabi, N., Tian, Y.: Stock price prediction leveraging reddit: the role of trust filter and sliding window. In: 2021 IEEE International Conference on Big Data (Big Data), pp. 1054–1060. IEEE (2021)
11. Memory, L.S.T.: Long short-term memory. Neural Comput. **9**(8), 1735–1780 (2010)
12. Noriega, L.: Multilayer perceptron tutorial. School of Computing. Staffordshire University, vol. 4, issue 5, p. 444 (2005)
13. Oetama, R.S., et al.: Sentiment analysis on official news accounts of twitter media in predicting Facebook stock. In: 2019 5th International Conference on New Media Studies (CONMEDIA), pp. 74–79. IEEE (2019)
14. Patil, R.: Financial news sentiment analysis using finBERT (2023). https://github.com/Raviraj2000/Financial-News-Sentiment-Analysis-using-FinBERT
15. Rogers, I.: The google pagerank algorithm and how it works (2002)
16. Sandeep, U., Vardhan, T.V., Yaswanth, B., et al.: Cracking the code: unleashing the power of sentiment analysis & ml for Moroccan stock market forecasting. In: 2023 8th International Conference on Communication and Electronics Systems (ICCES), pp. 1274–1278. IEEE (2023)
17. Sasank Pagolu, V., Nayan Reddy Challa, K., Panda, G., Majhi, B.: Sentiment analysis of twitter data for predicting stock market movements. arXiv e-prints arXiv:1610.09225 (2016)
18. Shah, D., Isah, H., Zulkernine, F.: Predicting the effects of news sentiments on the stock market. In: 2018 IEEE International Conference on Big Data (Big Data), pp. 4705–4708. IEEE (2018)
19. Thukral, S., et al.: Analyzing behavioral trends in community driven discussion platforms like reddit. In: 2018 IEEE/ACM International Conference on Advances in Social Networks Analysis and Mining (ASONAM), pp. 662–669. IEEE (2018)
20. Thukral, S., et al.: Understanding how social discussion platforms like reddit are influencing financial behavior. In: 2022 IEEE/WIC/ACM International Joint Conference on Web Intelligence and Intelligent Agent Technology (WI-IAT), pp. 612–619. IEEE (2022)

Transformative Intelligence in Data Analysis and Knowledge Exploration

Lidia Ogiela[1]($\boxtimes$), Makoto Takizawa[2], and Urszula Ogiela[1]

[1] AGH University of Krakow, 30 Mickiewicza Ave, 30-059 Kraków, Poland
{logiela,ogiela}@agh.edu.pl
[2] Research Center for Computing and Multimedia Studies, Hosei University,
3-7-2, Kajino-Cho, Koganei-Shi, Tokyo 184-8584, Japan
makoto.takizawa@computer.org

Abstract. Transformative Computing methods can be used to analyze and fuse sensor data for effective interpretation and classification. This work will describe new solutions based on Transformative Intelligence, which will allow for the semantic classification of analyzed data, their semantic categorization, as well as the extraction of knowledge from data, from various sources. This approach can significantly improve Big Data processing techniques and resources distributed in the cloud.

Keywords: Transformative intelligence · knowledge extraction · cognitive data analysis · information fusion

1 Introduction

The processes of data analysis nowadays are oriented not only to the description of the analyzed information sets, but primarily to the processes of interpretation, and understanding of the described data sets and the possibilities of semantic inference. The processes of understanding data have been widely described in the works of [1–5], and are based on semantic analysis based on the occurrence of the phenomenon of cognitive resonance.

On the basis of cognitive analysis, semantic information dedicated to the meaningful description of data is extracted in the process of automatic computer understanding [6–9]. Thus, this is an example of the implementation of the process of understanding data on the basis of the applied algorithms of meaningful interpretation and description of analyzed sets of information.

Semantic analysis makes it possible to extract the semantic information contained in a specific data set in order to evaluate, on its basis, the role that the semantic content plays in the process of evaluating the interpreted sets. The semantic content of data, also referred to as semantic decks, makes it possible to unambiguously interpret the analyzed sets due to its uniqueness. The unambiguity of the assignment of semantic information to a particular data set, is its unique feature and allows describing the data in an unambiguous way that distinguishes it from other sets.

L. Barolli (Ed.): AINA 2024, LNDECT 200, pp. 274–281, 2024.
https://doi.org/10.1007/978-3-031-57853-3_23

Semantics, therefore, allows for an unambiguous description and interpretation of data indicating the individual and characteristic features of the analyzed information sets.

Semantic analysis is based on the occurrence of the phenomenon of cognitive resonance, during which convergences are studied between the features that are characteristic of the analyzed data sets and certain expectations of these data, which the system generates automatically on the basis of the acquired knowledge.

Thus, it is a process that, on the one hand, serves to unambiguously and uniquely describe the data, on the basis of such features that are typical and characteristic only of these data sets.

On the other hand, it is used to indicate the optimal solution that converges to the greatest extent with the knowledge it has. This convergence is defined as the degree of similarity to that information which has been introduced in the system, or which the system has acquired, for example, in the process of learning based on AI algorithms.

The processes of teaching the system new solutions based on the AI algorithms used, allow even more complete matching of the analyzed data with the knowledge possessed by the system in this regard.

By attempting to compare solutions from both sides, it is possible to obtain a correspondence, or lack thereof, between the set of characteristics of the analyzed data sets and the knowledge that the system has in this regard. If the comparison process ends positively, that is, there will be a correspondence between the set of characteristics that uniquely describe the analyzed data and the system's expectations of the interpreted sets, then the system will correctly assign the analyzed data its description. In such a case, the phenomenon of cognitive resonance will occur, which will indicate full convergence between the analyzed data and the knowledge generated by the system in terms of its correct interpretation.

Otherwise, the system will not associate the knowledge it has with the features that have been indicated as characteristic of the analyzed datasets, due to the lack of knowledge regarding the described datasets. In such a case, the cognitive systems in which meaningful analysis is carried out will attempt to learn new solutions.

At this stage, the system uses AI algorithms to acquire knowledge regarding the analyzed sets, which influences the addition of the knowledge base that the system uses during the process of generating expectations from the analyzed data sets. This process can be repeated many times in case of a mismatch between a set of characteristics and the system's expectations.

However, in the case of repetition, a verification of the correctness of the description of the set of characteristics is ordered, due to the fact that the description may be incomplete, which may result in the inability of the system to match the correct pattern. If the verification is positive, the system will proceed with another attempt to learn a new solution until sufficient knowledge is acquired to enable the system to correctly match the knowledge to the pattern.

Thus, the cognitive analysis process can end with the first comparison attempt, or it can be repeated several times until the correct solution is identified.

The process of meaning analysis is shown in Fig. 1.

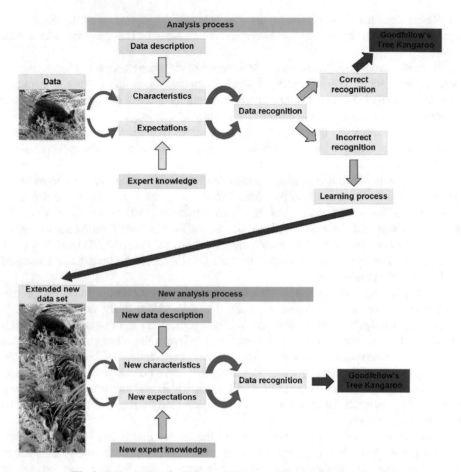

Fig.1. Meaning analysis in cognitive systems (own development).

Thus, the essence of semantic analysis is the realization of an automatic process of analyzing data, which allows its full semantic interpretation with a complete description of the meaning contained in the analyzed sets. It therefore has a special feature that distinguishes it from classical analysis processes. It is a description of meaning that also allows one to infer the causes of the current state as well as to predict changes in the data in the future. This feature distinguishes cognitive systems from other systems performing data analysis tasks.

2   Knowledge Extraction in Data Analysis

Knowledge extraction processes implemented during data analysis are based on the extraction of semantic knowledge. This knowledge is located in the analyzed sets and is often deeply embedded, and its extraction requires specialized knowledge regarding the analyzed data. Its description is possible on the basis of such features that will uniquely identify the analyzed data sets. It is therefore an example of deep knowledge possessed by experts (in the human world) and by machines (in the computer world).

The computer systems that are used to carry out the described tasks are cognitive systems that have expert knowledge extracted from a human – an expert in a specific field, creating a knowledge base, or from a machine – AI at the stage of teaching the system new solutions.

Thus, knowledge extraction processes are used to extract as much information as possible from the analyzed data sets, which will allow them to be unambiguously described and classified and distinguished from other sets. They should therefore provide answers to the question: *What distinguishes this data from others? Which of their characteristics are typical only for them?*

Knowledge extraction is the step of describing data based on its deep semantics. Semantic knowledge deposits can describe data sets both at a basic level (shallow), where semantic knowledge will be used to describe commonly known features, and at a specialized (expert) level, where semantic knowledge will identify those features that are known to a small number of specialists, and which will unambiguously allow the data to be properly classified using the indicated features.

The process of knowledge extraction is shown in Fig. 2.

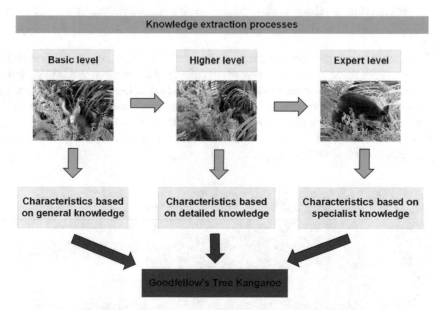

Fig. 2. Knowledge extraction process (own development).

Thus, the process of knowledge extraction is a contributor to the implementation of the actual process of meaningful analysis and the automatic attempt to understand them.

Knowledge extraction as one of the stages of meaningful data analysis allows for insightful interpretation of data and cognitive inference that enables deep understanding of data, that is, that allows predicting possible changes in the area of analyzed data.

3　Transformative Computing in Data Analysis Processes

Knowledge extraction processes can apply to different sets of analyzed data. Their advantage is that they represent an independent process of data analysis, which does not need to be implemented, and then only classical data analysis will take place, topped by the stage of its evaluation and description.

However, when knowledge extraction is applied, there will then be the possibility of realizing meaningful data analysis, which will culminate in the evaluation of the semantic content of the analyzed information. This process can be implemented for a variety of data sets, extracted from a variety of sources.

A special case of the data acquisition process is the possibility of recording data by a variety of sensors, which, recording the data each independently, allow their joint analysis. The realization of this process is served by transformative calculations, allowing to conduct analysis for different data, their diverse combinations – depending on the needs, and in real time. Transformative computing allows the use of various sensors that capture data, and then the data is processed in terms of the information needed. Thus, not all acquired data is analyzed, but only those that are relevant to the analysis process. The selection of relevant data is made possible through the use of algorithms of meaningful description of data, which already at the initial stage of the entire analysis process, allow the proper selection of relevant information. The meaningful description is intended to support the process of data extraction from the entire set of information collected by the sensors.

Then, using transformative computing on the analyzed dataset, it is possible to realize the process of knowledge extraction based on their deep description and interpretation. Thus, knowledge extraction enriches the implementation of transformative computing, in terms of the ability of cognitive systems to realize deep descriptions of data on the basis of specialized knowledge extracted from various sources [10–13].

A schematic view of the application of transformative computing in deep data analysis processes is shown in Fig. 3.

Fig. 3. Transformative computing in data analysis processes (own development).

4 Transformative Intelligence Based on Deep Knowledge Extraction

Transformative computing allows for insightful analysis of data acquired from different sources and recorded by different media. Thus, they are an amalgamation of the various data acquisition processes carried out by the various recorders, based on which the acquired data, can be freely processed and analyzed, and then based on their knowledge can be interpreted and semantically evaluated.

A modified solution based on transformative computing is transformative intelligence. Transformative intelligence represents a new approach to even deeper data analysis than previous solutions. This is because the field of transformative intelligence includes solutions developed on the basis of AI and cognitive systems, which bring new capabilities to transformative computing. These capabilities are based on the use of cognitive systems that implement semantic inference processes. Thus, this is a novel approach to insightful meaning analysis implemented at multiple levels.

Definition 1. *Transformative intelligence will therefore be understood as a solution based on the use of transformative computing in cognitive systems, enabling the implementation of semantic inference algorithms against data acquired through their fusion processes.*

This means that transformative intelligence is a solution that is dedicated to the implementation of semantic inference processes. Semantic inference is the most meaningful stage of data analysis, which indicates a full understanding of the analyzed data while recognizing that the process is entirely based on the knowledge possessed. Information systems that can carry out semantic inference processes are cognitive systems. Thus, the existing role of systems of meaningful interpretation of data is enriched with new capabilities that result from the implementation of semantic inference processes.

Transformative intelligence is based on transformative computing, which means that the data acquired for analysis comes from different sources and is acquired by different sensors. Transformative computing allows their unrestricted analysis in terms of their selection and selection of the information that is required at a given stage of analysis. Transformative intelligence, on the other hand, allows free selection of data and its multilevel analysis. This means that data can be processed at different levels, both at the level of the structure involved in the analyzed dataset and at the level of fog or cloud computing.

Thus, transformative intelligence applied to cognitive systems allows for the realization of semantic inference in large data sets acquired from their fusion.

Data fusion results from the cognitive system's processing of information acquired by various sensors. These data are described using linguistic algorithms applied to the processes of meaningful data analysis. At the same time, this analysis allows deep extraction of the knowledge contained in the analyzed sets.

Thus, transformative intelligence is a new direction in the development of data analysis issues, enabling the implementation of semantic inference steps on the basis of insightful knowledge extraction from specialized levels.

A schematic view of transformative intelligence processes is shown in Fig. 4.

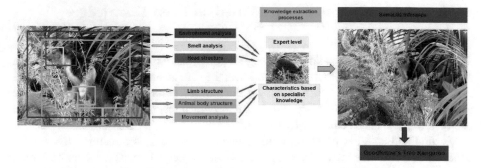

Fig. 4. Transformative intelligence – a schematic approach (own development).

Transformative intelligence can have a variety of applications, including such as:

- strategic data analysis,
- analysis of the economic and financial situation of enterprises, companies, organizations and business entities,
- analysis of diagnostic and therapeutic processes of patients,
- analysis of logistics and transport processes of shipping companies,
- analysis of defense and security processes of any entities,
- analysis of cyber security techniques,
- analysis of rock and geological structures,
- analysis of weather and climate change,
- analysis of air and water pollution,
- analysis of fuel states and fuel reserves,
- analysis of decision-making and optimization processes, etc.

The variety of application areas of transformative intelligence is due to the universal algorithms used by cognitive systems for data interpretation and semantic inference and knowledge extraction. These algorithms can be used in the processes of analyzing different types of data due to their ability to match the interpreted information.

5   Conclusions

Transformative intelligence is currently setting new directions in the development of data analysis issues. It is a solution that provides insightful, semantic interpretation of large data sets, and at the same time serves the tasks of inference. Inference processes are carried out on the basis of specialized knowledge, which is acquired by cognitive systems during the implementation of meaningful data analysis processes.

Transformative intelligence is therefore based on the use of transformative computing in cognitive systems, which allows the full implementation of semantic inference algorithms for information that is recorded during the combination of processes – data fusion – at the stage of the work of sensors that record the collected data.

Acknowledgments. The research project was supported by the program "Excellence initiative – research university" for the AGH University of Krakow.

References

1. Ogiela, L.: Cognitive systems for medical pattern understanding and diagnosis. In: Lovrek, I., Howlett, R.J., Jain, L.C. (eds.) KES 2008. LNCS (LNAI), vol. 5177, pp. 394–400. Springer, Heidelberg (2008). https://doi.org/10.1007/978-3-540-85563-7_51
2. Ogiela, L.: Cognitive informatics in automatic pattern understanding and cognitive information systems. In: 7th IEEE International Conference on Cognitive Informatics (ICCI 2008), Advances in Cognitive Informatics and Cognitive Computing, Studies in Computational Intelligence, vol. 323, pp. 209–226. Springer-Verlag Berlin Heidelberg (2010). https://doi.org/10.1007/978-3-642-16083-7_11
3. Ogiela, L.: Cognitive informatics in image semantics description, identification and automatic pattern understanding. Neurocomputing **122**, 58–69 (2013)
4. Ogiela, L.: Towards cognitive economy. Soft. Comput. **18**(9), 1675–1683 (2014)
5. Ogiela, L., Ogiela, M.R., Ko, H.: Intelligent data management and security in Cloud Computing. Sensors **20**(12), 3458 (2020). https://doi.org/10.3390/s20123458
6. Albus, J.S., Meystel, A.M.: Engineering of Mind – An Introduction to the Science of Intelligent Systems, a Wiley-Interscience Publication John Wiley & Sons Inc (2001)
7. Branquinho, J. (ed.): The Foundations of Cognitive Science. Clarendon Press, Oxford (2001)
8. Perconti, P., Plebe, A.: Deep learning and cognitive science. Cognition **203**, 104365 (2020). https://doi.org/10.1016/j.cognition.2020.104365
9. Zhang, H., Liu, F., Li, B., Zhang, L., Zhu, Y., Wang, Z.: Deep discriminative image feature learning for cross-modal semantics understanding. Knowl.-Based Syst. **216**, 106812 (2021). https://doi.org/10.1016/j.knosys.2021.106812
10. Ogiela, M.R., Ogiela, U.: Secure Information Splitting Using Grammar Schemes, Studies in Computational Intelligence 244, pp. 327–336, Springer-Verlag, Berlin-Heidelberg (2009). https://doi.org/10.1007/978-3-642-03958-4_28
11. Benlian, A., Kettinger, W.J., Sunyaev, A., Winkler, T.J.: The transformative value of cloud computing: a decoupling, platformization, and recombination theoretical framework. J. Manag. Inf. Syst. **35**, 719–739 (2018)
12. Gil, S., et al.: Transformative effects of IoT blockchain and artificial intelligence on cloud computing: evolution, vision, trends and open challenges. Internet Things **8**, 100118 (2019)
13. Nakamura, S., Ogiela, L., Enokido, T., Takizawa, M.: Flexible synchronization protocol to prevent illegal information flow in peer-to-peer publish/subscribe systems. In: Barolli, L., Terzo, O. (Eds.), Complex, Intelligent, and Software Intensive Systems, Advances in Intelligent Systems and Computing, vol. 611, pp. 82–93 (2018)

Enhancing Effectiveness and Efficiency of Customers Reviews Data Collection Through Multithreaded Web Scraping Approach

Dahlan Nariman[✉]

College of Sustainability and Tourism, Ritsumeikan Asia Pacific University (APU),
1-1 Jyumonjibaru Beppu-Shi, Oita, Japan
dahlan@apu.ac.jp

Abstract. In recent years, with the spread of the Internet and online applications, various changes have occurred in the tourism business and industries. One of them is the accommodation reservation site. These reservation sites allow users to leave evaluations and reviews of accommodation facilities. Those reviews have a great influence on the preferences of other users and the impact of the reviews become even stronger in the future. This review presents opportunities and challenges for tourism industry and consumer behavior researchers. Reviews provide an opportunity to understand consumer choices, preferences, and behavior. Also, the opportunity to promote products and services based on reviews. Hotels can also use this trend very seriously in their decision making. However, these reviews need to be collected, managed, and analyzed in an effective and efficient manner, and data collection from online platforms is one of the major challenges. We are developing a tool to collect hotels costumers' reviews called MULARS (Multi-languages Reviews Scrapers) with the aim of efficiently scraping multilingual online reviews. In our previous research, we made it possible to scrape multilingual data [1]. However, faster speed and more effective system of scraping is required for scraping large-scale data. Therefore, we are looking for ways to enhance the performance and effectiveness of the scraping process. In this research, we propose data scraping using multithreading. We implement a multithreaded scraping system and test its performance to demonstrate its improved speed and efficiency. This paper describes the proposed system, and the results of experiment of data collection performances will be presented. Finally, the paper discusses the improvement by adopting the multi-threading, the tool advantages, and future possibilities of the tool applications on the different areas.

1 Introduction

The rapid development of smartphone technology and social network services (SNS) has triggered an explosive increase in data volume, which continues to grow exponentially over time. This recent trend has ushered us in the era of big data [2, 3]. In the hospitality industry, Information technology has changed the face of the hospitality industry. The Internet also provides a way for travelers to express their opinions and recommendations. One of them is an accommodation reservation site where you can post ratings and reviews

L. Barolli (Ed.): AINA 2024, LNDECT 200, pp. 282–291, 2024.
https://doi.org/10.1007/978-3-031-57853-3_24

of accommodation facilities. Those reviews have a huge influence on the preferences of other users, and that influence will continue to grow. Studies show that the Internet in contrast to offline, became the main channel for travel reservations [4]. Therefore, this is important to collect and analyze the customers behaviors by utilizing the reviews. Hotel reviews have many benefits and opportunities such as understand the hotel's strengths and weaknesses and differentiate yourself from your competitors, potential to increase customer satisfaction and gain repeat customers, can improve brand image, even negative reviews can be turned into a positive image by responding appropriately. Hotels side can use guest ratings, complaints and requests written in reviews to improve operation and services. Hotel's managers can use trends characteristics of the reviews in their decision making. However, these reviews need to be collected, managed, and analyzed in an effective and efficient manner, and data collection from online platforms is one of the major challenges. In the tourism and hospitality sector in some areas like in Japan, the use of big data such as reviews on online platforms has not yet progressed. This is not yet ingrained and relies on the experience and intuition of local tourism stakeholders [5]. This is also caused by a lack of funding and human resources [6]. Therefore, it is important to create a model for the use of online reviews and demonstrate not only the analysis but also the impact and benefits for local tourism actors. It is also important to provide a tool which give high impact for local tourism actors such as reducing cost and time consuming and improving their business performances.

As a model of hotels reviews big data utilization, we are developing MULARS with the aim of efficiently scraping multilingual online reviews [1]. The Scraping is the process of creating a computer program to download, parse, and organize data from the web in an automated manner [7]. The MULARS have main features such as ability to collect data from TripAdvisor reviews and store multilingual reviews in a separate database with minimal computer operations, just enter the URL and click with one click. The tool also has possibility to analyze the reviews and others collected data [1]. Basically, collecting data using web scraping is a program who visits the main web page, and then visits the next web page linked to the main web page and so on. All accessed web pages are analyzed for their contents to be matched with the required query to collect the information contained in the web page. To do this, they have to download data from web pages and scrape it, which presents certain challenges. A web scraper should be able to visit the maximum number of pages, in minimum time and provide results that are relevant to the desired web content and web components.

In the current MULARS systems, a single thread is used to scrape data sequentially. The single thread has only one processing system, and all processes are performed one by one in sequence. For example, when scraping hotel reviews, each page of a hotel is scraped with reviews in turn. Once you have finished scraping reviews for that hotel, scrape reviews for the next hotel in the same order as the previous hotel. Since scraping could only be performed by a single process, the scraping latency was long. Since it is a linear relationship, the longer the number of hotel reviews and the number of hotels, the longer the wait time for scraping becomes. Faster speed and more effective system of scraping is required for scraping large-scale data.

2 Related Works

Web scraping systems have also been shown to be able to extract information from social networks [8]. A common and central component found in scraping systems is the crawl. Crawling can be defined as the process of automatically exploring and navigating a website by following hyperlinks [9]. The component or system responsible for crawling is commonly called a web crawler. However, there are several different categories of crawlers.

First type is a focused web crawler which is the process of finding pages that are related to some specific topics or satisfy some particular property. The focused web crawlers are also called as vertical web crawlers or specific web crawlers [10]. It collects the documents that are focused and relevant to a given topic. The focused web crawler determines the relevance of the document before crawling the page. It estimates if the given page is relevant to a particular topic and how to proceeds. The main advantage of this kind of crawler is that it requires less hardware resources. The focused crawlers use internet bandwidth and time efficiently, to dynamically browse the web by selecting the most advantageous links and retrieving the maximum number of relevant pages.

Second type of crawler is a deep web crawler [11]. Deep web or hidden web is part of World Wide Web that search engines cannot or will not index. Deep web consists of proprietary sites, sites with scripts, dynamic sites, ephemeral sites, sites blocked by search engine policy, sites with special format etc. The deep web crawler starts from a seed URL, identifies hyperlinks in the web page obtained from the URL, and recursively visits the hyperlinks to retrieve a new web page. The system does not consider the relationships between web pages and typically uses a breadth-first crawling strategy to retrieve web pages. In contrast to deep web crawlers, focused crawlers only crawl web pages that satisfy certain properties or topics. This type of crawler also able to performs site-based searching and its ranking of relevant sites [11].

Third type of crawler is an incremental crawler [12] which prioritizes and revisits URLs. This is concerned with maintaining the freshness of crawled content on web pages where new content may be modified. It adopts a new mechanism to update the data based on the previous results, it marks the existing collection. To have an updated content of downloaded web pages, an incremental web crawler links the review of previously downloaded pages to the first visit to new pages. The goal of the incremental deep web crawling is to select appropriate queries to retrieve as many incremental records as possible. The incremental crawling can be used in many scenarios, such as data source has not been crawled before, a record store instance does not contain at least one record generation, and seeds URLs have been removed from the crawl configuration. The advantage of incremental web crawlers is that only valuable data is provided to the user, thus saving network bandwidth and achieving data enrichment.

Fourth type is a distributed crawler. Since it is difficult to capture all Internet data in a single crawler process, the crawler process must be parallelized to complete in the shortest possible time. The distributed crawler is typically applied to multiple independent computers working on a single task. The distributed crawler is able to improve the performance of regular crawling systems by using a distributed architecture [13]. A distributed crawler can use one or more other classes of crawlers. The main goal of the distributed crawler design is to maximize performance and minimize time consumption.

It can make full use of the computing resources and improve the efficiency of the crawling system effectively [14]. Distributed crawlers are also an active area of research to improve the performance of crawler systems due to the huge amount of data that needs to be processed [9, 15].

3   Proposed System Architecture

We show the architecture of Multi-threaded Proposed System in Fig. 1, which is implemented in Python. We consider Beautiful Soup and Request module for HTML analysis and data collection from the website. In order to organize and manage data frame is used Pandas module. While, Openpyxl module is used to enter the collected data into an excel file and arrange the data to the specified format.

Basically, collecting reviews with the MULARS use a program to visits the master web page, and then visits the next web page linked to the main web page and so on. All accessed web pages are analyzed for their contents to be matched with the required query to collect the information contained in the web page. To do this, the system have to download data from web pages and scrape it, which presents certain challenges. Data collection using MULARS should be able to visit the maximum number of pages, in minimum time and also provide results that are relevant to the desired web content and web components. In the tool we previously developed, we used a single thread. This tool is able to collect data accurately, but the collection speed needs to be improved to enhance performance and time efficiency. To overcome this, data collection process may need to run in parallel, through multi-threading and with each data collection process running simultaneously to realize more faster and massive data collection. It is also necessary to ensure that in this way the collected data remains relevant to the query being processed. This will enable the creation of web scrapers for data collection with more optimal performance and accurate results.

In the single thread system, it has only one processing system, and all processes are performed one by one in sequence. Two or more processes cannot be executed at the same time. On the other hand, with multithreading, other threads branch off from the main thread and perform multiple processes in parallel as necessary. This branched thread is also called a sub-thread. In this way, multi-threading has the advantage of having multiple processing systems, and can be expected reduce the load on the main thread and processing speed faster than single-threading. Adopting multithreading may has the following advantages, such as improved processing speed and accuracy, improving application responsiveness, minimal system resource usage, simplify program structure, and enhancement of data communication in scraping process. This increases processing efficiency. Multi-threading can also be useful in the many situations such as performs scraping processing in the background, exploiting CPU wait time for slow devices, and make efficient use of multiprocessor environments. This increases the processing speed of web scraping, making it possible to collect large amounts of data in a short period of time. Multithreading is especially effective when working with large amounts of data or complex web scraping operations.

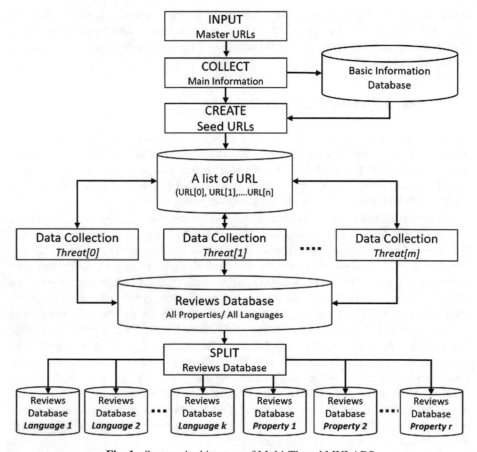

Fig. 1. System Architecture of Multi-Thread MULARS

4   Demonstration of System Performance Testing

The program is executed on windows Surface notebook PC with 8GB RAM memory, Intel(R) Core (TM) i5-1035G7 CPU @ 1.20GHz, 1498 MHz, 4 quad core and 8 logical processor. We collected the customers reviews of several 5 stars hotels in Bali Island, Indonesia. The list of hotels and the number of collected reviews are shown in Table 1. For performance testing, we collected a total of 8006 reviews from 7073 hotels.

Figure 2 shows the PC performance during data collection testing. Approximately 66% of memory and 58% of CPU is used while the system is running. The data received from web approximately is 11.7 Mbps during the data collection. Figure 3 shows example of collected reviews and collected hotel responses on excel formatted files, respectively. The collected data in the review's datasets are containing the reviewer's name, date of stay, review title of the reviews and the contents of the text reviews. In the hotel responses datasets, there are property/hotel name, date of publish, title and content of text of the responses. From the case of this data collection testing of some 5 starts hotels reviews in Bali Island, the MULARS is very effective because it is able to collect mostly all of the customers reviews and hotel side responses.

Table 2 shows data recorded in the performance tests. There are 5 experiments of the performance test conducted between 3rd to 5th of September 2023. The tests are performed on same Windows PC described above. Number of threads changed from single thread to 2, 3, 4, 5 and the maximum threads was 18. The performance test show that the minimum time required the data collection are when number of threads between 10 and 13. Averagely 13 threads are optimum number of the system performance running on the PC described above. Figure 4 shows visualization of the system performance depend on the number of threads. The performance of the data collection is a close to a logarithmically decreasing. The performance was drastically enhanced when the number of threads increased from 1 to 2, 3, 4 and 5. The performance will be slightly improved when the number of threads was 6 to 10. However, the performance was mostly same when the number of threads more than 13.

Table 1. Data collection of hotels reviews used for experiment

Hotel Name	English Reviews	Collected Reviews	Collected Responses from Hotels
The Apurva Kempinski Bali	828	826	739
The Mulia	1996	1874	1475
The St. Regis Bali Resort	1847	1813	1659
Mulia Resort	2885	2593	899
Total	8493	8006	7073

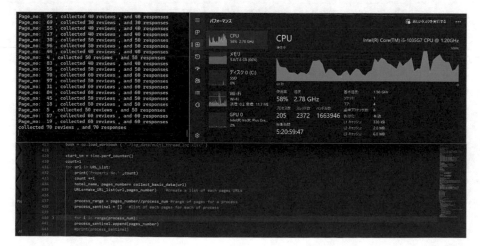

Fig. 2. Program execution preview and display CPU performance

Fig. 3. Example of collected reviews on excel formatted file

Table 2. Data recorded in the performance tests (Number of threads and time required for data collection)

Threads number	Experiment 1	Experiment 2	Experiment 3	Experiment 4	Experiment 5	Average (sec)
1	734.8750	695.9740	724.2122	944.3824	702.1263	760.3140
2	385.9066	383.6494	466.5637	575.9542	374.9049	437.3957
4	235.1432	262.1096	236.8364	300.2783	218.4101	250.5555
6	169.6887	188.5745	176.7358	219.8736	141.1294	179.2004
8	157.3062	176.5373	156.6163	172.0189	116.6956	155.8349
10	149.9628	172.9894	137.2897	167.3950	106.5771	146.8428
11	144.7232	140.6972	121.6185	160.9336	108.8133	135.3572
12	137.1413	134.2763	145.0731	151.7088	108.5831	135.3565
13	120.2206	162.6041	133.4608	128.4090	107.4658	130.4321
14	121.8990	158.6673	132.2702	137.7070	119.5523	134.0192
15	127.1503	151.5330	141.1532	142.1017	133.6477	139.1172
16	150.3905	148.2336	138.0036	125.1602	137.6838	139.8943
18	152.0680	158.4454	138.0875	134.2882	142.3681	145.0515
Max	734.8750	695.9740	724.2122	944.3824	702.1263	760.3140
Min	120.2206	134.2763	121.6185	125.1602	106.5771	130.4321
Enhanced efficiency	611.3%	518.3%	595.5%	754.5%	658.8%	582.9%
Fastest number of threads	13	12	11	16	10	13

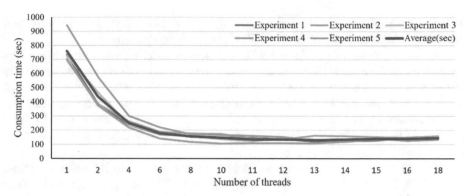

Fig. 4. Relation between number of thread(s) and its performances

5   Advantages and Future Possibilities

Multi-thread adoption in the MULARS give possibilities to enhance performance and effectivity in customers reviews data collection. Adoption of multi-thread was able to visit the huge number of web pages, collect specific dataset 5–6 times faster than conventional single-thread process method. As shown in Table 2, by adopting multithreads, the speed of data collection was improved by 582.9% on average. The system also provided data collection results which are relevant to desired web components and necessary content.

Based on the result of experiment testing, our system has the following advantages for customers data collection. The MULARS can provide the reviews data collection with a simple computer operation and it allows users to quickly collect a large number of hotel reviews without operating with complex menus. It takes about 0.013 to 0.017 s to collect each review. The implemented system is simple and has a low cost. Everyone can simply use the MULARS and collect ratings in real time, this allows MULARS to be used to monitor and perform sentiment analysis of thousands of customers reviews.

The MULARS is an efficient and effective tool for costumers' data collection and doing statistical analysis. The system can be extended with required new function in the future. Every day, there is a vast amount of data published by consumers. They are open to the public and can be easily extracted by the MULARS, which can analyze and obtain information about business, competitors, possible opportunities, and the latest trends. In the future, if the data collection is integrated with NLP (Natural Language Processing), it will be possible to know how customers react to their products and services, as well as their feedback on campaigns and products. Because the system was implemented in Python, it is modifiable, compatible with other python modules and scripts. The system can be used for many different tasks by making only small changes.

6   Conclusion and Future Work

The adoption of multi-threading in the MULARS has made it possible to improve the performance and effectiveness of customer review data collection. By running multi-threaded processes simultaneously, we were able to achieve faster data collection with high reliability. The multi-threaded processing enabled data collection to be performed 5.8 times faster than a conventional single-thread processing method.

Our work provides a simple tool with high performance in collecting online hotels reviews using the Python language and its modules. As our goal is to make hospitality researchers more comfortable collecting data online, to contemplate convenient and an effective technique to collect multi-language reviews data, cleaning, changing, storing on separate database, comparing, analyzing, and demonstrating information to find helpful data for decision making. We focus on how scraping processing functions on the main travel website such as TripAdvisor.

MULARS is able to present data collection quickly and effectively with very simple interactions. It is very easy for anyone to operate even if they do not understand python programming. In addition, the effectiveness of the data collection competition is very high, because the MULARS is able to collect mostly all of the reviews data. The system has ability to collect quickly and in real time, it is allowing to analyze reviews in a more up to date manner.

Because it is implemented in Python, the system will make it easier to integrate with other modules for the future development of some required features. This system requires improvements in the future, such as creating a user interface that makes it more possible for this system to run on various device platforms. The system still uses a single threat to collect data. In the future work, it is required to use multi-thread to improve speed in downloading huge number a hotel's reviews data or collecting thousands of specific hotels reviews effectively. Finally, the future work also requires integration with the required function layers such as, smart database layer for multi-languages, the data analysis layer, visualization layer, etc.

References

1. Nariman, D.: An effective technique for collecting multi-languages hotel reviews: a case study of 5 stars hotels in Bali Island. In: Lecture Notes on Data Engineering and Communications Technologies, vol. 186, pp. 209–217. Springer, Cham (2024). https://doi.org/10.1007/978-3-031-46784-4_19
2. Sagiroglu, S., Sinanc, D.: Big data: a review. In: International Conference on Collaboration Technologies and Systems (CTS), pp. 42–47. San Diego, CA, USA (2013)
3. Volk, M., Bosse, S., Turowski, K.: Providing clarity on big data technologies: a structured literature review. In: Proceeding IEEE Conference on Business Informatics (CBI 17). IEEE Press, vol. 01, pp. 388– 397 (2017)
4. Igor, H.C., Jadranka, G., Saša, K.: ETourism: a comparison of online and offline bookings and the importance of hotel attributes. J. Info. Organ. Sci. **34**, 41–54 (2010)
5. Fuji, T., Fujiu, M.: Data utilization in tourism industry. Hokuriku Econ. Res. J. **468**, 10–19 (2019)
6. Satoshi, W.: Efforts Toward Inbound Tourism Using ICT By Local Governments and Their Effects. InfoCom Research Inc., Tokyo (2018)
7. Broucke, S., Baesens, B.: Practical Web Scraping For Data Science Best Practices and Examples with Python. Apress, New York (2018). https://doi.org/10.1007/978-1-4842-3582-9
8. Erlandsson, F., Nia, R., Boldt, M., Johnson, H., Wu, S.F.: Crawling online social networks. In: Second European Network Intelligence Conference, pp. 9–16. Karlskrona (2015)

9. Raj, S., Krishna, R., Nayak, A.: Distributed component-based crawler for AJAX applications. In: Second International Conference on Advances in Electronics, Computers and Communications (ICAECC), pp. 1–6. Bangalore (2018)
10. Gjoka, M., Kurant, M., Butts, C.T., Markopoulou, A.: Practical recommendations on crawling online social networks. IEEE J. Sel. Areas Commun. **29**(9), 1872–1892 (2011)
11. Arun Patil, T., Chobe, S.: Web crawler for searching deep web sites. In: International Conference on Computing, Communication, Control and Automation (ICCUBEA), pp. 1–5. Pune, India (2017)
12. Gao, K., Wang, W., Gao, S.: Modelling on web dynamic incremental crawling and information processing. In: 5th International Conference on Modelling, Identification and Control (ICMIC), pp. 293–298. Cairo (2013)
13. Achsan, H.T.Y., Wibowo, W.C.: A fast distributed focused-web crawling. Procedia Eng. **69**, 492–499 (2014)
14. Ye, F., Jing, Z., Huang, Q., Chen, Y.: The research of a lightweight distributed crawling system. In: IEEE 16th International Conference on Software Engineering Research, Management and Applications (SERA), pp. 200–204. Kunming, China (2018)
15. Ying, Z., Zhang, F., Fan, Q.: Consistent hashing algorithm based on slice in improving Scrapy-Redis distributed crawler efficiency. In: IEEE International Conference on Computer and Communication Engineering Technology (CCET), pp. 334–340. Beijing (2018)

Investigating BERT Layer Performance and SMOTE Through MLP-Driven Ablation on Gittercom

Bathini Sai Akash[1]([✉]), Vikram Singh[2], Aneesh Krishna[3],
Lalita Bhanu Murthy[1], and Lov Kumar[2]

[1] BITS Pilani Hyderabad Campus, Hyderabad, India
akashbathinis@gmail.com, bhanu@hyderabad.bitspilani.ac.in
[2] National Institute of Technology, Kuruksehtra, India
{viks,lovkumar}@nitkkr.ac.in
[3] Curtin University Perth, Perth, Australia
A.Krishna@curtin.edu.au

Abstract. For software development teams, communication is necessary to preserve growth consciousness, streamline the management of projects, and avoid misunderstandings. Among the features that chat rooms offer to help and satisfy the interaction requirements of software-development groups are instant personal messaging, group conversations, and code knowledge exchange. This is all capable of occurring in real time. As a result, developers are increasingly using chat rooms. One of these prominent forums is Gitter, and the chats it includes might be a goldmine of information for researchers studying open-source software systems. The GitterCom dataset, the biggest repository of Gitter developer messages that have been meticulously labeled and organized, was used in this study to conduct a multi-label categorization for the dataset's Purpose Category. 9 MLP machine learning classifiers, six feature selection methods, and the layered architecture of the BERT transformer are all subjected to thorough empirical research and evaluation. As a consequence, our research process shows competent results with a maximum AUC score of 0.97 with MLP variants using Adam optimizer(MLP2 and MLP3). Additionally, the research process might be used to text data from software development forums for general multi-label text categorization. The insights of the research, which give a holistic understanding of the Machine learning pipeline driven by BERT, shall serve the research community for preferential selection of Feature selection techniques, BERT layers, and classification model selection, among others for text classification.

Keywords: BERT · Gittercom · MLP · Feature Selection · SMOTE

1 Introduction

Through the years, there has been an increasing number and quality of software infrastructure, which has raised the need for developer collaboration and

L. Barolli (Ed.): AINA 2024, LNDECT 200, pp. 292–302, 2024.
https://doi.org/10.1007/978-3-031-57853-3_25

led to the creation of large groups of software developers that are typically scattered, spanning numerous locations. Cooperative system design has various benefits, such as knowledge exchange, enhanced cooperation synergy, increased efficiency, and higher-quality programs. In order to coordinate their efforts, the group members have to interact with each other on a regular basis (e.g., by exchanging road maps, status and updates, problems or roadblocks, due dates, etc.). The Programming and development groups have traditionally employed email lists, message boards, and Internet Relay Chat for communication.

Developers chat rooms are now available in "private" and "public" versions. Public chat rooms are frequently used in open-source tasks, and the development of a cordial community that communicates and exchanges knowledge is highly valued. A networking tool created for such communities and developers, GitteCom (referred to as Gitter) provides both public and private chat rooms. Developers may interact with compatible people in open chat rooms, exchange expertise, and participate in conversations.

Furthermore, since conversations posted to open Gitter streams are forever saved within chat room records, all members are entitled to every single chat room talk dating to when the channel was originally created. The possibility of a Gitter room being connected to a GitHub archive is another distinctive feature of Gitter. Ten thousand communications from 10 Gitter groups for open-source developmental ventures are collected in GitterCom [5]. It offers insightful information for analyzing developer behaviors. However, professionals explicitly labeled the Purpose Category for every communication in the dataset. Our study is principally concerned with developing a method that analyses BERT, feature selection, and machine learning classification performance in-depth layer performance on the aforesaid dataset comprehensively and in comparison. As a result, we also suggest an approach that performs effectively for multi-label categorization, with an AUC of 0.96. GitterCom messages can utilize this structure to have their Purpose Categories automatically and instinctively labeled. The extrapolation of our method might help development teams organize talks into their appropriate assignments and objective groups while saving valuable resources and time. The study would afterward benefit the research community by providing exposure and viewpoints on everything from feature selection to BERT performance for text categorization in the context of forum discussions about software development. The method used in the study may also be used generally for text categorization in the domain and for applications similar to ours. To the best of our ability, we believe our work to be unique, and we carefully examine the accuracy of the data used in the study.

2   Related Work

The related research work is described in the part that follows. Supervised learning has often been used in the broad field of text categorization. It is a broad area of research that includes deep learning techniques, context-based language models, pre-trained models, and Markov-based language models [3]. However, it is crucial to clarify how these strategies work in conjunction. Significant research has included Gitter, notably in text and empirical analysis. The research in [6] identifies when developers connect, analyses developer interactions, and pinpoints the topics discussed. It also understands the Gitter community dynamics. The objective of the research is to close the knowledge and understanding gap in this area. In 2022, Esteban et al. [4] developed automated methods for evaluating message intent and, based on the findings, suggested possible topics for additional research. They compared Gitter's communication behaviors with those of other platforms. To effectively comprehend their capacity as effective communication platforms for developers, research in [1] aims to provide an understanding of how Slack and Gitter's forums are used in software development teams, perceive their impact on projects, and propose ways to enhance them. Our research addresses this gap by developing an efficient infrastructure that not only conducts thorough ablation studies but also proposes a competent design for the automated labeling of The *Purpose* Category on the dataset.

3   Dataset Description

GitterCom [5] comprises data from 10,000 messages collected from 10 Gitter[1] communities for open-source development projects (each community has 1000 messages). Each interaction (message) in the collection was manually labeled to classify the purpose of communication in accordance with the categories identified by Lin et al. [2]. Table 1 displays the distribution of the dataset by *Purpose Category*. To classify texts using multiple labels, we created the following problem by translating the category labels(string) into the numerical class labels as per 1.

Table 1. Dataset distribution by *Purpose category*

Category	Message count	Label
Communication	5274	2
Dev-ops	2776	1
Customer support	1431	3
Team collaboration	188	4
Participation in communities of practice	153	4
Fun	42	4
Discovery and aggregate news and information	97	4
Networking and social activities	39	4

---

[1] https://gitter.im/.

4   Research Framework

The research framework used for the study is explained in this section. The steps are explained below:

Step 1: Prior to analysis, the multi-labeled dataset with numerical labels (Sect. 3) was pre-processed. English stopwords, brackets, hashtags, URLs, and other punctuation were all removed as part of the pre-processing, which also included changing the capitalization to lowercase. Since pre-trained language models like BERT already had considerable prepossessing, further pre-processing steps like stemming and lemmatization were not used. They are already suited to processing unprocessed full-text entries, which is the cause. Its multi-head self-attention mechanism has been tailored to use all of the data in the document. Next, BERT and sentence embeddings were generated for all 12 transformer layers and the input encoding layer. We generated 13 outputs in this step, which are stored in the data warehouse.

Step 2: We applied six different feature selection techniques on the generated BERT embedding datasets. This was done to verify the Section analysis of the dimensional reduction and performance increase on BERT embeddings, which is done in Sect. 6.1. This resulted in 91 combinations. 13 layers, and six feature selection techniques were performed. Once again this is stored back into the data warehouse.

Step 3: We observed a significant class imbalance in the dataset for the multi-label issue. To solve this issue and protect ourselves from the class imbalance, we used SMOTE. The original BERT embedding dataset is included in the feature-selected datasets to which SMOTE is applied. After this phase, there are twice as many combinations as there were in the previous step, totaling 182.

Step 4: Next, the nine variants of MLP were trained on the 182 dataset combinations. The metric used for training and validation was the Area Under the curve. Finally, Ablation studies were performed on BERT transformer layer performance, classification models, Feature selection, and SMOTE performance enhancement (Section 6).

5   Results Analysis

In this work, we applied BERT and sentence embeddings for all 12 transformer layers and the input encoding layer. We generated 13 outputs in this step, which are stored in the data warehouse. We have also applied six feature selection techniques to the generated BERT embedding datasets. This was done to verify the Section analysis of the dimensional reduction and performance increase on BERT embeddings, which is done in Sect. 6.1. Thus, a total of 1638 [1 dataset * 13 BERT embedding layers * (6 sets generated using feature selection techniques + 1 set of original data) * (1 set generated using a class balancing technique + 1 original dataset) * 9 classifiers] distinct predictive models were built in this comparative analysis study. The predictive ability of the developed models was evaluated using the AUC metrics. These models were validated using the 5-fold cross-validation method. Table 2 reports the AUC obtained for the various models developed by applying different classifiers to original data and sampled data on different sets of features. By analyzing the information in the table, we observed that the model with the highest AUC is obtained by applying the MLP with the Adam Optimization Algorithm. Similarly, Table 2 reports the AUC values obtained for the various models developed by applying different classifiers to original data and sampled data on different sets of features. By analyzing the information in the table, we observed that the model with the highest AUC is obtained by applying the SMOTE technique to address the class imbalance problem.

Table 2. Performance of ML classifiers

| Layer | FR1 | | | | | | | | FR2 | | | | | | | |
| | OD | | | | SMOTE | | | | OD | | | | SMOTE | | | |
	MLP1-A	MLP2-A	MLP1-L	MLP2-L	MLP1-A	MLP2-A	MLP1-L	MLP2-L	MLP1-A	MLP2-A	MLP1-L	MLP2-L	MLP1-A	MLP2-A	MLP1-L	MLP2-L
LY1	0.55	0.47	0.57	0.52	0.55	0.54	0.64	0.59	0.33	0.34	0.48	0.48	0.53	0.53	0.60	0.57
LY2	0.58	0.56	0.59	0.58	0.71	0.73	0.69	0.68	0.56	0.54	0.56	0.56	0.65	0.66	0.64	0.64
LY3	0.57	0.58	0.60	0.60	0.71	0.72	0.67	0.67	0.55	0.57	0.55	0.55	0.64	0.64	0.63	0.62
LY4	0.55	0.55	0.59	0.59	0.71	0.73	0.67	0.66	0.55	0.57	0.56	0.55	0.64	0.66	0.64	0.63
LY5	0.57	0.57	0.57	0.58	0.72	0.72	0.65	0.65	0.57	0.57	0.56	0.57	0.67	0.68	0.66	0.65
LY6	0.57	0.57	0.57	0.57	0.73	0.72	0.67	0.67	0.55	0.58	0.56	0.57	0.63	0.63	0.63	0.62
LY7	0.56	0.58	0.57	0.57	0.72	0.72	0.67	0.66	0.56	0.59	0.57	0.56	0.65	0.66	0.62	0.62
LY8	0.60	0.57	0.57	0.56	0.74	0.75	0.69	0.69	0.57	0.58	0.57	0.57	0.67	0.68	0.65	0.64
LY9	0.56	0.57	0.57	0.57	0.75	0.74	0.69	0.69	0.59	0.59	0.57	0.57	0.66	0.67	0.64	0.64
LY10	0.58	0.58	0.57	0.56	0.75	0.76	0.70	0.69	0.55	0.56	0.56	0.58	0.66	0.68	0.64	0.65
LY11	0.57	0.57	0.57	0.57	0.75	0.75	0.67	0.67	0.56	0.55	0.54	0.54	0.68	0.70	0.68	0.68
LY12	0.58	0.58	0.58	0.58	0.73	0.73	0.68	0.68	0.57	0.56	0.54	0.57	0.71	0.71	0.70	0.71
LY13	0.59	0.58	0.58	0.58	0.75	0.75	0.70	0.70	0.58	0.57	0.57	0.55	0.69	0.69	0.67	0.69

6   Comparative Analysis

We conducted an extensive comparison of metrics Area Under Curve (AUC). To address the class imbalance issue that existed in the dataset, we additionally used SMOTE. In our data warehouse, we produced a total of 182 datasets. Each dataset relates to a particular set of analysis methods 4. To get results, all of the datasets underwent classification model training. Two tests were used: the *Wilcoxon Signed Rank* [7] and the *Friedman Test* [8] to apply statistical studies other than AUC investigations. This would assess the significance and reliability of the many techniques employed.

6.1   Feature Selection

Feature ranking algorithms AUC performance analysis utilizing box-plots: Figure 1 provides specifics on the AUC metrics for the six Feature selection approaches employed. In this section, the Original dataset (OD) without feature selection shall be referred to as FR-1, following FS procedures numbered 2 to 6. We evaluate mean values in descriptive statistics to assess performance. The information shown in the box plots Fig. 1 leads to the following conclusions:

- All the feature selection techniques perform on par or better with the Original dataset with all features. FR-2, FR-3, FR-6, FR-7 seem to perform similar to the original dataset.

– Chi Value Ranking(FR-5) seems to outperform other feature selection techniques with a mean AUC of 0.83

Statistical Hypothesis testing for comparison of Feature selection performance:

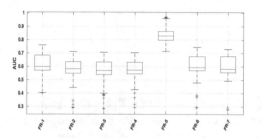
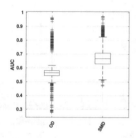

Fig. 1. Box Plot for Feature selection techniques

Fig. 2. Box Plot for SMOTE performance

As aforementioned, we used the Friedman and Wilcoxon Signed Rank Test as our two methods for assessing our hypotheses. The findings of the hypothesis test are shown in Table 3. On inspection of Table 3, we see that for most entity pairs $p \geq 0.05$, the null hypothesis is not rejected, and there is no significant difference in performance over the original dataset. However, in the case of FR-7, $p=0.03$, which is less than 0.05, the null hypothesis is rejected, and there is a significant difference in performance. The FR-7 performs differently as compared to FR-5, FR-4, and FR-6. However, there is no significant difference as compared to the original dataset(FR-1).

Additionally, we use the Friedman Test to assess the quantitative effectiveness of the methods. As stated above, the performance of the methods increases with decreasing rank. From Table 3, we note that FR-5 has the lowest rank and performs better than the original dataset without feature selection. However, the mean rank for the other FR techniques appears to be more than the original dataset. **We conclude that the feature selection Chi-Squared test(FR-5) could potentially be used on BERT embeddings to increase performance since lower dimensionality enhances the asymptotic performance of the classification models.**

6.2 Effect of Oversampling

AUC performance analysis of SMOTE using descriptive statistics and Boxplots: Fig. 2 shows the AUC values on the application of SMOTE. The information shown in box plots Fig. 2 leads to conclusions like the deployment of SMOTE; the mean AUC performance seems to have improved considerably. The mean AUC value notably improved from 0.58 to 0.68.

Table 3. Feature selection p values AUC

	FR1	FR2	FR3	FR4	FR5	FR6	FR7	Rank
FR-1	1.0	0.62	0.8	0.38	0.38	0.46	0.1	2.91
FR-2	0.62	1.0	1.0	0.16	0.46	0.8	0.07	5.14
FR-3	0.8	1.0	1.0	0.16	0.38	0.54	0.07	6.04
FR-4	0.38	0.16	0.16	1.0	0.38	0.21	0.03	5.61
FR-5	0.38	0.46	0.38	0.38	1.0	0.62	0.03	1.0
FR-6	0.46	0.8	0.54	0.21	0.62	1.0	0.04	3.18
FR-7	0.1	0.07	0.07	0.03	0.03	0.04	1.0	4.11

Table 4. SMOTE p values AUC

	OD	SMD	Mean-Rank
OD	1.00	0.33	2.00
SMD	0.33	1.00	1.00

Statistical Hypothesis testing to check for SMOTE enhancement: The findings of the hypothesis test are shown in Table 4. From Table 4, we see that p≥0.05, and note that there is no significant performance gain due to SMOTE. We see from Table 4 that the Mean rank decreases noticeably from OD to SMD. Thus, we conclude that the implementation of SMOTE greatly enhanced the efficacy of classification. **Thus, we conclude that the performance of multi-class prediction utilizing machine learning models improved notably but not significantly by using SMOTE over BERT embeddings of text dataset.**

6.3   Performance of Deep Learning Classifiers

AUC performance analysis of Classification Models using descriptive statistics and Boxplots: Figure 3 provides data regarding the AUC values for the nine machine learning models employed. The data presented in box plots as shown in Fig. 3 leads to subsequent observations:

- The mean AUC values for all the classification models range from 0.6 to 0.65, indicating that they might have similar performance.
- Models MLP1 and MLP2 slightly perform better regarding mean AUC. MLP7, MLP8, and MLP9 underperform relatively. The maximum performance of most models crosses 0.95, showing proficient results (Fig. 4).

Comparative statistical hypothesis testing for model classification performance: The findings of the hypothesis test are shown in Table 5. By statistically evaluating the hypothesis, if the value of p≤0.05, we rejected the null hypothesis. On inspection of Table 5, we see that MLP1 has p<0.05 for all entity pairs and is significantly different from all other MLP variants.MLP2 and MLP3 have significant differences, which is a vital inference considering both MLP2 and MLP3 have the same solver (Adam); MLP has three layers compared to just 2 for MLP2. MLP6 also has significant differences from MLP9, MLP3, and MLP1. In general, MLP1, MLP2, MLP6, and MLP3 have multiple entity pairs with p≤0.05. From Table 5, we also note that MLP2 and MLP3 achieve the lowest ranks and perform significantly better than all other models.

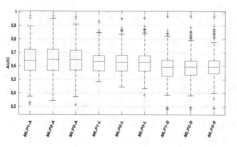 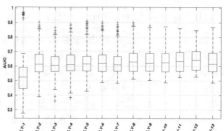

Fig. 3. Box-Plot for Classification Model

Fig. 4. Performance Parameter Box Plot for ML classifier

This justification demonstrates that, despite the possibility that applying too many model variations won't substantially improve or harm performance, it's still beneficial to look into all of the models above, which have various underlying architectures. **Hence, we conclude that MLP2 and MLP3, which are variants of Adam in our scenario of text multi-label classification utilizing BERT phrase embeddings, greatly beat all other models.**

Table 5. ML classifier p-values AUC

	MLP1	MLP2	MLP3	MLP4	MLP5	MLP6	MLP7	MLP8	MLP9	Mean-Rank
MLP1	1.0	0.0	0.0	0.0	0.0	0.0	0.0	0.0	0.0	3.38
MLP2	0.0	1.0	0.05	0.34	0.34	0.06	0.86	0.67	0.67	2.91
MLP3	0.0	0.05	1.0	0.39	0.02	0.0	0.08	0.02	0.05	2.68
MLP4	0.0	0.34	0.39	1.0	0.3	0.06	0.44	0.34	0.39	5.26
MLP5	0.0	0.34	0.02	0.3	1.0	0.06	0.34	0.93	1.0	4.91
MLP6	0.0	0.06	0.0	0.06	0.06	1.0	0.0	0.3	0.02	4.85
MLP7	0.0	0.86	0.08	0.44	0.34	0.0	1.0	0.26	0.39	7.73
MLP8	0.0	0.67	0.02	0.34	0.93	0.3	0.26	1.0	0.55	6.93
MLP9	0.0	0.67	0.05	0.39	1.0	0.02	0.39	0.55	1.0	6.35

6.4 BERT Layer Performance Comparison

AUC performance analysis of BERT layers using descriptive statistics and Boxplots: The AUC values for the 12 BERT layers are detailed in Fig. 2. LR-1 corresponds to the input encoding layer, and each consecutive label represents each BERT layer. The data presented in box plots shown in Fig. 2 leads to subsequent observations:

- No definitive conclusions can be drawn on the performance variability of BERT layers. Apart from the initial encoding layer(LR-1), every other layer performs alike.

- All the layers perform with a median AUC in the range of 0.61 to 0.62

Statistical Hypothesis testing for comparison of BERT Layer performance: The results from hypothesis testing(p-values) are presented in Table 6 for AUC. By statistical hypothesis testing, if the value of $p \leq 0.05$, we reject the null hypothesis and henceforth state that there is a significant difference in the performance of the layers. On inspection of Table 6, we notice that LR-1 has $p \leq 0.05$ with LR-8 and LR-9, respectively. LR-13 significantly differs in performance with every other layer apart from LR-7. LR-10, LR-8, and LR-12 also have some entity pairs with $p \leq 0.05$. Most other layer entity pairs do not show significant differences, as shown in Table 6. From Table 6, we see that LR-8 and LR-12 perform the best with a rank close to 5. LR-11, LR-13, and LR-6 perform similarly with ranks slightly higher than best-performing layers. Hence, While we analyze the 12 layers of BERT, we notice that it is not guaranteed that the final layer will collect the most pertinent data relating to the classification objective.

From this explanation, we elucidate that different layers in BERT seem to capture various information. The specific type of information that could be most vital for a particular embedding classification task determines which layer is the best performer. **Hence, we conclude that although BERT layer 11(LR-12) and layer 7(LR-8) outperform all other layers, layers 12(LR-13) and 10(LR-11) and 5(LR-6) perform similarly in our case of text multi-label classification using BERT sentence embeddings. Furthermore, LR-13 performs significantly differently from LR-8 and LR-12, respectively.**

Table 6. BERT layer p-values AUC

	LR1	LR2	LR3	LR4	LR5	LR6	LR7	LR8	LR9	LR10	LR11	LR12	LR13	Mean-Rank
LR-1	1.0	1.0	0.88	0.38	0.68	0.64	0.88	0.33	0.1	0.11	0.64	0.76	0.03	10.75
LR-2	1.0	1.0	1.0	0.64	0.64	0.61	1.0	0.12	0.11	0.12	0.8	0.8	0.01	7.75
LR-3	0.88	1.0	1.0	0.64	0.8	0.64	0.92	0.26	0.11	0.14	0.64	1.0	0.01	8.92
LR-4	0.38	0.64	0.64	1.0	0.84	1.0	0.47	0.54	0.18	0.02	0.31	0.54	0.0	7.65
LR-5	0.68	0.64	0.8	0.84	1.0	0.88	0.72	0.33	0.18	0.06	0.44	0.88	0.0	7.34
LR-6	0.64	0.61	0.64	1.0	0.88	1.0	0.57	0.41	0.23	0.04	0.41	0.84	0.0	6.31
LR-7	0.88	1.0	0.92	0.47	0.72	0.57	1.0	0.22	0.1	0.36	0.84	0.72	0.06	7.28
LR-8	0.33	0.12	0.26	0.54	0.33	0.41	0.22	1.0	0.64	0.02	0.17	0.33	0.0	5.06
LR-9	0.1	0.11	0.11	0.18	0.18	0.23	0.1	0.64	1.0	0.01	0.06	0.11	0.0	5.94
LR-10	0.11	0.12	0.14	0.02	0.06	0.04	0.36	0.02	0.01	1.0	0.44	0.09	0.54	6.83
LR-11	0.64	0.8	0.64	0.31	0.44	0.41	0.84	0.17	0.06	0.44	1.0	0.54	0.11	5.76
LR-12	0.76	0.8	1.0	0.54	0.88	0.84	0.72	0.33	0.11	0.09	0.54	1.0	0.02	5.06
LR-13	0.03	0.01	0.01	0.0	0.0	0.0	0.06	0.0	0.0	0.54	0.11	0.02	1.0	6.33

7 Conclusion

This study delivers a significant improvement in the evaluation of GitterCom text classification and for general multi-label text classification on software developer forum text data by successfully resolving the class imbalance and feature redundancy difficulties. By creating 182 datasets with various feature selection methods, SMOTE, and BERT layers, the efficacy of the models developed using nine unique MLP classifiers was assessed. The research makes use of SMOTE and six feature selection strategies in order to handle the problems of data coherence, dependability, feature irrelevance, and imbalanced data. Four study topics are thoroughly compared using the Wilcoxon sign rank test and Friedman test, and the results lead us to the following conclusions:

- In our application of text multi-label classification utilizing BERT sentence embeddings, layers 12(LR-13), 10(LR-11), and 5(LR-6) perform comparably, despite layer 11(LR-12) and layer 7(LR-8) outperforming all other layers. LR-13 also operates quite differently from LR-8 and LR-12, respectively.
- In our scenario of text multi-label classification using BERT phrase embeddings, Adam versions MLP2 and MLP3 significantly outperformed all other models.
- The usage of SMOTE over BERT embeddings of text dataset considerably but not significantly enhanced the performance of multi-class prediction using machine learning models.
- While reduced dimensionality improves the asymptotic performance of the classification models, the feature selection Chi-Squared test(FR-5) might possibly be applied on BERT embeddings to boost performance.

We also show superior performance in text classification with a maximum AUC of around 0.97 using our research pipeline. This was achieved by multiple combinations of MLP with BERT layers, for instance, FR5 on LY1 with MLP1 and MLP2. These findings may help the research community make judicious decisions about the performance evaluation of each particular BERT transformer layer and text categorization using BERT. Added to this, there is little research performed on feature selection of BERT features, Deep Learning classifiers, and efficacy of SMOTE inclusion, which are dealt with in a holistic mode in the research. Finally, our study process utilizing MLP2 and MLP3 classifiers produces competent results with an AUC max performance of 0.97. Furthermore, the study methodology and approach might be used to broaden multi-label text classification on text data from software development forums.

References

1. El Mezouar, M., da Costa, D.A., German, D.M., Zou, Y.: Exploring the use of chatrooms by developers: an empirical study on slack and gitter. IEEE Trans. Softw. Eng. **48**(10), 3988–4001 (2021)
2. Lin, B., Zagalsky, A., Storey, M.A., Serebrenik, A.: Why developers are slacking off: understanding how software teams use slack. In: Proceedings of the 19th ACM Conference on Computer Supported Cooperative Work and Social Computing Companion, pp. 333–336 (2016)

3. Lu, Z., Du, P., Nie, J.-Y.: VGCN-BERT: augmenting BERT with graph embedding for text classification. In: Jose, J.M., et al. (eds.) ECIR 2020. LNCS, vol. 12035, pp. 369–382. Springer, Cham (2020). https://doi.org/10.1007/978-3-030-45439-5_25
4. Parra, E., Alahmadi, M., Ellis, A., Haiduc, S.: A comparative study and analysis of developer communications on slack and gitter. Empir. Softw. Eng. **27**(2), 40 (2022)
5. Parra, E., Ellis, A., Haiduc, S.: GitterCom: a dataset of open source developer communications in gitter. In: Proceedings of the 17th International Conference on Mining Software Repositories, pp. 563–567 (2020)
6. Shi, L., et al.: A first look at developers' live chat on gitter. In: Proceedings of the 29th ACM Joint Meeting on European Software Engineering Conference and Symposium on the Foundations of Software Engineering, pp. 391–403 (2021)
7. Woolson, R.F.: Wilcoxon signed-rank test. Wiley Encyclopedia of Clinical Trials, pp. 1–3 (2007)
8. Zimmerman, D.W., Zumbo, B.D.: Relative power of the Wilcoxon test, the Friedman test, and repeated-measures ANOVA on ranks. J. Exp. Educ. **62**(1), 75–86 (1993)

Detecting Signs of Depression in Social Networks Users: A Framework for Enhancing the Quality of Machine Learning Models

Abir Gorrab[1], Nourhène Ben Rabah[2], Bénédicte Le Grand[1], Rébecca Deneckère[2(✉)], and Thomas Bonnerot[2]

[1] RIADI Laboratory, National School of Computer Science, University of Manouba, Manouba, Tunisia
Abir.Gorrab@riadi.rnu.tn, Benedicte.Le-Grand@univ-paris1.fr
[2] Centre de Recherche en Informatique, Université Paris1 Panthéon-Sorbonne, Paris, France
{Nourhene.Ben-Rabah,Rebecca.Deneckere}@univ-paris1.fr

Abstract. Depression is widely recognized as a contributor to global disability and a significant factor in the emergence of suicidal tendencies. On social networks, individuals openly share their thoughts and emotions through posts, comments, and other forms of communication. The use of Artificial Intelligence, particularly Machine Learning methods, holds great potential for analyzing this data. However, it is imperative to exercise caution in the application of these methods to avoid biases and overfitting, two problems that could compromise the quality of Machine learning models. In this paper, we present a framework for detecting signs of depression among users of the X social network. This framework is based on four phases aimed at minimizing both biases and overfitting, resulting in models that generalize well to new data, thereby enhancing their applicability by healthcare professionals and patients. To validate our framework, we present the results of three detailed experiments using nine Machine Learning algorithms.

1 Introduction

Depression is a serious issue that profoundly affects mental health [30]. According to the French National Institute of Health and Medical Research (INSERM), it is defined as a major personal suffering that can lead to chronic illnesses, health problems, and, in the most severe cases, the risk of suicide. The World Health Organization (WHO) estimates that 3.8% of the global population suffers from depression [28], with rates of 5% among adults and 5.7% among individuals over 60 years old. Overall, approximately 280 million people are affected by depression [12].

The detection of depression through the exploration of social networks represents a continuously expanding research field [3, 4, 7, 8, 25]. This approach is justified by the fact that people spend a lot of time on social networks, where they openly share their thoughts, emotions, and experiences [6, 29]. The analysis of these online messages can be used to detect potential signals associated with depression [2, 15, 23]. Researchers are exploring various approaches by combining Artificial Intelligence (AI) techniques, such as Machine Learning (ML) [7, 13, 21], Natural Language Processing (NLP) [14], and sentiment analysis methods [16, 18], to identify indicators of depression within social media posts.

However, due to the interdisciplinary nature of this issue, not all disciplines possess the same level of expertise, and their focus is not uniformly directed toward the challenges inherent in the generalization of machine learning algorithms. Moreover, Machine Learning should be used with care, as each algorithm may introduce potential biases and compromise the validity of results, thereby limiting its applicability by healthcare professionals and patients.

In this context, we propose a depression detection framework based on the analysis of users' tweets on the X social network. We start (1) with the data collection and annotation phase where we collect data and label it using Valence Aware Dictionary and sEntiment Reasoner (VADER) [11]. We then present (2) a data pre-processing phase consisting of (2.1) cleaning data by removing stopwords, data analyzing through tokenization and lemmatization; and data modeling with N-gram, to finally (2.2) encode data through BoW and TF-IDF methods. We then (3) train nine ML models, with a crucial phase of hyperparameter tuning, to end (4) with the model evaluation phase.

This paper presents several contributions:

- A comprehensive framework in which each step of the four phases is explained in detail, aiming to minimize both overfitting and bias issues. To our knowledge, no detailed study on depression detection while minimizing overfitting and bias has been conducted so far (see Related Works section). Moreover, our framework may also be useful for researchers in the healthcare area, who are not necessarily experts in Machine Learning.
- Detailed experimental results obtained from real data, comparing the performance of nine ML algorithms.

The remainder of this paper is organized as follows: Sect. 2 describes the proposed framework, while Sect. 3 presents three experiments to validate it. In Sect. 4, we discuss related works, before presenting our conclusions in Sect. 5.

2 Proposed Framework

The proposed framework relies on four key phases: (i) data collection and annotation, (ii) data pre-processing, (iii) hyperparameter tuning and model training, and (iv) model evaluation. Figure 1 describes these phases.

2.1 Phase 1: Data Collection and Annotation

In this first phase, our objective is to collect various data describing tweets from users in the X social network. The main aim of this collection is to obtain a balanced representation of *alarming messages*, i.e. reflecting a depressive state, and *normal messages*, i.e. containing no sign of depression. To achieve this objective, we build our database, following the steps described below:

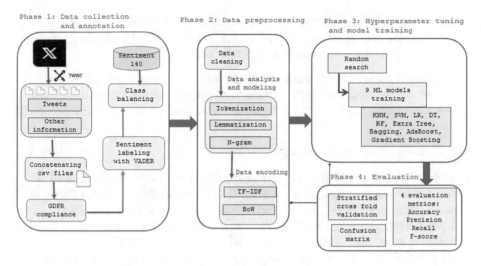

Fig. 1. Depression detection proposed framework.

- **Step 1: Data Collection with Twint.** We use Twint [24], a web scraping tool for tweets on X, to search for tweets containing one of these five specific keywords: 'depressed', 'depressive', 'hopeless', 'lonely', and 'suicide'. Each keyword-based query generates a distinct CSV file, which is concatenated with the others into a single one. The resulting file consists of 25 columns, including 'conversation_id', 'created at', 'date', 'time', 'timezone', 'user-id', 'user_name', 'name', 'place', 'tweet', 'mentions', 'urls', 'photos', 'replies_count', 'retweets_count', 'likes_count', 'hashtags', 'cashtags', 'links', 'retweet', 'quote_url', 'video', 'user_rt_id', 'near' and 'geo'.
- **Step 2: GDPR Compliance and removal of sensitive data.** Respecting GDPR rules is essential when dealing with personal data. To ensure confidentiality and privacy protection, sensitive columns such as 'user_name', 'name', 'place', 'near', 'geo', and 'user_rt_id' are removed. This deletion is a key measure to avoid any privacy violations. Non-exploitable data, typically null values, are also eliminated, leaving only 'user_id' and tweets. Moreover, only tweets in English are retained.
- **Step 3: Application of VADER for sentiment labeling.** VADER (Valence Aware Dictionary and sEntiment Reasoner) is used to label tweets based on sentiments. Each tweet can contain positive, negative, and neutral elements. VADER aggregates these elements to calculate an overall score that reflects the overall sentiment of the tweet. From VADER scores which range from -1 (very negative) to 1 (very positive), we assign *class 0* to negative score values, corresponding to 'alarming tweets' and *class 1* to positive score values – normal tweets. As a result, our dataset comprises 52,139 alarming tweets and 17,821 normal tweets.
- **Step 4: Class balancing.** Due to the imbalance between both classes and to prevent overfitting, we decided to retain 30,000 alarming tweets and we supplemented the 17,821 tweets with positive VADER scores (normal tweets) by adding positive tweets from an existing dataset called Sentiment140. Sentiment140 [9] is a dataset containing tweets labeled with polarity (negative, neutral, positive). From this dataset,

we selected positively polarized tweets containing other depressive keywords than those used in our initial dataset ('depressed', 'depressive', 'hopeless', 'lonely', and 'suicide'), in order to add diversity in the dataset.

Our final dataset consists of three columns: 'user_id', 'tweet', and 'label'. It includes 30,000 normal tweets containing terms related to depression and positive sentiment (class 1) and 30,000 alarming tweets with depressive words and negative sentiment (class 0).

2.2   Phase 2: Data pre-processing.

- **Step 1: Data cleaning:** Stop words refer to common words in a language that lack substantial meaning. To clean data, we remove stop words using the NLTK library (Natural Language Toolkit) [19]. We then eliminate unnecessary characters such as non-alphanumeric characters using Regex. We also remove short lines (< 2 characters), to finally keep 29,671 alarming tweets and 29,238 normal ones.
- **Step 2: Data analysis and modeling.** We proceed with a textual analysis composed of two stages: **tokenization** [26] and **lemmatization**. Tokenization consists of dividing a text into a collection of individual words named tokens. In the second step, lemmatization reduces words to their base or root form, known as the lemma. This allows us to consider the different forms of a term as a unique concept (eg. 'depressed', 'depression', 'depressing'). We then split our lemmas into uni-grams using N-gram modeling [17]. An N-gram is a sequence of N elements in data, such as characters or words, in the context of Natural Language Processing.
- **Step 3: Data encoding:** When dealing with textual data, it is necessary to encode it into numerical format for ML algorithms to function properly. For this, we use two encoding methods such as **TF-IDF** (Term Frequency-Inverse Document Frequency) and **BoW** (Bag of Words). TF-IDF is a term weighting measure in a document, text, or other content. A term that appears frequently in a document but rarely in the entire set of documents will have a high TF-IDF score, indicating its relative importance in the specific document. In our context, a term is a word, and a document is a tweet. Therefore, we calculate the score of each word in a tweet relative to its frequency in the corresponding tweet and in the overall set of tweets. The Bag of Words (BoW) model transforms arbitrary text into fixed-length vectors by counting the number of occurrences of each word.

2.3   Phase 3: Hyperparameter Tuning and Model Training

In this phase, we train and test nine ML models for depression detection from X data. These models are the k-nearest neighbors (**KNN**), support vector machine (**SVM**), logistic regression (**LR**), Decision Tree (**DT**), Random Forest (**RF**), Extra Tree (**ET**), **Bagging**, **AdaBoost**, and **Gradient Boosting**. In this study, we deliberately choose not to use deep learning algorithms and start with simpler models, thus facilitating the interpretation of results. Additionally, the selected algorithms do not require the access to graphics processing units (GPUs), unlike deep learning models for an efficient training. We use the implementations of these algorithms provided by the Scikit-learn library. To reduce the risk of bias when evaluating model performance, we use stratified 10-fold cross-validation; this method divides the dataset into 10 folds with the same distribution of classes, thus ensuring that the relative frequencies of the classes are approximately preserved in each training and test set.

Before training these models, it is important to select the right values of hyper-parameters as they have a significant impact on the performance of machine learning models. Optimal hyperparameters values contribute to preventing overfitting and enable the models to generalize correctly to new data. For time efficiency purposes, we apply an automatic method called Random Search, which randomly selects combinations of hyperparameters.

2.4 Phase 4: Model Evaluation

The performance of each model is measured using the following performance metrics: accuracy, precision, recall, and F1-score. These metrics are calculated from four values represented in a confusion matrix as shown in Table 1:

- True Positive (TP) refers to alarming tweets that are correctly predicted.
- True Negative (TN) refers to normal tweets that are correctly predicted.
- False Negative (FN) refers to alarming tweets that are predicted as normal tweets.
- False Positive (FP) refers to normal tweets that are incorrectly predicted as alarming tweets.

Table 1. Confusion matrix.

	Predicted Alarming tweets	Predicted Normal tweets
Actual Alarming tweets	TP	FN
Actual Normal tweets	FP	TN

In the following, we present the evaluation metrics:

- **Accuracy** represents the proportion of correct predictions:

$$Accuracy = \frac{TP + TN}{TP + FN + FP + TN} \tag{1}$$

- **Precision** is the proportion of actual alarming tweets among the tweets predicted as alarming. A low precision means that a high proportion of normal tweets are detected as alarming tweets (false positive).

$$Precision = \frac{TP}{TP + FP} \tag{2}$$

- **Recall** is the proportion of actual alarming tweets with regard to the actual number of alarming tweets. A low recall indicates that a large proportion of alarming tweets have been classified as normal (false negative).

$$Recall = \frac{TP}{TP + FN} \tag{3}$$

- **F_score** is the harmonic mean of precision and recall values, reaching its best value at 1 and its worst value at 0. It is calculated as follows:

$$F\_score = \frac{2 * Recall * Precision}{Recall + Precision} \tag{4}$$

3 Experiments

This section outlines three experiments to validate our proposed framework. In the initial experiment (Sect. 3.1), we compare tweet encoding methods, namely BoW and TF-IDF, by presenting the top 30 and 100 most frequent unigrams for each method. The second experiment (3.2) employs the Random Search method to identify optimal hyperparameter values associated with the nine selected ML algorithms: Random Forest, Extra-Tree, KNN, SVM, Logistic Regression, Decision Tree, AdaBoost, GradientBoosting, and Bagging. These adjustments are executed using both TF-IDF and BoW. The Third experiment (3.3) trains, tests and evaluates the different models with these optimized hyperparameters for both TF-IDF and BoW.

3.1 The Most Frequent Unigrams in Our Dataset

To compare the tweet encoding methods, namely BoW and TF-IDF, we selected the top 30 and 100 unigrams for each method. In Fig. 2, (a) and (b) illustrate word clouds corresponding to the top 30 and 100 unigrams (most frequent terms) in the dataset using the BoW method. Similarly, (c) and (d) represent word clouds of the top 30 and 100 unigrams using the TF-IDF method. These visualizations allow for a visual comparison of the most important terms according to each method. These terms are displayed with a larger font size.

The BoW word cloud highlights the frequently used terms in our dataset. Looking at (a) and (b), we can see terms that describe a depressive state, such as "lonely", "suicide", "depression", "depressed", "hopeless", as well as terms associated with verbs used to express feelings or needs, such as "need", "want", "know", "make", "feel". It also includes normal terms that don't express depression, such as "love", "people", "lol", "good", and "better".

Looking at (c) and (d), we essentially see the same terms, but we note a difference in the frequency of these terms. We can see that the terms "people" and "know" are larger in (c) than in (a), since they appear more frequently in the dataset.

3.2 Hyperparameter Tuning Using Random Search

A hyperparameter is a configuration variable of the machine learning algorithm and does not depend on the specifically trained model. The meaning of each hyperparameter is provided in the documentation of scikit-learn [22]. Tuning the hyperparameters of an algorithm allows to find the optimum values for a given set of data. Machine learning algorithms have several hyperparameters that can influence their performance. The hyperparameters of the nine algorithms we are comparing are shown in Table 2. It is

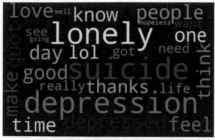

(a) The word clouds for the top 30 unigrams using the **BoW** method.

(b) The word clouds for the top 100 unigrams using the **BoW** method.

(c) The word clouds for the top 30 unigrams using the **TF-IDF** method

(d) The word clouds for the top 100 unigrams using the **TF-IDF** method.

Fig. 2. Word clouds for top 30 and 100 unigrams using BoW and TF-IDF methods.

generally accepted that these algorithms perform reasonably well with the default values of the hyperparameters specified in the software packages. However, adjusting the hyperparameters can improve their performance. One approach to choosing an optimal combination of values for our hyperparameters is to build a model for each possible combination of hyperparameter values. This method can be expensive and slow. To overcome these limitations, we have opted for a Random Search method for both TF-IDF and BoW. Table 2 shows the optimal hyperparameters for both methods. We may note that the encoding method impacts the optimal values of certain hyperparameters (represented in bold and italics).

3.3 Models Training, Testing and Evaluation for TF-IDF and BoW Encodings

Tables 3 and 4 show the performances of the nine ML models trained and tested with BoW and TF-IDF encoding methods respectively. As shown in Table 3, for BoW encoding, LR model gives the highest accuracy in the test set (95.48%), followed by SVM (95.12%), AdaBoost (95.04%) and DT (94.94%). For TF-IDF encoding, LR also presents the highest accuracy in the test set (95.52%). We then find SVM (95.41%), AdaBoost (95.34%) and then RF (95.09%).

We note that no overfitting occurs for LR and SVM, as the accuracy on the test set does not decrease significantly as compared to the accuracy on the training set. Moreover, we observe the TF-IDF encoding method provides better results than BoW.

Table 2. Hyperparameter tuning.

Algorithms	Hyperparameters with their possible values	Hyperparameters optimized for TF_IDF	Hyperparameters optimized for BoW
KNN	n_neighbors: [3, 5, 7, 9] weights: [uniform, distance]	n_neighbors = 9 weights = distance	n_neighbors = 9 weights = distance
SVM	C: [0.1, 1, 10] Kernel: [linear, rbf, poly] gamma: [scale, auto, 0.1, 1, 10]	*C = 1* *Kernel = rbf* *Gamma = 1*	*C = 0.1* *Kernel = linear* *Gamma = 10*
LR	Penalty: [l1, l2] C: [0.1, 1, 10]	Penalty = l2 *C = 10*	Penalty = l2 *C = 1*
DT	max_depth: [None, 10, 20, 30] min_samples_split: [2, 5, 10] min_samples_leaf: [1, 2, 4]	max_depth = 30 *min_samples_split = 10* *min_samples_leaf = 4*	max_depth = 30 *min_samples_split = 5* *min_samples_leaf = 2*
RF	n_estimators: [50, 100, 200] max_depth: [None, 10, 20, 30] min_samples_split: [2, 5, 10] min_samples_leaf: [1, 2, 4] bootstrap: [True, False]	n_estimators = 50 max_depth: None min_samples_split = 5 min_samples_leaf = 1 bootstrap = False	n_estimators = 50 max_depth: None min_samples_split = 5 min_samples_leaf = 1 bootstrap = False
ET	n_estimators: [50, 100, 200] max_depth: [None, 10, 20, 30] min_samples_split: [2, 5, 10] min_samples_leaf: [1, 2, 4] bootstrap: [True, False]	n_estimators = 50 max_depth: None min_samples_split = 5 min_samples_leaf = 1 bootstrap = False	n_estimators = 50 max_depth: None min_samples_split = 5 min_samples_leaf = 1 bootstrap = False
Bagging	base_estimator: RF n_estimators: [10, 20, 30] max_samples: [0.5, 0.7, 1.0] max_features: [0.5, 0.7, 1.0]	base_estimator: RF n_estimators = 10 max_samples = 0.7 max_features = 1	base_estimator: RF n_estimators = 10 max_samples = 0.7 max_features = 1

(continued)

Table 2. (*continued*)

Algorithms	Hyperparameters with their possible values	Hyperparameters optimized for TF_IDF	Hyperparameters optimized for BoW
Ada-Boost	base_estimator: RF n_estimators: [1, 5]	base_estimator: RF *n_estimators: 1*	base_estimator: RF *n_estimators: 5*
Gradient-Boosting	n_estimators: [1, 5] learning_rate: [0.01, 0.1, 0.2] max_depth: [3, 5, 7] min_samples_split: [2, 5, 10] min_samples_leaf: [1, 2, 4]	n_estimators = 1 *learning_rate = 0.01* max_depth = 7 min_samples_split = 2 min_samples_leaf = 1	n_estimators = 1 *learning_rate = 0.2* max_depth = 7 min_samples_split = 2 min_samples_leaf = 1

Table 3. Performances of ML models trained and tested on BoW encoding.

	Training set				Test set			
	Precision	Recall	F1	Accuracy	Precision	Recall	F1	Accuracy
KNN	99.97	99.70	99.84	99.84	96.28	85.02	90.30	90.94
SVM	99.87	91.29	95.38	95.61	99.72	90.42	94.84	95.12
LR	99.79	94.70	97.18	97.27	98.85	91.97	95.28	**95.48**
DT	99.82	90.39	94.87	95.15	99.49	90.27	94.65	94.94
RF	99.94	99.48	99.71	99.71	96.36	93.15	94.73	94.86
ET	99.95	99.56	99.75	99.75	95.10	93.51	94.30	94.39
Bagging	99.12	98.50	99.11	99.74	93.25	92.69	92.97	93.05
AdaBoost	99.93	99.74	99.84	99.84	97.11	92.78	94.89	95.04
Gradient-Boosting	100	89.03	94.19	94.55	99.98	89.03	94.18	94.54

In addition, Tables 5 and 6 represent the confusion matrices for the training and test sets respectively of the LR model using TF-IDF encoding. In our framework for detecting depression signs from social networks' users messages, the priority is to maximize the number of True positive (TP) and minimize the number of False Negative (FN), that is, tweets predicted as normal although being actually alarming. This is exactly what we can see in the confusion matrices, which shows the efficiency and relevance of our framework.

Table 4. Performances of ML models trained and tested on TF-IDF encoding.

	Training set				Test set			
	Precision	Recall	F1	Accuracy	Precision	Recall	F1	Accuracy
KNN	99.97	99.70	99.83	99.84	86.06	77.69	81.67	82.70
SVM	99.95	97.60	98.76	98.78	99.61	91.11	95.17	95.41
LR	99.67	97.81	98.73	98.75	98.14	92.73	95.36	**95.52**
DT	99.79	90.35	94.83	95.12	99.58	90.07	94.58	94.88
RF	99.90	99.75	99.83	99.83	97.01	92.96	94.94	95.09
ET	99.95	99.70	99.82	99.83	96.06	93.44	94.74	94.85
Bagging	99.85	98.12	98.98	99.00	97.91	91.78	94.74	94.95
AdaBoost	99.94	99.72	99.83	99.83	98.03	92.47	95.17	95.34
Gradient-Boosting	100	89.03	94.55	94.20	99.97	89.02	94.18	94.54

Table 5. LR confusion matrix for the training set with TF-IDF.

	Predicted Alarming tweets	Predicted Normal tweets
Actual Alarming tweets	266181 (TP)	858 (FN)
Actual Normal tweets	5761 (FP)	257381 (TN)

Table 6. LR confusion matrix for the test set with TF-IDF

	Predicted Alarming tweets	Predicted Normal tweets
Actual Alarming tweets	29156 (TP)	515 (FN)
Actual Normal tweets	2125 (FP)	27113 (TN)

4   Related Works and Discussion

To examine existing work on depression detection in social networks using artificial intelligence, specifically machine learning algorithms, we conducted queries based on keywords such as 'depression,' 'social network,' and 'machine learning.' We excluded articles published before 2018 to focus on recent works. In the Scopus bibliographic database, these keyword-based queries returned over 513 research articles. We read their abstracts and selected 16 articles that appeared as the most relevant for in-depth analysis. In this section, we present the observations from these studies for each step of our proposed framework, including the phases of (1) data collection and annotation, (2) preprocessing, (3) hyperparameter tuning and (4) model training and evaluation.

Regarding data collection and annotation, we observed a relative scarcity of publicly available datasets, leading many researchers to create their own datasets using web-scraping techniques [1, 8, 16, 20, 21]. However, these approaches are often limited by annotation issues and class imbalance problems, raising the risk of overfitting and resulting in poor model generalization.

The data preprocessing phase, including cleaning and encoding steps, varies according to the dataset. For example, cleaning is typically done through tokenization, stopword elimination, and stemming [5, 7, 23, 25]. For data encoding, various methods such as TF-IDF, bag-of-words, and LIWC (Linguistic Inquiry and Word Count) are employed [5, 21, 23].

Concerning hyperparameter tuning and models training, authors trained and tested various models [5, 7, 10]. Commonly used models include DT, LR, SVM, RF, AdaBoost, and MLP. We noted a complete absence of the hyperparameter tuning phase in these works. Authors present their results without addressing this phase, using default hyperparameter values specified in software packages. However, hyperparameter tuning can enhance model performance and contribute to preventing overfitting, allowing models to generalize well to new data. For evaluating model performance, researchers typically use four main metrics: accuracy, precision, recall, and F1 score [1, 14, 21]. However, the confusion matrix, despite being a powerful tool for evaluating model performance, is underutilized. As we showed in this paper, the confusion matrix provides a detailed understanding of how a model classifies instances into different categories.

Moreover, various evaluation methods are employed, including the division into training and test sets [1, 2, 5], as well as k-fold cross-validation [27]. It is important to note that these methods can introduce overfitting issues, especially with imbalanced datasets. For example, a heavily underrepresented class can result in training or test sets that do not adequately capture the variability of that class.

5 Conclusion and Future Work

In this paper, we proposed a framework using Machine Learning techniques in order to detect signs of depression in tweets from the social Network X users. We have overcome several limits of existing works in terms of bias and overfitting. To this end, we have introduced phases to specifically address class imbalance issues, to optimize hyperparameters and to perform stratified cross-fold validation. The experiments conducted on the dataset we collected showed that the TD-IDF was better than BoW for encoding data and that Linear Regression (LR) and Support Vector Machine (SVM) models presented the best performances than the others in terms of accuracy. One significant advantage of LR is its explanability, which is essential for our future work that will include collaboration with healthcare professionals for further experiments and evaluation.

References

1. Angskun, J., Tipprasert, S., Angskun, T.: Big data analytics on social networks for real-time depression detection. J. Big Data **9**(1), 69 (2022)
2. Ansari, L., Ji, S., Chen, Q., Cambria, E.: Ensemble hybrid learning methods for automated depression detection. IEEE Trans. Comput. Soc. Syst. **10**(1), 211–219 (2022)
3. Al Asad, N., Pranto, M.A.M., Afreen, S., Islam, M.M.: Depression detection by analyzing social media posts of user. In 2019 IEEE International Conference on Signal Processing, Information, Communication & Systems, pp. 13–17. IEEE (2019)
4. Benamara, F., Moriceau, V., Mothe, J., Ramiandrisoa, F., He, Z. Automatic detection of depressive users in social media. In: Conférence francophone en Recherche d'Information et al. Applications (CORIA) (2018)
5. Cacheda, F., Fernandez, D., Novoa, F.J., Carneiro, V.: Early detection of depression: social network analysis and random forest techniques. J. Med. Internet Res. **21**(6). e12554 (2019)
6. Chancellor, S., De Choudhury, M.: Methods in predictive techniques for mental health status on social media: a critical review. NPJ Dig. Med. **3**(1). 43 (2020)
7. Chiong, R., Budhi, G.S., Dhakal, S., Chiong, F.: A textual-based featuring approach for depression detection using machine learning classifiers and social media texts. Comput. Biol. Med. **104499**, 135 (2021)
8. Ghosh, T., Al Banna, M.H., Al Nahian, M.J., Uddin, M.N., Kaiser, M.S., Mahmud, M.: An attention-based hybrid architecture with explainability for depressive social media text detection in Bangla. Expert Syst. Appl. **119007**, 213 (2023)
9. Go, A., Bhayani, R., Huang, L.: Twitter sentiment classification using distant supervision. CS224N project report, Stanford **1**(12) (2009)
10. Govindasamy, K.A., Palanichamy, N.: Depression detection using machine learning techniques on twitter data. In: 2021 5th International Conference on Intelligent Computing and Control Systems (ICICCS), pp. 960–966. IEEE (2021)
11. Hutto, C., Gilbert, E.: Vader: a parsimonious rule-based model for sentiment analysis of social media text. In: Proceedings of the International AAAI Conference on Web and Social Media, vol. 8, issue 1, 216–225 (2014)
12. Institute of Health Metrics and Evaluation. GHDx: https://vizhub.healthdata.org/gbd-results/
13. Islam, M.R., Kabir, M.A., Ahmed, A., Kamal, A.R.M., Wang, H., Ulhaq, A.: Depression detection from social network data using machine learning techniques. Health Inf. Sci. Syst. **6**, 1–12 (2018)
14. Kabir, M.K., Islam, M., Kabir, A.N.B., Adiba Haque, Md., Rhaman, K.: Detection of depression severity using Bengali social media posts on mental health: study using natural language processing techniques. JMIR Formative Res. **6**(9), e36118 (2022). https://doi.org/10.2196/36118
15. Kour, H., Gupta, M.K.: An hybrid deep learning approach for depression prediction from user tweets using feature-rich CNN and bi-directional LSTM. Multimedia Tools Appl. **81**(17), 23649–23685 (2022)
16. Lin, C., et al.: Sensemood: depression detection on social media. In: Proceedings of the 2020 International Conference on Multimedia Retrieval, pp. 407–411 (2020)
17. Majumder, P., Mitra, M., Chaudhuri, B.B.: N-gram: a language independent approach to IR and NLP. In: International Conference on Universal Knowledge and Language, vol. 2 (2002)
18. Musleh, D.A., Alkhales, T.A., Almakki, R.A., Alnajim, S.E., Almarshad, S.K., Almuqhim, A.A.: Twitter Arabic sentiment analysis to detect depression using machine learning. Comput. Mater. Continua **71**(2) (2022)
19. Bird, S.: NLTK: the natural language toolkit. In: Proceedings of the COLING/ACL 2006 Interactive Presentation Sessions, pp. 69–72 (2006)

20. Ríssola, E.A., Bahrainian, S.A., Crestani, F.: A dataset for research on depression in social media. In: Proceedings of the 28th ACM Conference on User Modeling, Adaptation and Personalization, pp. 38–342 (2020)
21. Safa, R., Bayat, P., Moghtader, L.: Automatic detection of depression symptoms in twitter using multimodal analysis. J. Supercomput. **78**(4), 4709–4744 (2022)
22. Scikit-learn. https://scikit-learn.org/stable/
23. Tadesse, M.M., Lin, H., Xu, B., Yang, L.: Detection of depression-related posts in Reddit social media forum. IEEE Access **7**, 44883–44893 (2019)
24. Twint. https://github.com/twintproject/twint
25. Vasha, Z.N., Sharma, B., Esha, I.J., Al Nahian, J., Polin, J.A.: Depression detection in social media comments data using machine learning algorithms. Bull. Electr. Eng. Inf. **12**(2), 987–996 (2023)
26. Vijayarani, S., Ilamathi, M.J., Nithya, M.: Preprocessing techniques for text mining-an overview. Int. J. Comput. Sci. Commun. Netw. **5**(1), 7–16 (2015)
27. Wong, T.T.: Performance evaluation of classification algorithms by k-fold and leave-one-out cross validation. Pattern Recogn. **48**(9), 2839–2846 (2015)
28. World Health Organization (WHO). https://www.who.int/fr/health-topics/depression
29. Yang, K., Zhang, T., Ananiadou, S.: A mental state knowledge–aware and contrastive network for early stress and depression detection on social media. Inf. Process. Manage. **59**(4), 102961 (2022)
30. Zeberga, K., Attique, M., Shah, B., Ali, F., Jembre, Y.Z., Chung, T.S.: A novel text mining approach for mental health prediction using Bi-LSTM and BERT model. Comput. Intell. Neuroscience (2022)

Sweeping Knowledge Graphs
with SPARQL Queries to Palliate Q/A
Problems

Wiem Baazouzi[1]([⊠]), Marouen Kachroudi[2], and Sami Faiz[3]

[1] Université de la Manouba, Ecole Nationale des Sciences de l'Informatique,
Laboratoire de Recherche en génIe logiciel, Application Distribuées, Systèmes
décisionnels et Imagerie intelligente, LR99ES26, 2010 Tunis, Tunisie
wiem.baazouzi@ensi-uma.tn
[2] Université de Tunis El Manar, Faculté des Sciences de Tunis, Informatique
Programmation Algorithmique et Heuristique, LR11ES14, 2092 Tunis, Tunisie
marouen.kachroudi@fst.rnu.tn
[3] Université de Tunis El Manar, Ecole Nationale d'Ingénieurs de Tunis,
Laboratoire de Télédétection et Systèmes d'Information à Référence Spatiale,
99/UR/11-11, 2092 Tunis, Tunisie
sami.faiz@insat.rnu.tn

Abstract. Question answering over knowledge graphs (QA-KG) seeks
to leverage the information stored within a knowledge graph (KG) to
respond to questions posed in natural language. This approach assists
end users in accessing the extensive and valuable information contained
in the KG more efficiently and effortlessly, without requiring them to be
familiar with its underlying data structures. QA-KG presents a significant challenge, as extracting the semantic meaning of natural language
is a complex task for machines. Simultaneously, numerous techniques for
embedding knowledge within knowledge graphs have been introduced.
Knowledge Graph-based Question Answering (KGQA) still faces difficult challenges when transforming natural language (NL) into SPARQL
queries. Most systems answer simple questions referring only to a triple,
but more complex questions requiring complex queries containing subqueries or multiple functions are still a difficult challenge in this research
area. This article presents an approach that addresses these challenges
by providing several key features. First, it facilitates data annotation
to address concerns about misspelled and incomplete metadata. Second,
it allows data repair to handle missing values in the dataset. Third, it
provides data augmentation capabilities, allowing the dynamic addition
of meaningful columns and their corresponding cell values. Finally, he
learns to answer questions based on SPARQL queries within the framework of Knowledge Graphs. The effectiveness of this approach has been
evaluated using benchmark data sets with promising results.

Keywords: SPARQL Query · Knowledge Graph · Data annotation ·
Question Answering

© The Author(s), under exclusive license to Springer Nature Switzerland AG 2024
L. Barolli (Ed.): AINA 2024, LNDECT 200, pp. 316–330, 2024.
https://doi.org/10.1007/978-3-031-57853-3_27

1   Introduction

The aim of Knowledge Graph Question Answering (KGQA) research is to create an interaction approach that enables users to access extensive knowledge stored in a graph model using natural language queries. KGQA systems can be intricate pipelines with multiple task components or streamlined end-to-end solutions, often utilising deep neural networks, all presented through an intuitive and user-friendly interface. The task of developing successful KGQA systems has become more challenging due to the increasing volume and diversity of data within the Semantic Web and Linked Open Data cloud. New systems must efficiently manage larger quantities and a wider range of knowledge. Furthermore, there is an urgent need for enhanced multilingual capabilities to enhance the global accessibility of KGQA systems. The paper is thus organized as follows: In Sect. 2, the existing literature on question-answer systems based on Knowledge Graphs is reviewed. Section 3 present the proposed system and provides an illustration of the process. The experimentation, results, and system performance are covered in Sect. 4. Finally, the study is summarized in Sect. 5.

2   Background and Limitations

This section offers a concise overview of the fundamental ideas essential to comprehending the Knowledge Graph Question Answering (KGQA) field. It encompasses a structured explanation of Knowledge Graphs (KGs), an outline of SPARQL inquiries, and a delineation of the KGQA objective along with its related subtasks.

2.1   Knowledge Graphs

A Knowledge Graph (KG) serves as a structured depiction of factual information pertinent to a specific domain, encompassing both general and specialized areas. It consists of entities signifying the subjects of focus within the domain and relations (also interchangeably referred to as predicates or properties) signifying the connections between these entities. Formally, a Knowledge Graph is a multi-graph, symbolized as $\mathcal{KG} = (V, G)$, where V represents a collection of vertices and $E \subseteq (\mathcal{E} \cup \mathcal{B} \cup \mathcal{C}) \times R \times (\mathcal{E} \cup \mathcal{B} \cup \mathcal{C} \cup \mathcal{L})$ is a set of edges expressed as RDF triples. Here, $\mathcal{R}$ denotes the set of relations, $\mathcal{E}$ signifies the set of entities, $\mathcal{B}$ encompasses a set of blank nodes denoting the presence of a resource with a unique identifier, $\mathcal{C}$ stands for the set of classes, and $\mathcal{L}$ represents the set of non-unique literal values such as strings and dates. The Knowledge Graph can be further divided into an ontology segment, often referred to as the schema or T-Box denoted by $\mathcal{O} \subseteq \mathcal{C}) \times \mathcal{R} \times \mathcal{C}$, and a collection of facts, instances, or the so-called A−Box $\mathcal{M} \subseteq (\mathcal{E} \cup \mathcal{B}) \times \mathcal{R} \times (\mathcal{E} \cup \mathcal{B} \cup \mathcal{E} \cup \mathcal{L})$, as defined by Gu et al. in their work [5].

2.2   SPARQL Query

Formal query languages, such as SPARQL, -DCS, or FunQL, play a central role in retrieving and manipulating data stored within a Knowledge Graph (KG). These languages are characterized by precisely defined grammatical rules and structures, enabling the retrieval of intricate facts involving logical operators (like disjunction, conjunction, and negation), aggregation functions (such as grouping or counting), conditional-based filtering, and other ranking mechanisms. Among these languages, SPARQL stands out as a widely utilised option. The term "SPARQL" is a recursive acronym signifying "SPARQL Protocol and RDF Query Language." It serves as one of the most prevalent query languages for KGs and is supported by numerous publicly accessible KG repositories like DBPedia, Wikidata, and Freebase. For an in-depth grasp of SPARQL's syntax and semantics, readers seeking detailed information are directed to the W3C Technical Report[1]. At its core, a SPARQL query comprises a graph model, which involves KG resources (entities and relationships) and variables that can be associated with multiple KG resources.

2.3   Question Answering over KG

Utilised the introduced concepts, we proceed to establish a formal definition for the Knowledge Graph Question Answering (KGQA) task. In straightforward terms, KGQA involves the retrieval of responses to natural language questions (NLQs) or directives from a Knowledge Graph (KG). While the anticipated answers can assume intricate structures, a considerable portion of KGQA research operates under the assumption that responses generally adopt simpler forms: a collection of entities and/or literals, or binary Boolean outcomes. We can present this latter definition in a more rigorous manner as follows. Suppose $\mathcal{K}$ represents a KG and q denotes an NLQ. We then define the set of all conceivable answers $\mathcal{A}$ as the compilation of (a) the power set[2] $\mathcal{P}$ ($\mathcal{E} \cup \mathcal{L}$) comprising entities $\mathcal{E}$ and literals $\mathcal{L}$ within $\mathcal{K}$, (b) the outcome set of all possible aggregation functions $f\colon \mathcal{P}(\mathcal{E} \cup \mathcal{L}) \to \mathbb{R}$ (such as SUM or COUNT), and (c) the set of potential Boolean outcomes, True and False, required for addressing yes/no inquiries. Consequently, the KGQA task is formally defined as the provision of the accurate answer $\dashv \in \mathcal{A}$ for a given question q. A common approach employed by KGQA systems is to devise a structured query that mirrors the semantic arrangement of the question q. When executed on $\mathcal{K}$, this query yields the answer $\dashv$. As such, KGQA can be perceived as a domain within the sphere of semantic analysis. In a broader context, semantic analysis encompasses the conversion of a natural language statement q into a formal, executable representation of its meaning $f \in \mathcal{F}$. For a contemporary overview of semantic analysis, interested readers can refer to [6]. Queries addressed by KGQA systems are often classified into two categories: *simple* and *complex*. Questions that can be resolved by considering a

---

[1] https://www.w3.org/TR/rdf-sparql-query/#introduction.

[2] A collection of all subsets of a given set of elements.

single factual piece, devoid of supplementary constraints, are generally categorized as simple inquiries. Conversely, a question is deemed complex if answering
it necessitates awareness of multiple facts, as it entails a more intricate interplay
of information.

2.4   KGQA Subtasks

Question answering (QA) systems customarily break down the process of answering questions into three distinct stages: **(I)** *question comprehension,* which
involves extracting entities and relationships from the question to formulate an
abstract representation; **(II)** *binding,* which involves associating the abstract representation with corresponding vertices and predicates within the target Knowledge Graph (KG) to construct a properly structured SPARQL query; and **(III)**
filtering, which entails the selection of the most pertinent responses, considering
specific criteria. In the following section, we delve into the discussion of three
established QA systems: gAnswer, EDGQA, and NSQA . These systems serve
as exemplars of highly effective methodologies, representing the cutting-edge in
terms of precision (measured by F1 score) within the QALD challenge[3].

1. **Question Understanding:**

   The phase of question understanding involves recognizing the entities, variables, and connections referenced within the natural language question. Additionally, it produces a conceptual structured depiction, usually in graphical
   form, illustrating how these elements are interconnected.

   gAnswer system employs Stanford's dependency parser[4] to translate a question into a syntactic dependency tree. This tree is then traversed to produce
   an abstract visual depiction of the query known as the semantic query graph.
   To achieve this, a collection of fixed heuristic linguistic rules is applied, structured based on the QALD-9 training questions. For instance, question words
   that initiate with "wh*" (such as "what") are presumed to indicate unknowns.
   The procedure for constructing and utilising these models bears resemblance
   to our KEPLER-ASI+QA approach. Furthermore, gAnswer relies on a predefined set of synonyms[5].

   EDGQA utilising Stanford's basic NLP parser[6] to construct the question's
   constituency parse tree. Subsequently, it generates a rooted acyclic graph
   known as an entity description graph by sequentially breaking down the parse
   tree. This process is guided by rules tailored to the QALD-9 database. In the
   given instance, EDGQA erroneously identified[7] the Danish strait as a relation
   and the flow as an entity, resulting in the generation of this triple pattern:

---

[3] http://qald.aksw.org.

[4] https://nlp.stanford.edu/software/lex-parser.shtml.

[5] Predefined Synonyms 2022: https://drive.google.com/file/d/1hmqaftrToOqQNRAp
   CuxFXaBx7SosNVy/view.

[6] CoreNLP. 2022. https://stanfordnlp.github.io/CoreNLP/.

[7] EDGQA Log. 2021:https://raw.githubusercontent.com/HXX97/EDGQA/main/.

⟨unknown, Danish strait, flow⟩.

NSQA employs a deep neural symbolic framework to examine the question through a directed rooted acyclic graph, which is termed as an abstract meaning representation. This analysis is carried out by training a neural network on verb-oriented relations extracted from Propbank. The training dataset encompasses a variety of organized tree structures designed for different question types. The effectiveness of the model relies on the diversity inherent in these tree structures. Whether the model is capable of identifying relationships solely based on noun phrases, such as "city to land," remains uncertain10. The resulting graph exhibits a distinct level of detail compared to SPARQL, necessitating potential adjustments [7] to adapt the model for use in new domains.

2. **Entity and Relation Linking:**

Within Question-Answering (QA) systems, the process of binding, as pointed out by Lin et al.[8], presents a formidable challenge. This task involves the intricate endeavor of associating entities and relationships extracted from a given question with their corresponding nodes and predicates within a Knowledge Graph (KG). Current approaches within this domain formalize this association process in the following manner: **(i)** a *lookup* in an inverted index that maps synonyms to relevant vertices and predicates of the target KG, or **(ii)** a *keyword search* on a database of tagged documents constructed from the vertices and predicates of the target KG, or **(iii)** a *deep learning-based transformation* defined on the vector space of the target KG. These methods are KG-specific and require expensive pre-processing for each target KG:

gAnswer achieves entity linkage through the using of the crossWikis dictionary[8], which is constructed from a comprehensive web crawl. This dictionary establishes a connection between each entity and its corresponding vertex within the designated Knowledge Graph (KG), accompanied by a confidence score that quantifies the reliability of the association. In the realm of relational linkage, gAnswer develops a separate dictionary that connects relational mentions to their corresponding predicates within the target KG. For instance, a relational phrase such as "starred by" is mapped to the specific "starred" predicate within a particular KG. The process of generating these mappings involves employing general Natural Language Processing (NLP) techniques, or alternatively, utilising n-grams.

EDGQA employs a trio of indexing mechanisms, specifically Falcon [13], EARL [4], and Dexter [14], to establish indices for the vertices and predicates encompassed within the target Knowledge Graph (KG). The process within EDGQA combines the outcomes of these three systems to create associations between entities and KG vertices. Subsequently, this information is harnessed to narrow down the scope of exploration for relational connections. Eventually, a BERT-based model is employed to rank the array of predicates. The mentioned indexing mechanisms operate by dissecting the target KG,

---

[8] https://github.com/nitishgupta/crosswikis.

extracting vertices and predicates, and capturing their accompanying descriptions. For instance, Falcon utilises conventional #label and #altLabel predicates to amass descriptions and synonyms. Within this context, Falcon undertakes parts-of-speech tagging and n-gram extraction to produce tagged documents, facilitating subsequent keyword searches during the linking process. In scenarios where a KG lacks the #label predicates, a suitable indexing predicate must be manually designated for each category of vertex. As an illustration, for vertices of the "author" type, "authorName" could be chosen as the indexing predicate.

NSQA differs slightly in its approach, focusing on the identification of entities and relationships during the linking process. It commences by utilising Blink [16] to create embeddings of the targeted Knowledge Graph (KG) vertices within a vector space. Subsequently, it takes in the abstract graph obtained from the preceding phase and leverages a BERT-based neural model, which is trained on 3,651 questions from LC-QuAD 1.0. These questions possess manually annotated mentions, aiding in the detection of entities. These identified entities are then seamlessly integrated into the same vector space as the KG and are connected to their nearest neighboring vertices. When it comes to relational linking, NSQA adopts SemRel [9], a neural model based on transformers. SemRel is trained to predict relationships, thereby generating a list of prospective relation candidates from the pool of relations existing within the KG, subsequently arranging them in order of relevance. It's worth noting that within NSQA, the most resource intensive aspect of the preprocessing phase revolves around constructing the vector space, encompassing the integration of target vertices and KG predicates.

3. **Filtering:**
Some question answering (QA) systems, like EDGQA, incorporate a step where potential answers are screened using an RDF engine. During this filtering process, these systems gather prior information in advance, in a preprocessing phase, about various types or classes of nodes in a Knowledge Graph (KG). The filtering is carried out by associating an unknown value (e.g., ?sea) with a specific vertex type within the target KG, provided that such a type exists. The SPARQL query is then adjusted to include the type constraint and executed, leading to the retrieval of pre-filtered responses. Additionally, alternative approaches employ different models, such as those based on neural networks. However, our current work does not delve into exploring this particular research direction. Readers interested in other models are encouraged to explore relevant resources.

3   The KEPLER-ASI+QA Approach

The proposed approach KEPLER-ASI+QA implements three distinct phases: question understanding, binding, and execution via KEPLER-ASI with filtering, as flagged by Fig. 1.

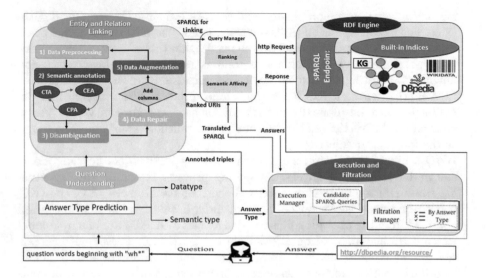

Fig. 1. KEPLER-ASI+QA Architecture

3.1   Question Understanding

KEPLER-ASI+QA introduces a formalization of question comprehension in the form of a text generation challenge. The process involves examining the input question to create a SPARQL query, with this analysis being conducted without dependence on any specific Knowledge Graph (KG). Furthermore, KEPLER-ASI+QA forecasts both the projected data and the semantic category of the resulting response. This phase takes the English question as input and yields the anticipated data type for the answer, encompassing categories like date, numeric, boolean, and string as summarized by Table 1. In the latter scenario,

Table 1. Question Types and Expected Answer Type.

Question Type	Answer Type
What	Thing
Why	Reason
Which	Choice
Whose	Possession
Where	Location or Place
When	Time or Date or Year
Who	Person
"How Much" or "How Many"	Number
Name, List, Show	Set or List or Group
Does, Has, Was, Did, Do	Yes/No or True/False

we additionally make predictions about the semantic type based on contextual cues. The nature of a question is determined by its syntactic structure, which grants us insights into extracting question characteristics and anticipating diverse potential answers. Question analysis is conducted to deduce these question attributes, facilitating the acquisition of pertinent responses. These attributes assist in establishing the question type associated with a query, thus guiding the search for an answer and specifying the required answer type. In the scope of this study, we concentrate on "Wh" questions, which exhibit distinct features, structures, and traits that enable their categorization based on question types. The proposed approach efficiently identifies answers to a range of question types, including *What, Where, How Much, Who, When, Why, Which, Whose,* and *List of Questions.* The QA system forwards the gathered details to the DBpedia projector for recognizing named entities. Leveraging Stanford CoreNLP, we pinpoint keywords and extract verbs, adjectives, and nouns, which in turn guide the identification of DBpedia properties and classes. The QA system takes measures to distinguish and exclude entities that might also be linked to classes, minimizing potential confusion. Subsequently, the system compiles a list of comparatives and superlatives to further refine the selection of queries and identify the most suitable one. These identified properties are then employed to construct the SPARQL query. The ensuing patterns outline the framework for formulating queries, where the '+' symbol indicates concatenation, the '|' symbol signifies options, 'JJS' represents an adjective in superlative form, and '*' denotes any word.

1. $[Where] + [*]+?$
2. $[What] + [JJS] + [*]+?$
3. $[Where] + [JJS] + [*]+?$
4. $[When] + [*]+?$
5. $[Who] + [JJS] + [*]+?$
6. $[How] + [much] + [*]+?$
7. $[How] + [many] + [*]+?$
8. $[List|Name|Show|Give] + [*]+?$

Entities detected are annotated with the corresponding question tags and subsequently forwarded to the SPARQL query process. The process of selecting the appropriate SPARQL query is now elucidated based on the nature of the question. The SPARQL Query is equipped with pre-established templates catering to various question types, illustrating how to construct a SPARQL query for each specific type. The Question Answering (QA) system then supplements properties, classes, and entities to the chosen Query, derived from the question processing segment, in accordance with the question type. These queries are ultimately executed in the DBpedia endpoint utilising SPARQL queries custom-tailored for a range of question types, encompassing comparisons and superlatives. The outcome of these queries is ranked based on their attributes, pinpointing the highest count of triples within DBpedia. Post-ranking, the foremost query in the list predominantly holds the correct answer to the user's inquiry. The proposed algorithm outlining the construction of the SPARQL query from the input

---

Algorithm 1: Query Formulation Algorithm

Data: Question q, Question type t

Result: SPARQL Query

1   BoolQuestion=Arrays.asList('Did','Was','Have','Does','Were','Is','Are','Be','Do');
AnswerType=Arrays.asList('Boolean','Date','Number',
'String','Literal','Resource'); builder=new SparqlQueryBuilder(q); result=null;

2   **switch** t **do**

3   **case** *'WHO'* **do**

4   **if** *q.contains(superlative)* **then**

5   result=builder.sparqlWho();

6   break;

7   **case** *'HOW'* **do**

8   **if** *q.contains(superlative)* **then**

9   result=builder.sparqlHow();

10   break;

11   **case** *'WHERE'* **do**

12   **if** *q.contains(superlative)* **then**

13   result=builder.sparqlWhere();

14   break;

15   **case** *'WHAT'* **do**

16   **if** *q.contains(superlative)* **then**

17   result=builder.sparqlWhat();

18   break;

19   **case** *'WHICH'* **do**

20   **if** *q.contains(superlative)* **then**

21   result=builder.sparqlWhich();

22   break;

23   **case** *'WHEN'* **do**

24   result=builder.simpleSparql(",' FILTER ((datatype(?answer)
xsd:date)|| (datatype(?answer) xsd:gYear))');

25   result= builder.sparqlList(); **if** *(BoolQuestion.contains(t))* **then**

26   q.questionType='ASK'; result=builder.simpleASK();

27   **else**

28   result=builder.simpleSparql();

29   break;

30   **while** *(result=null||result.isEmpty())and (q.entityList!=null) and
(q.entityList.size()>1) and (builder.getEntityIndex()<q.entityList.size()))*
do

31   container.setAnswers(result);
container.setSparqlQuery(builder.getLastUsedQuery()); return
container; builder.incrementIndex();

32   **return** (SPARQL Query)

---

question is outlined in Algorithm 1. The KEPLER-ASI+QA system transmits the details assembled by the SPARQL Query Builder Components, such as extracted nouns (representing classes), verbs, and adjectives (indicating properties), to the QA projector to discern named entities. All identified entities are then affixed with corresponding question labels and subsequently relayed to the QA Match with KG component. The ensuing subsection delves into the selection process of the Annotation, Repair, and Augmentation Components within the Knowledge Graph (KG) DBpedia context.

3.2 Entity and Relation Linking

In a intuitive sense, binding refers to the meaningful connection between the elements present in the graph and the actual information within the target Knowledge Graph (KG). Since we have no preliminary information about the target KG, the matching process must occur at the moment of answering questions, in a just-in-time fashion. KEPLER-ASI+QA is a system designed to recognize entities, classes, and relationships in order to complete placeholders within a SPARQL model. Due to the brevity of questions compared to longer texts, conventional semantic entity linking tools like DBpedia Spotlight [3] or MAG [10] are ineffective due to the absence of sufficient contextual information. To overcome this limitation, KEPLER-ASI+QA employs an approach termed KEPLER-ASI, which involves annotating, repairing, and enhancing the Knowledge Graph to identify potential entity, property, and class candidates within the KG, independent of specific questions. KEPLER-ASI+QA follows a three-phase process established prior to the execution of filtering steps.

1. **Data Annotation:** The primary objective of the KEPLER-ASI [1] approach is to automatically provide annotations to tabular data elements through integration with the DBpedia Knowledge Graph (KG). This stage builds upon the KEPLER-ASI approach [2] and involves the generation of three distinct tasks[9]:

   (a) Column Type Annotation (CTA): This task focuses on the assignment of ontology classes to the columns present in the table.
   (b) Column Entity Annotation (CEA): This task involves linking table cells (values) to corresponding entities within the DBpedia KG.
   (c) Column Property Annotation (CPA): The goal of this task is to establish mappings between pairs of columns and ontology properties.

   To enhance the matching process, KEPLER-ASI employs a variety of services, including DBpedia search, interaction with the DBpedia endpoint, and using of WikiData search. This multifaceted approach contributes to an improved and comprehensive matching mechanism.

---

[9] https://sem-tab-challenge.github.io/2023/.

2. **Data Repair:**

The KEPLER-ASI approach is designed to automate the inclusion of missing cell features (values) or undefined 'nan' values. Our algorithm employs SPARQL queries that take into account the cell entity annotations (CEA) from the subject of the column (Col Subject) and the column property annotations (CPA) to retrieve the absent cell entities (i.e., CEA). An illustrative SPARQL query serves to exemplify the process of obtaining the missing cell feature for the first row in the given dataset.

3. **Data Augmentation:**

The KEPLER-ASI approach empowers users to supplement the annotated dataset with pertinent columns. This is achieved by applying a straightforward SPARQL query to the applied column property annotations (CPA). Multiple Uniform Resource Identifiers (URIs) for column properties (CPA) might be present, such as http://dbpedia.org/ontology/birthDate and http://dbpedia.org/ontology/birthDeath. The user has the option to select a single CPA, resulting in the addition of KEPLER-ASI as a new column in the dataset. Alternatively, they can choose to add a column with the exact name, such as 'birthDate'. Consequently, the system provides the user with the flexibility to incorporate the chosen CPA, such as http://dbpedia.org/ontology/birthDate, into the dataset.

3.3   Filtering

Our primary emphasis is on the associations between semantic types, which are contingent on the contextual nuances of a question, and the entities/relations present in the DBpedia Knowledge Graph. We evaluate the similarity of the question against all DBpedia types and identify types with high similarity scores as potential candidates. To generate the set of candidate answers, denoted as C_R, for a given question Q, we construct it by amalgamating the entities, relations, and constraints derived from Query S. Specifically, we merge the triples derived from the descriptions into a query S tailored to address the question Q. This query S can be envisioned as a collection of triplets signifying potential answers to question Q. In order to assess the score for each candidate, we compute the product of the similarity scores associated with all components within S. This aggregated score at the candidate level, denoted as C_R, is calculated by multiplying the relevant similarity scores. The final score for a candidate is determined by a combination of the aforementioned two scores, as presented in Eq. (1):

$$P_{candidate-Answer_{(C_R)}} = \lambda \times P_{comp(S)} + (1 - \lambda) \times P_{comp(Q)} \tag{1}$$

In conclusion, with λ representing a hyper-parameter, a prioritized inventory of candidate subqueries is formulated for block b. From this collection, we choose the highest-ranked top-k candidate-Responses (C_R) to serve as the outcome of the filtering stage.

4   Evaluation

In this section, we will showcase the outcomes achieved by Kepler-aSI+QA through a comparison with alternative systems across various Benchmark datasets. These outcomes underscore the commendable performance of Kepler-aSI+QA, even in the face of numerous challenges and complications.

4.1   Compared Systems

We assess the performance of KGQAn against benchmark systems, namely gAnswer, EDGQA, and NSQA. EDGQA has demonstrated advanced capabilities in the QALD-9 and LC-QuAD 1.0 contexts. Notably, gAnswer attained the top ranking in the QALD-9 challenge [11]. NSQA, leveraging deep learning models for question comprehension and linking, surpassed gAnswer's performance in the QALD-9 setting. While the code for gAnswer[10] and EDGQA[11] is accessible, we have successfully reproduced the outcomes of gAnswer and EDGQA. In the case of NSQA, which employs a logical neural network [12] and specialized datasets for AMR model training, both resources are unavailable. Consequently, we present the results solely as provided in [7].

4.2   Benchmarks and Question Sets

QA benchmarks consist of English questions that have been annotated alongside corresponding SPARQL queries and the correct answer set, tailored to a specific version of a Knowledge Graph (KG). QALD-9 and LC-QuAD 1.0 are widely employed for evaluating QA systems across various iterations of the DBpedia KG. QALD-9 comprises 408 questions for training and 150 questions for testing, with an average question length of 7.5 words. LC-QuAD 1.0, on the other hand, encompasses a training set of 4000 questions and a separate test set containing 1000 questions. These questions are constructed using diverse templates to capture a range of query types. In addition, our evaluation will encompass the using of QALD-10[12]. This more recent benchmark involves a training set comprising 412 questions. Its primary objective is to assess QA systems in tackling questions that bear similarities to those in QALD-9, within the realms of both well-established and less conspicuous domains within the KG.

4.3   Kepler-aSI+QA implementation

In adherence to the FAIR principles (Findable, Accessible, Interoperable, and Reusable) outlined in Wilkinson et al.'s work [15], we uphold these principles concerning our experimental findings by leveraging GERBIL QA[13], an accessible and open-source online evaluation tool.

---

[10] GAnswer code. 2022. https://github.com/pkumod/gAnswer.
[11] EDGQA code. 2021. https://github.com/HXX97/EDGQA..
[12] https://www.nliwod.org/challenge.
[13] https://gerbil-qa.aksw.org/gerbil.

Table 2. Results for KEPLER-ASI+QA

System	LC-QuAD 1.0			QALD-9[a]			QALD-10[b]		
	P	R	F1	P	R	F1	P	R	F1
NSQA	0.447	0.458	0.444	0.318	0.320	0.316			
gAnswer	0.394	0.398	0.318	0.294	0.326	0.298	0.453	0.457	0.453
EDGQA	0.505	0.56	0.531	0.313	0.403	0.320	0.419	0.408	0.414
KEPLER-ASI+QA	0.578	0. 520	0.547	0. 554	0.456	0.500	0. 580	0.471	0.520

[a]https://github.com/KGQA/QALD-9
[b]https://github.com/KGQA/QALD-10

Within this framework, we adhere to established evaluation metrics for Knowledge Graph Question Answering (KGQA), which encompass key parameters such as Precision, Recall, and the Macro F1 measures as defined in the QALD benchmarks. A concise overview of the three benchmarks we employ is presented in Table 2. Our approach was implemented using the Google Colaboratory platform[14], which provides a fully integrated Jupyter notebook environment.

5 Conclusion and Outlooks

Current QA systems are tailored to specific domains and Knowledge Graphs, necessitating a costly preprocessing phase. This article introduces KEPLER-ASI+QA, which proposes a fresh conceptualization for question comprehension involving triple extraction. Our model has the capacity to generalize its comprehension across diverse domains. Furthermore, KEPLER-ASI+QA introduces an innovative just-in-time binding and filtering approach that executes entity and relationship binding through semantic search queries. Our assessment of prevailing QA systems, using widely-accepted benchmarks and three distinct real Knowledge Graphs, demonstrates that KEPLER-ASI+QA get an $F1$ score comparable to other participants in standard benchmarks, while surpassing QALD participants. Our forthcoming endeavors encompass the expansion of our question comprehension training to encompass more intricate queries, as well as the provision of support for questions with multiple intents in a purely interactive framework.

---

[14] https://colab.research.google.com/?utm_source=scs-index.

References

1. Baazouzi, W., Kachroudi, M., Faiz, S.: A matching approach to confer semantics over tabular data based on knowledge graphs. In: Fournier-Viger, P., Hassan, A., Bellatreche, L. (eds.) Model and Data Engineering. MEDI 2022. LNCS, vol. 13761. Springer, Cham (2023). https://doi.org/10.1007/978-3-031-21595-7_17
2. Baazouzi, W., Kachroudi, M., Faiz, S.: Yet another milestone for kepler-aSI at SemTab 2022. In: Proceedings of the Semantic Web Challenge on Tabular Data to Knowledge Graph Matching (SemTab 2022) co-located with the 21th International Semantic Web Conference (ISWC 2022), Virtual Event, 23–27 October 2022, Proceedings, pp. 80–91. Springer (2022)
3. Daiber, J., Jakob, M., Hokamp, C., Mendes, P.: Improving efficiency and accuracy in multilingual entity extraction. In: Proceedings of the 9th International Conference On Semantic Systems, pp. 121–124 (2013)
4. Dubey, M., Banerjee, D., Chaudhuri, D., Lehmann, J.: EARL: joint entity and relation linking for question answering over knowledge graphs. In: Vrandečić, D., et al. (eds.) ISWC 2018. LNCS, vol. 11136, pp. 108–126. Springer, Cham (2018). https://doi.org/10.1007/978-3-030-00671-6_7
5. Gu, Y., et al.: Beyond IID: three levels of generalization for question answering on knowledge bases. In: Leskovec, J., Grobelnik, M., Najork, M., Tang, J., Zia, L., (Eds.) Beyond I.I.D. Three Levels of Generalization for Question Answering on Knowledge Bases. In WWW 2021: The Web Conference 2021, Virtual Event/Ljubljana, Slovenia, 19–23 April (2021)
6. Kamath, A., Das, R.: A survey on semantic parsing (2018). arXiv preprint arXiv:1812.00978
7. Kapanipathi, P., et al.: Leveraging abstract meaning representation for knowledge base question answering (2020). arXiv preprint arXiv:2012.01707
8. Lin, X., Li, H., Xin, H., Li, Z., Chen, L.: KBPearl: a knowledge base population system supported by joint entity and relation linking. Proc. VLDB Endowment **13**(7), 1035–1049 (2020)
9. Naseem, T., et al.: A semantics-aware transformer model of relation linking for knowledge base question answering. In: Proceedings of the 59th Annual Meeting of the Association for Computational Linguistics and the 11th International Joint Conference on Natural Language Processing (Volume 2: Short Papers), pp. 256–262 (2021)
10. Ngonga Ngomo, A.-C., et al.: Entity linking in 40 languages using MAG. In: Gangemi, A., et al. (eds.) ESWC 2018. LNCS, vol. 11155, pp. 176–181. Springer, Cham (2018). https://doi.org/10.1007/978-3-319-98192-5_33
11. Ngomo, N.: 9th challenge on question answering over linked data (QALD-9). Language **7**(1), 58–64 (2018)
12. Riegel, R., et al.: Logical neural networks (2020). arXiv preprint arXiv:2006.13155
13. Sakor, A., et al.: Old is gold: linguistic driven approach for entity and relation linking of short text. In: Proceedings of the 2019 Conference of the North American Chapter of the Association for Computational Linguistics: Human Language Technologies, Volume 1 (Long and Short Papers), pp. 2336–2346 (2019)
14. Trani, S., Ceccarelli, D., Lucchese, C., Orlando, S., Perego, R.: Dexter 2.0: an open source tool for semantically enriching data. In: Proceedings of the 2014 International Conference on Posters & Demonstrations Track. vol. 1272, pp. 417–420. Citeseer (2014)

15. Wilkinson, M., et al.: The fair guiding principles for scientific data management and stewardship. Sci. data **3**(1), 1–9 (2016)
16. Wu, L., Petroni, F., Josifoski, M., Riedel, S., Zettlemoyer, L.: Scalable zero-shot entity linking with dense entity retrieval (2019). arXiv preprint arXiv:1911.03814

Trustworthiness of 𝕏 Users: A One-Class Classification Approach

Tanveer Khan[✉], Fahad Sohrab, Antonis Michalas, and Moncef Gabbouj

Tampere University, Tampere, Finland
{tanveer.khan,fahad.sohrab,antonios.michalas,moncef.gabbouj}@tuni.fi

Abstract. 𝕏 (formerly Twitter) is a prominent online social media platform that plays an important role in sharing information making the content generated on this platform a valuable source of information. Ensuring trust on 𝕏 is essential to determine the user credibility and prevents issues across various domains. While assigning credibility to 𝕏 users and classifying them as trusted or untrusted is commonly carried out using traditional machine learning models, there is limited exploration about the use of One-Class Classification (OCC) models for this purpose. In this study, we use various OCC models for 𝕏 user classification. Additionally, we propose using a subspace-learning-based approach that simultaneously optimizes both the subspace and data description for OCC. We also introduce a novel regularization term for Subspace Support Vector Data Description (SSVDD), expressing data concentration in a lower-dimensional subspace that captures diverse graph structures. Experimental results show superior performance of the introduced regularization term for SSVDD compared to baseline models and state-of-the-art techniques for 𝕏 user classification.

1 Introduction

Online Social Networks (OSNs) have become an essential tool for modern communication, enabling people to interact, while spending significant time on these platforms. It is now becoming an integral part of our lives, and people are using it for different purposes, including connection with friends and family, participation in online communities, brand promotion, finding and sharing information, and much more [1]. The most popular OSNs include Facebook, 𝕏, Instagram, and LinkedIn. This work focuses on 𝕏 – an OSN platform that allows users to share and discover short messages or tweets limited to 280 characters. 𝕏 has 310 million active users publishing 500 million tweets per day [2]. Also, 𝕏 has become a valuable tool enabling users to share information with a wide audience quickly and easily. It allows users to see tweets relevant to their interests, retweet or like other users' tweets, or post their own tweets. While this makes it easy for 𝕏 users to share updates and information with their followers in real-time, it also makes it easier for fake account users to carry out malicious activities such as sharing unverified information [3]. Additionally, it has been observed that fake news spreads more rapidly on 𝕏 than real news, damaging the reputation and reliability of

---

Tanveer Khan, Fahad Sohrab : The first two authors contributed equally to this work.

© The Author(s), under exclusive license to Springer Nature Switzerland AG 2024
L. Barolli (Ed.): AINA 2024, LNDECT 200, pp. 331–343, 2024.
https://doi.org/10.1007/978-3-031-57853-3_28

the X. Various techniques have been proposed [4] to tackle the spread of false information, with one approach being the classification of X users as trusted or untrusted [5]. This classification is of significant importance in maintaining the reputation and reliability of the X platform. For example, identifying a trusted and reliable X user ensures the continued success and usefulness of the X platform as a trusted and valuable social media tool.

There are various ways for X user classification such as using Machine Learning (ML) [6], and Natural Language Processing (NLP) [7]. Among these, ML models have been widely used in various research to classify X users into different categories based on their profiles, activity, and content of the tweets. The process involves collecting and preprocessing large amounts of data, including user profiles and tweets, and then training an ML model to classify users into different categories based on the available features. The model can then be used to classify new, unseen users. The ML algorithms used for X user classification are supervised [8], unsupervised [9], and semi-supervised [10, 11]. Classifying X users as trusted or untrusted using only ML models can be challenging due to high-dimensional and variable characteristics of big data [12]. Despite the *curse of dimensionality* and the imbalanced nature of the data, the appropriate techniques and models hold the potential to address these challenges successfully. In our approach, we rely on OCC models, where the decision function is inferred using training data from a single class only. It is used when a large amount of data is available for the class of interest but little or no data is available for other classes [13]. OCC differs from traditional binary classification models, which are trained using data from both categories. We use a manually labeled dataset obtained from Khan *et al.* [10] research, which involved gathering data for 50,000 X users, with manual labeling for 1,000 of them. By applying different OCC models to the labeled dataset, our goal is to answer the following **research questions (RQs)**:

RQ 1: How effective is the OCC in accurately identifying political X users as trusted or untrusted, and what are the comparative strengths and weaknesses among different OCC models in this context?

RQ 2: What are the key challenges OCC faces when classifying political users on X, and can the performance of OCC be optimized for political user identification through subspace learning for OCC?

RQ 3: Can we encode the relationships between the training data points in a lower-dimensional subspace optimized for OCC while capturing and preserving the local structure of target class data?

Contributions: The main contributions of this work can be summarized as follows:

C1. We propose using subspace-learning-based OCC for X user identification.

C2. We propose a novel regularizer for Subspace Support Vector Data Description (SSVDD) expressing the concentration of the data in a lower-dimensional subspace that captures different graph structures.

C3. In the proposed regularization term, any suitable graph can be used to encode the corresponding graph structure, and we evaluate its effectiveness by comparing it with different OCC models.

1.1   Organization

The rest of the paper is organized as follows. In Sect. 2, we provide necessary background information about different OCC models. In Sect. 3, we provide important published works in the area of X user credibility, accompanied by a detailed discussion of our proposed approach in Sect. 4. The data collection and experimental results are presented in Sect. 5. Finally, we conclude the paper in Sect. 6.

2   Preliminaries

In ML, OCC refers to an approach to building a model by considering data from a single class only. OCC is appropriate for scenarios where it is critical to identify one of the categories, but the examples from that specific category are scarce or statistically so diverse that they cannot be used during the training process. OCC has found application in different areas, such as early detection of myocardial infection [14], rare insect classification [15], and credit card fraud detection [16]. These applications present data scarcity challenges from one of the categories to be modeled.

Among the widely-used OCC approaches, One-class Support Vector Machine (OCSVM) and Support Vector Data Description (SVDD) have been proven as powerful data description methods over time. These methods identify the so-called *support vectors* as crucial for determining the decision boundary. In OCSVM, a hyperplane is created to separate the target class in a way that maximizes the distance of the hyperplane from the origin [17]. The classification of a new data point is determined by its location relative to the hyperplane: if it falls on the positive side, it is considered normal; otherwise, it is flagged as abnormal. SVDD, on the other hand, creates a hyperspherical boundary around the target class data within the original feature space by minimizing the volume of the hypersphere.

Let us denote the target class training samples to be encapsulated inside a hypersphere by a matrix $\mathbf{X} = [\mathbf{x}_1, \mathbf{x}_2, \ldots, \mathbf{x}_N], \mathbf{x}_i \in \mathbb{R}^D$, where N is the number of samples and D is dimensionality of data. The formulation of SVDD is expressed as follows:

$$\min \quad F(R, \mathbf{a}) = R^2 + C \sum_{i=1}^{N} \xi_i \quad \text{s.t.} \quad \|\mathbf{x}_i - \mathbf{a}\|_2^2 \leq R^2 + \xi_i, \quad \xi_i \geq 0, \, \forall i \in \{1, \ldots, N\}, \quad (1)$$

where R represents the radius, $\mathbf{a} \in \mathbb{R}^D$ is the center of the hypersphere, and slack variables $\xi_i, i = 1, \ldots, N$ are introduced to enable the possibility of target data being outliers. The hyperparameter $C > 0$ controls the trade-off between the volume of the hypersphere and the presence of data points outside the hypersphere. A test sample is assigned to the positive class if its distance from the center of the hypersphere is equal to or less than the radius R.

A distinct category in OCC, Graph Embedded One-Class Classifiers, refers to methods that integrate generic graph structures expressing relevant geometric relationships in their optimization processes. Graph Embedded One-Class Support Vector Machine (GEOCSVM) is an example that incorporates graph-based information and enhances the traditional OCSVM approach. By leveraging graph information, GEOCSVM compares favorably to the standard OCSVM. In GEOCSVM, the relationship between training patterns can be described locally and globally using a single graph or a combination of fully connected and kNN graphs [18]. Similarly, Graph Embedded Support Vector Data Description (GESVDD) is a type of OCC that combines the SVDD approach with graph-based information. In GESVDD, the graph-based information is incorporated into the optimization process of the SVDD. Like SVDD, GESVDD also creates a hypersphere around the target class data to separate the target class data from the outliers in an OCC problem. However, graph-based information in GESVDD provides additional information that can help to improve the separation of target class data from outliers [18]. Other extensions of graph-based OCC include Graph Embedded Subspace Support Vector Data Description (GESSVDD) [19] that poses the subspace learning for OCC as a graph embedding problem.

Traditional boundary-based OCC methods primarily find a data description in the given feature space. However, a contemporary paradigm shift is evident in the form of subspace learning-based techniques that not only form a data description but also optimize a subspace simultaneously. A leading technique in this paradigm is the SSVDD [20], which defines a data description along with data mapping to low-dimensional feature space optimized for OCC. To define a concise representation of the target class, the method repeatedly optimizes data mapping and data description. The optimization function of SSVDD is as follows:

$$\min \quad F(R,\mathbf{a}) = R^2 + C\sum_{i=1}^{N} \xi_i \quad \text{s.t.} \quad \|\mathbf{Q}\mathbf{x}_i - \mathbf{a}\|_2^2 \le R^2 + \xi_i, \quad \xi_i \ge 0, \ \forall i \in \{1,\ldots,N\}, \quad (2)$$

where $\mathbf{Q} \in \mathbb{R}^{d \times D}$ is the projection matrix for mapping the data from the original D-dimensional feature space to an optimized lower d-dimensional space. In SSVDD, an augmented version of the Lagrangian with a regularization term ψ is optimized:

$$L = \sum_{i=1}^{N} \alpha_i \mathbf{x}_i^\mathsf{T} \mathbf{Q}^\mathsf{T} \mathbf{Q} \mathbf{x}_i - \sum_{i=1}^{N}\sum_{j=1}^{N} \alpha_i \mathbf{x}_i^\mathsf{T} \mathbf{Q}^\mathsf{T} \mathbf{Q} \mathbf{x}_j \alpha_j + \beta \psi, \quad (3)$$

where α represents the Lagrange multipliers, and β is used to control the importance of the regularization term. The regularization term ψ expresses the class variance in the d-dimensional space and it is denoted as

$$\psi = \text{Tr}(\mathbf{Q}\mathbf{X}\lambda\lambda^\mathsf{T}\mathbf{X}^\mathsf{T}\mathbf{Q}^\mathsf{T}), \quad (4)$$

where $\text{Tr}(\cdot)$ is the trace operator and $\lambda \in \mathbb{R}^N$ is a vector used to select the contribution of certain data points in the optimization process, leading to different variants of SSVDD. The different variants are as follows.

- SSVDDψ1: In this variant, the regularization term becomes obsolete and is not used during the data description.
- SSVDDψ2: In this case, all the training samples describe the class variance in the regularization term.
- SSVDDψ3: In this case, the samples belonging to the boundary and outside the boundary are used in the regularization term.

- SSVDDψ4: In this variant, only the support vectors that belong to the class boundary are used to describe the class variance in the regularization term.

The selection of different data instances in the regularization term is carried out by replacing the λ value accordingly with the α values. The updating of the projection matrix $\mathbf{Q}$ is carried out by utilizing the gradient of (3), expressed as:

$$\mathbf{Q} \leftarrow \mathbf{Q} - \eta \Delta L. \tag{5}$$

Here, η denotes the learning rate parameter. This work primarily focuses on subspace learning-based OCC and proposes a graph-based regularization for SSVDD.

3   Related Work

A lot of research has looked into different aspects of $\mathbb{X}$, such as bot detection, analysis of the spread of fake news, and assessing the credibility of $\mathbb{X}$ users. Bots can be helpful for tasks such as posting information about news and providing assistance during emergencies, etc [21], but some bots can be used for malicious purposes such as influencing public opinion or spreading malware [22]. Hence, identifying bots is vital for $\mathbb{X}$ to enforce its platform terms and conditions. Hence, researchers have proposed different methods [23] to create accurate models for bot detection.

Apart from bot detection, another important area of research is the detection of fake news, which is rampant – tends to be retweeted faster than true ones [24]. Various ML models, particularly of the supervised classification, have been used for fake news detection [25]. For example, Hassan *et al.* [26] extracted features from the sentences and used a support vector machine to detect fake news. Despite the popularity of the topic there has been limited progress in fake news detection. This is partly due to the ongoing controversy surrounding the term 'fake news' and the lack of a universally accepted definition thereof [24]. Nevertheless, several works have delved into fake news detection by assessing the credibility of tweets or classifying the $\mathbb{X}$ users as trusted or untrusted. Another study proposes an automated ranking technique to evaluate tweet credibility. Gupta *et al.* worked on assigning a credibility score to each tweet. Another interesting work in this domain is the work conducted by Tanveer *et al.* [10, 11], presented a model that analyzes $\mathbb{X}$ users, assigning a score to each user based on their social profile, tweet credibility, and h-index score. While there has been considerable research in this domain, it is important to note that only a limited number of studies utilize an OCC to classify $\mathbb{X}$ user as trusted or untrusted. This underscores a critical gap in the existing body of knowledge. Adopting OCC becomes particularly valuable when the task involves identifying a specific category with limited or diverse training instances.

4   Methodology

This research aims to develop a regularization strategy for training OCC models, specifically tailored for identifying political $\mathbb{X}$ users, categorizing them as either trusted or untrusted, as shown in Fig. 1. For this reason, we use the manually labeled dataset of 1000 political $\mathbb{X}$ users from the paper [10]. For each user, a unique profile is created, containing various features. Some are basic features extracted for each $\mathbb{X}$ user linked to their account. More specifically, these features are (i) Number of friends, (ii) Number of followers, (iii) Number of retweets, (iv) Number of likes, (v) URLs, (vi) Lists, (vii) Status and (viii) Mention by others.

The basic features are used to calculate more advanced features like a social reputation score, an h-index score, a sentiment score, and tweet credibility. Below, we provide a brief description of these advanced features:

- Social reputation score: It provides the number of users interested in the updates of an $\mathbb{X}$ user.
- H-index score: The h-index is used to measure how impactful an $\mathbb{X}$ user is. This is measured by considering the number of likes and retweets of a $\mathbb{X}$ user.
- Sentiment score: The tweets of a $\mathbb{X}$ user are classified as positive, negative, and neutral, based on which sentiment score is assigned to each $\mathbb{X}$ user.
- Tweet credibility: It is calculated by considering the retweet ratio, liked ratio, URL ratio, user hashtag ratio, and original content ratio.
- Influence score: The influence score of a $\mathbb{X}$ user is calculated by considering the social reputation, h-index score, sentiment score, and tweet credibility.

Details on calculating influence scores from basic features and using advanced features are beyond this paper's scope. For more information, refer to the previous article on this topic [10]. All political $\mathbb{X}$ users are classified as trusted or untrusted based on social reputation, tweet credibility, sentiment score, h-index score, and influence score. All $\mathbb{X}$ accounts with abusive and harassing tweets, a low social reputation, h-index, and influence score are grouped as untrusted users, while those who are more reputable among users with a high h-index score, more credible tweets and a high influence score are grouped as trusted users.

Having a dataset for political $\mathbb{X}$ users as either trusted or untrusted based on various criteria, we then focus on inferring a model based on using information only from trusted users. We train different OCC models, including SVDD, ESVDD, OCSVM, SSVDDrψ_1, SSVDDrψ_2, SSVDDrψ_3, SSVDDrψ_4 GEOCSVM and GESVDD. We also propose a novel regularization term for SSVDD. The newly proposed regularization term considers the graph information, which measures the concentration of the data in a lower-dimensional subspace and captures the essential features of the training set while preserving the local structure of the data. The proposed regularization term is defined as

$$\gamma = \mathrm{Tr}(\mathbf{Q}\mathbf{X}\mathbf{L}_x\mathbf{X}^\mathsf{T}\mathbf{Q}^\mathsf{T}), \tag{6}$$

where $\mathbf{L}_x$ is the Laplacian matrix of the graph. The subscript x denotes the adopted graph type. The Laplacian is defined as

$$\mathbf{L}_x = \mathbf{D}_x - \mathbf{A}_x, \quad [\mathbf{D}_x]_{ii} = \sum_{j \neq i} [\mathbf{A}_x]_{ij}, \forall i \in \{1, \dots, N\}, \tag{7}$$

where $\mathbf{D}_x$ is the degree matrix and $\mathbf{A}_x \in \mathbb{R}^{N \times N}$ serves as the graph's weight matrix. In what follows, we drop the subscript X for notation simplicity.

We investigated the three different graph Laplacians in the proposed regularization term γ. In the first experiment, we exploit the local geometric information by employing k-Nearest Neighbor (kNN) and setting the Laplacian matrix to

$$\mathbf{L}_{kNN} = \mathbf{D}_{kNN} - \mathbf{A}_{kNN},\tag{8}$$

where $[\mathbf{A}_{kNN}]_{ij} = 1$, if $\mathbf{x}_i \in \mathcal{N}_j$ or $\mathbf{x}_j \in \mathcal{N}_i$ and 0, otherwise. $\mathcal{N}_i$ denotes the nearest neighbors of $\mathbf{x}_i$. Adjusting the k numbers of neighbors in kNN allows the neighborhoods $\mathcal{N}_i$ to be defined accordingly. In the second experiment, we use within-cluster Laplacian information.

$$\mathbf{L}_w = \mathbf{I} - \sum_{c=1}^{\mathcal{C}} \frac{1}{N_c}\mathbf{1}_c\mathbf{1}_c^T,\tag{9}$$

where $\mathbf{I}$ is an identity matrix, $\mathcal{C}$ denotes the total numbers of clusters, $\mathbf{1}$ is a vector of ones, N_c is the total number of instances belonging to cluster c and $\mathbf{1}_c$ represents a vector with ones corresponding to instances that belong to cluster c and zeros elsewhere. In the third experiment, we use the between-cluster scatter information:

$$\mathbf{L}_b = \sum_{c=1}^{\mathcal{C}} N_c \left(\frac{1}{N_c}\mathbf{1}_c - \frac{1}{N}\mathbf{1}\right)\left(\frac{1}{N_c}\mathbf{1}_c - \frac{1}{N}\mathbf{1}^T\right).\tag{10}$$

In this paper, we denote the three variants of the proposed regularization strategies for SSVDD as SSVDDγL_{kNN}, SSVDDγL_w, and SSVDDγL_b, respectively. For non-linear data description, we employed non-linear projection trick (NPT) [27]. NPT is equivalent to employing the widely recognized kernel trick while enabling the use of the method's linear variant. The kernel matrix is obtained as

$$\mathbf{K}_{ij} = \exp\left(\frac{-\|\mathbf{x}_i - \mathbf{x}_j\|_2^2}{2\sigma^2}\right),\tag{11}$$

where σ is a hyperparameter scaling the distance between $\mathbf{x}_i$ and $\mathbf{x}_j$. We followed similar steps for non-linear data description as adapted in recent variants and extensions of SSVDD [28,29].

Fig. 1. One-class classification categorizes X platform users as trusted or untrusted

5 Experimental Results and Model Evaluation

To extract the features from X and generate the dataset, we used Python 3.5. The Python script was executed locally on a machine with the following configuration: Intel

Core i7, 2.80*8 GHZ, 32GB, Ubuntu 16.04 LTS 64 bit. For the training and evaluation of the OCC models, we switched to Matlab and performed the experiments on Intel(R) Xeon(R) CPU E5-2650 v3 2.30GHz 64GB RAM. We provide the open-source implementation of our work on Github[1].

A comprehensive set of evaluating metrics is reported over the test set to compare different OCC models. Accuracy (Accu) provides the ratio of correctly classified instances to the total number of instances, True Positive Rate (TPR) represents the proportion of positive instances correctly classified, while True Negative Rate (TNR) indicates the ratio of true negatives to the total number of negative samples. Precision (Pre) measures the proportion of instances classified as positive that are truly positive, and the F1-score is defined as the harmonic mean of precision and TPR. Additionally, Geometric Mean is employed to discern the best-performing parameters on training set, calculated as square root of product of TPR and TNR.

5.1   Preprocessing the Data

In this work, we chose to analyze the $\mathbb{X}$ account of 1,000 politicians[2] and the main reason for evaluating the profiles of politicians is their intrinsic potential to influence public opinion as their content originates and exists in a sphere of political life, which is, unfortunately, often surrounded by controversial events and outcomes. We selected 70 percent of the data for training and 30 percent for testing. The train and test sets are randomly selected by keeping the proportions of the two classes similar to the collected dataset. We perform the random selection five times; hence, we use five different train-test sets for the experiments to check the robustness of the OCC methods. We normalize the data by subtracting the mean and dividing it by the Standard Deviation (STD). These are both computed using only the target class samples from the training set. During the training, a 5-fold cross-validation technique is used over the training set to select the hyperparameters of the models. More details on the hyperparameters can be found in the GitHub.

5.2   Results and Discussions

In Table 1, we report the average performance measures of various OCC methods on the five data splits of the dataset. The classifiers are divided into two categories: linear OCC and non-linear OCC. In Table 1, we also report the STD of evaluating metrics for the linear and non-linear methods over the five data splits of the dataset.

Considering the GM values, the linear OCC generally have lower performance measures than non-linear OCC. This indicates that non-linear OCC are more adept at correctly predicting both positive and negative classes than linear OCC. For example, in non-linear OCC, SSVDDγL_{kNN} achieves the highest GM value, which is 0.80, surpassing the 0.64 obtained by SSVDDγL_w, a linear OCC. Conversely, non-linear SSVDD$\psi 1$ OCC achieve the lowest GM value which is 0.43, as opposed to 0.19 recorded by ESVDD linear OCC.

---

[1] https://github.com/fahadsohrab/xssvdd.

[2] https://zenodo.org/records/7014109.

Regarding Accu, most non-linear OCC models consistently outperform their linear counterparts. As shown in Table 1, the highest Accu, reaching 0.80, is achieved by non-linear OCC SSVDDγL_{kNN}. In contrast, three linear OCC models – SSVDDγL_w, SSVDDγL_b and SSVDDγL_{kNN} – received a slightly lower Accu of 0.74. The lowest Accu among non-linear OCC models is 0.48, attributed to OCSVM, while the linear OCSVM achieves 0.44. For the other evaluation metrics, Pre and F1 remain stable, while TPR consistently remains high, indicating the model's effectiveness in identifying positive instances.

To summarize, linear OCSVM has the lowest Accu (0.44) and F1-score (0.41) among all classifiers, while SVDD and ESVDD have very low TNR and GM values in linear cases. Linear SSVDD classifiers with regularization terms $\psi 1$ and $\psi 4$ have similar performance measures and are somewhat better than OCSVM, SVDD, and ESVDD. GEOCSVM has the highest Accu (0.78) and GM (0.78) scores, indicating its superior performance in identifying positive and negative instances.

The superior performance of non-linear OCC models in terms of GM and Accu can be attributed to the inherent complexity of the data distribution. Non-linear classifiers are more flexible in capturing intricate relationships and patterns within the data, especially when the decision boundary is non-linear.

Examining the SSVDD variants and their performance metric, GM, concerning the regularization term ψ, reveals that $\psi 3$ yields the most favorable outcomes for both linear and non-linear classifiers followed closely by $\psi 2$, then $\psi 1$, with $\psi 4$ performing the least effectively. The superiority of SSVDD$\psi 3$ can be attributed to its consideration of samples inside and outside the class boundary during the training in the regularization term, providing a more comprehensive understanding of the class variance. Conversely, SSVDD$\psi 4$ performs poorly as it only considers support vectors on the class boundary in the regularization term, potentially missing crucial information about class distribution.

The best results for linear and non-linear OCC models are obtained by appending our new regularization term γ to SSVDD (see Table 1). The performance of all three linear OCC models, namely: SSVDDγL_{kNN}, SSVDDγL_b and SSVDDγL_w, is nearly identical, bearing limited impact on performance metrics. On the other hand, among the non-linear OCC models, SSVDDγL_{kNN} demonstrates superior performance, outperforming the other two counterparts, namely SSVDDγL_w and SSVDDγL_b, where the latter ranks the lowest in performance.

Additional information about the use of γ for SSVDD and its impact on performance metric can be found in Fig. 2 (a) for linear classification using kNN, and in Fig. 2 (b) for non-linear classifiers. Looking at the linear OCC models in Fig. 2 (a), the GM value remains steady at 0.57 to 0.63, and Accu falls within the range of 0.70 to 0.74 across various values of k for kNN, showing a stable performance level. All the performance metrics show stability, except TNR fluctuates, which tend to be on the lower side. Unlike the stable results in the linear classifier, the non-linear classifier shows a distinct pattern (see Fig. 2 (b)). The non-linear OCC displays variable performance, with GM values from 0.54 to 0.80 and Accu ranging from 0.67 to 0.80.

We also present the performance metrics for linear and non-linear SSVDD with the proposed regularization term γ, focusing on the between-cluster scatter Laplacian (L_b).

Table 1. Measuring the performance of linear and non-linear OCC models averaged over five test splits with ± STD

	Accu	TPR	TNR	Pre	F1	GM
	Linear OCC					
SSVDDψ1	0.68 ± 0.02	**0.98** ± 0.01	0.27 ± 0.05	0.65 ± 0.01	0.78 ± 0.01	0.51 ± 0.05
SSVDDψ2	0.69 ± 0.03	0.97 ± 0.02	0.31 ± 0.07	0.66 ± 0.02	0.78 ± 0.02	0.54 ± 0.06
SSVDDψ3	0.73 ± 0.06	0.97 ± 0.02	0.41 ± 0.16	0.70 ± 0.06	0.81 ± 0.03	0.62 ± 0.12
SSVDDψ4	0.69 ± 0.03	**0.98** ± 0.01	0.29 ± 0.08	0.66 ± 0.02	0.79 ± 0.02	0.53 ± 0.07
OCSVM	0.44 ± 0.10	0.34 ± 0.10	0.59 ± 0.36	0.59 ± 0.13	0.41 ± 0.04	0.39 ± 0.16
SVDD	0.58 ± 0.01	0.96 ± 0.01	0.05 ± 0.03	0.59 ± 0.01	0.73 ± 0.00	0.22 ± 0.06
ESVDD	0.57 ± 0.01	0.96 ± 0.02	0.04 ± 0.02	0.58 ± 0.00	0.72 ± 0.01	0.19 ± 0.06
SSVDDγL_{kNN}	0.74 ± 0.04	0.97 ± 0.01	0.41 ± 0.12	0.70 ± 0.04	0.81 ± 0.02	0.63 ± 0.09
SSVDDγL_b	0.74 ± 0.07	**0.98** ± 0.01	0.42 ± 0.19	0.71 ± 0.07	**0.82** ± 0.04	0.63 ± 0.14
SSVDDγL_w	0.74 ± 0.02	0.97 ± 0.01	0.42 ± 0.04	0.70 ± 0.02	0.81 ± 0.01	0.64 ± 0.03
	Non-Linear OCC					
SSVDDψ1	0.66 ± 0.09	0.88 ± 0.15	0.35 ± 0.39	0.68 ± 0.12	0.75 ± 0.04	0.43 ± 0.31
SSVDDψ2	0.70 ± 0.08	0.80 ± 0.16	0.55 ± 0.35	0.75 ± 0.14	0.75 ± 0.04	0.61 ± 0.19
SSVDDψ3	0.74 ± 0.09	0.80 ± 0.11	0.67 ± 0.31	0.80 ± 0.12	0.79 ± 0.06	0.70 ± 0.20
SSVDDψ4	0.65 ± 0.09	0.85 ± 0.19	0.36 ± 0.41	0.69 ± 0.13	0.73 ± 0.06	0.40 ± 0.33
OCSVM	0.48 ± 0.04	0.53 ± 0.04	0.40 ± 0.12	0.56 ± 0.04	0.54 ± 0.02	0.45 ± 0.06
SVDD	0.71 ± 0.12	0.84 ± 0.11	0.53 ± 0.43	0.76 ± 0.16	0.78 ± 0.05	0.56 ± 0.34
ESVDD	0.56 ± 0.02	0.58 ± 0.17	0.54 ± 0.26	0.66 ± 0.09	0.60 ± 0.07	0.52 ± 0.04
GEOCSVM	0.78 ± 0.02	0.74 ± 0.06	0.83 ± 0.06	0.86 ± 0.03	0.79 ± 0.03	0.78 ± 0.01
GESVDD	0.70 ± 0.12	0.61 ± 0.27	0.84 ± 0.13	0.87 ± 0.09	0.67 ± 0.24	0.68 ± 0.17
SSVDDγL_{kNN}	**0.80** ± 0.05	0.76 ± 0.11	**0.85** ± 0.08	**0.88** ± 0.05	0.81 ± 0.06	**0.80** ± 0.05
SSVDDγL_b	0.73 ± 0.09	0.77 ± 0.10	0.68 ± 0.21	0.78 ± 0.10	0.77 ± 0.07	0.71 ± 0.12
SSVDDγL_w	0.78 ± 0.02	0.84 ± 0.07	0.70 ± 0.08	0.80 ± 0.03	**0.82** ± 0.03	0.76 ± 0.02

As can be seen in Fig. 2 (c), and Fig. 2 (d), the choice of hyperparameter $\mathcal{C}$ in L_b significantly impacts the performance of both linear and non-linear OCC models. The linear OCC models show varied performance across different values of $\mathcal{C}$ for L_b. For example, Accu ranges from 0.61 to 0.74, with the highest performance achieved at $L_{b=4}$, while precision falls within the range of 0.61 to 0.71. TPR fluctuates between 0.96 and 0.98, showing a reliable identification of positive instances. GM and TNR show a similar pattern, both displaying lower values. In contrast, non-linear OCC models show distinct behavior, with TPR at the top, varying between 0.66 and 0.86, followed by F1-score and Accu, both of which show a stable performance. However, GM and TNR, position at the last, show high variation. Following, we will analyze the results for within-cluster Laplacian (L_w). As can be seen in Fig. 2 (e), linear classifiers show varying performance across different $\mathcal{C}$ values for L_w. Accu ranges from 0.64 to 0.74 (at $L_w = 7$), and TPR fluctuates between 0.96 and 0.98, indicating a consistent ability to identify positive instances correctly. Precision ranges from 0.65 to 0.70, and GM shows variations but generally remains between 0.56 and 0.64. The non-linear classifiers in Fig. 2 (f), show different behavior, with Accu ranging from 0.68 to 0.78. TPR varies between 0.73 and 0.89 and precision fluctuates between 0.71 and 0.81. It is noteworthy that at $L_w = 3$, the non-linear OCC models achieve their highest F1-score and GM value.

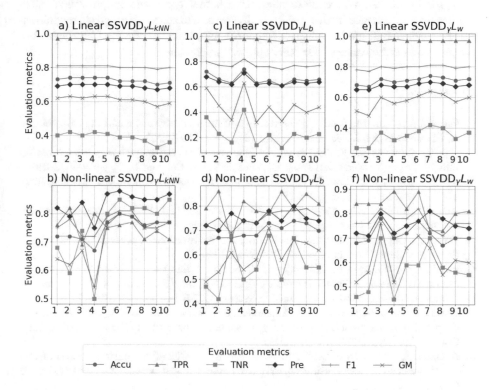

Fig. 2. Evaluating SSVDD performance with the proposed regularization term γ, varying k for kNN, and different cluster values ($\mathcal{C}$) for L_b and L_w.

6   Conclusion

Considering the significant impact of information sharing on social media, specifically on platform **X**, our goal was to identify the trusted or untrusted **X** users. Our study provided insights into the effectiveness of OCC models in classifying political users on platform **X**, through exploring OCC models. In addition, it included a novel regularization term for SSVDD.

In response to the research questions *RQ 1–3*, our findings demonstrate the effectiveness of OCC models in identifying political **X** users as trusted or untrusted. The results consistently demonstrate that non-linear OCC classifiers outperform their linear counterparts. This paper provided brief insights on the recent improvements in OCC, notably the new paradigm of subspace learning for SVDD used to tackle the curse of dimensionality. Our study confirmed the potential of OCC performance optimization for political user identification through subspace learning. The proposed subspace-learning-based approach, particularly with the introduced regularization term for SSVDD, showcased superior performance compared to baseline models.

In the future, we will explore alternative kernel types and graph structures to enhance the performance further. Additionally, we aim to adapt the proposed regularization term to the Multi-modal Subspace Support Vector Data Description [30] framework and analyze its effectiveness over other application domains.

References

1. Gazi, M.A., Çetin, M., Çaki, C.: The research of the level of social media addiction of university students. Int. J. Soc. Sci. Educ. Res. **3**(2), 549–559 (2017)
2. Das, R., Karmakar, G., Kamruzzaman, J.: How much i can rely on you: measuring trustworthiness of a twitter user. IEEE Trans. Dependable Secure Comput. **18**(2), 949–966 (2019)
3. Vosoughi, S., Roy, D., Aral, S.: The spread of true and false news online. Science **359**(6380), 1146–1151 (2018)
4. Khan, T., Michalas, A., Akhunzada, A.: Fake news outbreak 2021: can we stop the viral spread? J. Netw. Comput. Appl. **190**, 103112 (2021)
5. Zhang, Z., Gupta, B.B.: Social media security and trustworthiness: overview and new direction. Futur. Gener. Comput. Syst. **86**, 914–925 (2018)
6. Pritzkau, A., Winandy, S., Krumbiegel, T.: Finding a line between trusted and untrusted information on tweets through sequence classification. In: 2021 International Conference on Military Communication and Information Systems (ICMCIS), pp. 1–6. IEEE (2021)
7. Devarajan, G.G., Nagarajan, S.M., Amanullah, S.I., Mary, S.S.A., Bashir, A.K.: AI-assisted deep NLP-based approach for prediction of fake news from social media users. IEEE Trans. Comput. Soc. Syst. **PP**(99), 1–11 (2023)
8. Asfand-e Yar, M., Hashir, Q., Tanvir, S.H., Khalil, W.: Classifying misinformation of user credibility in social media using supervised learning. Comput. Mater. Continua **75**(2), 2921 (2023)
9. Ahmad, F., Rizvi, S.A.M.: Information credibility on twitter using machine learning techniques. In: Singh, P.K. (ed.) FTNCT 2019. CCIS, vol. 1206, pp. 371–381. Springer, Singapore (2020). https://doi.org/10.1007/978-981-15-4451-4_29
10. Khan, T., Michalas, A.: Trust and believe-should we? evaluating the trustworthiness of twitter users. In: 2020 IEEE 19th International Conference on Trust, Security and Privacy in Computing and Communications (TrustCom), pp. 1791–1800. IEEE (2020)
11. Khan, T., Michalas, A.: Seeing and believing: evaluating the trustworthiness of twitter users. IEEE Access **9**, 110505–110516 (2021)
12. Wang, J., Jing, X., Yan, Z., Fu, Y., Pedrycz, W., Yang, L.T.: A survey on trust evaluation based on machine learning. ACM Comput. Surv. (CSUR) **53**(5), 1–36 (2020)
13. Alam, S., Sonbhadra, S.K., Agarwal, S., Nagabhushan, P.: One-class support vector classifiers: a survey. Knowl.-Based Syst. **196**, 105754 (2020)
14. Degerli, A., Sohrab, F., Kiranyaz, S., Gabbouj, M.: Early myocardial infarction detection with one-class classification over multi-view echocardiography. In: 2022 Computing in Cardiology. vol 498, pp. 1–4. IEEE (2022)
15. Sohrab, F., Raitoharju, J.: Boosting rare benthic macroinvertebrates taxa identification with one-class classification. In: 2020 IEEE Symposium Series on Computational Intelligence, pp. 928–933. IEEE (2020)
16. Zaffar, Z., Sohrab, F., Kanniainen, J., Gabbouj, M.: Credit card fraud detection with subspace learning-based one-class classification. In: 2023 IEEE Symposium Series on Computational Intelligence (SSCI), pp. 407–412. IEEE (2023)

17. Schölkopf Ü, B., Williamson, R.C., Smola Ü, A., Shawe-Taylor Y, J.: SV estimation of a distribution's support. Adv. Neural Inf. Process. Syst. **41**, 582–588 (2000)
18. Mygdalis, V., Iosifidis, A., Tefas, A., Pitas, I.: Graph embedded one-class classifiers for media data classification. Pattern Recogn. **60**, 585–595 (2016)
19. Sohrab, F., Iosifidis, A., Gabbouj, M., Raitoharju, J.: Graph-embedded subspace support vector data description. Pattern Recogn. **133**, 108999 (2023)
20. Sohrab, F., Raitoharju, J., Gabbouj, M., Iosifidis, A.: Subspace support vector data description. In: 2018 24th International Conference on Pattern Recognition (ICPR), pp. 722–727. IEEE (2018)
21. Haustein, S., Bowman, T.D., Holmberg, K., Tsou, A., Sugimoto, C.R., Larivière, V.: Tweets as impact indicators: examining the implications of automated "bot" accounts on Twitter. J. Assoc. Inf. Sc. Technol. **67**(1) 232–238 (2016)
22. Qiang, F., Feng, B., Guo, D., Li, Q.: Combating the evolving spammers in online social networks. Comput. Secur. **72**, 60–73 (2018)
23. Rodríguez-Ruiz, J., Mata-Sánchez, J.I., Monroy, R., Loyola-Gonzalez, O., López-Cuevas, A.: A one-class classification approach for bot detection on twitter. Comput. Secur. **91**, 101715 (2020)
24. Meyers, M., Weiss, G., Spanakis, G.: Fake news detection on twitter using propagation structures. In: van Duijn, M., Preuss, M., Spaiser, V., Takes, F., Verberne, S. (eds.) MISDOOM 2020. LNCS, vol. 12259, pp. 138–158. Springer, Cham (2020). https://doi.org/10.1007/978-3-030-61841-4_10
25. Pedro Henrique Arruda Faustini and Thiago Ferreira Covoes: Fake news detection in multiple platforms and languages. Expert Syst. Appl. **158**, 113503 (2020)
26. Hassan, N., Arslan, F., Li, C., Tremayne, M.: Toward automated fact-checking: detecting check-worthy factual claims by ClaimBuster. In: Proceedings of the 23rd ACM SIGKDD International Conference on Knowledge Discovery and Data Mining, pp. 1803–1812 (2017)
27. Kwak, N.: Nonlinear projection trick in kernel methods: an alternative to the kernel trick. IEEE Trans. Neural Networks Learn. Syst. **24**(12), 2113–2119 (2013)
28. Sohrab, F., Raitoharju, J., Iosifidis, A., Gabbouj, M.: Ellipsoidal subspace support vector data description. IEEE Access **8**, 122013–122025 (2020)
29. Sohrab, F., Laakom, F., Gabbouj, M.: Newton method-based subspace support vector data description. In: 2023 IEEE Symposium Series on Computational Intelligence (SSCI), pp. 1372–1379. IEEE (2023)
30. Sohrab, F., Raitoharju, J., Iosifidis, A., Gabbouj, M.: Multimodal subspace support vector data description. Pattern Recogn. **110**, 107648 (2021)

Entity Co-occurrence Graph-Based Clustering for Twitter Event Detection

Bundit Manaskasemsak[(✉)], Natthakit Netsiwawichian, and Arnon Rungsawang

Massive Information and Knowledge Engineering Laboratory, Department of Computer Engineering, Faculty of Engineering, Kasetsart University, Bangkok 10900, Thailand
{bundit.m,natthakit.ne,arnon.r}@ku.th

Abstract. Social streams from existing online platforms such as Twitter or X have consistently shown the most up-to-date information about current events. Detecting and categorizing Twitter content connected to events is one of the most challenging tasks. Many researchers have therefore been interested in this issue for many years. In this paper, we propose an approach to identifying events that occurred from tweet text. To accomplish this goal, (1) tweets are first decomposed into sets of representative word-entities using NER technology; (2) a graph representing relationships between co-occurring entities is created; (3) several clustering methods are investigated on the graph; and finally, (4) tweets are assigned to the locally dense subgraphs (i.e., clusters) as identified events. Experiments conducted on two standard Twitter datasets demonstrate that the proposed strategy outperforms state-of-the-art methods. The results also show the achievement of accurately detecting significant events at the top of the ranking results.

1 Introduction

Twitter (now known as X) is one of the most popular online social media platforms that allows users to create short messages of up to 280 characters, called tweets. By using mobile devices and the Internet, people can quickly send and receive their status and share information and news. As a result, today, users are easily more informed of events around the world. However, as the number of online users increases, the amount of data generated also increases rapidly. In fact, active Twitter users may write multiple tweets in a single day. This makes them and other users experience information overload. Automated event detection is therefore an important tool but also a challenge for researchers.

To address the issue of event detection on Twitter, existing research efforts have introduced several techniques in various aspects. For example, some studies focus on identifying events of a particular type [1, 2]. Most of these methods typically use domain-specific knowledge along with machine learning processes; however, they would fail to detect a broad range of events. While other studies focus on domain-independent events [3, 4], these methods mostly determine event characteristics and use an unsupervised learning technique, especially clustering, to infer groups of events. Moreover, some studies propose probabilistic topic modeling for tweet content [5, 6]. The basic idea is to create a representation of tweets based on words or entities and group them according to similar event topics. Meanwhile, some studies [3, 4, 7–9] create a correlation graph

© The Author(s), under exclusive license to Springer Nature Switzerland AG 2024
L. Barolli (Ed.): AINA 2024, LNDECT 200, pp. 344–355, 2024.
https://doi.org/10.1007/978-3-031-57853-3_29

that represents the relationships of words or named entities and then employ clustering techniques on the graph. Each subgraph is therefore defined as an event.

In this paper, we concentrate on domain-independent event detection from Twitter using graph clustering techniques. Our approach relies on the assumption that tweets mentioning the same event should contain the same set of keywords (i.e., word-entities) related to that event. This is achieved by first generating an entity co-occurrence graph that represents the relationships between word-entities. Locally dense subgraphs that infer events are then specified. And tweets mentioning those detected events are finally identified. Our contributions are summarized as follows:

- We used the Named Entity Recognition (NER) technique to generate representations for tweets. Here, a number of NERs have been examined in order to select the best entity representation.
- We proposed an entity co-occurrence graph, defined as a weighted and undirected graph. Within the graph, word-entities (vertices) are connected to others with weights determined by the number of co-occurrences in tweets.
- We applied various clustering algorithms to partition the graph into locally dense subgraphs to infer events.
- We conducted experiments on two gold-standard Twitter corpora and compared the event detection performance of our approach with that of state-of-the-art methods. Moreover, we also demonstrated its effectiveness in identifying the most discussed events at the top of the ranking results.

The remainder of this paper is organized as follows: Sect. 2 reviews research studies related to ours. Section 3 details the proposed method for Twitter event detection. Section 4 reports our performance in comparison with state-of-the-art methods. Finally, Sect. 5 concludes the paper.

2 Related Work

Due to tweets on Twitter containing a lot of rich human-generated data, many studies are interested in using tweet text for social media analysis. The study in the field of event detection is also the same. Currently, there are many approaches to dealing with the problem of automatic event detection.

Several early studies on event detection are based on topic modeling [10–14]. Latent Dirichlet Allocation (LDA) is the most commonly used topic modeling model [15]. By pre-setting the number of events, this model clusters the documents based on the similarity probabilities of the topic in each document. The disadvantage of LDA is that it causes incorrect clustering by selecting an inappropriate number of events in advance. To solve this problem, Zhou et al. [5] suggested a simple Bayesian model to extract representations of events as a tuple of a non-location-named item, date, location, and event-related keyword. Later, they proposed another Twitter event extraction method, a non-parametric Bayesian mixture model [6], based on their prior research.

For the graph-based event detection model, Weng and Lee [7] proposed event detection with clustering of wavelet-based signals to filter the trivial words and cluster the words by the modularity-based graph partitioning technique to form events. Hromic et al. [8] proposed a model for mining the structure of interaction between users to find particular topics and events by building two graph representations: the user-to-user graph and the user-to-tweet graph. Katragadda et al. [3] proposed an event detection model by using a graph where the vertex is the word that appeared in the tweet and the weight is the number of co-occurring pairs of words in the graph. They then pruned and clustered the resulting graph to detect the events.

There have been many studies that use NER instead of word recognition in tweets and employ graphs for event detection. McMinn and Jose [9] examined the named entity and used it for an event detection model by using the TF-IDF weight cosine similarity score to cluster the tweet. Edouard et al. [4] proposed an approach that used named entity context to create a temporal event graph and used a simple graph technique to detect clusters of tweets that correspond to the same events. Pandya et al. [16] proposed an event detection model called MaTED. They used NER and harnessed DBPedia to extract event-related phrases from tweets and the text titles of webpages that URL links in the tweets point to. After creating temporal profiles based on tweets' occurrences, they then use a graph-based clustering algorithm to group those bursty phrases into events. Rather than using the NER, Dhiman and Toshniwal [17] proposed a model using a graph of sentences where nodes represent the tweets' word embeddings and edge weights represent the magnitude of semantic and temporal similarity between the tweet pairs. They then applied an uncertain graph clustering algorithm to extract the clusters of those tweets as events.

3   Methodology

To discover domain-independent events in tweets, we present an entity co-occurrence graph-based clustering (ECoGC) approach. The method is based on the idea that individuals tend to utilize the same set of keywords in the content of their tweets when publishing events. Our goal is to represent the relationships between word-entities as a graph and then partition it into tightly connected clusters (i.e., subgraphs), with each cluster inferring an event.

Figure 1 presents the workflow of our event detection model. Given a collection of tweets. The workflow consists of five main tasks. Firstly, the *tweet preprocessing task* aims to tokenize the input tweet text, clean it, and transform it into a normalized format (Sect. 3.1). Secondly, the *word-entity extraction task* extracts potential keywords (i.e., word-entities) from the tweet content (Sect. 3.2). Thirdly, the *graph construction task* generates a graph by linking related word-entities (Sect. 3.3). Fourthly, the *graph clustering task* identifies groups of tightly connected word-entities (Sect. 3.4). Lastly, the *tweet assignment task* assigns the tweets into clusters and infers them as detected events (Sect. 3.5).

3.1   Tweet Preprocessing

From the tweet collection, we perform the general task of preprocessing the content of tweets. Here, we first tokenize the tweet text into tokens and then remove retweets, non-ASCII characters, and emoticons. For URLs and user mentions, we follow the method

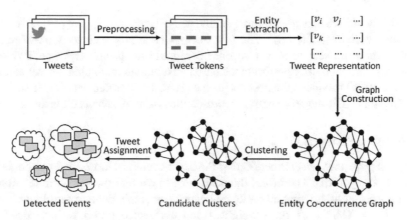

Fig. 1. Workflow of the proposed ECoGC method.

suggested by Ushio et al. [18] to transform them into a normalized format. That is, the URLs are converted to the specific token {{URL}}. The mentions of non-verified usernames are converted to {{USERNAME}}, while verified ones are replaced with the display name along with the @ symbol. In this work, we define verified usernames as popular persons, companies, organizations, etc. whose names can be found on the Wikipedia website. We believe that the names extracted from the crowd-sourced encyclopedia have a certain level of credibility. In the following, a sample tweet.

```
Hey @jamie142. U gotta give PSY credit. That vid is ultra creative

& frickin hilarious. RT @CNN: 'Gangnam' is most watched video ever.

http://t.co/eg1uBsp1
```

 will be processed as:

```
Hey {{USERNAME}} U gotta give PSY credit That vid is ultra creative

frickin  hilarious  {@CNN@}  Gangnam  is  most  watched  video  ever

{{URL}}
```

3.2 Word-Entity Extraction

The main objective of this task is to develop a representation for each tweet. The meaning of a tweet content is usually determined by the words it contains. However, we believe that using almost all the terms in the tweet text will make the model complex, as many of these terms are not event related. Therefore, we employ the Named Entity Recognition (NER) technique to extract potential keywords (namely, word-entities) from tweets and thereby hypothesize that such keywords might be used to establish the context of the event.

In this work, we considered three candidate NER models: spaCy [19], TweetNER7 [18], and tweebank-NER [20]. The first one has been proposed for general text processing, while the second and third ones contribute to specific tasks on Twitter. We afterwards examined their performance based on a human-annotated dataset and found that TweetNER7 produced the best results (see Sect. 4.1 for more details). Therefore, we employed this NER model to extract word-entities from tweets used for the later steps.

3.3 Graph Construction

We hypothesize that tweets mentioning the same event are more likely to contain the same set of keywords. Therefore, this task aims to express the semantic relationships between word-entities using an entity co-occurrence graph, defined as follows.

Let $\mathcal{G} = (\mathcal{V}, \mathcal{E}, \mathcal{W})$ be a weighted and undirected graph with vertices $\mathcal{V} = \{v_1, v_2, \ldots, v_n\}$ as the set of n word-entities in consideration and edges $\mathcal{E} \subseteq \mathcal{V} \times \mathcal{V}$ as the set of m relations between those word-entities. We define $w_{ij} \in \mathcal{W}$ as the weight on edge $e(v_i, v_j) \in \mathcal{E}$, which is represented by the total number of times the word-entities v_i and v_j appear together. In other words, it can be the number of tweets that contain v_i and v_j. This indicates that the higher the weight, the more relevant two word-entities are. Otherwise, zero weight will be assigned if they are not related (or no graph edge).

Figure 2 visualizes the sample entity co-occurrence graph generated from 400 excerpted tweets in the experimental dataset. As shown in the figure, there are a total of 10 possible entities extracted from those tweets. The value labeled each edge represents the number of tweets (out of 400) that contain that entity pair.

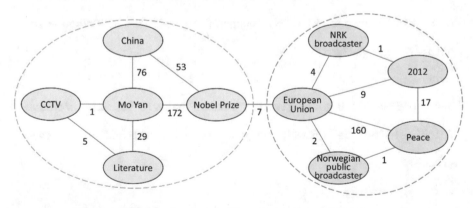

Fig. 2. A sample entity co-occurrence graph.

3.4 Graph Clustering

As previously stated, tweets related to the same event often share a common set of keywords, while tweets related to another event or not related to any event are displayed with different keywords [4, 9]. In addition, the frequency of the co-occurring keywords can reflect their relevance to the event.

Consider an entity co-occurrence graph obtained from the previous stage. Here, we will use a partitioning technique (also known as community detection) on the graph to cluster tightly connected word-entities, and further infer that each subgraph identifies an event on its own. From the example graph shown in Fig. 2, we would expect to obtain two clusters of word-entities: one for the green-colored entities on the left represents the event that Mo Yan, a Chinese writer known for his hallucinator realism, won the Nobel Prize in Literature; and the other for orange-colored entities on the right represents the event when the Norwegian Nobel Committee decided that the 2012 Nobel Peace Prize would be awarded to the European Union.

To do so, in this work, we apply four different state-of-the-art graph clustering algorithms: the Greedy Modularity [21], Louvain [22], Leiden [23], and Infomap [24] algorithms, respectively. The performance comparison of using these algorithms in event detection is then presented in Sect. 4.2.

3.5 Tweet Assignment

At this stage, we assume that events discovered using distinct clusters (i.e., subgraphs) are unrelated. Each cluster is therefore examined separately in the event detection process.

Taking each cluster into account, we now have a set of word-entities that constitute an event description. We would, however, prefer to know which tweets mentioned the incident. Based on the assumption discussed earlier, the meaning of tweet content can be determined by the words it contains. Therefore, each representative tweet will be assigned to a cluster if it contains the most closely matching word-entities. Note that a tweet can be assigned to more than one cluster if it matches all of them equally.

4 Experiments

In this section, we present the performance evaluation results divided into two parts. The first one examines the performance of candidate NERs (Sect. 4.1), and the second one evaluates the performance in event detection (Sect. 4.2).

4.1 Performance of Candidate NERs

As outlined in Sect. 3.2, we employ the NER model to extract probable word-entities to create a representative set of tweets. For this reason, we would like to investigate which candidates from spaCy [19], TweetNER7 [18], and tweebank-NER [20] perform better under our criteria.

Our trials were conducted on the WNUT16 dataset [25]. The collection contains 3,856 tweets with human-annotated named entities. The performance of those three models is measured in terms of precision, recall, and F1-score, respectively. Here, precision is defined as the fraction of tagged entities that are correct, regardless of type. While recall is defined as the fraction of actual entities that are tagged regardless of type as well; the F1-score summarizes both. The overall performances of the three candidate NER models are concluded in Table 1.

Table 1. Effectiveness of the candidate NER models.

Models	Precision	Recall	F1-score
spaCy [19]	0.39	0.60	0.47
TweetNER7 [18]	**0.59**	**0.79**	**0.68**
Tweebank-NER [20]	0.48	0.60	0.53

Bold text emphasizes the highest value.

As can be observed, the TweetNER7 model outperforms all others. As a result, we decided to incorporate it in our word-entity extraction process.

4.2 Event Detection Performance

In this subsection, we evaluate the resulting set of detected events and compare our performance with other state-of-the-art approaches.

1) Datasets
We conducted our experiments on two gold standard Twitter corpora: EVENT2012 [26] and RepLab2013 [27].

EVENT2012. The corpus consists of 120 million tweets collected over a 28-day period from October to November 2012. It also contains 506 labeled events generated from the Wikipedia Current Events Portal [28]. Each tweet was thus annotated for a specific type of event, such as politics, sports, art, and other topics. Of those tweets, 159,952 were identified as being related to such an event. However, due to the Twitter restrictions, the body text of the tweets is not included in the dataset; only the IDs are reported. We therefore used the Tweepy library [29] to fetch tweets through their IDs. As the corpus ages, many tweets are no longer available and must be excluded. Finally, we received 61,739 event-related tweets.

RepLab2013. The corpus originally consisted of more than 140,000 tweets annotated by trained annotators, curated and verified by reputation experts. However, the dataset used in this work was taken from [30]. The dataset was filtered to include only tweets belonging to topics with a size of at least 100 tweets. Thus, there were 2,567 tweet IDs with 13 different labeled topics, such as for sale, Suzuki Cup, user comments, money laundering/terrorism finance, etc. In the same way, we need to retrieve the tweet content later. After pulling those tweets, we received 2,275 relevant ones.

2) Results on the EVENT2012 Dataset
We evaluate the proposed ECoGC method using four graph clustering algorithms: Greedy Modularity [21], Louvain [22], Leiden [23], and Infomap [24] in event detection. For the EVENT2012 dataset, we also compare it with three state-of-the-art approaches: NEED [6], EDO [3], and the method proposed by Edouard et al. [4].

The qualitative measure is performed in terms of precision, recall, and F1-score. Within our work, each detected event is considered correct if the majority of tweets in that cluster belong to the same event in the ground truth. Table 2 reports the experimental results conducted on the EVENT2012 dataset. As can be seen, ECoGC using Infomap clustering provides the best performance, while ECoGC using Greedy Modularity gives the second-best one. Furthermore, all ECoGC methods except using Leiden can improve F1-scores above the baselines.

Table 2. Evaluation results on the EVENT2012 dataset.

Models	Precision	Recall	F1-score
ECoGC$_{\text{Greedy}}$	0.773	0.729	0.751
ECoGC$_{\text{Louvain}}$	0.761	0.694	0.726
ECoGC$_{\text{Leiden}}$	0.744	0.512	0.607
ECoGC$_{\text{Infomap}}$	**0.786**	**0.757**	**0.771**
NEED [6]	0.636	0.383	0.478
EDO [3]	0.754	0.512	0.638
Edouard et al. [4]	0.750	0.668	0.710

Bold text emphasizes the highest value.

In general, an event's significance could be inferred from the number of tweets that mention it. Consequently, we investigate further how well our method works to identify the most discussed events at the top of the ranking results. We thus report this in term of the number of correctly assigned events in decreasing rank order P@k (i.e., the precision at the first k rank). The performance comparison of the first 10 identified events is shown in Fig. 3. ECoGC employing Leiden clustering yields the greatest P@k values in all ranking orders although it appears to be the worst of the three clustering algorithms in Table 2. This suggests that this method might be more appropriate if we only have limited time to consider the first few interesting events. In addition, Louvian-based ECoGC offers the second best.

3) Results on the RepLab2013 Dataset
We have repeated the earlier studies on the RepLab2013 dataset. Rather, we now contrast our ECoGC with the state-of-the-art approach suggested by Dhiman and Toshniwal [17]. Table 3 reports the findings of the experiment. It is evident that ECoGC using Greedy Modularity clustering yields the best results. In terms of F1-scores, both the Louvain- and Leiden-based models perform marginally better.

In Fig. 4, we also report the precision values of the first 10 events that were found. It is evident that ECoGC based on Louvain and Leiden clustering methods yield comparable positive outcomes. This suggests that the two methods can be used to identify the initial few significant events in the dataset.

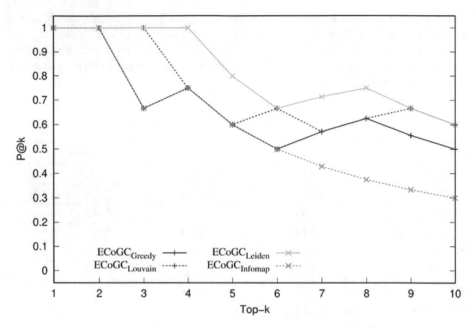

Fig. 3. Detection performance at the first k events for the EVENT2012 dataset.

Table 3. Evaluation results on the RepLab2013 dataset.

Models	Precision	Recall	F1-score
ECoGC$_{Greedy}$	**0.846**	0.769	**0.806**
ECoGC$_{Louvain}$	0.769	0.769	0.769
ECoGC$_{Leiden}$	0.769	0.769	0.769
ECoGC$_{Infomap}$	0.769	0.692	0.729
Dhiman and Toshniwal [17]	0.710	**0.833**	0.766

Bold text emphasizes the highest value.

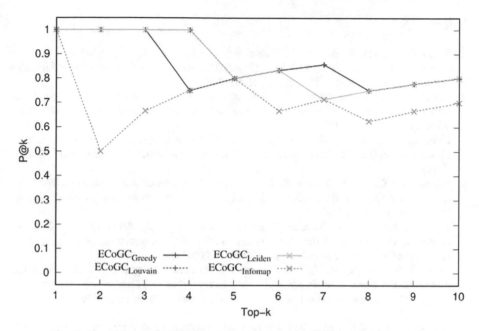

Fig. 4. Detection performance at the first k events for the RepLab2013 dataset.

5 Conclusion

In this paper, we propose a graph-based clustering method for domain-independent event detection from Twitter data. We have demonstrated the effectiveness of NER techniques in generating representations for tweets. Then, an entity co-occurrence graph that represents semantic entity (i.e., keyword) relationships for events was introduced, and several clustering techniques have been applied to the graph to identify those events. Experimental results on two benchmark datasets reveal that the proposed approach outperforms the baseline with better F1-scores. Moreover, the proposed method using either Leiden or Louvain clustering technique can accurately detect the most discussed events at the top of the ranking results.

Creating representations for tweets is an early and important phase of our work. Therefore, we intend to improve the entity extraction performance in the future with the expectation that it will enhance the overall detection performance. In addition, we also intend to broaden the experiments to detect real-time events from Twitter stream data.

Acknowledgments. In this work, Natthakit Netsiwawichian was supported by the Master's Degree Education Promotion Program under a grant of Faculty of Engineering, Kasetsart University.

References

1. Sankaranarayanan, J., Samet, H., Teitler, B.E., Lieberman, M.D.: TwitterStand: news in tweets. In: Proceedings of the 17th ACM SIGSPATIAL International Conference on Advances in Geographic Information System, pp. 41–51 (2009)
2. Sakaki, T., Okazaki, M., Matsuo, Y.: Earthquake shakes Twitter users: real-time event detection by social sensors. In: Proceedings of the 19th International Conference on World Wide Web, pp. 851–860 (2010)
3. Katragadda, S., Virani, S., Benton, R., Raghavan, V.: Detection of event onset using twitter. In: Proceedings of the 2016 International Joint Conference on Neural Networks, pp. 1539–1546 (2016)
4. Edouard, A., Cabrio, E., Tonelli, S., Le-Thanh, N.: Graph-based event extraction from twitter. In: Proceedings of the International Conference Recent Advances in Natural Language Processing, pp. 222–230 (2017)
5. Zhou, D., Chen, L.Y., He, Y.: A simple Bayesian modelling approach to event extraction from twitter. In: Proceedings of the 52nd Annual Meeting of the Association for Computational Linguistics (Volume 2: Short Papers), pp. 700–705 (2014)
6. Zhou, D., Zhang, X., He, Y.: Event extraction from twitter using non-parametric Bayesian mixture model with word embeddings. In: Proceedings of the 15th Conference of the European Chapter of the Association for Computational Linguistics (Volume 1: Long Papers), pp. 808–817 (2017)
7. Weng, J., Lee, B.-S.: Event detection in twitter. In: Proceedings of the 5th International AAAI Conference on Weblogs and Social Media, vol. 5, no. 1, pp. 401–408 (2011)
8. Hromic, H., Prangnawarat, N., Hulpuş, I., Karnstedt, M., Hayes, C.: Graph-based methods for clustering topics of interest in twitter. In: Cimiano, P., Frasincar, F., Houben, G.J., Schwabe, D. (eds.) Engineering the Web in the Big Data Era: The 15th International Conference on Web Engineering, LNCS, vol. 9114, pp. 701–704. Springer, Cham (2015)
9. McMinn, A.J., Jose, J.M.: Real-time entity-based event detection for twitter. In: Mothe, J., et al. (eds.) Experimental IR Meets Multilinguality, Multimodality, and Interaction: The 6th International Conference of the Cross-Language Evaluation Forum for European Languages, LNCS, vol. 9283, pp. 65–77. Springer, Cham (2015)
10. Cui, L., Zhang, X., Zhou, Z., Salim, F.: Topical event detection on Twitter. In: Cheema, M.A., Zhang, W., Chang, L. (eds.) Databases Theory and Application: The 27th Australasian Database Conference, LNCS, vol. 9877, pp. 257–268. Springer, Cham (2016)
11. Sokolova, M., Huang, K., Matwin, S., Ramisch, J., Sazonova, V., Black, R., Orwa, C., Ochieng, S., Sambuli, N.: Topic modelling and event identification from Twitter textual data. arXiv preprint, arXiv:1608.02519 (2016)
12. Hong, L., Davison, B. D.: Empirical study of topic modeling in Twitter. In: Proceedings of the First Workshop on Social Media Analytics, pp. 80–88 (2010)
13. Yin, H., Cui, B., Lu, H., Huang, Y., Yao, J.: A unified model for stable and temporal topic detection from social media data. In: Proceedings of the IEEE 29th International Conference on Data Engineering, pp. 661–672 (2013)
14. Diao, Q., Jiang, J., Zhu, F., Lim, E.-P.: Finding bursty topics from microblogs. In: Proceedings of the 50th Annual Meeting of the Association for Computational Linguistics (Volume 1: Long Papers), pp. 536–544 (2012)
15. Blei, D.M., Ng, A.Y., Jordan, M.I.: Latent dirichlet allocation. J. Mach. Learn. Res. **3**, 993–1022 (2003)
16. Pandya, A., Oussalah, M., Kostakos, P., Fatima, U.: MaTED: metadata-assisted twitter event detection system. In: Lesot, M.-J., et al. (eds.) Information Processing and Management of Uncertainty in Knowledge-Based Systems, CCIS, vol. 1237, pp. 402–414. Springer, Cham (2020)

17. Dhiman, A., Toshniwal, D.: An approximate model for event detection from twitter data. IEEE Access **8**, 122168–122184 (2020)
18. Ushio, A., Barbieri, F., Sousa, V., Neves, L., Camacho-Collados, J.: Named entity recognition in twitter: a dataset and analysis on short-term temporal shifts. In: Proceedings of the 2nd Conference of the Asia-Pacific Chapter of the Association for Computational Linguistics and the 12th International Joint Conference on Natural Language Processing (Volume 1: Long Papers), pp. 309–319 (2022)
19. spaCy – Industrial-Strength Natural Language Processing. https://spacy.io. Accessed 31 Dec 2023
20. Jiang, H., Hua, Y., Beeferman, D., Roy, D.: Annotating the Tweebank corpus on named entity recognition and building NLP models for social media analysis. In: Proceedings of the 13th Conference on Language Resources and Evaluation, pp. 7199–7208 (2022)
21. Clauset, A., Newman, M.E.J., Moore, C.: Finding community structure in very large networks. Phys. Rev. E **70**(6), 066111 (2004)
22. Blondel, V.D., Guillaume, J.-L., Lambiotte, R., Lefebvre, E.: Fast unfolding of communities in large networks. J. Stat. Mech.: Theory Exp. 2008(10), P10008 (2008)
23. Traag, V.A., Waltman, L., van Eck, N.J.: From Louvain to Leiden: guaranteeing well-connected communities. Sci. Rep. **9**, 5233 (2019)
24. Edler, D., Holmgren, A., Rosvall, M.: The MapEquation software package, https://mapequation.org. Accessed 31 Dec 2023
25. Strauss, B., Toma, B., Ritter, A., de Marneffe, M.-C., Xu, W.: Results of the WNUT16 named entity recognition shared task. In: Proceedings of the 2nd Workshop on Noisy User-generated Text, pp. 138–144 (2016)
26. McMinn, A.J., Moshfeghi, Y., Jose, J.M.: Building a large-scale corpus for evaluating event detection on twitter. In: Proceedings of the 22nd ACM International Conference on Information and Knowledge Management, pp. 409–418 (2013)
27. Amigó, E., et al.: Overview of RepLab 2013: Evaluating online reputation monitoring systems. In: Forner, P., Müller, H., Paredes, R., Rosso, P., Stein, B. (eds.) CLEF 2013. LNCS, vol. 8138, pp. 333–352. Springer, Heidelberg (2013). https://doi.org/10.1007/978-3-642-40802-1_31
28. Wikipedia – Portal:Current events. https://en.wikipedia.org/wiki/Portal:Current_events. Accessed 31 Dec 2023
29. Tweepy – An easy-to-use Python library for accessing the Twitter API. https://www.tweepy.org. Accessed 31 Dec 2023
30. Curiskis, S.A., Drake, B., Osborn, T.R., Kennedy, P.J.: An evaluation of document clustering and topic modelling in two online social networks: Twitter and Reddit. Inf. Process. Manage. **57**(2), 102034 (2020)

Anomaly Detection with Generalized Isolation Forest

Brett E. Downey, Carson K. Leung$^{(\boxtimes)}$ ⓘ, Adam G. M. Pazdor, Ryan A. L. Petrillo,
Denys Popov, and Benjamin R. Schneider

University of Manitoba, Winnipeg, MB, Canada
Carson.Leung@UManitoba.ca

Abstract. The existing Isolation Forest algorithm detects anomalies by using binary trees. However, a few limitations can be observed. To resolve this problem, we present a systematic approach to generalize and improve the Isolation Forest algorithm in this paper. The resulting Generalized Isolation Forest algorithm is based on the identification of common properties that characterize successful variations of Isolation Forest. It is designed to detect different kinds of anomalous data. Our analytical evaluation shows that the generalization conforms to the identified properties. Our experimental evaluation on real-world data demonstrates that our generalized method can significantly improve the performance of Isolation Forest when compared to the basic implementation. Overall, our work provides a foundation for understanding the fundamentals of Isolation Forest and introduces novel ways to extend the algorithm for improved performance in specific application domains.

Keywords: advanced information applications · data mining · anomaly detection · Isolation Forest algorithm · anomalous data · anomaly patterns · performance improvement · clustering · anomaly scores

1 Introduction

Nowadays, big data are everywhere due to advancements in technology so that huge amounts of data can be easily generated and collected at a rapid rate from a wide variety of real-life application domains. Embedded in these big data are implicit, previously unknown but useful information. Data science solutions [1–3]—which make good use of data mining [4–7], machine learning [8–11], mathematical and statistical techniques, and visualization [12, 13]—help discover the useful information that was previously hidden in the big data. These solutions can be applicable for tasks like bioinformatics and health informatics [14–17], social media mining [18] (e.g., web mining [19]), social network analysis [20–22], sports analytics [23, 24], and transportation analytics [25], which can be for social good. In this paper, we focus on the data mining task of anomaly detection.

Anomaly detection is a problem that arises in many contexts. These application areas include cybersecurity, finance, healthcare, and manufacturing [26]. There are a variety of approaches to anomaly detection. Most approaches try to represent the dataset and

L. Barolli (Ed.): AINA 2024, LNDECT 200, pp. 356–368, 2024.
https://doi.org/10.1007/978-3-031-57853-3_30

report anomalies as data that are dissimilar from the generated model's representation of the distribution. However, these approaches often require significant information about the distribution of the dataset. For example, k-means based anomaly detection methods still require knowledge about the number of clusters in the dataset to work effectively. In contrast, the *Isolation Forest* algorithm [27] utilizes the fundamental scarcity of anomalous data to characterize what data are anomalous. Since its introduction, there have been various ways to extend Isolation Forest. Most variations of Isolation Forest create an ensemble of *iTree* estimators that are used to generate anomaly scores. In this paper, we present a more systematic approach to improving Isolation Forest. We start by inspecting the common properties that characterize successful variations of Isolation Forest. They provide an important foundation for understanding the fundamentals of the algorithm. Isolation Forest also has very fast convergence, resulting in training that requires minimal data compared to other methods. However, Isolation Forest has limited hyper-parameters that can be used to customize the model on specific use cases. Once the model has sufficient ensemble size and tree height, the only parameter a user can tune is the anomaly score threshold. This hyper-parameter is a value representing the model's sensitivity. We stick with the original Isolation Forest's objective to be anomaly focused. However, we introduce hyper-parameters to help users apply knowledge of anomalies within their domain to create performant anomaly detection models. Observing that non-anomalous data are highly regular, we expect that patterns exist within anomalous data in application domains. Intuitively, we expect an anomaly (e.g., a fraudulent credit card transaction) to have similarities with other fraudulent credit card transactions. By introducing ways for the Isolation Forest model to leverage information about the nature of anomalies in specific application domains, we achieve better performance compared to generic implementations of Isolation Forest. Our *key contributions* of this paper include:

- Presentation and explanation of several properties required for effective generalizations of the *iTree* algorithm.
- Design of a new generalization on the *iTree* algorithm for detecting different kinds of anomalous data
- Evaluation of the resulting generalized method on its applications to real-world data to improve performance over the basic implementation of Isolation Forest.

The reminder of this paper is organized as follows. The next section gives background and related works. Section 3 describes our generalized Isolation Forest algorithm. Evaluation and conclusions are shown in Sects. 4 and 5, respectively.

2 Background and Related Works

2.1 Background

The original version of the Isolation Forest algorithm [27] first chooses a feature uniformly randomly, and then chooses the split point uniformly randomly in the minimum to the maximum range of that feature. This process is performed recursively until (i) all points are isolated at the node, or (ii) the maximum height is reached. See Algorithm 1.

To elaborate, the Isolation Forest algorithm first constructs an *iTree*, which is a data structure that represents this spatial partitioning method by having each internal node of the tree represents each split around a feature. See Algorithm 1. The leaves in an *iTree* represent isolated points that were used to construct the tree. For computational simplicity, the *iTree* is often set to have a maximum height *l*. Crucially, due to the inherent scarcity of anomalous data, the spatial partitioning in each *iTree* tends to create shallow leaves corresponding to anomalous data. This is a property that Isolation Forest exploits to detect anomalies. Our paper uses the same ensemble of *iTrees* as the original Isolation Forest to construct the *iForest* [28] (as shown in Algorithm 2). As a preview, we will modify the *iTree* subroutine that is called by *iForest*.

Algorithm 1 *iTree*(input data X, current tree height e, height limit l)
Output: *iTree*
1   **if** $(e \geq l)$ or $(|X| \leq 1)$
2   **then return** exNode{size$\leftarrow|X|$}
3   **else** $Q \leftarrow$ a list of attributes in X
4   randomly select an attribute $q \in Q$
5   randomly select a split point p from min & max values of attribute q in X
6          $X_L \leftarrow$ filter(X, $q < p$)
7          $X_R \leftarrow$ filter(X, $q \geq p$)
8   **return** inNode{left$\leftarrow$*iTree*(X_L, $e+1$, l), right$\leftarrow$*iTree*(X_R, $e+1$, l),
                  splitAttr$\leftarrow q$, splitVal$\leftarrow p$}

Algorithm 2 *iForest*(input data X, number of trees t, sub-sampling size ψ)
Output: a set of t *iTrees*
1   **initialize** *Forest*
2   height limit $l \leftarrow$ ceiling($\log_2 \psi$)
3   **for** $i \leftarrow 1$ to t **do**
4          $X' \leftarrow$ sample(X, ψ);
5   *Forest* $\leftarrow$ *Forest* $\cup$ *iTree*(X', 0, l)
6   **return** *Forest*

We also adopt the recommended ensemble size $t = 100$ and sub-sample size $\psi = 256$ for all examples and applications as set by Liu et al. [27]. Anomaly scores will be calculated using the same *pathLength* algorithm and anomaly scoring calculation as in the original version of Isolation Forest [28]. In Algorithm 3, $c(\cdot)$ is the average path length of an unsuccessful binary search in an *iTree*:

$$c(n) = 2(\mathrm{H}(n-1)) - \frac{2}{n} \tag{1}$$

where $\mathrm{H}(n)$ is an approximation for the n-th harmonic number:

$$\mathrm{H}(n) \approx \ln(n) + 0.57721566492 \tag{2}$$

with 0.57721566492 being the Euler's constant or Euler-Mascheroni constant. In the first call of Algorithm 3, the current path length e is initialized to 0. Then, anomaly scores for a given point are calculated as follows:

$$S(x) = 2^{-\frac{E(h(x))}{c(\psi)}} \tag{3}$$

where

- $h(x)$ is the height of point x in an *iTree* given by Algorithm 3;
- $E(\cdot)$ is the average across all *iTrees* in the *forest*; and
- $S(x)$ is the output anomaly score for a point x that is compared to a threshold value to decide whether a point is anomalous or nominal:

  - We use the term *anomalous* to refer to points that are irregular when compared to the distribution of the dataset the model is trained on. It is important to note that what is considered anomalous is domain dependent.
  - We use the term *nominal* to refer to data that matches the distribution of the greater dataset.

Algorithm 3 *PathLength*(an instance x, an *iTree* T, current tree height e)
Output: path length of x
1   **if** T is an external node
2   **then return** e + c(T.size) // c() is defined in Eq. (1)
3   $a \leftarrow T$.splitAttr
4   **if** $x_a < T$.splitVal
5   **then return** PathLength(x, T.left, e+1)
6   **else return** PathLength(x, T.right, e+1)

We use several basic facts about continuous distributions in the proofs in this paper. In particular, we use the fact that:

$$P(a \leq X \leq b) = \int_a^b f(x)dx \qquad (4)$$

For any continuous distribution $f(x)$ and its corollary [29]:

$$P(a \leq X \leq a) = \int_a^a f(x)dx = 0 \qquad (5)$$

That is, the probability of generating a value that is exactly equal to some value a is 0. This result [30] follows from applying the fundamental theorem of calculus to Eq. (4):

$$P(a \leq X \leq a) = \int_a^a f(x)dx = F(a) - F(a) = 0$$

This is used in Sect. 3 to discard cases where two randomly generated points are exactly equal, as those cases are 0 probability events.

2.2 Related Works

Some works have been done on a variety of methods for extending Isolation Forest. Other versions of Isolation Forest modify how data are partitioned at each node during the construction of an *iTree*. For example, *Extended Isolation Forest* (EIF) [31] modifies Isolation Forest to create a uniform anomaly scoring along all feature directions. There have also been other variations on Isolation Forest that use statistical methods to improve performance by increasing sampling in low-density areas of the distribution.

For example, *Probabilistic Generalization of Isolation Forest* (PGIF) [32] uses information about the distribution and density to choose split points between clusters. Due to the versatility of the algorithm, there are many ways in which Isolation Forest has been modified. Combining Isolation Forest with techniques from mathematics and clustering has produced several novel and effective variations of Isolation Forest. For example, Karczmarek et al. [33] and Paweł et al. [34] found that combining Isolation Forest with *k*-means and fuzzy *c*-means to be an effective way to construct *iTrees* in Isolation Forest. As Isolation Forest is fundamentally an ensemble of simple *iTrees*, there are many ways to construct different variations of Isolation Forest by changing how *iTrees* are constructed. Other modifications entirely replace the way an *iTree* is constructed. For example, Galka et al. [35] proposed an Isolation Forest modification that merges *iTree* nodes by making use of minimal spanning trees.

As many applications require real-time monitoring of anomalies, there has also been work to modify Isolation Forest to process and detect anomalies in live data streams, and also to improve the efficiency of Isolation Forest data stream anomaly detection. Additionally, streaming data are infinite and can vary over time, which presents a challenge for the original Isolation Forest as a predetermined contamination parameter or anomaly threshold will not work. To overcome this problem, Laskar et al. [36] applied Isolation Forest in an industrial big data stream environment searching for network security anomalies using *k*-means to eliminate the requirement for a contamination parameter. Other techniques have been developed to allow Isolation Forest to handle data streams, such as using Mondrian decision trees to create a hybrid Isolation Mondrian Forest in order to handle data stream batches [37]. These data stream modifications allow Isolation Forest to be effectively applied to more application domains that require real-time anomaly detection on continuous streams of data.

Isolation Forest is an isolation-based anomaly detection algorithm [28]. Other anomaly detection approaches include statistical, depth-based, distance-based, clustering-based, deviation-based, etc. [26]. Isolation-based anomaly detection is unique from all these approaches by focusing on isolated data points. Some versions of Isolation Forest can incorporate aspects of other anomaly detection approaches such as statistical methods when generating *iTrees*. However, the key concept of isolated data points is what differs isolation-based anomaly detection from other anomaly detection methods.

3   Our Generalized Isolation Forest

To improve Isolation Forest, let us understand the properties that create effective Isolation Forest methodologies. Here, we list four important properties for creating effective partitioning methods in the Isolation Forest algorithm:

1. $|X_L| > 0$ *and* $|X_R| > 0$: The idea behind *iForest* is to split the data until each data point is isolated in a leaf or maximal height is reached [38]. If this property does not hold, it could allow for empty branches in the constructed *iTree*. As the scoring of points corresponds to the depth of a point in ensembles of *iTrees*, empty branches distort the evaluation of points in an *ITree*.
2. *For any two points* $x_1 \neq x_2$, *there exists an iTree such that the leaf containing* x_1 *is not same as the leaf that contains* x_2: This property shows that our scheme can isolate all points. That is, if we continue our partitioning process, any two points would eventually end up isolated in leaves. If this property does not hold, the algorithm could not distinguish between certain sets of points [39]. As *iTrees* are spatial partitioning data structures, we need to be able to partition between any two non-identical points in the feature space.
3. *Path length of anomalous data is statistically shorter compared to nominal data*: As path length dictates anomaly scoring in Isolation Forest, it is important that data that are anomalous tend to have shorter paths in *iTrees*. This results in higher anomaly scores as calculated by Eq. (3).
4. *The partitioning method splits the feature where the distribution of feature is sparse with high probability:* This property allows for generalization of the model to multi-cluster datasets. If a dataset has multiple clusters, then it is important that *iTrees* create partitions between the clusters. If an algorithm does not do this, then the distribution of larger clusters will dictate what points are considered anomalous in smaller clusters. This results in poor generalization to multi-cluster distributions. Anomalies should be calculated with respect to the clusters that the point is close to. That is, we want our anomaly detection method to operate *locally* not *globally*.

These properties form the basis for creating novel ways for generating *iTrees*. Any method that does not satisfy these properties will struggle to create *iTrees* that work as individual estimators in the *iForest* ensemble. Using these properties as a foundation for what makes effective *iTree* estimators, we designed the Generalized Isolated Forest with feature weighted (as described in the next section).

3.1   Generalized iTree Algorithm Feature Weighted

By statistically weighting which features are used to generate partitions at each node in the *iTree*, we have found that we can control which features are most impactful when the model is scoring inputs. A feature q weighted with higher than uniform probability will result in the model being more sensitive to outlying inputs in feature q. Intuitively, the average constructed *iTree* will have more nodes where q is selected as the partition feature, leading to inputs that are outliers in q to be isolated in shallower paths in the *iTree*. This leads to higher anomaly scores for these inputs when they are scored against the trained model. The addition of weighted features allows for a much higher degree of customization when training the model. Depending on the application domain, certain features are more impactful when deciding whether inputs are anomalies. Through the introduction of a weighting hyper-parameter $w = (w_1, w_2, \ldots, w_N)$ such that $\sum_{i=1}^{N} w_i = 1$ *and* $w_i > 0$, we create a much more versatile version of Isolation Forest. Our version of Generalized Isolation Forest allows for features with different levels of importance to be included in the model. In Isolation Forest, all features are equally weighted, this results in poor performance if features in the model have varying importance for identifying anomalies.

Algorithm 4 *feature Weighted iTree*(input data X, current tree height e, height limit l, feature weight $w=(w_1, w_2, \ldots, w_N)$)

Output: iTree
1   **if** $(e \geq l)$ or $(|X| \leq 1)$
2   **then return** exNode{size$\leftarrow |X|$}
3   **else** $Q \leftarrow$ a list of attributes in X
4   select an attribute $q \in Q$ such that Prob$(q=q_i) = w_i$ where q_i is the i-th feature of Q
5   randomly select a split point p from min & max values of attribute q in X
6        $X_L \leftarrow$ filter$(X, q < p)$
7        $X_R \leftarrow$ filter$(X, q \geq p)$
8   **return** inNode{left$\leftarrow iTree(X_L, e+1, l)$, right$\leftarrow iTree(X_R, e+1, l)$,
              splitAttr$\leftarrow q$, splitVal$\leftarrow p$}

Consider the case of a dataset with two features, X and Y. Suppose we know that inputs that are outliers in feature X are more likely to be anomalies. In conventional implementations of Isolation Forest, there is no easy way to represent this knowledge within the model. To use this knowledge would require pre-processing of the dataset. However, in our model, we can represent this knowledge by setting $w = (0.75, 0.25)$. Furthermore, as w becomes a model hyper-parameter we can use hyper-parameter search to find the ideal values for w, we generate the data for this experiment.

4   Evaluation

4.1   Analytical Evaluation of Our Generalized Isolation Forest Algorithm

In this section, we show how our introduced algorithm satisfies Properties 1 and 2 as defined in Sect. 3. In the following proofs, we assume that the dataset X is a set of vectors drawn from a continuous distribution. That is, for any two $x_i, x_j \in X$ (where $i \neq j$) the probability that $x_i = x_j$ along a feature is 0. This follows from the fact that in a continuous distribution, the probability of a given point being generated is 0.

Lemma 1. *Property 1* (i.e., $|X_L| > 0$ *and* $|X_R| > 0$) *holds.*

Proof. We assume that *iTree* is called with valid parameters, i.e., $|X| > 0$, $e \leq l$, and $w_i > 0$ for all i in w. On the one hand, if line 1 (of Algorithm 4) is true, then the proof is trivial, because X_L and X_R are never generated. On the other hand, if line 1 is false, then there exists $x_1, x_2 \in X$ such that x_1 and x_2 are respectively the minimum and maximum values of q in X. As x_1 and x_2 are drawn from a continuous distribution, $x_1 \neq x_2$ in feature q. With *min* < *max* from line 4, it implies that p $\in$ (*min, max*) as Prob(p = *min*) = Prob(p = *max*) = 0 with p being a uniformly sampled real number from the interval [*min, max*]. Hence, $x_1 \in X_L$ and $x_2 \in X_R$ because $x_1 = min < p < max = x_2$ in feature q. Consequently, $|X_L| > 0$ and $|X_R| > 0$.

Lemma 2. *Property 2 (i.e., for any two points $x_1 \neq x_2$, there exists an iTree such that the leaf containing x_1 is not same as the leaf that contains x_2).*

Proof. Let $x_1 \neq x_2 \in X$. As x_1 and x_2 are drawn from a continuous distribution, $x_1 \neq x_2$ in feature q is selected on line 4 (of Algorithm 4). Without loss of generality, assume that $x_1 < x_2$ in feature q. As the split point p is chosen randomly uniformly, there exists an *iTree* such that $x_1 < p < x_2$. Hence, $x_1 \in X_L$ and $x_1 \in X_R$ implies x_1 and x_2 belong to different leaves in the *iTree* by the recursion on line 7.

Lemmas 1 and 2 prove that our Generalized Isolation Forest with feature weighted can generate *iTrees* that differentiate between any two differing points given sufficiently large maximum tree height.

4.2 Experimental Evaluation of Our Generalized Isolation Forest Algorithm on the HTTP Dataset

In this section, we apply both the original Isolation Forest partition scheme and the feature weighted partition scheme on the HTTP KDD CUP 1999 dataset. The HTTP KDD CUP 1999 is a reduced dataset containing 567,479 data points and 2211 (0.4%) outliers that was derived from the original KDD CUP 1999 simulated military environment network intrusion dataset [40]. This reduced dataset contains three basic attributes, two of which will be analyzed to detect anomalies: source bytes (src_bytes) and destination bytes (dst_bytes). Source bytes refer to the number of bytes transmitted from the web client to the web server, whereas destination bytes are the number of bytes transmitted from the web server to the web client. These basic attributes can indicate what the typical request and response sizes are for a web server. Depending on what types of operations the server supports, different clusters can exist. For example, one cluster with low values for the source and destination bytes may indicate a small GET request, whereas a different cluster with high values may indicate more computationally demanding operations (such as a download request). Anomalies that occur outside of operation clusters will have different source and destination bytes than any operation that is known to be supported by the server, which could indicate a potential problem.

An evaluation of HTTP networking data can reveal a variety of use cases for detecting anomalies. These anomalies may be caused by security threats, system errors, or oversights in system design, and their detection is essential for ensuring the secure and efficient operation of networks. For example, a software bug that leads to the submission of malformed requests in specific edge cases can be difficult to identify and resolve, as it may not be immediately apparent that there is a networking issue. However, if the anomalous request is detected, it can be thoroughly investigated to determine the root cause and implement a solution. Security threats can come in the form of repeated network requests, unusually large requests that are not typically supported by the web server, and more. Lastly, anomalies may be found after implementing a system, which can indicate problems in the overall architecture and design, such as redundant or repeated network requests.

HTTP networking data are complex, with a wide range of information transmitted for each request, including request type, headers, and body. When examining anomalies in these data, it is important to consider a wide range of features and not just focus on individual features. It can be useful to apply weights to different features. Hence, we will compare our results with the original Isolation Forest algorithm, and the known outliers of the data set. Our evaluation results show that the original Isolation Forest may not detect clusters of outliers, whereas our implementation can correctly identify the groupings of outliers by applying specific weights to the feature that the outliers are most prominent in.

We consider the comparison of the actual known outliers in the HTTP KDD CUP 1999 dataset (Fig. 1(a)), compared to the outliers detected by scikit-learn's implementation of Isolation Forest[1] (Fig. 1(b)). Here, the contamination parameter was set to 0.004 = 0.4%. This value represents the percentage of outliers in the dataset and controls how many outliers will be identified by the original Isolation Forest.

As observed from Fig. 1(a), a large nominal cluster that contains 99% of the overall data exists in the source byte destination range [5.1, 6.3]. We note that there are two nominal data points far away outside of this cluster. The first data point is (10.90669, 9.02570), which is located inside of the cluster of anomalies near $X = 10$ and has 2121 instances in the HTTP KDD CUP 1999 dataset. The second data point is $(-2.30258, -2.30258)$, which has 303 instances located as the most bottom left point.

Examining the result indicates a large cluster of anomalies that are undetected by the original Isolation Forest, which are surrounded by a dense cluster of identical nominal data points. Scikit-learn implementation of Isolation Forest detects approximately 2207 outliers, of which the most anomalous data point (10.90669, 9.02570) was incorrectly predicted to be anomaly for a total of 2121 times. This also accounts for every nominal instance of that data point in the HTTP KDD CUP 1999 dataset. Additionally, this data point is the most repeated row in the HTTP KDD CUP 1999 dataset (0.4% of the dataset), followed by the data point $(-2.30258, -2.30258)$, which occurs much less frequently for a total of 303 times. As every instance of this data point was predicted to be an anomaly, this leaves approximately 100 other data points that are permitted to be predicted as anomalous, limited by the supplied contamination parameter. As the dense nominal cluster accounts for a very small percentage of the total data compared to the entire data set, it is reasonable that this data point was marked anomalous. However, this indicates that Isolation Forest is susceptible to detecting a large amount of identical nominal data as anomalous but failing to predict actual anomalous points. Increasing the contamination score would allow Isolation Forest to detect these anomalies because the total amount of anomalies allowed are limited to approximately 2200 (around 4% of 567479). However, other large nominal groupings would incorrectly be flagged as anomalous as well and would then require an additional contamination score increase to overcome. In addition, this increases the likelihood that the destination bytes feature will cause more anomalous values to be marked (instead of the source destination bytes, which contains the real anomalies).

Instead, we show that the difference in anomalies detected by applying the feature weighted method—using (i) the same value for alpha and (ii) a weight on the source

---

[1] https://scikit-learn.org/stable/modules/generated/sklearn.ensemble.IsolationForest.html

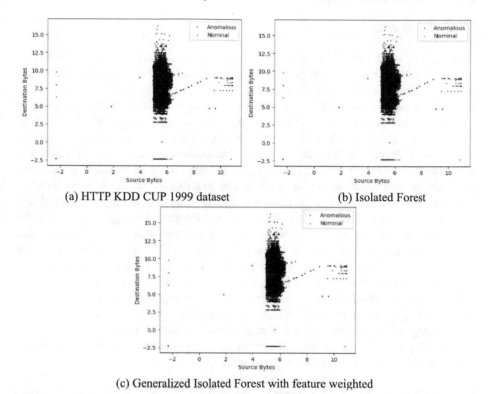

(a) HTTP KDD CUP 1999 dataset   (b) Isolated Forest

(c) Generalized Isolated Forest with feature weighted

Fig. 1. (a) Ground truth for the HTTP KDD CUP 1999 dataset, (b) (Scikit-learn implementation of) Isolation Forest with contamination $= 0.004$, (c) Our Generalized Isolation Forest with feature weighted using $w = (0.67, 0.33)$ and anomaly threshold $= 0.66$

destination attribute—can contain more anomalies being detected close to dense clusters of nominal data. Applying the feature weighted partitioning scheme increases the total amount of outliers (that are detected), by a total of approximately 2700. By applying (i) the same alpha value of 0.61, (ii) a feature weight on the source bytes of 67%, and (iii) a weight on the destination bytes of 33%, we can use the feature weighted partitioning scheme to identify the large source destination anomalies more accurately that exist near the dense cluster of nominal data (Fig. 1(c)). Although this partitioning scheme detects the dense cluster of nominal data at (10.90669, 9.02570), it successfully predicts the cluster of anomalies surrounding the nominal cluster. We observed that (i) 98% of all data points occur within the large cluster ranging from approximately [5.1, 6.3], and (ii) a wide range of destination bytes are accepted as nominal in this range Hence, applying a weight on the source destination can assist with the detection of anomalies that outside of the normally accepted range even when there exist smaller clusters of nominal data near the anomalous data. These operations may indicate far less used supported network operations, which require a large difference in the source destination bytes. One example use case of this can be uploading files.

5   Conclusions

In this paper, we introduced properties for generalizing modifications of Isolation Forest. We justified why each of these properties are required for creating Isolation Forest variations that produce meaningful anomaly scores and by extension, effective anomaly detection models. We applied these properties to create generalized versions of Isolation Forest. We also introduced the feature weighted method and integrated it into Isolated Forest. By weighting which features are sampled, we demonstrated how to control the importance of features in the resulting model. This method is very powerful for encoding information about the underlying anomalies in the Isolation Forest model. Through leveraging information about the nature of anomalies in a dataset, we created a generalized Isolation Forest algorithm. During the evaluation, we show that our algorithm conforms to the properties we described. Finally, we applied the feature weighted method to the HTTP dataset. We compared the performance of the feature weighted method to scikit-learn's implementation of Isolation Forest on using the HTTP dataset. This comparison showed that our algorithm performed better than scikit-learn's implementation when compared to the ground truth for the dataset. Overall, our findings suggest that by leveraging domain-specific knowledge about the nature of anomalies, the Generalized Isolation Forest algorithm is more effectively customized for specific application domains. As *ongoing and future work*, we would expand our validation techniques. Developing a numeric way to score models would allow for more automated hyper-parameter search, which would further increase model performance.

Acknowledgement. This work is partially supported by Natural Sciences and Engineering Research Council of Canada (NSERC) and University of Manitoba.

References

1. Anderson-Grégoire, I.M., et al.: A big data science solution for analytics on moving objects. In: Barolli, L., Woungang, I., Enokido, T. (eds.) AINA 2021, Volume 2, LNNS, vol. 226, pp. 133–145. Springer, Cham (2021). https://doi.org/10.1007/978-3-030-75075-6_11
2. Atif, F., Rodriguez, M., Araújo, L.J.P., Amartiwi, U., Akinsanya, B.J., Mazzara, M.: A survey on data science techniques for predicting software defects. In: Barolli, L., Woungang, I., Enokido, T. (eds.) AINA 2021, Volume 3. LNNS, vol. 227, pp. 298–309. Springer, Cham (2021). https://doi.org/10.1007/978-3-030-75078-7_31
3. Dierckens, K.E., et al.: A data science and engineering solution for fast k-means clustering of big data. In: IEEE TrustCom-BigDataSE-ICESS 2017, pp. 925–932 (2017)
4. Alam, M.T., Ahmed, C.F., Samiullah, Md., Leung, C.K.-S.: Discovering interesting patterns from hypergraphs. ACM Trans. Knowl. Disc. Data (TKDD) **18**(1), 32:1–32:34 (2024). https://doi.org/10.1145/3622940
5. Leung, C.K.-S., Jiang, F.: Frequent pattern mining from time-fading streams of uncertain data. In: Cuzzocrea, A., Dayal, U. (eds.) DaWaK 2011. LNCS, vol. 6862, pp. 252–264. Springer, Heidelberg (2011). https://doi.org/10.1007/978-3-642-23544-3_19
6. Leung, C.K.-S., Tanbeer, S.K.: Fast tree-based mining of frequent itemsets from uncertain data. In: Lee, S.-G., Peng, Z., Zhou, X., Moon, Y.-S., Unland, R., Yoo, J. (eds.) DASFAA 2012, Part I. LNCS, vol. 7238, pp. 272–287. Springer, Heidelberg (2012). https://doi.org/10.1007/978-3-642-29038-1_21

7. Liu, C., Li, X.: Mining method based on semantic trajectory frequent pattern. In: Barolli, L., Woungang, I., Enokido, T. (eds.) AINA 2021, Volume 2. LNNS, vol. 226, pp. 146–159. Springer, Cham (2021). https://doi.org/10.1007/978-3-030-75075-6_12

8. Abahussein, S., Zhu, T., Ye, D., Cheng, Z., Zhou, W.: Protect trajectory privacy in food delivery with differential privacy and multi-agent reinforcement learning. In: Barolli, L. (ed.) AINA 2023, Volume 3. LNNS, vol. 655, pp. 48–59. Springer, Cham (2023). https://doi.org/10.1007/978-3-031-28694-0_5

9. Leung, C.K., et al.: Machine learning and OLAP on big COVID-19 data. In: IEEE BigData 2020, pp. 5118–5127 (2020)

10. Lu, W.: Applied machine learning for securing the internet of medical things in healthcare. In: Barolli, L. (ed.) AINA 2023, Volume 2. LNNS, vol. 654, pp. 404–416. Springer, Cham (2023). https://doi.org/10.1007/978-3-031-28451-9_35

11. Olawoyin, A.M., Leung, C.K., Hryhoruk, C.C.J., Cuzzocrea, A.: Big data management for machine learning from big data. In: Barolli, L. (ed.) AINA 2023, Volume 1. LNNS, vol. 661, pp. 393–405. Springer, Cham (2023). https://doi.org/10.1007/978-3-031-29056-5_35

12. Di Giacomo, E., Di Martino, B., Didimo, W., Esposito, A., Liotta, G., Montecchiani, F.: Design of a process and a container-based cloud architecture for the automatic generation of storyline visualizations. In: Barolli, L. (ed.) AINA 2023, Volume 3. LNNS, vol. 655, pp. 90–99. Springer, Cham (2023). https://doi.org/10.1007/978-3-031-28694-0_9

13. Leung, C.K., et al.: Big data visualization and visual analytics of COVID-19 data. In: IV 2020, pp. 415–420 (2020)

14. Faiz, M.F.I., Iqbal, M.Z.: XceptionUnetV1: a lightweight DCNN for biomedical image segmentation. In: Barolli, L., Hussain, F., Enokido, T. (eds.) AINA 2022, Volume 3. LNNS, vol. 451, pp. 23–32. Springer, Cham (2022). https://doi.org/10.1007/978-3-030-99619-2_3

15. Leung, C.K.: Biomedical informatics: state of the art, challenges, and opportunities. BioMedInformatics 4(1), 89–97 (2024)

16. Matsui, T., et al.: Analysis of visualized bioindicators related to activities of daily living. In: Barolli, L., Woungang, I., Enokido, T. (eds.) AINA 2021, Volume 1. LNNS, vol. 225, pp. 731–744. Springer, Cham (2021). https://doi.org/10.1007/978-3-030-75100-5_62

17. Souza, J., Leung, C.K., Cuzzocrea, A.: An innovative big data predictive analytics framework over hybrid big data sources with an application for disease analytics. In: Barolli, L., Amato, F., Moscato, F., Enokido, T., Takizawa, M. (eds.) AINA 2020. AISC, vol. 1151, pp. 669–680. Springer, Cham (2020). https://doi.org/10.1007/978-3-030-44041-1_59

18. Cabusas, R.M., Epp, B.N., Gouge, J.M., Kaufmann, T.N., Leung, C.K., Tully, J.R.A.: Mining for fake news. In: Barolli, L., Hussain, F., Enokido, T. (eds.) AINA 2022, Volume 2. LNNS, vol. 450, pp. 154–166. Springer, Cham (2022). https://doi.org/10.1007/978-3-030-99587-4_14

19. Lee, W., et al.: Mobile web navigation in digital ecosystems using rooted directed trees. IEEE Trans. Ind. Electron. (TIE) 58(6), 2154–2162 (2011)

20. Cameron, J.J., et al.: Finding strong groups of friends among friends in social networks. In: IEEE DASC 2011, pp. 824–831 (2011)

21. Choudhery, D., Leung, C.K.: Social media mining: prediction of box office revenue. In: IDEAS 2017, pp. 20–29 (2017)

22. Tanbeer, S.K., et al.: Interactive mining of strong friends from social networks and its applications in e-commerce. J. Organ. Comput. Electron. Commer. 24(2–3), 157–173 (2014)

23. Anuraj, A., et al.: Sports data mining for cricket match prediction. In: Barolli, L. (ed.) AINA 2023, Volume 3. LNNS, vol. 655, pp. 668–680. Springer, Cham (2023). https://doi.org/10.1007/978-3-031-28694-0_63

24. Isichei, B.C., et al.: Sports data management, mining, and visualization. In: Barolli, L., Hussain, F., Enokido, T. (eds.) AINA 2022, Volume 2. LNNS, vol. 450, pp. 141–153. Springer, Cham (2022). https://doi.org/10.1007/978-3-030-99587-4_13

25. Leung, C.K., Braun, P., Hoi, C.S.H., Souza, J., Cuzzocrea, A.: Urban analytics of big transportation data for supporting smart cities. In: Ordonez, C., Song, I.-Y., Anderst-Kotsis, G., Tjoa, A.M., Khalil, I. (eds.) DaWaK 2019. LNCS, vol. 11708, pp. 24–33. Springer, Cham (2019). https://doi.org/10.1007/978-3-030-27520-4_3
26. Mehrotra, K., et al.: Anomaly Detection Principles and Algorithms. Springer, Cham (2017). https://doi.org/10.1007/978-3-319-67526-8
27. Liu, F., et al.: Isolation forest. In: IEEE ICDM 2008, pp. 413–22 (2008)
28. Liu, F.T., Ting, K.M., Zhou, Z.-H.: Isolation-based anomaly detection. ACM Trans. Knowl. Disc. Data (TKDD) 6(1), 3:1–3:39 (2012). https://doi.org/10.1145/2133360.2133363
29. Sinai, Y.G.: Probability Theory: An Introductory Course. Springer, Heidelberg (1992). https://doi.org/10.1007/978-3-662-02845-2
30. Stewart, J.: Calculus: Early Transcendentals, 9th edn. Cengage (2021)
31. Hariri, S., et al.: Extended isolation forest. IEEE Trans. Knowl. Data Eng. (TKDE) 33(4), 1479–1489 (2021)
32. Tokovarov, M., Karczmarek, P.: A probabilistic generalization of isolation forest. Inf. Sci. 584, 433–449 (2022)
33. Karczmarek, P., et al.: K-means-based isolation forest. Knowl. Based Syst. 195, 105659:1–105659:15 (2020)
34. Pawel, K., et al.: Fuzzy c-means-based isolation forest. Appl. Soft Comput. 106, 107354 (2021)
35. Galka, L., et al.: Isolation forest based on minimal spanning tree. IEEE Access 10, 74175–74186 (2022)
36. Laskar, M.T.R., et al.: Extending isolation forest for anomaly detection in big data via k-means. ACM Trans. Cyber-Phys. Syst. (TCPS) 5(4), 41:1–41:26 (2021). https://doi.org/10.1145/3460976
37. Ma, H., et al.: Isolation Mondrian forest for batch and online anomaly detection. In: IEEE SMC 2020, pp. 3051–3058 (2020)
38. Lesouple, J., et al.: Generalized isolation forest for anomaly detection. Pattern Recogn. Lett. 149, 109–119 (2021)
39. Zambon, D., et al.: Graph iForest: isolation of anomalous and outlier graphs. In: IJCNN 2022, pp. 5153–5160 (2022)
40. Kelly, M., et al.: The UCI machine learning repository. https://archive.ics.uci.edu

Toward a Deep Multimodal Interactive Query Expansion for Healthcare Information Retrieval Effectiveness

Sabrine Benzarti[1,3(✉)], Wafa Tebourski[2,3], and Wahiba Ben Abdessalem Karaa[2,3]

[1] Esprit School of Business, Tunis, Tunisia
sabrine.benzarti@esprit.tn
[2] High Institute of Management of Tunis, Tunis, Tunisia
wafatebourskiisg@yahoo.com, wahiba.bak@gmail.com
[3] RIADI Laboratory, National School of Computer Science, Manouba, Tunisia

Abstract. The emerging trend of Health Information Retrieval (HIR) aims to efficiently address users' specific information requirements. However, a beyond challenge arises in the healthcare domain due to the necessity for specialized dictionaries, which influence the outcomes of HIR. The Vocabulary Mismatch (VM) phenomenon necessitates more robust efforts in the Health Information Retrieval domain, mainly in query formulation. It is crucial to support clinicians, biomedical scientists, and non-specialists in their daily retrieval endeavors. In this paper, we propose an innovative approach that combines deep image captioning for clinical diagnosis generation then MetaMap normalization, and finally a user involvement step. This combination aims to optimize query formulation task for an enhanced query and an efficient HIR process. Experimental results, conducted on widely used search engines such as Google and Bing, reveal that our approach has demonstrated its effectiveness by enhancing result quality and delivering documents from reliable sources. It has significantly improved the user experience, ensuring relevant results appear within the top five ranking links. The findings demonstrate that the integration of various techniques is a valuable enhancement to the Query Expansion (QE) process, resulting in a notable increase in weighted precision rates by around the twofold a MAP to over 70% and an apparent reduction in Vocabulary Mismatch for Healthcare Information Retrieval.

Keywords: Information retrieval · query formulation · Query expansion · deep learning · MetaMap · user interaction · search engine

1 Introduction

The field of information retrieval's main purpose is the extraction and organization of information for later retrieval. This process studies the selection of information that best meets the user's requirements by integrating different models and techniques of information access [1, 2].

Health information retrieval (HIR) is a critical task in healthcare, allowing practitioners and researchers to access relevant and reliable information quickly and efficiently

L. Barolli (Ed.): AINA 2024, LNDECT 200, pp. 369–379, 2024.
https://doi.org/10.1007/978-3-031-57853-3_31

[3, 4]. However, the specialized terminology used in the health domain and the existence of multiple lexical variants for keywords used in user queries is decisive and can lead to vocabulary mismatch (VM) and significantly influence the effectiveness of the retrieval process [5]. In recent years, breakthroughs in multimodal deep learning have enabled the development of more sophisticated approaches for Query Expansion(QE), which aim to enhance retrieval effectiveness by generating more relevant and diverse queries [6].

In this paper, we propose a novel approach for HIR query expansion (QE), called Deep Multimodal interactive Query expansion for healthcare information retrieval. Our method combines clinical diagnosis generation via deep image captioning, MetaMap normalization, and UMLS concept detection to generate appropriate queries. We also introduce an interactive feedback mechanism to allow users to refine their queries and boost retrieval effectiveness further. To evaluate the effectiveness of our approach, we conducted experiments using widespread search engines such as Google and Bing.

Our paper is structured as follows: the Sect. 1 will present a brief literature review, followed by an explanation of our approach in the Sect. 2. The conducted experiments will be presented in the Sect. 3. Finally, we will conclude by discussing our funding and presenting perspectives for future work.

2 Literature Review

Several recent studies have shown the potential of multimodal deep learning for HIR Query Expansion. For a broader review on the application of deep learning in medical informatics we vouch for [7] where medical image analysis is briefly touched upon. The authors draw attention to fundamental applications of deep learning in the areas of translational bioinformatics, medical informatics, pervasive sensing, and public Healthcare. One dedicated review on the Application of Multi-modal Deep Learning in Medicine, we recommend, Fei Jiang and his team [3] present the worth of the Artificial Intelligence in Healthcare from prior use to newest research. However, most existing information retrieval approaches spotlight on single-modal input as text or only Image and do not consider the complex and heterogeneity nature of healthcare data [8]. Our method tries to overcome these constraints by leveraging multiple modalities and incorporating an interactive feedback mechanism, which has been proven to enhance retrieval effectiveness significantly.

The field of health information retrieval has been a surge of interest for scientists and practitioners who attempt to develop effective and efficient systems for excerpting and arranging information for later retrieval. With the rising reliance on the web as a source of information, the challenge of satisfying the information requirements of Healthcare Information seekers has become more complicated. The heterogeneity of information seekers' profiles, their altering needs which depend on their circumstances, background studies and context of their quest, make the retrieval process more thought-provoking. Other raison that leads to this complexity is the emergence of new technologies and the shift in user behavior from passive information consumers to active information generators (such as social networks and blogs) have contributed to the radical evolution of the IR field [9, 10]. The main purpose of HIR is that results should be pertinent and achieve a high level of precision. For instance, a survey writer might be interested in a broad range of subdomains related to their overall current study, considering the information requirements expressed in their query. However, a medical expert requires precise, pertinent, and refined latest specialized documents. To address these challenges, various methods have been proposed, such as Query Expansion, which leverages external resources like domain knowledge and relevant query terms to improve the representation and enrichment of the query content [11].

Medical data are heterogeneous. It entails different types of data such as text, image, video, etc. The use of medical images as queries called Content Based Image Retrieval (CBIR) [12] is advantageous in terms of excluding textual ambiguity and saving time. Image-based query formulation involves visual features (pixel) to formulate a query and usually return a set similar image [13]. As the work of Ghosh et al. which propose a literature review of medical image retrieval systems [14]. However, restricting the query formulation to just visual data can lead to semantic gap phenomena. In such case, Annotation is assumed to be the best solution, however, it is usually subjective. The human image captioning process is time-consuming and requires experts in specific domains, especially in the medical field. In the same context, Text-based query formulation called Content Based Text Retrieval (CBTR) involves key words to formulate a query unfortunately it suffers essentially from ambiguity and polysemy [15].

Therefore, an evident solution is an eventual combination of both visual and textual features for effective Health Information Retrieval. Integrating both features type into HIR query formulation poses a challenge, as there is a lack of studies assessing models that include both aspects. Therefore, it is commanded to develop effective methods, such as medical image captioning or diagnosis generation, to annotate such data, contributing considerably to the improvement of research in the field of Information Retrieval [16].

Sumbal [17] proposes a hybrid query expansion framework for boosting the retrieval of biomedical literature. The framework combines two approaches: a co-occurrence-based method and a semantic-based approach, to enhance the precision and recall of retrieval. The co-occurrence-based method excerpts the utmost frequent co-occurring terms in the relevant documents of the initial query, while the semantic-based method uses the Unified Medical Language System (UMLS) [18] to develop the query with semantically related terms. The fundings of the research proof that the proposed hybrid approach overtook the traditional baseline approach, both in terms of precision and recall. The study also found that the hybrid method was further effective in retrieving highly relevant documents and decreasing the number of irrelevant documents. Overall, the hybrid Query Expansion method provides an effective tool for biomedical researchers and clinicians to optimize their literature retrieval [17]. Another, recent study [19] utilizes an automated approach for recommending MeSH [20] terms in systematic reviews to conduct in an efficient query formulation for literature search. The proposed method is useful to significantly improve the effectiveness of search strategies by suggesting further relevant MeSH [21] terms that were not initially included in the query search. The values of this approach bring in a remarkable saving in time, and effort required for query formulation, improving the precision and recall of retrieval results, and eventually leading to high-quality systematic reviews [19].

Overall, our results demonstrate that our method is a promising approach for improving HIR effectiveness, which might assist healthcare practitioners and researchers in accessing relevant, pertinent, and reliable information.

This paper aims to propose a Deep Multimodal Interactive Healthcare Query Expansion approach for web search engines retrieval effectiveness. To achieve this goal, the paper is structured as follows. First, a literature review of related works on health information retrieval is presented to provide the necessary background knowledge. Then, the proposed approach is described in detail, including its three main components: clinical diagnosis generation via deep image captioning, MetaMap [22] normalization for UMLS concept detection and finally the user involvement step.

3 Method

The proposed Deep Multimodal Interactive Healthcare Query expansion approach involves three main steps: i) *Deep image captioning using Transformer*, ii) *MetaMap normalization*, and iii) *user interaction and involvement*. The final query could be launched directly on the web via any search engine to seek relevant information.

In the first step, we use a pre-trained Transformer-based model for deep image captioning to generate textual descriptions of medical images. This involves encoding the image using a Convolutional Neural Network (CNN) [23] and decoding the encoded image into a natural language annotation using the Transformer model [24, 25]. The generated textual captions serve as an initial query for the retrieval process. Next, we apply MetaMap [22] normalization to the initial generated key words query. MetaMap is a tool that maps text to Unified Medical Language System (UMLS) Metathesaurus Concepts [18], which provides a standardized vocabulary for biomedical concepts. This helps to ensure that the query is formulated using appropriate medical terminology and trim down the risk of Vocabulary Mismatch (VM). Finally, the user interacts with the system to refine the query based on their specific information requirements.

The resulting query is then used to retrieve pertinent documents from a search engine, which are ranked based on their relevance to the introduced query. This approach merges deep image captioning, MetaMap [22] normalization, and user involvement to generate an expanded query that is more effective at retrieving relevant information from web search engines.

Procedure 1. The proposed approach Procedure steps.

Procedure: Proposed approach steps

Deep Textual Query Generation (DTQG step)

Input: User information needs as an image

Output: Deep textual query keywords

Steps:

Apply natural language processing techniques to extract semantic and syntactic features from user input.

Generate deep textual queries that capture the essence of the user's information needs.

Textual Query Mapping (TQM step)

Input: Deep textual query keywords

Output: Structured query language

Steps:

Map deep textual queries to a structured query language that can be used to retrieve relevant data from various sources.

Consider user preferences and constraints in the query mapping process.

User Involvement/ Interaction (UI step)

Input: Structured query language

Output: Refined queries based on user feedback

Steps:

Allow for user feedback and preferences in the query formulation process.

3.1 Deep Textual Query Generation

The transformer architecture is a neural network that depend on self-attention mechanisms to depict global dependencies between input tokens. The transformer model [24, 25] has upsurge widespread usage in various medical domains due to its versatility and effectiveness. It has been well applied in clinical Natural Language Processing (NLP) for tasks such as text classification, Name Entity Recognition (NER). In medical image analysis and annotation, transformers have excelled in tasks like classification, segmentation, and caption generation [26]. Transformers have also been widely used in electronic health records analysis for patient risk prediction and treatment recommendation [27]. The transformer's ability to deal with large-scale datasets, high adaptability with different modalities, and detection of complex patterns has made it a very valuable tool in medical research progress.

We propose a deep learning model for automatic generation of medical image captions. Our model is based on the transformer architecture, which has shown state-of-the-art performance on several Natural Language Processing (NLP) tasks. We use the shared code Of Zhihong Chen et al. [28]. They use two well-known benchmark datasets: IU X-RAY [29] and MIMIC-CXR [30]. The IU X-RAY dataset consists of 7,470 chest X-ray images and 3,955 corresponding reports. The MIMIC-CXR dataset is the largest publicly available radiography dataset, containing 473,057 chest X-ray images and 206,563 reports. Those datasets provide valuable resources for evaluating and training models in radiology report generation. Zhihong Chen et al., utilize it for the visual extractor, they choose the ResNet101 [31] pre-trained on ImageNet [32] to extract patch features, each represented by 512 dimensions. As for the encoder-decoder backbone, they use a Transformer structure with 3 layers, 8 attention heads, and hidden states of 512 dimensions, initializing it randomly. In the Cross-Modal Memory Networks (CMN), the memory matrix has a dimension of 512 and consists of 2048 memory vectors, also initialized randomly. For memory querying and responding, they established the thread number to 8 and the K value to 32. During training, they utilize cross-entropy loss with the Adam optimizer [33]. The learning rates for the visual extractor and other parameters are set to 5×10^{-5} and 1×10^{-4}, respectively, and decomposed by a rate of 0.8 per epoch for all datasets. In the report generation step, they set the beam size to 3 to stabilize effectiveness and efficiency. Overall, the model provides a promising approach for medical image captions, which has the potential to assist healthcare professionals in medical diagnosis and treatment recommendation.

3.2 Textual Query Mapping

MetaMap [22] is a well-used Natural Language Processing (NLP) tool for mapping free text to Unified Medical Language System (UMLS) Metathesaurus concepts [18]. It is usually used in the biomedical and clinical domain for tasks such as information extraction, information retrieval, and data mining. One common application of MetaMap is in query formulation, where it is used to map user queries to UMLS concepts, which can then be utilized to retrieve relevant information from biomedical literature databases such as PubMed. The Textual Query Mapping step involves the use of MetaMap annotation system to identify and extract biomedical concepts from the initial query. These biomedical concepts encompass various terms with associated meanings, and they can be single or multi-word phrases linked to a unique identifier called Concept Unique Identifier (CUI). MetaMap is a flexible annotation system widely used to map biomedical text to the Unified Medical Language System (UMLS) Metathesaurus. In the UMLS Metathesaurus, each medical concept is associated with a specific semantic type, using Mesh ontology for a given concept such as "*Pulmonary Disease*" for the concept "*Pulmonary Embolism*" or "*Pulmonary Thromboembolisms*". In our approach, we use MetaMap to extract medical concepts as well synonyms from the generated query based on the predefined semantic types. These semantic types are recurrently faced in clinical notes and medical literature and afford valuable information about patients' diseases and symptoms. Through the medical concept recognition phase, the generated query is fed into MetaMap, which uses a four-step processing algorithm. This algorithm divides the generated query in a list of keywords, performs variant generation, candidate retrieval, candidate evaluation, and builds mappings to the UMLS Metathesaurus.

3.3 User Involvement Interaction

Query formulation and user involvement are essential characteristics of information retrieval systems that intend to improve the user search experience and enhance the pertinence of search results. Involving users in the process of queries formulation and refining search strategies can lead to more precise and customized results. The interactive aspect is ensured by the user participation that might update the query and personalize it to have more effective results that will amend the user experience rating. Query formulation is a crucial step in information retrieval, and implying users in the query formulation process can lead to more effective and personalized search results.

Relevance feedback mechanisms empower users to reveal their assessments of search result relevance. In response to this feedback, the system regulates and fine-tunes subsequent search queries to align more closely with the user's intent. Additionally, interactive query formulation enables users to actively shape and develop their queries in real-time, leveraging features such as suggestions, autocomplete, and interactive interfaces. Personalization, on the other hand, customizes search results based on individual user preferences, search history, and behavior. Through adaptive learning, systems can continually enhance relevance by adjusting insights from user interactions. Crucially, the ongoing collection of user feedback and the systematic evaluation of usability has critical roles in the upgrade of information retrieval systems. This iterative aspect serves to discover and adopt user needs, preferences, and challenges (Table 1).

Table 1. Proposed approach Procedures steps results

Initial query		DTQG step	TQM step	UI step
User 1	Image #1	pulmonary thromboembolism emphysema	Pathological accumulation of air in tissues	Pathological accumulation of air in tissues *treatment*
User 2	Image #1	pulmonary thromboembolism emphysema	Pathological accumulation of air in tissues	Pathological accumulation of air in tissues *symptoms*
expert	Image #1	pulmonary thromboembolism emphysema	pulmonary thromboembolism emphysea	Pulmonary thromboembolism emphysema *diagnosis*

4 Results and Discussion

Retrieval effectiveness measures are just one part of the overall evaluation of search engine quality [35]. Calculating the effectiveness of search engine results is a basic part of such frameworks and is usually based on a TREC-style setting [36]. However, these methods have been condemned as being inadequate for evaluating Web search engines, where results present differs from other information retrieval systems [34].

Furthermore, retrieval effectiveness tests have been complained about not considering actual user behavior, i.e., they are based on the query-response paradigm and do not consider interaction in the searching process. While these criticisms should not be ignored, continued enhancement of query-response-based methods is still a valuable pursuit [37].

To assess the proposed methodology, we conducted a series of 50 queries for each step, resulting in a full dataset of 150 queries launched through Bing and Google. For each given query, we gathered 25 initial results, accumulating a total of 3750 links for evaluation. Precision@K, a measure in ranked Information Retrieval (IR), refers to the precision computed at cut-off K. It represents the percentage of top K documents that are deemed relevant. If R relevant articles are retrieved, precision@K can be computed accordingly. In this study, we considered K values of 5, 10, 15, 20, and 25, as illustrated in Table 2.

Table 2. Evaluation of the proposed approach

Bing							Google					
Query	P@5	P@10	P@15	P@20	P@25	MAP	P@5	P@10	P@15	P@20	P@25	MAP
Image	0,48	0,54	0.58	0.62	0,64	0.56	0,32	0,40	0,44	0.46	0.52	0.43
DTQG	0.46	0.54	0.6	0.63	0.68	0.51	0.64	0.68	0.74	0.70	0.64	0.65
TQM	0.52	0.58	0.61	0.65	0.7	0.62	0.68	0.72	0.75	0.78	0.68	0.70
UI	0.64	0.78	0.76	0.74	0.73	0.73	0.82	0.85	0.87	0.85	0.83	0.85

Our outcomes, as presented in Table 2, underscore the efficiency of the proposed approach, showcasing a nearly twofold improvement in precision rate. The incorporation of deep learning, medical concept detection via MetaMap mapping, and user involvement in query formulation emerges as a focal consideration. This incorporation proves effectives in not only bolstering precision but also in attenuating query mismatches. Remarkably, it excels in retrieving highly relevant literature and medical documentation. Furthermore, our results indicate that while the second query combination TQM aids in identifying more biomedical contextual terms, it falls short in ensuring the optimal precision for biomedical literature retrieval. This observation aligns with our findings that user involvement (UI) outperformed domain-agnostic medical concept mapping (DTQG) in the biomedical context.

Our approach aims to limit the Vocabulary Mismatch (VM) phenomenon by integrating automatic query generation and then UMLS concept detection. Our initial query was an image; when we use it as a query, Google Images and Bing try to generate some annotations. We noticed that image captioning won't give more than 3 words in the best case, which influences the retrieval effectiveness.

Our approach will generate at least 5 words concatenated with a set of UMLS concepts and a set of possible words that the user could add, which will increase the final numbers of the query keywords and normally improve the overall retrieval effectiveness

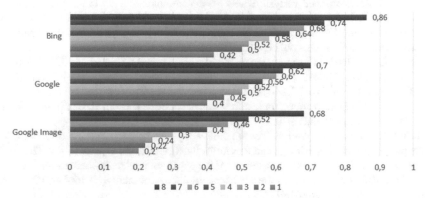

Fig. 1. A figure represents the influence of the number of keywords in a query on the precision rate.

rate, as shown in Fig. 1 above. The number of key words introduced as a query by a user to explain his information requirements is crucial to returning a relevant result from the fifth first returned link or document.

5 Conclusion

In summary, this study not only contributes to the field of healthcare information retrieval but also lays the groundwork for future endeavors in refining and expanding upon the presented approach to meet the growing challenges in the realm of biomedical literature exploration and knowledge discovery .

In conclusion, this study introduces an innovative meth that integrates deep learning, medical concept detection (MetaMap mapping), and user involvement to significantly boost precision in biomedical literature retrieval. Notably, the second query combination excels in identifying additional biomedical contextual terms, but it falls short of ensuring optimal precision, highlighting the crucial need for a balanced approach that considers both comprehensiveness and precision in information retrieval strategies.

Numerous promising avenues for further research emerge. The exploration of advanced deep learning architectures as incorporate medical word embedding or the fine-tuning of models to adapt to evolving medical terminologies and contexts holds potential to enhance the overall effectiveness of the proposed approach. Acknowledging the dynamic nature of biomedical research, continuous user feedback and iterative improvements are crucial for sustaining the system's efficacy over time.

In summary, this research not only contributes to the field of biomedical information retrieval but also lays a foundation for future works to refine and expand upon the presented approach.

References

1. Wang, J.: Mean-variance analysis: a new document ranking theory in information retrieval. In: Advances in Information Retrieval: 31th European Conference on IR Research, ECIR 2009, Toulouse, France, April 6–9, 2009. Proceedings 31, pp. 4–16. Springer Berlin Heidelberg (2009). https://doi.org/10.1007/978-3-642-00958-7_4
2. Salton, G., Michael, J.: Introduction to Modern Information Retrieval. McGill (1986)
3. Wang, Y., et al.: Clinical information extraction applications: a literature review. J. Biomed. Inform.Biomed. Inform. **77**, 34–49 (2018). https://doi.org/10.1016/j.jbi.2017.11.011
4. Ravi, D., et al.: Deep learning for health informatics. IEEE J. Biomed. Health Inform. **21**(1), 4–21 (2017). https://doi.org/10.1109/JBHI.2016.2636665
5. Yadav, N., di Bari, A., Wei, M., et al.: Mitigating vocabulary mismatch on multi-domain corpus using word embeddings and thesaurus. In: ICAART, no. 1, pp. 441–445 (2020)
6. Pei, X., Zuo, K., Li, Y., et al.: A review of the application of multi-modal deep learning in medicine: bibliometrics and future directions. Int J Comput Intell Syst **16**, 44 (2023). https://doi.org/10.1007/s44196-023-00225-6
7. Jiang, F., et al.: Artificial intelligence in healthcare: past, present and future. Stroke Vasc. Neurol. **2**(4), 230–243 (2017). https://doi.org/10.1136/svn-2017-000101
8. Mutabazi, E., Ni, J., Tang, G., Cao, W.: A review on medical textual question answering systems based on deep learning approaches. Appl. Sci. **11**(12), 5456 (2021). https://doi.org/10.3390/app11125456
9. Agichtein, E., Brill, E., Dumais, S. : Improving web search ranking by incorporating user behavior information. In: Proceedings of the 29th Annual International ACM SIGIR Conference on Research and Development in Information Retrieval, pp. 19–26 (2006). https://doi.org/10.1145/1148170.1148177
10. Lambert, S.D., Loiselle, C.G.: Health information—seeking behavior. Qual. Health Res. **17**(8), 1006–1019 (2007). https://doi.org/10.1177/1049732307305199
11. Wei, Y., Croft, W.B., Allan, J.: Effective query formulation with multiple information sources. In: Proceedings of the 18th ACM Conference on Information and Knowledge Management, pp. 563–572 (2009). https://doi.org/10.1145/2124295.2124349
12. Tagare, H.D., Jaffe, C.C., Duncan, J.: Medical image databases: a content-based retrieval approach. J. Am. Med. Inf. Assoc. **4**(3), 184–198 (1997). https://doi.org/10.1136/jamia.19
13. Wang, J., Yang, J., Yu, K., Lv, F., Huang, T., Gong, Y.: Locality-constrained linear coding for image classification. In: 2010 IEEE Computer Society Conference on Computer Vision and Pattern Recognition, San Francisco, CA, USA, pp. 3360–3367 (2010). https://doi.org/10.1109/CVPR.2010.5540018
14. Ghosh, P., Antani, S., Long, L.R., Thoma, G.R.: Review of medical image retrieval systems and future directions. In: 2011 24th International Symposium on Computer-Based Medical Systems (CBMS), Bristol, UK, pp. 1–6 (2011). https://doi.org/10.1109/CBMS.2011.5
15. Sharma, S., Dudeja, R.K., Aujla, G.S., et al.: DeTrAs: deep learning-based healthcare framework for IoT-based assistance of Alzheimer patients. Neural Comput. Applic.Comput. Applic. (2020). https://doi.org/10.1007/s00521-020-05327-2
16. Beddiar, D.R., Oussalah, M., Seppänen, T.: Automatic captioning for medical imaging (MIC): a rapid review of literature. Artif. Intell. Rev.. Intell. Rev. **56**(5), 4019–4076 (2023)
17. Malik, S., Shoaib, U., Bukhari, S.A.C., El Sayed, H., Khan, M.A.: A hybrid query expansion framework for the optimal retrieval of the biomedical literature. Smart Health **23**, 100247 (2022). https://doi.org/10.1016/j.smhl.2021.100247
18. Bodenreider, O.: The unified medical language system (UMLS): integrating biomedical terminology. Nucleic Acids Res. **32**(90001), 267D – 270 (2004). https://doi.org/10.1093/nar/gkh061

19. Wang, S., Scells, H., Koopman, B., Zuccon, G.: Automated MeSH term suggestion for effective query formulation in systematic reviews literature search. Intell. Syst. Appl. **16**, 200141 (2022). https://doi.org/10.1016/j.iswa.2022.200141

20. Zieman, Y.L., Bleich, H.L.: Conceptual mapping of user's queries to medical subject headings. In: Proceedings of the AMIA Annual Fall Symposium, p. 519. American Medical Informatics Association (1997)

21. Richter, R.R., Austin, T.M.: Using MeSH (medical subject headings) to enhance pubmed search strategies for evidence-based practice in physical therapy. Phys. Ther.Ther. **92**(1), 124–132 (2012)

22. Aronson, A.R.: MetaMap: mapping text to the UMLS metathesaurus. Bethesda, MD: NLM, NIH, DHHS **1**, 26 (2006)

23. Krizhevsky, A., Sutskever, I., Hinton, G.E.: ImageNet classification with deep convolutional neural networks. In: Advances in Neural Information Processing Systems, vol. 25 (2012)

24. Vaswani, A., et al.: Attention is all you need. In: Advances in Neural Information Processing Systems, vol. 30 (2017)

25. Devlin, J., Chang, M.W., Lee, K., Toutanova, K.: Bert: pre-training of deep bidirectional transformers for language understanding. arXiv preprint arXiv:1810.04805 (2018)

26. Shamshad, F., et al.: Transformers in medical imaging: a survey. Med. Image Anal. **88**, 102802 (2023). https://doi.org/10.1016/j.media.2023.102802

27. Selivanov, A., Rogov, O.Y., Chesakov, D., et al.: Medical image captioning via generative pretrained transformers. Sci. Rep. **13**, 4171 (2023). https://doi.org/10.1038/s41598-023-312 23-5

28. Chen, Z., Shen, Y., Song, Y., et al.: Cross-modal memory networks for radiology report generation. In: Proceedings of the Joint Conference of the 59th Annual Meeting of the Association for Computational Linguistics and the 11th International Joint Conference on Natural Language Processing (2021)

29. Demner-Fushman, D., et al.: Preparing a collection of radiology examinations for distribution and retrieval. J. Am. Med. Inform. Assoc. **23**(2), 304–310 (2016)

30. Johnson, A.E.W., et al.: MIMIC-CXR: a large publicly available database of labeled chest radiographs. arXiv preprint arXiv:1901.07042 (2019)

31. He, K., Zhang, X., Ren, S., Sun, J.: Deep residual learning for image recognition. In: Proceedings of the IEEE Conference on Computer Vision and Pattern Recognition, pp. 770–778 (2016)

32. Deng, J., Dong, W., Socher, R., Li, L.J., Li, K., Li, F.F.: ImageNet: a large-scale hierarchical image database. In: 2009 IEEE Conference on Computer Vision and Pattern Recognition, pp. 248–255 (2009)

33. Kingma, D.P., Ba, J.: Adam: a method for stochastic optimization. CoRR, abs/1412.6980 (2015). https://doi.org/10.48550/arXiv.1412.6980

34. Lewandowski, D., Höchstötter, N.: Web searching: a quality measurement perspective. In: Web Search: Multidisciplinary Perspectives, pp. 309–340. Springer, Berlin Heidelberg (2008). https://doi.org/10.1007/978-3-540-75829-7_16

35. Lewandowski, D.: A framework for evaluating the retrieval effectiveness of search engines. In: Next Generation Search Engines: Advanced Models for Information Retrieval, pp. 456–479. IGI Global (2012)

36. Lewandowski, D., Sünkler, S.: Designing search engine retrieval effectiveness tests with RAT. Inf. Serv. Use **33**(1), 53–59 (2013)

37. Lewandowski, D.: Challenges for search engine retrieval effectiveness evaluations: universal search, user intents, and results presentation. In: Pasi, G., Bordogna, G., . Jain, L.C. (eds.) Quality issues in the management of web information, pp. 179–196. Springer, Heidelberg (2013). https://doi.org/10.1007/978-3-642-37688-7_9

Multi-level Frequent Pattern Mining on Pipeline Incident Data

Connor C. J. Hryhoruk, Carson K. Leung$^{(\boxtimes)}$, Jingyuan Li,
Brandon A. Narine, and Felix Wedel

University of Manitoba, Winnipeg, MB, Canada
`Carson.Leung@UManitoba.ca`

Abstract. Pipeline incidents occur throughout the world and can have devastating financial and ecological impacts. A large amount of data is collected and made publicly available for each pipeline incident. Although some important information is explicitly visible in such datasets, a lot of implicit information remains hidden. In this paper, we explore frequent pattern mining approaches—specifically, multi-level frequent pattern mining, which help discover some of this implicit information that was previously hidden in the data. The resulting frequent patterns and their corresponding association rules provides some new insights into pipeline incidents that may help improve the safety and reliability of pipelines. Evaluation results show the effectiveness and practicality of our multi-level frequent pattern mining solution.

Keywords: advanced information applications · data mining · frequent pattern mining · multi-level mining · association rules · pipeline accident

1 Introduction

Nowadays, big data are everywhere due to advancements in technology so that huge amounts of data can be easily generated and collected at a rapid rate from a wide variety of real-life application domains. Embedded in these big data are implicit, previously unknown but useful information. Data science solutions [1–3]—which make good use of data mining [4–7], machine learning [8–11], mathematical and statistical techniques, and visualization [12,13]—help discover the useful information that was previously hidden in the big data. These solutions can be applicable for tasks like health informatics [14], social media mining [15] (e.g., web mining [16]), social network analysis [17–19], sports analytics [20,21], and transportation analytics [22], which can be for social good. In this paper, we focus on the data mining task of frequent pattern mining from pipeline incident data.

In the year of 2021 alone, 99 Canadian companies utilized approximately 69,400 km of government-regulated pipelines to transport natural gas, oil and other substances [23]. While often considered as a safer alternative to vehicular transport, pipelines are still susceptible to accidents and failures. Due to the

L. Barolli (Ed.): AINA 2024, LNDECT 200, pp. 380–392, 2024.
https://doi.org/10.1007/978-3-031-57853-3_32

nature of the substances that are transported through these pipelines, these incidents can often be extremely costly both financially and environmentally [24]. While relatively rare, some pipeline incidents result in serious injuries and sometimes fatalities. Research and development of ways to mitigate or prevent these incidents has been an ongoing endeavor. However, pipeline incidents remain a constant threat, with the average number of annual pipeline incidents in Canada from 2011 to 2021 sitting at 117 [23]. Recent data also suggest that the number of annual pipeline incidents in Canada has been relatively stable within the last couple of years. With the demand for energy increasing, it seems likely that we may see an increase in the construction and use of natural gas and oil pipelines in the near future. Because of this, it is important to find ways to prevent pipeline incidents from happening and to find ways to mitigate the negative impacts of pipeline incidents.

Many organizations—including government agencies such as the Canada Energy Regulator (CER), or formerly National Energy Board (NEB)—collect data related to pipeline incidents that occur within their country [23]. These data are often publicly available and contains detailed information about each incident. These include information like severity, causes of the incident and some maintenance history details [25]. These historical data contain a lot of useful information that could potentially be used to prevent these incidents or predict when they occur. As is often the case with large and complex datasets, not all of the useful information is explicitly listed, nor can it be easily extracted through conventional means. This becomes increasingly apparent that datasets (e.g., pipeline incident dataset compiled and managed by CER) may contain data columns with implicit hierarchies that may hide certain features and relationships between the data. At present, CER's pipeline incident dataset contains records for 1736 pipeline incidents that have occurred across Canada since the year 2008. Each record contains 102 different columns of data. These data range from dates to text descriptions and various counts. While this data lends itself well too many statistical models and machine-learning techniques (e.g., linear regression modeling, k-means clustering), our work involves discovering interesting relationships and patterns from the dataset and its implicit data hierarchies by using frequent pattern mining approaches that stem from the realm of data mining.

Frequent pattern mining [26] involves discovering relationships between different items in a database entry that occur often enough to be considered frequent. From these relationships or frequent patterns, association rules can be generated, which allow us to make predictions about future events and to predict the likelihood of sets of items being present in a database entry, given the presence of another item. Frequent pattern mining alone could be used to obtain many interesting frequent patterns and association rules from pipeline incident data. These data could then be used for predictions and classifications related to pipeline incidents. As previously mentioned, however, CER's pipeline incident data include natural data hierarchies embedded within certain data columns. Using a basic frequent pattern mining approach may fail to uncover certain patterns within the different levels of these hierarchies. In order to uncover interesting frequent patterns and association rules at the different data-hierarchy levels, we consider other techniques such as multi-level mining. Specifically, multilevel frequent pattern mining allows us to uncover associations between different variables that have natural data hierarchies, such as spatial and temporal data. This can be applied to the pipeline incident data to discover patterns between the different specificity levels contained within a single date or location related column [27]. For example, by using multi-level mining, we can uncover patterns related to not only a specific year, month, day and combination, but also between years and months separately.

Our *key contribution* of this paper is to provide useful insights into pipeline incidents and the events that led up to them. By designing and applying the technique of multi-level mining to pipeline incident data, we uncover previously unknown or poorly understood relationships embedded within the data. These insights have the potential to help with predictions of when and where pipeline incidents may occur, as well as the severity and cause of the incidents. With these predictions, new informed decisions can be made about incident prevention, new safety measures and damage control measures. These new insights are expected to help reduce the frequency and impact of pipeline incidents and perhaps prevent future injuries or fatalities from occurring.

The remainder of this paper is organized as follows. The next section provides background and discusses related works. Section 3 describes our multi-level frequent pattern mining solution for analyzing pipeline incident data. Evaluation results are shown in Sect. 4. Finally, conclusions are drawn in Sect. 5.

2   Background and Related Works

2.1   Frequent Pattern Mining

Much of the foundation for our frequent pattern and multi-level mining approaches come from two influential data-mining papers. Agrawal and Srikant [26] proposed Apriori algorithm for mining frequent patterns (at a single level). They also compiled a plethora of techniques, concepts and algorithms for various types of frequent pattern mining in their book on frequent pattern mining. Eavis and Zheng [27] extended the Apriori to both naive approaches to multi-level mining as well as novel algorithms that are specialized and efficient at multi-level mining. As a preview, our work makes use of some of these techniques and algorithms (e.g., Apriori algorithm and iterative multi-level mining) and applies them to pipeline incident data in order to extract useful information.

2.2   Pipeline Incidents

Other research has gone into understanding and preventing pipeline accidents using methods other than multi-level mining. For example, the concepts of natural language processing and text mining [28] have gone towards understanding pipeline incidents and what causes them. These approaches focused mainly on the text data collected and were fundamentally different approach from multi-level mining.

Some research [29] has also gone into improving the understanding of pipeline incident datasets and analyzing the causal factors listed in the incident reports using neural networks. This research could serve to improve data collection for pipeline incidents, which may result in better results when applying techniques like multi-level mining or natural language processing to these datasets.

Aroh et al. [24] investigated the negative effects of pipeline incidents. Their work provided a better understanding of the impacts that pipeline incidents may have. It also provided motivation to work towards solutions and preventative measures. Without research into the consequences of pipeline incidents, this project and similar projects may have never been undertaken.

3   Our Mining Solution

In order to extract relationships and frequent patterns from the different hierarchies embedded within the spatial and temporal data columns within CER's pipeline incident dataset, our solution first separates the hierarchies into separate columns. This allows us to include only the levels of the hierarchy that we wish to mine at a given time, using the multi-level mining. The spatial and temporal hierarchies within the data set are distinct from each other, containing different levels and data types, and therefore require different processes to separate.

3.1   Data Preprocessing

Preprocess Temporal Data. Our solution handles both temporal and spatial data. For the temporal data in CER's pipeline incident dataset, they are in the form of simple date strings. One inconsistency between columns containing temporal data is that some columns—namely, the columns relating to when the incident was discovered and when the incident occurred—contain information as specific as the hour. Other columns containing temporal data only have information as specific as the day.

Another inconsistency is that some date columns contain dates in the form of MM/DD/YYYY, whereas other columns have dates in the form of YYYY-MM-DD. Despite these inconsistencies, each date column could be separated into multiple columns, where each column contains one piece of the date. This was done by performing the following steps (say, via a Python script):

– Read the dataset into a data frame.
– Extract the date columns & store them in separate one-column data frames.
– For each date column,
  • create a new one-column data frame for each piece of information contained in the date string,
  • read each row,

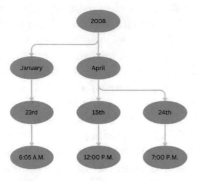

2007-12-27 14:00	2007	Dec-07	2007-12-27
2008-01-02 11:30	2008	Jan-08	2008-01-02
2008-01-23 6:05	2008	Jan-08	2008-01-23
2008-01-26 10:45	2008	Jan-08	2008-01-26
2008-01-27 0:01	2008	Jan-08	2008-01-27
2008-02-22 0:00	2008	Feb-08	2008-02-22
2008-02-25 0:00	2008	Feb-08	2008-02-25
2008-03-02 10:30	2008	Mar-08	2008-03-02
2008-03-01 16:45	2008	Mar-08	2008-03-01
2008-03-14 13:30	2008	Mar-08	2008-03-14
2008-03-24 9:50	2008	Mar-08	2008-03-24
NA	NA	NA	NA
2008-04-16 14:30	2008	Apr-08	2008-04-16
2008-04-21 7:00	2008	Apr-08	2008-04-21

(a) Temporal hierarchy (with year at level 1, month at level 2, day at level 3, and time at level 4)

(b) Processed temporal hierarchical representation

Fig. 1. Processed hierarchical temporal data representation.

- split the date string on either "/" or "-" characters, and
- insert each resulting string after the split into one of the new data frames.

For a date column containing year, month, day and hour, this results in four new one-column data frames, each representing one piece of the date/time. For example, Fig. 1(b) shows that the original date "2007-12-27 14:00" is processed into year "2007", month "Dec" (of 2007), day "27" (of 2007-12/Dec), and hour "14:00".

The original date format already represents the first level of our data hierarchy. However, to represent the second, third and, the fourth (in the case that the column contains an hour) level of the hierarchy, we combine certain columns. See Fig. 1(a) for an example of temporal hierarchy. For date columns as specific as hour, the columns representing the data hierarchies look as in Fig. 1(b). Apart from data pertaining to when the incident occurred and when it was discovered, the time between these events also seemed potentially interesting. CER's pipeline incident dataset does not explicitly contain this information. However, when using the occurrence and discovery information, a value representing the time between the two was easily obtained. This was done by converting the date/time in the occurrence and discovery columns into the Portable Operating System Interface for Unix calendar time (POSIXct) date format, which stores time as number of seconds since 01 January 1970. This allows the use of simple arithmetic on the resulting date values to get a floating point value representing the hours between the two events.

Preprocess Spatial Data. While CER's pipeline incident dataset already contained two separate levels of spatial data, we make use of a reverse geocoding application programming interface (API) in order to convert latitude and longitude values into addresses. By scanning through the latitude and longitude

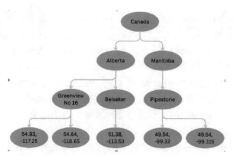

province	municipality	coordinates
Alberta	Greenview_No_16, Alberta	54.84, -118.65
Manitoba	Pipestone, Manitoba	49.753, -101.239
Manitoba	Pipestone, Manitoba	49.754, -101.237
British_Columbia	Northern_Rockies, British_Columbia	58.012, -122.694
British_Columbia	Peace_River, British_Columbia	56.143, -120.516
Alberta	Beiseker, Alberta	51.385, -113.535
Saskatchewan	Moosomin_No_121, Saskatchewan	50.207, -101.487
Ontario	Dryden, Ontario	49.789, -92.765
Ontario	Fauquier-Strickland, Ontario	49.303, -82.043
Manitoba	Oakview_RM, Manitoba	50.076, -99.998
Saskatchewan	Progress_No_351, Saskatchewan	51.92, -109.17
Ontario	Timiskaming, Ontario	47.446, -79.762
Ontario	Kenora_District, Ontario	49.808, -94.189
Saskatchewan	Regina, Saskatchewan	50.48, -104.566
British_Columbia	Peace_River, British_Columbia	56.279, -120.963
British_Columbia	Peace_River, British_Columbia	55.338, -121.871
Manitoba	Pipestone, Manitoba	49.754, -101.239

(a) Spatial hierarchy (with country at level 1, province at level 2, city/town at level 3, and GPS coordinates at level 4)

(b) Processed spatial hierarchical representation

Fig. 2. Processed hierarchical spatial data representation.

values and sending them individually to the HERE reverse geo-coding API, a list of addresses pertaining to the pipeline incidents is compiled. For each latitude and longitude pair, the JavaScript Object Notation (JSON) object returned by the API was parsed to extract the relevant information so that it could be stored in a data frame and then exported as a comma-separated values (CSV) file.

It was tempting to use the street value returned by the API. However, since pipelines often run through unpopulated and undeveloped areas, a large number of coordinates resulted in NA values for the nearest street. Hence, it is better to exclude this information. While very few of the coordinate pairs resulted in NA values for the city portion of the address, many of these values referred to municipalities or districts within a given province. This was again likely due to the fact that pipelines do not only run through populated areas. The municipality strings were then combined into one column with the corresponding province, which was already present in the dataset. This was done so that two entries could be distinguished in the case that two provinces happened to have a municipality or city of the same name. The three resulting levels of the spatial data hierarchy are province, municipality and coordinates. See Fig. 2(a) for an example of spatial hierarchy. For example, Fig. 2(b) shows that the original Global Positioning System (GPS) coordinates "54.84°N, 118.65°W" is processed into province "Alberta", Municipal District of "Greenview No. 16" (in Alberta), and GPS coordinates "(54.84, −118.65)".

After extracting the data hierarchies from the spatial and temporal data columns, original versions of these columns in the pipeline incident data set were replaced by a column representing a specific level of the hierarchy. As the occurrence date and discovery date columns had four hierarchy levels, the end result is four different datasets. Each of these data sets contains spatial and temporal data at a specific level of their respective data hierarchies. For columns only containing three levels, the same values are carried over from the third to fourth level data set. Columns that contain comma-separated values (e.g., the "detailed what happened" column can contain a list of descriptions) were also separated into multiple columns in preparation for encoding. This was done so that each piece of the comma-separated value could be mined individually. For example, an original value "cracking, material loss" would be separated so that "cracking" and "material loss" showed up separately in the data set entry. This allows the mining algorithm to count the values individually rather than counting the combination of the two values. In other words, if two columns originally contained "cracking, material loss" and "cracking, corrosion," after being separated, our mining algorithm would be able detect the "cracking" value in both entries, which was not the case before separation.

Another problem with keeping the comma-separated values in a single column is that entries containing the same set of values in different orders would have also been counted separately, which is not desirable. The last step before encoding the data set was to get rid of unnecessary mostly empty columns. Unique data columns (e.g., pipeline incident number column) were removed because they do not provide any real insight when performing frequent data mining and they would increase the amount of columns in the encoded data set by the number of entries in the column. Columns that contained over 50% empty values were also dropped in order to focus on columns that were more likely to produce frequent patterns.

3.2   One-Hot Encoding

To one-hot encode each of the datasets that are to be mined, our solution creates dummy variables for each of the possible values in a given column. For a column with k possible values, there are a total of k dummy variables that are each represented as columns in the one-hot encoded dataset. Each entry in these new columns is a binary value that represents whether or not that dummy variable was present in the original data entry. This allows the frequent pattern mining algorithm to look at binary values rather than having to look at and compare the various data types contained in each column of the original data set and only having to look at the string name of the dummy variable if it reads in a value of one for that entry. For the purposes of frequent pattern mining, one-hot encoding also takes care of null or missing values because they will simply contain 0 s for each of the dummy variables generated from the original column, and will therefore be ignored by the algorithm. While one-hot encoding vastly increases the number of columns in the dataset, it also allows all entry values to be stored as simple binary values and makes frequent pattern mining much simpler and more efficient in the long run.

3.3   Multi-level Frequent Pattern Mining

To discover frequent patterns, our solution applies frequent pattern mining to the pipeline incident data. The mining can be conducted in a similar fashion as in algorithms like Apriori for a single level (e.g., time and GPS coordinates at level 4). At the same time, our solution also mines the data at multiple levels (e.g., day and city/town at level 3).

Knowing that mining at lower minimum support values may cause some memory issues because the binary values in each column of the dataset were being stored in memory as 64-bit integers. We resolve this problem by explicitly reading them in as 8-bit integers, which greatly reduced the memory required.

3.4   Formation and Visualization of Association Rules

Once frequent patterns are mined, our solution then forms association rules. This algorithm would return the antecedents and consequences of each association rule separately in the form of frozensets, as well as a number of metrics to evaluate and filter out rules. Along with the support of the consequences, antecedent and both combined, each association rule is returned with its confidence, lift, leverage, conviction and Zhang's metric. These metrics are important when filtering out interesting and useful association rules, especially when there could potentially be thousands of rules generated.

In combination with the encoded datasets for each level of the data hierarchies within the pipeline incident dataset, the frequent pattern mining and association rule mining algorithms could be used to mine each dataset individually to obtain frequent patterns and association rules. Based on the hierarchy level of the dataset, the minimum support of frequent patterns could be adjusted, since patterns at lower levels of the hierarchy are much less likely to appear because of their specificity.

To visualize the association rules, we created a grouped matrix using a balloon plot. The balloon plot is a scatter plot with the antecedent and consequence on the x and y-axes respectively. Each data point in the plot represents an antecedent and consequence pair, and the size of the balloon corresponds to the confidence level of the pair. Larger balloons indicate a higher confidence level. In addition to size, the color of the balloons represents the lift of the association rule. Lift is a measure of how much more often the antecedent and consequence occur together compared to what would be expected by chance. We used a shade of blue for the balloon color, with darker shades indicating higher lift values [30].

To highlight the most important associations, we reordered the antecedents and consequences by lift value. This ensured that the association rules with the highest lift values were plotted in the top right corner of the plot. By doing so, we could quickly identify the most significant associations in the dataset.

4   Evaluation

To evaluate our solution, we used real-life CER's pipeline incident dataset [25]. We applied our solution to preprocess data—especially, both temporal and spatial data. Figures 1(b) and 2(b) show processed hierarchical temporal and spatial representations, respectively.

We implemented our solution with Python libraries such as MLxtend (for machine learning extension), which contain well-implemented versions of the Apriori algorithm. Despite Apriori being generally less efficient than algorithms like FP-growth, all versions of the preprocessed and one-hot encoded pipeline incident dataset could be mined in less than a second, even with a minimum

Data Hierarchy Level	Minimum Support	Number of Frequent Patterns	Minimum confidence	Minimum lift	Minimum conviction	Number of Association rules
Level 1	0.85	483	99%	>1	>4	465
Level 2	0.75	2845	99%	>1	>4	2022
Level 3	0.65	9247	99%	>1	>10	10519
Level 4	0.55	34137	99%	>1.5	>15	27270

Fig. 3. Parameters for frequent pattern mining

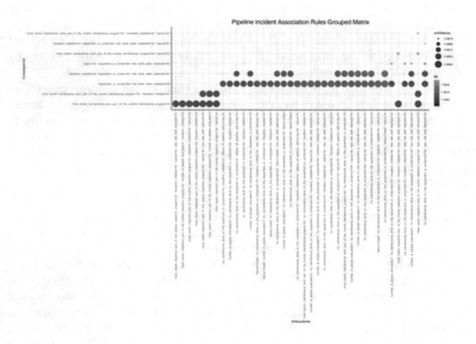

Fig. 4. Top-15 association rules with the highest lift values on the first level of the hierarchy

support value of 50%, which was more than good enough for the purposes of this work.

Based on the hierarchy level of the dataset, the minimum support of frequent patterns could be adjusted, since patterns at lower levels of the hierarchy are much less likely to appear because of their specificity. After some experimentation, the parameters shown in Fig. 3 were settled upon for the final run of frequent pattern mining and association rule generation.

After finding frequent patterns, our solution also forms and visualizes association rules. Figure 4 shows the top-15 association rules with the highest lift values on the first level of the analysis. The most interesting association rule in our analysis was determined based on the highest lift value. This association rule has the following antecedent and consequence:

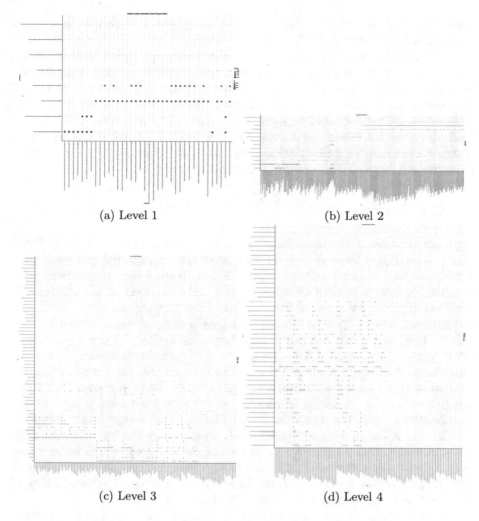

(a) Level 1

(b) Level 2

(c) Level 3

(d) Level 4

Fig. 5. Association rules at multiple levels of hierarchy

- Antecedent: {"Was the most recent inspection part of the routine inspection program? No", "Was NEB staff deployed? No"}
- Consequent: {"Was the most recent maintenance work part of the routine maintenance program? No", "Was insulation installed? No", "Rupture? No"}

The text after the question in each item represents the value of the column variable. For example, "Was the most recent inspection part of the routine inspection program? No" refers to the column variable "most recent inspection part of the routine inspection program" and its value "No". This convention was used to make it easier to identify the column variable associated with each value in the association rules. Since ruptures in the dataset were a rare occurrence, it makes sense that the "No" value for that variable is frequent, therefore we can ignore that part of the consequence.

Furthermore, there is a strong relationship between pipeline incidents where the pipeline was not inspected recently as part of the routine inspection program, no NEB staff were deployed, no recent maintenance work was done as part of the routine maintenance program, and no insulation was installed. This could be a useful insight for pipeline maintenance and safety practices, as it suggests that these variables may be interrelated and could be worth considering as a group when identifying potential risk factors. Visualizations of the association rules derived from all four datasets can be seen in Fig. 5.

5   Conclusions

Pipeline incidents occur throughout the world and can have devastating financial and ecological impacts. A large amount of data is collected and made publicly available for each pipeline incident. Although some important information is explicitly visible in such datasets, a lot of implicit information remains hidden. In this paper, we explored frequent pattern mining approaches—specifically, multi-level frequent pattern mining, which help discover some of this implicit information that was previously hidden in the data. Our solution preprocesses both temporal and spatial data, incorporates temporal and spatial hierarchies, and conducts one-hot encoding. Then, it mines the preprocessed one-hot encoded data at multiple levels to discover frequent patterns. These patterns form building blocks for the formation of association rules, which reveal important factors leading to pipeline incidents. To help users easily comprehend the mined results, our solution also provides visualization of association rules. As such, some new insights into pipeline incidents can be observed, which in turn may help improve the safety and reliability of pipelines. Evaluation results on real-life CER's pipeline incident dataset show the effectiveness and practicality of our multi-level frequent pattern mining solution.

As ongoing and future work, we explore further enhancement to our solution. For instance, we explore cross-level mining and their visualization. We would also apply this data science solution to other real-life application domains.

Acknowledgements. This work is partially supported by Natural Sciences and Engineering Research Council of Canada (NSERC) and University of Manitoba.

References

1. Anderson-Grégoire, I.M., et al.: A big data science solution for analytics on moving objects. In: Barolli, L., Woungang, I., Enokido, T. (eds.) AINA 2021, Volume 2. LNNS, vol. 226, pp. 133–145. Springer, Cham (2021). https://doi.org/10.1007/978-3-030-75075-6_11
2. Atif, F., Rodriguez, M., Araújo, L.J.P., Amartiwi, U., Akinsanya, B.J., Mazzara, M.: A survey on data science techniques for predicting software defects. In: Barolli, L., Woungang, I., Enokido, T. (eds.) AINA 2021, Volume 3. LNNS, vol. 227, pp. 298–309. Springer, Cham (2021). https://doi.org/10.1007/978-3-030-75078-7_31
3. Dierckens, K.E., et al.: A data science and engineering solution for fast k-means clustering of big data. In: IEEE TrustCom-BigDataSE-ICESS 2017, pp. 925–932 (2017)
4. Alam, M.T., et al.: Discovering interesting patterns from hypergraphs. ACM Trans. Knowl. Discov. Data (TKDD) $18(1)$, 32:1-32:34 (2024)
5. Leung, C.K.-S., Jiang, F.: Frequent pattern mining from time-fading streams of uncertain data. In: Cuzzocrea, A., Dayal, U. (eds.) DaWaK 2011. LNCS, vol. 6862, pp. 252–264. Springer, Heidelberg (2011). https://doi.org/10.1007/978-3-642-23544-3_19
6. Leung, C.K.-S., Tanbeer, S.K.: Fast tree-based mining of frequent itemsets from uncertain data. In: Lee, S., Peng, Z., Zhou, X., Moon, Y.-S., Unland, R., Yoo, J. (eds.) DASFAA 2012, Part I. LNCS, vol. 7238, pp. 272–287. Springer, Heidelberg (2012). https://doi.org/10.1007/978-3-642-29038-1_21
7. Liu, C., Li, X.: Mining method based on semantic trajectory frequent pattern. In: Barolli, L., Woungang, I., Enokido, T. (eds.) AINA 2021, Volume 2. LNNS, vol. 226, pp. 146–159. Springer, Cham (2021). https://doi.org/10.1007/978-3-030-75075-6_12
8. Abahussein, S., et al.: Protect trajectory privacy in food delivery with differential privacy and multi-agent reinforcement learning. In: Barolli, L. (ed.) AINA 2023, Volume 3. LNNS, vol. 655, pp. 48–59. Springer, Cham (2023). https://doi.org/10.1007/978-3-031-28694-0_5
9. Leung, C.K., et al.: Machine learning and OLAP on big COVID-19 data. In: IEEE BigData 2020, pp. 5118–5127 (2020)
10. Lu, W.: Applied machine learning for securing the internet of medical things in healthcare. In: Barolli, L. (ed.) AINA 2023, Volume 2. LNNS, vol. 654, pp. 404–416. Springer, Cham (2023). https://doi.org/10.1007/978-3-031-28451-9_35
11. Olawoyin, A.M., et al.: Big data management for machine learning from big data. In: Barolli, L. (ed.) AINA 2023, Volume 1. LNNS, vol. 661, pp. 393–405. Springer, Cham (2023). https://doi.org/10.1007/978-3-031-29056-5_35
12. Di Giacomo, E., et al.: Design of a process and a container-based cloud architecture for the automatic generation of storyline visualizations. In: Barolli, L. (ed.) AINA 2023, Volume 3. LNNS, vol. 655, pp. 90–99. Springer, Cham (2023). https://doi.org/10.1007/978-3-031-28694-0_9
13. Leung, C.K., et al.: Big data visualization and visual analytics of COVID-19 data. In: IV 2020, pp. 415–420 (2020)
14. Souza, J., Leung, C.K., Cuzzocrea, A.: An innovative big data predictive analytics framework over hybrid big data sources with an application for disease analytics. In: Barolli, L., Amato, F., Moscato, F., Enokido, T., Takizawa, M. (eds.) AINA 2020. AISC, vol. 1151, pp. 669–680. Springer, Cham (2020). https://doi.org/10.1007/978-3-030-44041-1_59

15. Cabusas, R.M., et al.: Mining for fake news. In: Barolli, L., Hussain, F., Enokido, T. (eds.) AINA 2022, Volume 2. LNNS, vol. 450, pp. 154–166. Springer, Cham (2022). https://doi.org/10.1007/978-3-030-99587-4_14

16. Lee, W., et al.: Mobile web navigation in digital ecosystems using rooted directed trees. IEEE TIE **58**(6), 2154–2162 (2011)

17. Cameron, J.J., et al.: Finding strong groups of friends among friends in social networks. In: IEEE DASC 2011, pp. 824–831 (2011)

18. Choudhery, D., Leung, C.K.: Social media mining: prediction of box office revenue. In: IDEAS 2017, pp. 20–29 (2017)

19. Tanbeer, S.K., et al.: Interactive mining of strong friends from social networks and its applications in e-commerce. J. Organ. Comput. Electron. Commer. **24**(2–3), 157–173 (2014)

20. Anuraj, A., et al.: Sports data mining for cricket match prediction. In: Barolli, L. (ed.) AINA 2023, Volume 3. LNNS, vol. 655, pp. 668–680. Springer, Cham (2023). https://doi.org/10.1007/978-3-031-28694-0_63

21. Isichei, B.C., et al.: Sports data management, mining, and visualization. In: Barolli, L., Hussain, F., Enokido, T. (eds.) AINA 2022, Volume 2. LNNS, vol. 450, pp. 141–153. Springer, Cham (2022). https://doi.org/10.1007/978-3-030-99587-4_13

22. Leung, C.K., Braun, P., Hoi, C.S.H., Souza, J., Cuzzocrea, A.: Urban analytics of big transportation data for supporting smart cities. In: Ordonez, C., Song, I.-Y., Anderst-Kotsis, G., Tjoa, A.M., Khalil, I. (eds.) DaWaK 2019. LNCS, vol. 11708, pp. 24–33. Springer, Cham (2019). https://doi.org/10.1007/978-3-030-27520-4_3

23. Statistical summary: pipeline transportation occurrences in 2021. Transportation Safety Board of Canada. https://www.tsb.gc.ca/eng/stats/pipeline/2021/ssep-sspo-2021.html

24. Aroh, K.N., et al.: Oil spill incidents and pipeline vandalization in Nigeria. Disaster Prev. Manag. Int. J. **19**(1), 70–87 (2010)

25. Pipeline incident data. National Energy Board. https://doi.org/10.35002/nb1p-vw48

26. Agrawal, R., Srikant, R.: Fast algorithms for mining association rules in large databases. In: VLDB 1994, pp. 487–499 (1994)

27. Eavis, T., Zheng, X.: Multi-level frequent pattern mining. In: Zhou, X., Yokota, H., Deng, K., Liu, Q. (eds.) DASFAA 2009. LNCS, vol. 5463, pp. 369–383. Springer, Heidelberg (2009). https://doi.org/10.1007/978-3-642-00887-0_33

28. Liu, G., et al.: Identifying causality and contributory factors of pipeline incidents by employing natural language processing and text mining techniques. Process Saf. Environ. Prot. **152**, 37–46 (2021)

29. Kumari, P., et al.: A unified causation prediction model for aboveground onshore oil and refined product pipeline incidents using artificial neural network. Chem. Eng. Res. Des. **187**, 529–540 (2022)

30. Hahsler, M., Karpienko, R.: Visualizing association rules in hierarchical groups. J. Bus. Econ. **87**(3), 317–335 (2016)

Real-Time Task Scheduling and Dynamic Resource Allocation in Fog Infrastructure

Mayssa Trabelsi[✉] and Samir Ben Ahmed

Faculty of Sciences of Tunis, University of Tunis El Manar, Tunis, Tunisia
{mayssa.trabelsi,samir.benahmed}@fst.utm.tn

Abstract. Efficient task scheduling in Fog Computing infrastructures necessitates adaptive solutions to dynamically manage real-time tasks and improve the utilization of resources. In this paper, we propose a real-time dynamic task scheduler that assigns incoming tasks to fog nodes based on maximizing resource utilization and satisfying task deadlines. The proposed approach leverages optimization technique and heuristic algorithm to allocate fog resources and schedule real-time tasks efficiently. The objective is an efficient use of resources while ensuring tasks are executed within their specified deadlines. Hence, the algorithm aims to maximize resource utilization in the architecture so that a large percentage of real-time tasks meet their deadline. The algorithm is implemented in the iFogSim simulator, and its performances are evaluated and compared to other algorithms.

Keywords: Internet of Things (IoT) · Fog Computing · Resource Allocation · Task Scheduling · Quality of Service (QoS) · Resource Utilization

1 Introduction

In recent years, the Internet of Things (IoT) and real-time applications have experienced explosive growth, becoming widely used in various domains such as smart healthcare, real-time manufacturing, smart homes, connected and autonomous cars, and smart cities [1]. Real-time applications, particularly those sensitive to delays, demand strict timing constraints and require processing capabilities that often exceed the limitations imposed by severely resource-constrained IoT devices [3]. Given the emergence of many IoT applications requiring real-time response and low latency, cloud computing is often unsuitable due to high communication delay between the IoT devices and the cloud servers [3,4]. To address these issues, Cisco has introduced a new technology called Fog Computing (FC). FC extends the Cloud Computing paradigm to the edge of the network [5]. It is a novel decentralized computing paradigm that provides resources such as computation, storage, and networking closer to IoT devices [5,7]. FC enables delay-sensitive IoT applications to improve their quality of service (QoS) by reducing the transmission time and making real-time processing of IoT requests [2].

FC faces three major scheduling challenges including Task Scheduling (TS), Resource Allocation (RA), and Workflow Scheduling (WS) [6,13]. The challenges associated with TS and RA are crucial, given the heterogeneity of resource-constrained Fog Nodes (FNs), as well as task characteristics such as release time, wait time, length,

© The Author(s), under exclusive license to Springer Nature Switzerland AG 2024
L. Barolli (Ed.): AINA 2024, LNDECT 200, pp. 393–403, 2024.
https://doi.org/10.1007/978-3-031-57853-3_33

and deadlines, in the highly unpredictable nature of fog environments. These challenges directly impact the overall performance enhancement of the system [6]. The efficient allocation of a set of heterogeneous tasks on a large number of geographically distributed FNs makes TS and RA NP-hard problems [6].

In recent years, many research has proposed methods for scheduling IoT tasks in fog environments to optimize resource utilization and various performance metrics along with task deadline constraints [1,3,9,11,12]. The authors in [1] proposed an approach to solving the Task Scheduling Problem (TSP) to minimize energy consumption while ensuring high QoS for IoT tasks. In [3], the authors proposed two semi-greedy-based algorithms to optimize FNs' total energy consumption while meeting IoT tasks' QoS requirements. A priority-based TS and RA mechanism is designed in [9] to reduce the total task processing time and the bandwidth cost as much as possible. A priority TS algorithm is introduced in [10] for improved RA and load balancing in fog computing. The algorithm has reduced the overall average response time and the total cost. The authors in [11] studied TS and RA problems to minimize the waiting time and end-to-end delay of tasks for dynamic vehicular fog networks. In [7], the authors proposed a TS algorithm and reallocation mechanism to improve the resource utilization of FNs and reduce task delays for IoT users and low resource loss. The authors in [12] proposed a heuristic algorithm, called Hybrid-EDF, for the dynamic scheduling of multiple real-time IoT workflows considering the optimization of cost.

In this work, we propose an online TS and dynamic RA algorithm called the Dynamic Heuristic Algorithm (DTS) in fog infrastructure. The main goal of this algorithm is to maximize both the resource utilization and the percentage of real-time tasks meeting their deadline.

The algorithm is implemented in the iFogSim simulator [8] and compared with Hybrid-Earliest Deadline First (Hybrid-EDF), Fog-Earliest Deadline First (Fog-EDF), First Come First Serve (FCFS), and Cloud-Only algorithms.

The main contributions of this paper can be summarized as follows:

- A scheduler for dynamically assigning tasks in real-time in a hierarchical fog infrastructure is developed.
- A mathematically formulation is proposed to address the task scheduling problem. The resource utilization of FNs is optimized while ensuring that tasks meet their deadlines.
- An efficient dynamic heuristic algorithm called DTS is proposed that dynamically assigns the IoT tasks to FNs with available resources to provide high QoS for IoT users and maximize resource utilization.
- A simulation of the proposed algorithm using iFogSim is performed. DTS is compared with other benchmark algorithms using parameters such as the percentage of IoT tasks that meet their deadline, the resource utilization at different levels in fog infrastructure, and the average response time.

The paper is organized as follows: Sect. 2 presents the proposed architecture and problem formulation. In Sect. 3, we describe the proposed algorithm. The evaluation of the proposed solution is presented in Sect. 4. Finally, Sect. 5 concludes the paper and suggests directions for future work.

2   Task Scheduling in Fog Infrastructure

In this section, we present the system architecture in the first subsection and mathematically formulate the TSP in the hierarchical fog infrastructure in the second subsection.

2.1   System Architecture

The proposed architecture is presented in Fig. 1. The fog infrastructure for TS consists of three layers: IoT devices, Fog, and Cloud. The IoT devices layer includes a large number of heterogeneous IoT devices such as wearable devices, smart home sensors, smartphones, cameras, smart meters, healthcare devices, and so on. The IoT devices assigned real-time tasks to the higher layers for real-time processing on fog/cloud nodes. The second layer in the fog architecture is called the fog layer, including a task scheduler, fog gateways, and FNs. These FNs have the capability of computation, storage, and network connectivity. The tasks scheduler receives incoming tasks from IoT devices in real time and is responsible for their scheduling. It allocates the tasks between the fog and cloud layers and selects suitable nodes to execute tasks in the fog infrastructure. Each fog node is used to compute and store the tasks assigned to it through the scheduler. Finally, the cloud layer consists of powerful servers with high computing power and storage capacity. These servers are responsible for processing tasks forwarded by the scheduler [4].

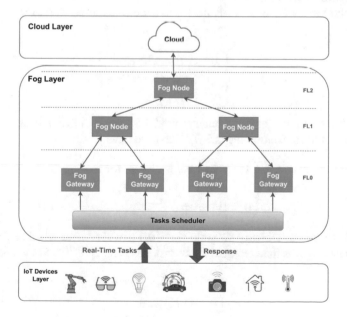

Fig. 1. Proposed system architecture.

2.2   Problem Formulation

This subsection presents the mathematical formulation for the TSP in a fog infrastructure. In this work, we assume that the scheduler receives independent and non-preemptive tasks.

Let us consider a set of n independent tasks offloaded from IoT devices to the DTS, as follows: $T = \{T_1, T_2, \ldots, T_n\}$. Each task $T_i \in T$ has a set of attributes $T_i = \left\{ T_i^{length}, T_i^{deadline}, T_i^{in}, T_i^{out} \right\}$, where T_i^{length} is the length of the task in MI (Millions of Instructions), $T_i^{deadline}$ is the deadline of the task in ms (millisecond), T_i^{in} is the input file size in KB (Kilo Byte), and T_i^{out} is the output file size in KB.

Suppose a fog network $G = (F, L)$ where $F = \{F_1, F_2, \ldots, F_m\}$ presents a set of m computing nodes including f FNs and c cloud nodes and $L = e_{uv} | u, v \in F$ presents the communication links between nodes in the fog infrastructure. Each computing node F_j has some characteristics $F_j = \left\{ F_j^{sc}, F_j^{cpu} \right\}$, where F_j^{sc} refers to storage capacity, F_j^{cpu} is the CPU processing rate in MIPS (Millions of Instructions Per Second). Also, each link $e_{uv} \in L$ is associated with some characteristics such as propagation delay $e_{u,v}^P$ in ms and bandwidth $e_{u,v}^B$ in Mbps (Mega bit per second).

In the following, we will present the decision variables and describe the system's response time and resource utilization.

2.2.1   Decision Variables

We define the binary decision variable $x_{i,j}$ to denote the allocation of task T_i on computing node F_j.

$$
x_{i,j} = \begin{cases} 1, & \text{if } T_i \text{ is allocated to } F_j \\ & (\forall T_i \in T, \forall F_j \in F) \\ 0, & \text{Otherwise} \end{cases} \tag{1}
$$

For each task T_i, we employ a binary variable $y_{uv} i \in \{0, 1\}$ to indicate whether link $e_{uv} \in L$ is selected for routing the task T_i. Thus, we have

$$
y_{u,v}^i = \begin{cases} 1, & \text{if link } e_{uv} \text{ is chosen for routing } T_i \\ & (\forall u, v \in F, \forall T_i \in T) \\ 0, & \text{Otherwise} \end{cases} \tag{2}
$$

Finally, we define a binary variable $z_i \in \{0, 1\}$ to represent whether the response time $RT_{i,j}$ must be less than or equal to the task deadline $T_i^{deadline}$. So, we have

$$
z_i = \begin{cases} 1, & \text{if } RT_{i,j} \le T_i^{deadline} \\ 0, & \text{Otherwise} \end{cases} \tag{3}
$$

2.2.2   Response Time

The TS challenge for sensitive delay applications is allocating real-time tasks on suitable FNs, ensuring the deadline is met. The response time of each task T_i in each fog node F_j must be calculated. The response time $RT_{i,j}$ for a given task T_i is the time required to send the task request from an IoT device to the connected fog node and get the results back. It comprises the propagation delay $PD_{i,j}$ from the source to the destination and vice versa, execution time $ET_{i,j}$, transmission time $TT_{i,j}$, and waiting time $W_{i,j}$ in the queue of node F_j. Therefore, the execution time $ET_{i,j}$ of task T_i in the fog node F_j can be obtained as follows:

$$ET_{i,j} = \frac{T_i^{length}}{F_j^{cpu}}, \forall T_i \in T \tag{4}$$

in which T_i^{length} and F_j^{cpu} are the length of ith task T_i and the CPU processing rate of node F_j, respectively. The transmission time of the task T_i can be represented as follows:

$$TT_{i,j} = \sum_{\forall e_{u,v} \in L} \frac{T_i^{in} + T_i^{out}}{e_{uv}^B} \times y_{u,v}^i, \forall T_i \in T \tag{5}$$

where T_i^{in} and T_i^{out} are the sizes of the input and output files of task T_i and e_{uv}^B is the bandwidth between two connected nodes F_u and F_v.

The propagation delay for the task T_i can be represented as follows:

$$PD_{i,j} = \sum_{\forall e_{u,v} \in L} (e_{uv}^P \times y_{u,v}^i), \forall T_i \in T \tag{6}$$

Therefore, the response time for the task T_i can be defined as follows:

$$RT_{i,j} = (2 \times PD_{i,j}) + ET_{i,j} + TT_{i,j} + W_{i,j}, \forall T_i \in T \tag{7}$$

By employing the variable NT to denote the number of tasks that meet their deadline, we determine the count of IoT tasks successfully satisfying their respective deadlines. Hence, we have:

$$NT = \sum_{\forall T_i \in T} z_i, \forall T_i \in T \tag{8}$$

The percentage of IoT tasks meeting the deadline is defined as $NT^\%$ that can be obtained as follows.

$$NT^\% = \frac{NT \times 100}{n} \tag{9}$$

2.2.3 Resource Utilization

To maximize resource utilization efficiency, we calculate the resource utilization of each fog node by summing the CPU lengths of current tasks (both running and waiting) assigned to this computing node and then dividing it by the total CPU capacity of this node. We determine the resource utilization using the following equation:

$$RU_j = \frac{1}{F_j^{cpu}} \times (\sum_{i=1}^{n} T_i^{length} \times x_{i,j}), \forall T_i \in T \tag{10}$$

2.2.4   Objective Function

A DTS is proposed to address the TSP in a fog infrastructure. The main goal of the DTS algorithm is to maximize both resource utilization (see Eq. 10) and the percentage of IoT tasks that meet their deadline (see Eq. 9). The deadlines of all the tasks need to be satisfied. Therefore, our overall objective function can be expressed as follows:

$$Max : (\sum_{j=1}^{m} RU_j + NT^{\%}) \tag{11}$$

Subject to the following constraints:

$$\sum_{j=1}^{m} x_{i,j} = 1, \forall T_i \tag{12}$$

$$x_{i,j}, y_{u,v}^i, z_i \in \{0,1\} \tag{13}$$

$$RT_{i,j} \leq T_i^{deadline}, \forall T_i \in T \tag{14}$$

Constraint 12 ensures that each task is assigned to only one computing node. The domain of the decision variables is defined by Constraint 13, while Constraint 14 guarantees the satisfaction of deadlines for each task.

3   Dynamic Heuristic Algorithm

In this section, we propose DTS, a dynamic heuristic algorithm, for solving the TSP in a fog infrastructure. Our primary objective is maximizing resource utilization while ensuring tasks meet their deadlines.

The scheduler operates in real-time, receiving tasks without prior knowledge of their arrival times. The main goal is to make dynamic decisions on the destination of each task, assigning them to appropriate fog nodes based on the system's current state. The tasks are independent and unpredictable, each with different characteristics, such as task size and deadlines. Thus, DTS takes these factors into account. First, the algorithm selects nodes that satisfy the time constraint. Second, it maximizes resource utilization by assigning tasks to fog nodes with minimal resource usage, thus reducing waiting time and achieving a more balanced task distribution across the network. Consequently, tasks with longer deadlines are assigned to higher-level nodes, while tasks with very short deadlines are allocated to nodes near the gateways. As a result, the number of real-time tasks successfully meeting their deadlines increases. This approach ensures the efficient use of fog resources and improves the system's overall performance in meeting real-time constraints.

In Algorithm 1, we present the pseudocode of the DTS algorithm.

The algorithm receives two inputs: a list of incoming tasks request $List_{T_i}$ with their attributes from IoT users at the same time t, and the list of the available computing nodes F in the architecture. The process starts in line 1 by initializing three empty lists, $L1$, $L2$, and $L3$.

As the dynamic task scheduler can receive more than one task at a time t from several IoT devices, we sort the list $List_{T_i}$ in ascending order based on the deadline of

---

Algorithm 1 DTS Algorithm

    Input: $List_{T_i}$, List F
    Output: $T_i \longrightarrow F$
1: $L1 := \emptyset$, $L2 := \emptyset$, $L3 := \emptyset$;
2: *sort $List_{T_i}$ in ascending order of deadline T_i;*
3: **for each** $T_i \in List_{T_i}$ **do**
4:     $L1 := \emptyset$;
5:     **for each** $F_j \in ListF$ **do**
6:        *calculate $RT_{i,j}$ using eq.(7);*
7:        $L1 := L1 \cup RT_{i,j}$;
8:     **end for**
9:     $L2 := \emptyset$;
10:    **for each** $RT_{i,j} \in L1$ **do**
11:      **if** $RT_{i,j} \le T_i^{deadline}$ **then**
12:        $L2 := L2 \cup F_j$;
13:      **end if**
14:    **end for**
15:    $L3 := \emptyset$;
16:    **if** $L2 \ne \emptyset$ **then**
17:      **for each** $F_j \in L2$ **do**
18:        *calculate RU_j using eq.(10);*
19:        $L3 := L3 \cup RU_j$;
20:      **end for**
21:      **if** $cloud \in L3$ **then**
22:        *allocate T_i on cloud;*
23:      **else if** $count\_min\_RU(L3) > 1$ **then**
24:        *allocate T_i on $F_j \in L3$ with the minimum $RT_{i,j}$;*
25:      **else**
26:        *allocate T_i on $F_j \in L3$ with the minimum RU_j ;*
27:      **end if**
28:    **else**
29:      *reject T_i ;*
30:    **end if**
31: **end for**

---

each task (line 2). The code (in lines 3 to 31) will be processed for each task in List T_i. In line 4, the list $L1$ is reinitialized inside the loop. Afterward, for each computing node in *List F*, we calculate the response time of each task T_i at node F_j. The response time $RT_{i,j}$ is then added to the list $L1$ (lines 5 to 6). Then, the list $L2$ is reinitialized (line 7), and the algorithm will iterate over the list $L1$ (line 8). If at least one node can meet the task deadline, it is added to the list $L2$ (lines 8 to 12).

After the algorithm reinitializes the list $L3$ (line 13), if the list $L2$ is not empty, the algorithm calculates the resource utilization of each fog node F_j in $L2$. It then adds each computing node's resource utilization to the list $L3$ (lines 14 to 18).

Next, if the cloud node is in list $L3$, the task T_i will be allocated on the cloud. Otherwise, task T_i will be allocated on a fog node according to the resource utilization RU_j in list $L3$ (lines 21 to 27).

The function *count_min_RU (L3)* is employed to identify the lowest resource utilization in the list *L3* and determine the number of occurrences of the minimum value. If there is more than one fog node with the same minimum resource utilization (i.e., *count_min_RU (L3)* returns a value greater than 1) (line 22), the task T_i will be allocated on the computing node with the shortest response time (line 23) among those with the same minimal resource utilization. Otherwise, task T_i will be assigned to the node with the least resource utilization to balance the load and maximize the number of tasks assigned in fog infrastructure (line 25).

Finally, if the list *L2* is empty (meaning no node meets the task's deadline), the task will be rejected (line 29).

4   Performance Evaluation

In this section, we present the performance evaluation and comparison of the proposed algorithm with other benchmarks in terms of (1) the percentage of IoT tasks that successfully meet their deadline (see Eq. 9), (2) the percentage of resource utilization in fog infrastructure, (3) the average resource utilization of CPU for each computing node for different strategies, and (4) the average response time. We evaluate the performance of the proposed algorithm with other heuristic and baseline scheduling algorithms: (1) Hybrid-EDF, (2) Fog-EDF, (3) FCFS, and (4) Cloud-Only algorithms.

4.1   Simulation Setting

We used iFogSim [8] to simulate and test our DTS strategy. Simulation experiments were conducted on a PC Intel with Core i7-2670 CPU 2.20 GHz, 3.7 GB RAM, and Ubuntu 20.04.5 LTS. The architecture shown in Fig. 1 models the simulated infrastructure. We considered eight IoT applications, each with one sensor with a distinct periodic frequency $Freq_i$ and one actuator.

Moreover, the tasks of the IoT applications are independent and have distinct deadlines and lengths. Table 1 presents the parameter settings for IoT tasks.

We present eight fog gateways in fog level 0 (FL0). Next, four FNs are considered in fog level 1 (FL1), followed by two in fog level 2 (FL2), and finally, one cloud server in the cloud layer.

Table 2 illustrates the Propagation Delays (PD) between the infrastructure components in *ms*.

Table 1. Characteristics of IoT Tasks.

Parameter	Value
Number of instructions (MI)	[80000–300000]
Deadline (ms)	[3500–4500]
Frequency (ms)	[50–70]
Input file size for task	[50000–150000] Kb
Output file size for task	3 Kb
Bandwidth of links	[100–1000] MB

Table 2. PD between nodes.

Network link	PD (ms)
Fog node - cloud	1200
Fog node - Fog node	500
Fog node - Fog Gateway	50
Sensor - Fog Gateway	10
Actuator - Fog Gateway	10

4.2   Results Analysis

In this subsection, we present the results of the simulation experiments with various settings that are outlined in Subsect. 4.1.

Evaluating the tasks to meet the deadlines is essential, as it represents the effectiveness of the real-time task scheduling algorithm in providing Quality of Service (QoS) to IoT users.

Figure 2 illustrates the percentage of tasks with met deadlines. It is evident that our proposed algorithm consistently attains a significantly higher percentage of tasks meeting their deadlines (100%) compared to other strategies. The Hybrid-EDF strategy achieves 80.44%, the Fog-EDF strategy reaches 75.74%, the Cloud-Only strategy reaches 47.07%, and the FCFS strategy obtained the lowest percentage with 44.99%. The results show that our proposed algorithm establishes a feasible schedule, ensuring that all real-time tasks consistently meet their deadlines.

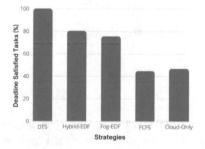 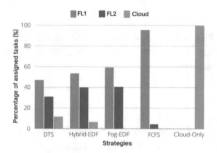

Fig. 2. Percentage of tasks that meet their deadline.

Fig. 3. Resource utilization in fog infrastructure.

The resource utilization is computed as the ratio of the number of IoT tasks allocated on fog nodes at the same layer to the total number of IoT tasks assigned. This metric provides insight into the utilization of various resources.

Figure 3 illustrates the resource utilization of the fog infrastructure for various policies. Our proposed algorithm, DTS, has allocated 47.07% of IoT tasks in FL1, 30.98% in FL2, and 11.622% in the cloud node. In contrast, the Hybrid-EDF strategy assigns more IoT tasks to FL1 (53.54% of tasks), 39.98% to FL2, and 6.47% to the cloud node. The Fog-EDF strategy allocates 59.39% of IoT tasks to FL1, 40.6% to FL2, and none to the cloud node. On the other hand, the FCFS strategy assigns 95.6% of IoT tasks to FL1, 4.39% to FL2, and none to the cloud node. Finally, the Cloud-Only strategy has assigned all IoT tasks to the cloud layer.

Figure 4 presents the average resource utilization of the CPU for each computing node under different strategies. There are seven computing nodes: four in fog level 1 (FL1) - Fog_Node1.1, Fog_Node1.2, Fog_Node1.3, and Fog_Node1.4, two in fog level 2 (FL2) - Fog_Node2.1 and Fog_Node2.2, and finally, one cloud node in the cloud layer. The results demonstrate that the proposed algorithm effectively allocates tasks to the available computational nodes, optimizing processor utilization at all levels. The Hybrid-EDF, Fog-EDF, and FCFS algorithms use more processors in the available FNs in FL1 than in FL2. The Cloud-Only strategy stands out for handling high-processing tasks comfortably due to its powerful processor resources. It has allocated all the tasks in the cloud.

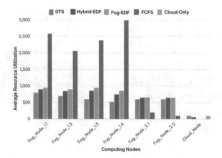 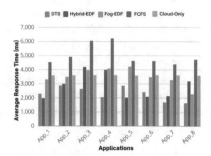

Fig. 4. Average resource utilization of CPU for each computing node for different strategies.

Fig. 5. Average response time.

Figure 5 illustrates the average response time of IoT tasks requested by real-time applications. Our proposed algorithm is expected to demonstrate a reduced average response time compared to other algorithms. Our algorithm focuses on distributing tasks in a balanced manner across all levels of the infrastructure, thus avoiding overloading the queues of FNs (as seen in Hybrid-EDF, Fog-EDF, and FCFS algorithms). Additionally, our algorithm prevents assigning all tasks to the cloud layer, except tasks that can meet their deadline on the cloud, which incurs propagation time from source to destination (as observed in the Cloud-Only algorithm).

5   Conclusion

This work focused on task scheduling and resource allocation problems using the Dynamic Heuristic Algorithm (DTS) in fog infrastructure. We mathematically formulated the task scheduling problem and solved it using a dynamic heuristic algorithm. DTS dynamically assigned incoming tasks to the suitable fog nodes in real-time in the fog infrastructure. The task scheduler first sorts the tasks by their deadline and then selects the fog nodes that maximize resource utilization while satisfying the time constraints. Finally, the proposed algorithm performance is evaluated according to the QoS criteria and compared to Hybrid-EDF, Fog-EDF, FCFS, and Cloud-Only algorithms. The results show that DTS has very encouraging results regarding the percentage of real-time tasks that meet their deadline compared with the other algorithms. In future work, we aim to expand this work by incorporating energy consumption and monetary cost as additional criteria in real-time task scheduling in fog infrastructure.

References

1. Salimi, R., Azizi, S. Abawajy, J.: A greedy randomized adaptive search procedure for scheduling IoT tasks in virtualized fog-cloud computing (2023)
2. Sultan Hajam, S.: Deadline-cost aware task scheduling algorithm in fog computing networks. Int. J. Commun. Syst. **37**, e5695 (2024)
3. Azizi, S., Shojafar, M., Abawajy, J., Buyya, R.: Deadline-aware and energy-efficient IoT task scheduling in fog computing systems: a semi-greedy approach. J. Netw. Comput. Appl. **201**, 103333 (2022)
4. Ali, H., Rout, R., Parimi, P., Das, S.: Real-time task scheduling in fog-cloud computing framework for iot applications: a fuzzy logic based approach. In: 2021 International Conference On COMmunication Systems & NETworkS (COMSNETS), pp. 556–564 (2021)
5. Bonomi, F., Milito, R., Zhu, J. Addepalli, S.: Fog computing and its role in the internet of things. In: Proceedings of the First Edition of the MCC Workshop on Mobile Cloud Computing, pp. 13–16 (2012)
6. Jamil, B., Ijaz, H., Shojafar, M., Munir, K., Buyya, R.: Resource allocation and task scheduling in fog computing and internet of everything environments: a taxonomy, review, and future directions. ACM Comput. Surv. (CSUR) **54**, 1–38 (2022)
7. Yin, L., Luo, J., Luo, H.: Tasks scheduling and resource allocation in fog computing based on containers for smart manufacturing. IEEE Trans. Ind. Inform. **14**, 4712–4721 (2018)
8. Gupta, H., Vahid Dastjerdi, A., Ghosh, S., Buyya, R.: iFogSim: a toolkit for modeling and simulation of resource management techniques in the Internet of Things, Edge and Fog computing environments. Softw. Pract. Exper. **47**, 1275–1296 (2017)
9. Sharif, Z., Jung, L., Ayaz, M., Yahya, M., Pitafi, S.: Priority-based task scheduling and resource allocation in edge computing for health monitoring system. J. King Saud Univ.-Comput. Inf. Sci. **35**, 544–559 (2023)
10. Choudhari, T., Moh, M., Moh, T.: Prioritized task scheduling in fog computing. In: Proceedings Of The ACMSE 2018 Conference, pp. 1–8 (2018)
11. Jamil, B., Ijaz, H., Shojafar, M., Munir, K.: IRATS: a DRL-based intelligent priority and deadline-aware online resource allocation and task scheduling algorithm in a vehicular fog network. Ad Hoc Netw. **141**, 103090 (2023)
12. Stavrinides, G., Karatza, H.: A hybrid approach to scheduling real-time IoT workflows in fog and cloud environments. Multimedia Tools Appl. **78**, 24639–24655 (2019)
13. Sharma, O., Rathee, G., Kerrache, C., Herrera-Tapia, J.: Two-stage optimal task scheduling for smart home environment using fog computing infrastructures. Appl. Sci. **13**, 2939 (2023)

Secure Lightweight Data Communication Between the IoT Devices and Cloud Service

Tran Viet Xuan Phuong[1], Tho Thi Ngoc Le[2(✉)], and Huy Le Ngoc[3,4]

[1] Department of Computer Science, University of Arkansas at Little Rock, Arkansas, USA
ptran@ualr.edu
[2] Faculty of Information Technology, HUTECH University, Ho Chi Minh, Vietnam
ltn.tho@hutech.edu.vn
[3] Faculty of Computer Networks and Communications, University of Information Technology,
Ho Chi Minh City, Vietnam
17520074@gm.uit.edu.vn
[4] Vietnam National University, Ho Chi Minh City, Vietnam

Abstract. The proliferation of Internet of Things (IoT) devices has been dramatically increasing and interacting with a large amount of data, such as human bio-meteorology [26]. IoT capacity is currently limited while the users need to securely back up these sensitive data during the processing/storage. Therefore, employing a cloud service can diminish the storage cost for IoT devices. However, the data storage can be malleable if the cloud is corrupted or is being maintained. To mitigate this problem, the IoT devices need to check/update the data periodically between the storage data in the cloud and the collecting data. An efficient secure protocol is desirable for checking/updating data between the cloud and IoT devices.

In this paper, we propose an efficient secure pattern matching algorithm between the cloud and the IoT devices. The main construction relies on the XOR operator and the counter mode of block ciphers, which achieves lightweight computation matching. In addition, the proposed algorithm enables the secure checking/updating between the storage data from the cloud and the IoT's data. Motivated by the work of Ateniese et al. [4], we strengthen our protocol that supports data verification before performing the pattern-matching algorithm. This protocol can prevent the attackers from interacting with our data illegally in the context of the corrupted cloud. Our experiments are demonstrated on a common machine as the cloud service and a Raspberry Pi 3 as the client. The performance shows that our proposed work is reasonable on constrained devices.

1   Introduction

Pattern matching is one of the primary and essential problems in computer science. For an example, fingerprints matching is an application for the authentication of users. Suppose that Alice wants to authenticate herself through IoT devices, she uploads her current fingerprints to the cloud server for the matching operation. Her data will be in jeopardy if her fingerprints are leaked out. Therefore, the data encryption and the secure pattern matching are necessary to avoid this leakage. In fact, pattern matching is usually

ⓒ The Author(s), under exclusive license to Springer Nature Switzerland AG 2024
L. Barolli (Ed.): AINA 2024, LNDECT 200, pp. 404–415, 2024.
https://doi.org/10.1007/978-3-031-57853-3_34

performed by a third party, such as a Cloud Service Provider (CSP) [27], where we need to ensure that the CSP can not learn any information from patterns. For convenience, most people are storing the data constantly on the cloud. Therefore, the data integrity must be checked before the pattern matching happens to assure the correctness of the result. The checking phase will eliminate the data malleable before sending it to the client. Based on the aforementioned problems, this research will investigate a pattern matching protocol that can protect the confidentiality and integrity of sensitive data.

Recently, the problem of "insecure data transfer and storage," one of the top 7 OWASP IoT 2018 [21], points out the importance of data encryption on IoT devices. Furthermore, the advent of the quantum computer caused many ciphers and cryptographic hash functions to impact the leakage data of the IoT devices [14]. To remedy the breakdown, the Advanced Encryption Standard (AES) is considered to be one of the cryptographic primitives that is resilient in quantum computations, but only when AES is used with key sizes of 192 or 256 bits [20].

In this work, we investigate the AES-256 and SHA-256, which are fast and quantum-resistant algorithms [20], to build our protocols on IoT devices supporting cloud storage and pattern matching related to security, integrity, and efficiency. Additionally, motivated by the work of Ateniese et al. [4], we realize that there is no previous work that considers data verification to perform secure pattern matching. Hence, we integrate [4] to support our pattern matching protocol. Our contributions are as follows:

- Firstly, we propose a secure pattern matching algorithm, which relies on the XOR operator and counter mode of block ciphers by using AES-256. Our proposed algorithm achieves secure, lightweight computation.
- Secondly, we utilize the work of Ateniese et al. [4] to support the data verification before performing the pattern-matching algorithm.
- Thirdly, using the Python 3 language, we build a prototype implementing our models on SOCOFing dataset [24] and some large random-generated files. The computational cost is evaluated on a common machine as the CSP and a Raspberry Pi 3 as the client.

2 Related Works

Secure Pattern Matching on Low-powered Devices. Kamara et al. [19] investigate garbled-circuit-based protocols, a set of mutually distrustful parties to evaluate a function of their joint inputs without revealing their inputs to each other. Later, Cater et al. [9] create a protocol that allows mobile devices to securely outsource the majority of computation required to evaluate a garbled circuit, which builds on the most efficient garbled circuit evaluation techniques. Research on privacy-preserving wildcard pattern matching has become a hot topic in recent years. Hazay and Toft [18] show that wildcard pattern matching can be converted to exact pattern matching using additively homomorphic encryption. Besides string matching, image matching is also one of the interesting problems to attract many works. Abduljabbar et al. [1,2] propose works of secure image matching in two articles "Towards Secure Private Image Matching" and "SEPIM: Secure and Efficient Private Image Matching." While Baron et al.'s technique [5] is an exact pattern matching, image matching is an approximate pattern matching. The approximate pattern matching requires a score representing the similarity between two patterns. Their protocols are helpful for the secure image matching scheme between two parties where images are confidential. In three articles [1, 2, 5], homomorphic encryption is applied to design the security schemes. Homomorphic encryption is a type of cryptographic algorithm that allows calculating some operators on encrypted data. However, due to the cumbersome computation of homomorphic encryption, their solution is impractical to deploy on resource-constrained IoT devices.

Provable of Data Possession. Ateniese et al. [3] firstly propose Provable of Data Possession (PDP) to verify the data integrity remotely. PDP model computes a probabilistic proof of a file at the server (provider) and sends it to the client (verifier). Then, the client verifies the file integrity by using that probabilistic proof. PDP model uses public key encryption, which is too hard to be implemented on IoT devices. Following up, Ateniese et al. publish "Scalable and Efficient PDP" [4], which uses the symmetric encryption and hash function instead of only asymmetric encryption in [3]. Similarly, Chen and Curtmola [10], Erway et al. [12] propose the dynamic storage, called Dynamic PDP (DPDP), which allows the client to update a file without performing the chain of actions: retrieving, modifying, and uploading. To deploy PDP in the mobile devices, Yang et al. [28] proposed a method, which uses Merkel Hash Tree, digital signature, and bilinear map. They advance previous work by enabling the clients to store only small pieces of metadata without verification. However, this approach requires the TPM-supported mobile devices and a trusted Third Party Agent (TPA), which is inapplicable. Recently, Ferri [15] proposes a secure storage to develop an efficient method for transmitting and storing data in a secure way. In addition, Ren et al. [23] utilizes the IoT security service architecture based on SIM card, and for secure data transmission with local secure storage.

Quantum-Resistance Cryptography. In Mavroeidis et al.'s article [20], the Advanced Encryption Standard (AES) is considered to be one of the cryptographic primitives that is resilient in quantum computations, but only when AES is used with key sizes of 192 or 256 bits. Brassard et al. [8] also utilize Grover's algorithm to create a quantum birthday attack that can solve the problem of short digest hash functions (such as SHA-1). But, both SHA-2 (including SHA-256) and SHA-3, with longer outputs, remain quantum resistant.

3   Problem Statement

3.1   System Model

As depicted in Fig. 1, our model comprises three parties: Alice, Bob, and the Cloud Service Provider (CSP).

1. *Alice and Bob.* The people work on IoT devices. They need to store their data on the cloud and verify the integrity of their remote data. They also request the pattern matching result from the stored data on CSP.

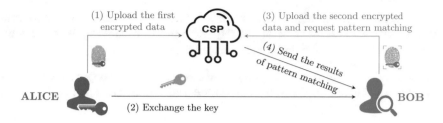

Fig. 1. Data flow in our model.

2. *Cloud Storage Provider (CSP).* An untrusted party works on a common machine and provides the storage service to Alice and Bob. Moreover, the CSP must perform pattern matching operations on stored data if it is requested.

3.2 Security Model

1. *Alice and Bob.* They agreed on their shared secret key. It means that Alice can share this key with Bob if she trusts him. We note that Bob can not download Alice's data from the CSP without her permission.
2. *Cloud Storage Provider (CSP).* We consider the CSP as the honest-but-curious. In addition, the CSP is assumed not to collude with the other parties (Alice or Bob).

3.3 Objectives

Our scheme aims four objectives:

1. The CSP can not learn anything about the data of Alice and Bob.
2. The remote data must be intact before performing pattern matching.
3. The CSP must perform pattern matching operations without decrypting the data.
4. All algorithms and protocols must be suitable for the capability of IoT devices.

4 Preliminaries

4.1 Integrity Checking with Replay-Attack Resistance

We apply the work of Ateniese et al. [4] to elaborate our integrity-checking phase. When Alice checks the data integrity locally, the data digest will be compared with the remote data digest computed by the CSP.

- Alice computes some data **digests** with some random **nonce** values before sending her data to the CSP by using [4]: $\text{nonce} \xleftarrow{R} \{0,1\}^l; \text{digest} = \text{hash}(\text{nonce} + \text{data})$, where the **nonce** is a random value with length l generated by the data owner, leading to the non-repeated data digests. The **nonces** and **digests** are kept secretly for future validation.
- Alice verifies the data integrity, she picks an arbitrary pair of **nonce** and **digest** and sends the **nonce** to the CSP.

- Alice receives the **digest** of concatenating of current data and the sent **nonce**. If this **digest** is the same as the local **digest**, the CSP proved the data integrity successfully. Otherwise, the client will detect the damaged data.

As **nonce** values are different for every validation request, this approach can counter replay attacks. Therefore, the values of **nonce** and **digest** are reinitiated for every request.

4.2 Probabilistic Integrity Checking

Ateniese et al. [3,4] propose a probabilistic proof approach by dividing the data into many blocks. Instead of the entire data verification, the client only needs to verify some blocks. This work can speed up the integrity checking operation many times. The authors prove that if we divide data into 10,000 blocks and verify 460 of them (representing 4.6% of the blocks), we can detect server misbehavior in 1% of the blocks with a probability greater than 99%.

In this work, we will apply 'Scalable and Efficient PDP" [4] in models as follows:

- D is the data, which is represented as a single contiguous file of d equal-sized blocks: $D[1], \ldots, D[d]$.
- **SHA(.)** is the cryptographic hash function SHA-256.
- **nonce** is the value generated randomly by the client.
- $I = i_1, \ldots, i_r$ is a list of block indices that the client asks the server to prove. Like **nonce**, these indices are also randomly generated.

Alice will compute the **digest** as following: $\mathsf{digest} = \mathsf{SHA}(\mathsf{nonce}, D[i_1], \ldots, D[i_r])$, where commas are concatenating operators. In this approach, the local secret group includes the **nonce** value, the indices I, and the **digest**. When Alice wants to verify the data integrity, she needs to send the list of indices I along with **nonce** to the server. Then, our protocol follows the steps in the previous Subsect. 4.1.

4.3 Pattern Matching with the XOR Operator

In many pattern matching algorithms, XOR operator is used to compare the corresponding bytes between the two patterns. If two bytes are the same, the XOR result is a zero; otherwise, a non-zero number. To calculate the similarity score, we need to count how many zero bytes in the XOR operation of two patterns and divide it by the size of the patterns. We present the formula as follows: $\mathsf{similarity} = \frac{\mathsf{count}(P_1 \oplus P_2)}{\mathsf{size}(P_1)}$, where P_1, P_2 are the two patterns of the same size, and the notation of $\mathsf{size}(P_1)$ is the size of the pattern P_1 and also equals to $\mathsf{size}(P_2)$. After obtaining the similarity score, we set a similar threshold to determine whether two patterns match or not.

5 Proposed Scheme

To satisfy the objectives in the aforementioned Subsect. 3.3, we propose our secure pattern matching model including three schemes/protocols, which are:

1. Data Uploading Sect. 5.2 describes how the data owner uploads her data to the CSP.
2. Data Integrity Checking Sect. 5.3 helps the data owner know whether her remote data is changed or not.
3. Secure Pattern Matching Sect. 5.4 describes how the parties perform the secure pattern matching algorithm.

5.1 Notation

- k is the secret key of Alice.
- D_1 is the data (or pattern) of Alice, and D_2 is the data (or pattern) of Bob. We assume that D_1, D_2 can be represented as single contiguous files of d equal-sized blocks: $D_1[1], \ldots, D_1[d]$ and $D_2[1], \ldots, D_2[d]$.
- SHA(.) is the cryptographic hash function SHA-256.
- $\text{AES}_k(.)$ is the AES-256 encryption operation in counter mode with key k.
- RNG() is a Random Number Generator.
- $\text{RPG}_d(r)$ is a Random Permutation Generator creating a r-combination of set $\{1, \ldots, d\}$ ($r \le d$).

5.2 Data Uploading

Fig. 2. Data uploading scheme.

When the data owner Alice wants to upload her data D_1, comprising r blocks to be proved, to the CSP, she generates the Proofs and follows the steps in Fig. 2 as:

(1a) Alice generates the key.
(1b) Alice generates the nonce and indices as:
  - RNG() $\rightarrow$ nonce,
  - $\text{RPG}_d(r) \rightarrow I = \{i_1, \ldots, i_r\}$.
(2) Encrypt the data : $\text{AES}_k(D_1) \rightarrow D'_1$.
(3) Compute the digest: SHA(nonce, $D'_1[i_1], \ldots, D'_1[i_r]) \rightarrow$ digest.
(4a) Alice uploads the encrypted data D'_1 to the CSP and delete the original data D_1.
(4b) Alice keeps the Proof $= \{$nonce, I, digest$\}$ locally to check data integrity of D'_1 in the future.

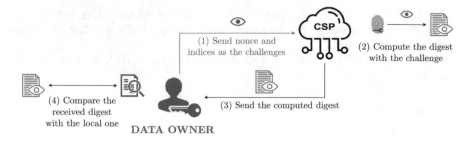

Fig. 3. Integrity checking scheme.

5.3 Data Integrity Checking

Data owner, Alice, in Fig. 3, firstly picks an arbitrary **Proof** stored locally; then she sends I and **nonce** in the **Proof** to the CSP. The CSP must compute the data **digest'** by combining stored data D'_1 with the parameters received from Alice:

$$\mathsf{SHA}(\mathsf{nonce}, D'_1[i_1], \ldots, D'_1[i_r]) \rightarrow \mathsf{digest'}.$$

After the CSP sends the **digest'** to Alice, she compares **digest'** with the local **digest** to determine if the data integrity is guaranteed or not.

$$\mathsf{Compare}(\mathsf{digest'}, \mathsf{digest}) \rightarrow \{\mathsf{success}, \mathsf{failure}\}.$$

5.4 Secure Pattern Matching

To match the patterns securely, patterns from both Alice and Bob must be *in the same size* and *encrypted by the same key*. To get the matching result, the CSP performs XOR operator on the two encrypted patterns directly, as illustrated in Fig. 4. Firstly, Alice uploads her encrypted pattern D'_1 to the CSP. When Bob wants to check whether his pattern D_2 matches with Alice's pattern or not, the protocols are followed:

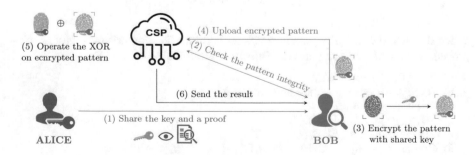

Fig. 4. Secure pattern matching scheme.

1. Alice shares her key k and a Proof with Bob.
2. Bob checks the integrity of Alice's pattern with shared Proof.
3. Bob encrypts his pattern D_2 by shared key and sends the encrypted pattern D_2' to the CSP: $\text{AES}_k(D_2) \rightarrow D_2'$.
4. The CSP performs the XOR operator on encrypted patterns: $\text{XOR}(D_1', D_2') = D_1' \oplus D_2' \rightarrow X$.
5. The CSP calculates the similarity score by dividing the number of zero bytes in X by X's size: $\frac{\text{count}(X)}{\text{size}(X)} \rightarrow$ similarity.

If the matching is valid, the CSP will send the similarity to Bob as the result of pattern matching between two patterns.

6 Security Analysis

6.1 Correctness of Secure Pattern Matching

In the Sect. 5.4, we claim that "the result of the XOR operator on encrypted patterns equals to the one on original patterns." Then, with:

$$0 \oplus a = a; \text{ and } a \oplus a = 0 \tag{1}$$

$$\text{AES}_k(\text{counter}) \rightarrow \text{keystream}; D_1' = D_1 \oplus \text{keystream}; D_2' = D_2 \oplus \text{keystream} \tag{2}$$

From (1) and (2), we have:

$$D_1' \oplus D_2' = D_1 \oplus \text{keystream} \oplus D_2 \oplus \text{keystream} = D_1 \oplus D_2 \oplus 0 = D_1 \oplus D_2.$$

6.2 Data Privacy

In our scheme, the data is encrypted by the AES-256 algorithm with CTR mode. The key k can be held only by Alice and Bob. Since Alice trusts Bob, he will not disclose the key k to the CSP. The CSP receives only encrypted data D_1' and D_2', so it only can get the pattern matching result $X = D_1 \oplus D_2$. In the CTR mode, the counter numbers are generated with every data by the data owner. Even if the CSP decrypts a certain data, it can not decrypt the others. Our scheme uses the key k of 256-bit length, which is proven to be resilient to quantum attacks [20]. Moreover, the CSP must run the algorithms whose computational complexity is $2^{254.4}$ for biclique attack [7] and 2^{176} for related-key attacks [6].

6.3 Security of Integrity Checking

Intuitively, nonce values should have at least a 256-bit length to bypass the replay attack. If the CSP re-sends the digest with the previous nonce, the probability for this attack to be successful is $1/2^{256}$. Moreover, if the CSP wants to pre-compute all possible proofs, it must spend $2^{256} \cdot 32$ (bytes) $\approx 3 \cdot 10^{70}$ (gigabytes), where 32 is the message digest length (in bytes) of hash function SHA-256. The above storage space is too large to store, which means that the CSP incurs incredible memory.

Probabilistically, there is a small chance that the CSP avoids being detected by the changes in data. However, according to Ateniese et al. [4], if we divide data into d blocks and request to prove r blocks, the probability that the CSP avoids detecting m changed blocks is: $P_{avoid} = (1 - \frac{m}{d})^r$. For example, if the fraction of corrupted blocks (m/d) is 1%, and we verify $r = 460$ blocks, the probability of avoiding detection is 0.98%.

7   Experiment

We conduct the following setups for experiment:

Table 1. The configuration of devices are experienced in our work.

Devices	CPU	RAM
Raspberry Pi 3 (as Alice or Bob)	1.4 GHz quad-core A53 64-bit	1 GB
Common Machine (as CSP)	Intel(R) Core(TM) i5-7200U CPU @ 2.50 GHz (4 CPUs) 64-bit	12 GB

1. Simulation Setup: Experiments are conducted using Python 3 language and *cryptography* library [22]. We simulate a common machine as the CSP and treat a Raspberry Pi 3 as the client. Table 1 shows the detail information in terms of CPU and RAM ratio.
2. Database Setup: We utilize the SOCOFing [24], which is a popular fingerprint dataset. SOCOFing contains fingerprints of 600 people, with three different levels of alteration for obliteration, central rotation, and z-cut. The ascending levels are *Altered-Easy*, *Altered-Medium*, and *Altered-Hard*. Each image has the size of 11 kB. To show the efficiency, we generated larger random-content files of 100 MB to 500 MB.

Cost of Encryption and Decryption: Firstly, we run the encryption and decryption on large files and compare their cost between constrained device and common machine. As shown in Fig. 5a for encryption cost and Fig. 5b for decryption cost, the operations on the constrained device are much slower than on the common machine. It is noted that this cost embraces the data privacy on IoT devices.

Cost of Proof Computing: Secondly, the clients and the CSP need to compute the proof for the data integrity checking phase. We show the cost of proof computing on large files in Fig. 6a, where the total number of blocks d is 10,000 and the number of proved blocks r is 460. The results show that the execution speed in the CSP and in the client are similar. Moreover, we also compare the cost of using probabilistic proof $(r = 460, d = 10,000)$ to deterministic proof $(r = d = 1)$ on the client.

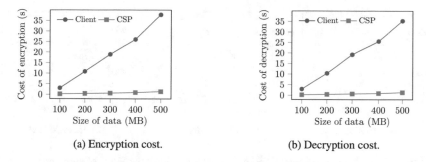

(a) Encryption cost.　　　　　　　(b) Decryption cost.

Fig. 5. Cost of encryption and decryption on the CSP and the Client.

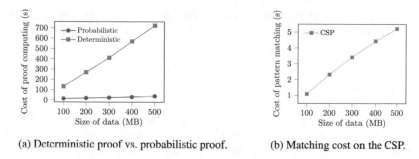

(a) Deterministic proof vs. probabilistic proof.　　(b) Matching cost on the CSP.

Fig. 6. Cost of computing.

We test the secure pattern-matching algorithm on the common machine of the CS with both the images of SOCOFing dataset and large self-generated files. The execution time is approximately 0.001 s. The result on large files, showing in Fig. 6b, indicated that our matching scheme is highly efficient. Our algorithm is 100–1,000x faster than other studies [1,5] in the same size input of 100 MB to 500 MB.

Table 2. Similarity score.

Easy				Medium				Hard			
Pat. 1	Pat. 2	Sim(%)	Match	Pat. 1	Pat. 2	Sim(%)	Match	Pat. 1	Pat. 2	Sim(%)	Match
		92.88	Yes			72.82	Yes			64.95	Yes
		84.64	Yes			76.48	Yes			62.07	Yes
		37.14	No			34.20	No			32.00	No

Result of Pattern Matching with XOR Operator: Thirdly, we demonstrate the results of the proposed pattern matching algorithm on the SOCOFing dataset, in Table 2, regarding the similarity scores of some pairs of patterns. Figures 7a, 7b, 7c visualize

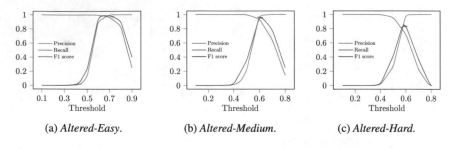

(a) *Altered-Easy.*   (b) *Altered-Medium.*   (c) *Altered-Hard.*

Fig. 7. Evaluation by different thresholds at different levels of Easy, Medium and Hard.

the precision, recall, and F1 score when we change the threshold. These figures are showing the performance which are conducted on 2,000 images of all three versions *Altered-Easy*, *Altered-Medium*, and *Altered-Hard*.

8   Conclusion

In this paper, we proposed a secure pattern-matching scheme between IoT devices and cloud service. Our scheme utilizes PDP to check data integrity before performing the matching for data validation. The proposed pattern-matching algorithm applies the XOR operator and the counter mode of block ciphers. Remarkably, the CSP cannot learn anything about the content of patterns due to the hashing mechanism. In addition, our scheme is quantum-resistance by using the AES-256 and SHA-256. We conduct the experiments on the Raspberry Pi 3, which is a common IoT device, thus is well-suitable to IoT environments.

Acknowledgement. This research was fully funded by the HUTECH University under grant number 2022/02/CNTT 62/HD-DKC.

References

1. Abduljabbar, Z.A., et al.: Towards efficient authentication scheme with biometric key management in cloud environment. In: Proceedings of the IEEE 2nd BigDataSecurity, pp. 146–151 (2016)
2. Abduljabbar, Z.A., et al.: SEPIM: secure and efficient private image matching. Appl. Sci. **6**(8), 213 (2016)
3. Ateniese, G., et al.: Provable data possession at untrusted stores. In: Proceedings of the ACM CCS 2007, pp. 598–609 (2007)
4. Ateniese, G., Di Pietro, R., Mancini, L.V., Tsudik, G.: Scalable and efficient provable data possession. In: Proceedings of the 4th EAI SecureComm, pp. 1–10 (2008)
5. Baron, J., El Defrawy, K., Minkovich, K., Ostrovsky, R., Tressler, E.: 5PM: secure pattern matching. In: Visconti, I., De Prisco, R. (eds.) SCN 2012. LNCS, vol. 7485, pp. 222–240. Springer, Heidelberg (2012). https://doi.org/10.1007/978-3-642-32928-9_13
6. Biryukov, A., Khovratovich, D.: Related-key cryptanalysis of the full AES-192 and AES-256. In: Matsui, M. (ed.) ASIACRYPT 2009. LNCS, vol. 5912, pp. 1–18. Springer, Heidelberg (2009). https://doi.org/10.1007/978-3-642-10366-7_1

7. Bogdanov, A., Khovratovich, D., Rechberger, C.: Biclique cryptanalysis of the full AES. In: Lee, D.H., Wang, X. (eds.) ASIACRYPT 2011. LNCS, vol. 7073, pp. 344–371. Springer, Heidelberg (2011). https://doi.org/10.1007/978-3-642-25385-0_19

8. Brassard, G., HØyer, P., Tapp, A.: Quantum cryptanalysis of hash and claw-free functions. In: Lucchesi, C.L., Moura, A.V. (eds.) LATIN 1998. LNCS, vol. 1380, pp. 163–169. Springer, Heidelberg (1998). https://doi.org/10.1007/BFb0054319

9. Carter, H., Mood, B., Traynor, P., Butler, K.: Secure outsourced garbled circuit evaluation for mobile devices. J. Comput. Secur. **24**(2), 137–180 (2016)

10. Chen, B., Curtmola, R.: Robust dynamic provable data possession. In: Proceedings ICDCS Workshops 2012, pp. 515–525 (2012)

11. Chen, L., et al.: Report on post-quantum cryptography. NIST Interagency/Internal Report (NISTIR), National Institute of Standards and Technology, Gaithersburg, MD, [online] (2016). https://doi.org/10.6028/NIST.IR.8105

12. Erway, C.C., Küpçü, A., Papamanthou, C., Tamassia, R.: Dynamic provable data possession. ACM Trans. Inf. Syst. Secur. (TISSEC) **17**(4), 1–29 (2015)

13. Erway, C.C., Papamanthou, C., Tamassia, R., et al.: Apparatus, methods, and computer program products providing dynamic provable data possession. US Patent No. US8978155B2 (2015)

14. Fernández-Caramés, T.M.: From pre-quantum to post-quantum IoT security: a survey on quantum-resistant cryptosystems for the internet of things. IEEE Internet Things J. **7**(7), 6457–6480 (2019)

15. Ferri, C.M.F.: The integrally secure storage protocol (2022)

16. Grover, L.K.: A fast quantum mechanical algorithm for database search. In: Proceedings of the STOC 1996, pp. 212–219 (1996)

17. Hak, T., Dul, J.: Pattern matching (2009). http://hdl.handle.net/1765/16203

18. Hazay, C., Toft, T.: Computationally secure pattern matching in the presence of malicious adversaries. In: Proceedings of the ASIACRYPT'10, pp. 195–212 (2010)

19. Kamara, S., Mohassel, P., Riva, B.: Salus: a system for server-aided secure function evaluation. In: Proceedings of the ACM CCS 2012, pp. 797–808 (2012)

20. Mavroeidis, V., Vishi, K., Zych, M.D., Jøsang, A.: The impact of quantum computing on present cryptography. arXiv preprint arXiv:1804.00200 (2018)

21. OWASP. OWASP Internet of Things. https://owasp.org/www-project-internet-of-things/

22. PyPI. cryptography 3.3.1. https://pypi.org/project/cryptography/

23. Ren, X., Yue, Z., Li, Z.: Security system of Internet of Things based on subscriber identity module. J. Phys: Conf. Ser. **1972**(1), 012004 (2021)

24. Shehu, Y.I., Ruiz-Garcia, A., Palade, V., James, A.: Sokoto coventry fingerprint dataset. arXiv preprint arXiv:1807.10609 (2018)

25. Shor, P.W.: Algorithms for quantum computation: discrete logarithms and factoring. In: Proceedings of the 35th Annual Symposium on Foundations of Computer Science, pp. 124–134 (1994)

26. Vanelli, B., et al.: IoT data storage in the cloud: a case study in human biometeorology. In: Proceedings of the IISSC'17 and CN4IoT'17, pp. 253–262 (2017)

27. Wang, B., Song, W., Lou, W., Hou, Y.T.: Privacy-preserving pattern matching over encrypted genetic data in cloud computing. In: Proceedings of the IEEE INFOCOM 2017, pp. 1–9 (2017)

28. Yang, J., Wang, H., Wang, J., Tan, C., Yu, D.: Provable data possession of resource-constrained mobile devices in cloud computing. J. Networks **6**(7), 1033–1040 (2011)

IoT Based Road State Sensing System Toward Autonomous Driving in Snow Country

Yoshitaka Sibata[1,3]([envelope]), Yasushi Bansho[2], and Shoichi Noguchi[3]

[1] Sendai Foundation for Applied Science, Iwate Prefectural University, Sugo, Takizawa, Iwate 152-89, Japan
shibata@iwate-pu.ac.jp
[2] Holonic Systems Ltd., Kami-Hirasawa 7-3, Shiwa, Iwate, Japan
bansho@holonic-systems.com
[3] Sedai Foundation for Applied Science, 1-5-1 Nishiki-Machi, Aoba-Ku, Sendai, Miyagi, Japan
Noguchi@sfais.or.jp

Abstract. In order to realize autonomous driving in snow country, IoT based road sate sensing system by combination of various environmental sensors is introduced. Using this system, not only various road states, such as dry, wet, slush, snowy, icy states are correctly identified, but the friction rates and roughness on road surface are quantitatively calculated so that the dangerous locations are identified and shown on the road map. Those road state data are also integrated and processed on MEC or cloud system to predicate the future road state in time and used for safe autonomous driving. In this paper, the proposed road state sensing system is designed and implemented as an experimental prototype to evaluate the performance and feasibility of autonomous driving in winter of a small snow city.

1 Introduction

As increase of traffic accidents by old ages and decrease of number of drivers for public transportation, introduction of autonomous driving has getting popular in those decay. There are actual and practical autonomous services, such as buses, taxies and logistics services on public roads are available in automobile-oriented countries, such as Japan in Level 2 and 3 in autonomous driving level. In level 2, the autonomous driving cars control their speed and steering angle by ECU based controller using various on-board sensors such as RGB camera, Li-DAR, millimeter wave radar and AI technology. There is a driver on the autonomous driving car who can control the driving mode, namely, automatic mode or manual mode. In level 3, autonomous driving car is controlled based on on-board sensors. If car has some troubles while driving, then autonomous driving mode is changed from autonomous mode to manual mode. Then the operator in remote operation center where the driving, traffic and road states is always monitored by multiple attached cameras in addition to various on-board environmental sensors in realtime through external communication network such as V2X and mobile network, after confirming the car's safety, the operator restarts and switches from manual to automatic mode.

L. Barolli (Ed.): AINA 2024, LNDECT 200, pp. 416–423, 2024.
https://doi.org/10.1007/978-3-031-57853-3_35

In level 4, there is no on-board operator and the car is completely autonomously controlled by AI based computer although the autonomous car runs on predetermined line or the limited road. Furthermore, in level 5, autonomous driving car is completely controlled on free road without any road restrictions.

So far, most of the current autonomous driving services are performed to the dedicated roads and highways in restricted urban city [1] and do not consider the mountain areas although those areas are occupied by heavy snows and ices and conducts serious accidents [2]. In addition, communication facility in those areas are not enough to support high level autonomous driving. Thus, a lot of problems inherent to snow country have to be challenged and overcome for reducing the influences of population and public traffic transportation shortages.

A lot of efforts have to be paid to those sensing and communication facility. In particular, high quality sensing and detecting technologies for autonomous driving including, sensing by camera, Li-DAR, millimeter wave radar, etc. must be developed. With Li-DAR, in the research [3], by applying machine learning, clustering and labelling techniques to recognize road obstacles such as bicycles are investigated and obstacles within 50 m are identified. However, the resolution for correct recognition of the objects beyond 50 m area is difficult. With millimeter wave radar, in the researches [4, 5], road traffic obstacles including pedestrians are dedicated even at night, rainy and foggy weather conditions. However, due to limitation of the resolution, decision of shape and classification of the obstacles at long distance is difficult. With RGB camera as image sensor, various application fields are existed due to its simple and cost-effective sensors to detect the objects. In the research [6], long distance object detection more than 200m is attained although the accuracy is influenced by night and weather conditions. Furthermore, in research [7, 8], deep convolutional neural networks are introduced to effectively detect the kindness and position of obstacles. The resolution of object at long distance is still maintained as compared with the LiDAR and millimeter wave radar. With road surface state identification, temperature sensor to detect the road surface temperature [9], 6-axis acceleration sensor to correctly measure road surface roughness [10] are used to decide the 6 road states such as dry, wet, rainy, snowy, ice and slush.

Thus, various sensors are developed to identify the obstacles and surface conditions on road. In this paper, to maintain the safety autonomous driving on challenged road conditions, those sensor technologies are integrated and attained as cloud sensing system to qualitatively and quantitively identify the correct road states and share with cars running on the road in wide area.

In the following, a general system of the proposed road state sensing system toward autonomous driving is explained in section two. Objective autonomous driving in challenged road environment for snow mountain area is explained in section three. The proposed road state sensing system prototype for autonomous vehicle system is precisely explained in section four. In final, concluding remarks are summarized in section five.

2 Road State Sensing System

In order to realize the autonomous driving in snow country, road states in all seasons must be observed in realtime using environmental sensors and managed widely. The sensor data are processed by on-board MEC using AI technology to recognize the road states. The recognized road states information is exchanged with other vehicles through SMB for V2X communication as shown in Fig. 1. At the same time, those recognized road states information and the collected sensor data are sent to local cloud at the remote control center using SRS for V2X facility and managed to monitor on the current road conditions. Based on current road state information, the driving schedule is adjusted. For example, if the road state is icy along the road, the running speed is controlled at lower speed or stopped for safety, while the road state is dry and the surface friction rate is low, then the running speed is set to higher value. Thus, the autonomous driving can be controlled based on various road conditions depending on the locations.

Those collected sensor data of the local cloud at the remote control center and the roadside servers along road are also sent to the global cloud and integrated into a bigdata and used to predict the wide area road conditions in temporal and geographical locations in future. Thus, the autonomous vehicle service can be realized even in challenged environment in snow country by introducing road sensing technology in addition to V2X communication facility.

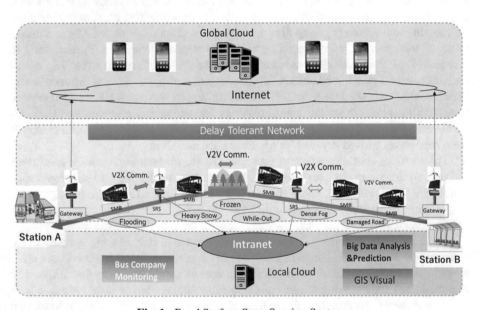

Fig. 1. Road Surface State Sensing System

3 Effective Autonomous Driving System

Currently, two type of autonomous driving vehicle systems including free road driving vehicle system and restricted road driving system are existed. In case of free road driving vehicle system, autonomous vehicle can run on any roads such as highway and general and dedicated, public and private roads with no restrictions and coexisted together with ordinal manual operated vehicles. For those reasons, many functions for autonomous driving including self-location estimation function, environmental recognition function such as existence of pedestrians, road surface and surround traffic conditions, traffic signal sensing and recognition function, route planning and following functions and more for safe and reliable autonomous driving. Therefore, since many intelligent sensing, decision and control devices and software are required, realization of autonomous driving involves very high cost and only urban area are objectives as autonomous driving area.

On the other hand, in the restricted road driving systems, since the driving roads is fixed and other manual operated vehicles are exclusively limited to the specific road, the required autonomous driving functions are simple and a few such as environmental recognition and traffic signal recognition functions. The traffic rule can be also simplified comparing with free road driving vehicle system. Therefore, the cost of the driving system can be realized with low cost.

Due to those simple functions with low-cost realization, this restricted driving system is used for small but heavy snow area as shown in Fig. 2. EV typed autonomous driving vehicle runs on the road where electric magnet lines are embedded under about 10 cm in depth. By detecting those magnetic field force using electro-magnetic sensor, the vehicle can always keep running on the right way. When there are obstacles on the way or pedestrians around the vehicle, the millimeter wave sensor and stereo vision camera sensors can detect those and automatically stop the vehicle or even remotely detect and immediately stop driving using V2X communication. After confirming the safety of driving, the on-board operator can restart the vehicle and continue the autonomous driving. In the same way, using road state censors, road states, surface friction and roughness data are always observed, then based on those road conditions, the vehicle is controlled by reducing its running speed or immediately stopping for safe and reliable driving. After confirming the safety of driving, the on-board operator can restart the vehicle and continue the autonomous driving. Thus, autonomous driving can be maintained even though in small town in snow country.

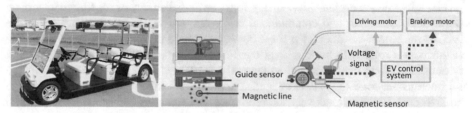

Fig. 2. Electro-Magnetic Field Typed Autonomous Driving Vehicle

4 Prototype System

In order to verify the effects and usefulness of the proposed IoT based sensing system in winter, a social experimental prototype system is constructed in Morioka castle park in Morioka city, Iwate prefecture which is located at northern part of Japan in winter season as shown in Fig. 3. First, we prepare autonomous driving vehicle, Koani-car which is currently under service at the small village with no accident for 4 years since the beginning the service. Then we installed electromagnetic wires on the ground about 1 km. Koani-car includes various sensors such as dynamic accelerator, gyro sensor, temperature sensor, humidity sensor, near infrared laser sensor, camera and GNSS, sensor server system and Communication server System as shown in Fig. 4. For experimental evaluation, a number of the citizens including handicapped are invited to this social experiment. As the tourist information service, the history of the castle and stone was explained by overlapping the VR contents of old castle and battle places on the current locations using tablet terminal as shown in Fig. 5.

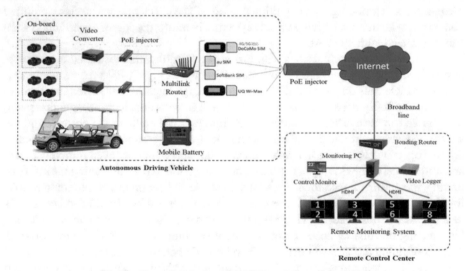

Fig. 3. Prototype System for V2X and Camera Sensing

Next, in order to evaluate the sensing function and accuracy of sensing road surface condition, both sensor and communication servers and various sensors are set to the EV. We ran Koani-car about 4 h for 2 days to evaluate decision accuracy in realtime on the winter road with various road conditions such as dry, wet, snowy, damp and icy condition along the leased line village. The experimental results of both qualitative and quantitative time series data of road states and road roughness are indicated in the Fig. 6 through Fig. 8, respectively. Currently, we are evaluating the road surface decision function using the video camera to compere the decision state and the actual road surface state. Thus, social experiment of autonomous driving in snow could be easily constructed and performed for sightseeing purpose even for handicapped in cold snow/iced weather condition (Fig. 7).

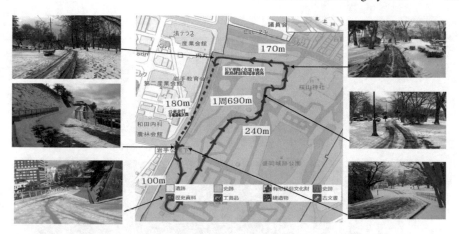

Fig. 4. Autonomous Driving Course in Morioka Castle Park

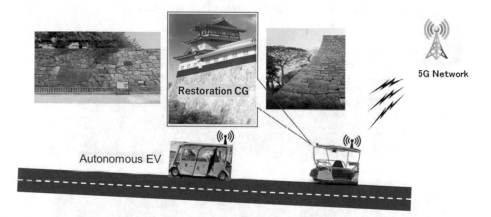

Fig. 5. VR based Tourist Information on Prototype

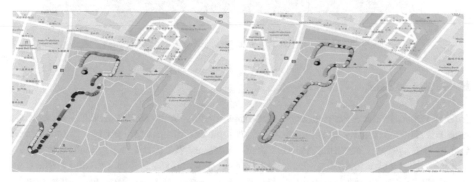

Fig. 6. Road State (Left) and Friction Rate (Right

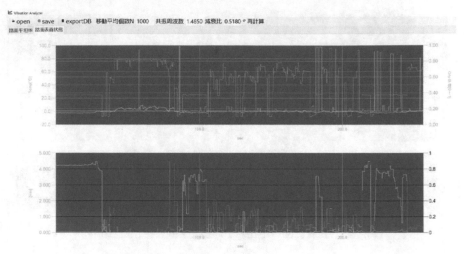

Fig. 7. Quantitative Time Series Data of Road States

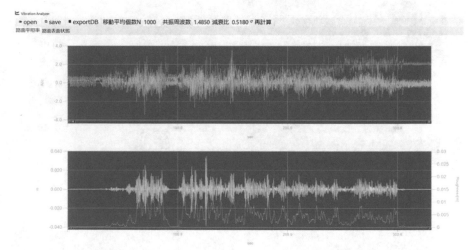

Fig. 8. Quantitative Time Series Data of Road Roughness

5 Conclusions

In this paper, in order to realize autonomous driving in snow country IoT based road sate sensing system by combination of various environmental sensors is introduced. Using this system, not only various road states, such as dry, wet, slush, snowy, icy states are correctly identified, but the friction rates and roughness on road surface are quantitatively calculated so that the dangerous locations are identified and shown on the road map. Those road state data are also integrated and processed on MEC or cloud system to predicate the future road state in time and used for safe autonomous driving. In this paper, the proposed road state sensing system is designed and implemented as an experimental prototype to evaluate the performance and feasibility of autonomous driving in winter of a small snow city.

Acknowledgments. The research was supported by Sendai Foundation for Applied Science, by Japan Keiba Association Grant Numbers 2021M-198, JSPS KAKENHI Grant Numbers JP 20K11773 and Ministry of Land Infrastructure, Transport and Tourism.

References

1. SAE International, "Taxonomy and definitions for terms related to driving automation systems", for On-Road Motor Vehicles. J3016_201806 (2018)
2. Police department in Hokkaido, "The actual state of winter typed traffic accidents". https://www.police.pref.hokkaido.lg.jp/info/koutuu/fuyumichi/blizzard.pdf (2018)
3. Himmelsbach, M., et al.: Real-time object classification in 3D point clouds using point feature histograms. In: 2009 IEEE/RSJ International Conference on Intelligent Robots and Systems, pp. 994–1000 (2009)
4. Sugimoto, S., Tateda, H., Okutomi, M.: Obstacle detection using millimeter-wave radar and its visualization on image sequence. In: The 17th International Conference on Pattern Recognition (ICPR2004), vol. 3, pp. 342–345 (2004)
5. Kaliyaperumal, K., Lakshmanan, S., Kluge, K.: An algorithm for detecting roads and obstacles in radar image. IEEE Trans. Veh. Technol. **50**(1), 170–182 (2001)
6. Pinggera, P., Franke, U., Mester, R.: High-performance long rang obstacle detecting using stereo vision. In IEEE/RSJ International Conference on Intelligent Robot and Systems (IROS2015), pp. 1308–1313 (2015)
7. Simonyan, K., and Zisserman, A.: Very deep convolutional networks for larger-scale image recognition, Comput. Vision Pattern Recogn. arXiv, pp.1409–1556, 2014
8. Ren, S., et al.: Faster R-CNN: toward realtime object detection with region proposal networks, In Advanced in neural information processing systems", Adv. Neural Inf. Process. Syst. 28 (NIPS 2015), 2015
9. Takahashi, N., Tokunaga, R.A., Sato, T., Ishikawa, N.: Road surface temperature model accounting for the effects of surrounding environment. J. Jpn. Soc. Snow Ice **72**(6), 377–390 (2010)
10. Fujita, S., Tomiyama, K., Abliz, N., Kawamura, A.: Development of a roughness data collection system for urban roads by use of a mobile profilometer and GIS. J. Japan Soc. Civ. Eng., 69(2), I_90–I_97 (2013)

An Approach to Attack Modeling for the IoT: Creating Attack Trees from System Descriptions

João B. F. Sequeiros[1,2(✉)], Francisco Tchissaquila Chimuco[1,2,3],
Tiago M. C. Simões[1,2], Mário M. Freire[1,2], and Pedro R. M. Inácio[1,2]

[1] Universidade da Beira Interior, Covilhã, Portugal
jbfs@ubi.pt
[2] Instituto de Telecomunicações, Lisboa, Portugal
[3] Instituto Superior de Ciências de Educação de Huíla, Lubango, Huíla, Angola

Abstract. This paper presents an Internet of Things (IoT) architecture and associated attack taxonomy, along with a tool named *Attack Trees in IoT* (ATIoT), which was designed to generate attack trees from a description of a system. The tool obtains the description via a series of questions about the IoT system. The proposed IoT architecture was developed with security into consideration, allowing to define security requirements that each component may need to fulfill with more granularity. The associated attack taxonomy provides a comprehensive overview of the different types of attacks that an IoT system may face, categorized from the components of the proposed architecture. The ATIoT tool leverages the IoT architecture and attack taxonomy to generate attack trees that can be used to identify potential attack vectors, and therefore aids in prioritizing security controls for an IoT system. The tool asks a series of questions about the IoT system, including its functionalities and characteristics, and generates an attack tree based on the responses. The tool is designed to be accessible to developers with little to no security expertise, providing a user-friendly interface and automated attack tree generation. Using the tool, developers can gain a better understanding of the security risks associated with their IoT systems and implement appropriate security controls to mitigate those risks.

1 Introduction

The Internet of Things (IoT) is changing the way we interact with technology, bringing new opportunities for innovation and automation [7]. However, with the rapid growth of IoT, security has become a major concern. As IoT devices become more prevalent in our daily lives, the attack landscape and potential security breaches increase, putting personal data, privacy, and even physical safety at risk [9]. The security challenges of IoT stem from the vast number and diversity of connected devices, the majority of which were not designed with security in mind. Furthermore, the complexity of IoT systems and the lack of standardized technologies and protocols makes securing IoT systems a significant challenge [1], which can only be consistently overcome by following a *security by design* approach.

Security by design is an approach to system design that incorporates security considerations throughout the development process, rather than adding security as an

L. Barolli (Ed.): AINA 2024, LNDECT 200, pp. 424–436, 2024.
https://doi.org/10.1007/978-3-031-57853-3_36

afterthought. By integrating security into the design process, developers can identify and address security risks at every stage of development, from concept to deployment, and ensure that the system is secure from the ground up [5]. There is also a lack of cohesive security frameworks to aid the development process [13], specially in those that focus on teams with little to no security expertise. Lack of expertise in cybersecurity is another major gap and obstacle to achieve secure systems.

Attack modeling is a proactive approach to IoT security that helps identify and address potential security vulnerabilities in IoT systems, falling into the early stages of development. Attack modeling involves creating an attack taxonomy, identifying potential attack paths, and determining the final targets and objectives of the attack, allowing for developers to then identify potential mitigations and security mechanisms that should be implemented and tested later on. Attack modeling can also be used to evaluate the effectiveness of existing security measures and identify areas where additional security controls or technologies may be needed.

The work proposed herein attempts to reduce the previously identified gaps, by further expanding on previous work to create an all-encompassing framework, composed of a set of tools aimed at providing developers with the necessary knowledge and support to create, design and implement secure IoT systems. The tool *Attack Trees in IoT* (ATIoT) is therefore proposed as a means for developers to quickly identify potential attacks their system may be exposed to. The main contributions of this work are: (i) the proposals for an IoT architecture with security as its main focus, which provides the fundamental blocks for the definition of security requirements for each of its components, (ii) an associated attack taxonomy for IoT, which applies directly to the components of the proposed architecture, and (iii) the ATIoT tool, which generates attack trees for a given system through its description, obtained via a guided set of questions that the user is required to answer, and outputting the attack trees directly, without requiring them to possess any intrinsic security expertise.

The remainder of this paper is organized as follows. Section 2 presents the background and related works. Section 3 presents the proposed IoT architecture and attack taxonomy based on the architecture. Section 4 presents the ATIoT tool. Section 5 presents the functionality testing and exemplifies a use case of the tool, and finally, Sect. 6 presents the main conclusions and future work.

2 Related Works and Background

Despite IoT being a relatively recent computing paradigm, there has been an attempt in standardizing some definitions and architecture of IoT systems, identifying layers and components of an IoT system. Regarding security, several of the traditional attack modeling approaches are also applicable and adaptable to IoT.

2.1 IoT Architectures

There have been attempts to create different IoT architectures that help describe the main components or layers that may define a generic IoT system. This is a challenge in itself, as heterogeneity is one of the main characteristics of IoT, with systems that wildly differ from one another in terms of characteristics and functionality. Some architectures for IoT are described below:

- **The 3-Layer Model** – the most commonly widespread IoT architecture is the three layers model [2]. In this architecture, IoT systems are described in terms of the three following layers: (i) *Perception*, which describes the layer responsible for sensing devices and data gathering, including both data collectors and actuators; (ii) *Network*, responsible for all communications inside the system and with the outside, all data transfer and processing; and (iii) *Application*, responsible for receiving data, and providing the end users analysis, processes and different services;
- **The IoT Reference Model** – the IoT Reference Model was proposed by Cisco, IBM and Intel [4] as a comprehensive IoT architecture that encompasses seven levels: the (i) *Physical Devices and Controllers* level, which includes sensors, devices, and other "things" that generate data and interact with the environment; the (ii) *Connectivity* level, responsible for providing the network connectivity for the devices to communicate with each other and with processing units; the (iii) *Edge Computing* level, responsible for a first data analysis and transmission; the (iv) *Data Accumulation* level, which pertains to data storage; the (v) *Data Abstraction* level, which processes, aggregates, structures and analyzes the data; the (vi) *Applications* level, which provides the reporting, analytics and controls that enable users to interact with the system; and the (vii) *Collaboration and Process* level, which entails all the business and people processes;
- **ETSI M2M Architecture** – the European Telecommunications Standards Institute (ETSI) has proposed the Machine-to-Machine (M2M) architecture [6] which places IoT devices in one out of two main domains. The (i) *Device and Gateway Domain* integrates devices, the local area network where they are included and the gateways that communicate between them and other parts of the system; and the (ii) *Network Domain*, which integrates applications, the core system network, and management and access functions of the system. From this architecture, ETSI has been creating technical specifications for enabling better interoperability between devices.

2.2 Attack Modeling Strategies

Attack modeling techniques attempt to identify attacks against a given system, through adversary techniques and possible attack paths. Through this modeling, it is possible to understand the behavior, tactics and objectives of an attacker, and therefore organize and decide on security mechanisms that should be implemented by the system. Herein we approach three main attack modeling mechanisms: *Kill Chain*, *Diamond Model*, and *Attack Trees*. Together with a proposal for an architecture and by defining an attack taxonomy, it is possible to later define the attack models that will apply to a given set of system characteristics:

- **Kill Chain** – the Kill Chain [15] can be used to describe the different stages of an attack. It is divided into several stages, namely, *Reconnaissance*, where the attacker gathers information about the target, *Weaponization*, the phase in which they create or acquire tools or exploits to carry out the attack, *Delivery*, when they deliver the weaponized payload to the target system, *Exploitation*, where they exploit a vulnerability to gain access to the system, *Installation*, where they install malware or other malicious tools on the system, *Command and Control*, the phase in which they establish communication with the system to control it, and *Actions on Objectives*, where they achieve their objective;

– **Diamond Model** – the Diamond Model [3] is used to analyze and visualize threats. The Diamond Model includes four elements: the *Adversary*, which is the person, group, or organization that is carrying out the attack; the *Capability*, which refers to the technical skills, tools, and resources that the adversary possesses; the *Infrastructure*, which is The technical infrastructure that the adversary is using to carry out the attack; and the *Victim*, who, as the name implies, is the person, group, or organization that is the target of the attack;

– **Attack Trees** – Attack Trees [12] are a visual representation of possible attack scenarios that are structured like a tree. The root of the tree represents the ultimate goal of the attack, while the branches represent different ways the attacker could achieve that goal. For each branch, the attack tree will continue to branch out further, showing the specific steps an attacker could take to achieve that objective. At each step, the attacker must overcome some sort of obstacle, such as a security control or a technical challenge. By visualizing the attack scenario in this way, one can identify where the efforts of the attacker can be thwarted.

2.3 Attack Modeling Tools

Attack modeling tools are essential to aid in the process of security engineering, aiding in the process of identifying and evaluating potential attacks a system may be subject to. The ADTool [8] aids in graphically modeling attack-defense scenarios through attack, protection and defense trees, and also allows the quantitative analysis of attacks through these trees. IoTSAT [10] presents a framework for formal security analysis of IoT systems, using device configurations, network topology, user policy, and attack surfaces in its modeling process, then using the model to measure resilience against attacks and identify threat vectors. AT-AT (Attack Tree Analysis Tool) [16] is a tool that aids in designing attack trees, calculates scenarios and calculates likelihood, impact, resources and time for each scenario through the values defined for these metrics for each step on the tree.

2.4 Framework for IoT Security by Design

The ATIoT tool proposed in this paper was developed in the context of the SECURIoTESIGN project [14], being one of several tools integrated in a framework whose main purpose is aiding the secure design and development process for IoT systems. The tools address the different stages of system development. These tools are designed with ease of use in mind, and should output recommendations, guidelines and directions so that developers can create more secure systems.

For these purposes, a framework integrates and connects the different tools, which include, e.g., Security Requirements Elicitation (SRE) and Security Best Practices and Guidelines (SBPG) [11]. The ATIoT tool, presented in this paper, is the most recent development for this framework. The web-based framework and its backend is named SAM (Security Advisory Modules), and integrates all tools in a modular fashion. This framework was also structured to enable a quicker integration of new tools, and updating of existing ones, in a way to improve the longevity of the platform, and keeping it up to date in a scenario that is always evolving. To use the tools in SAM, users are asked sets of multiple choice questions, from which the outputs are generated. The SRE tool, in particular, is of greater relevance for this work. From the answers given by the user, it presents a set of security requirements that the system should fulfill. The set of questions partially intersects with those necessary for the ATIoT tool (that will be described in the following sections), so the use of the SRE tool before the ATIoT tool will benefit the user, by giving them, together with a set of attack trees, a set of security requirements that should be fulfilled.

3   IoT Architecture and Attack Taxonomy

Most of the architectures proposed for IoT are centered or expand on the concept of three main layers: *Application*, *Network*, and *Perception/Physical*. While there are architectures that separate this into a greater number of layers or components, these three can be considered as the mainstay of any IoT system. However, most architectures focus on the working side of IoT and security is rarely taken into account (and even when it is, it is considered as a separate layer or component, instead of an integral part of the different parts that a system may be composed of). This causes issues, mainly when a system is being developed by a team without strong expertise in terms of security, which will have difficulty in defining what potential vulnerabilities their system may have. This cumulatively impacts the other IoT issues: its heterogeneity, complexity and limited resources.

Herein, we thus present an IoT architecture which intends to ensure security is embedded in each of its layers, so that developers can understand more precisely and given a set of security requirements, in which parts of a system these requirements apply to. This architecture is key for the development of the ATIoT tool, on par with the attack taxonomy proposed below. This architecture is composed of seven main components, which can be briefly described as follows:

- *Entity* - taking into account a physical and virtual existence, an *entity* identifies what the system directly interacts with, be it through sensing or actuating (e.g., the air temperature, a box to be picked up, or the cardiac rhythm of a person);
- *Device* - which refers to sensors or actuators, the actual hardware that is deployed, and refers to the perception and actuation the system performs on a given environment;
- *Data* - which concerns all data that is gathered, processed, transmitted or stored on the system, from its original collection up to its final destination;
- *Communications and Networking* - referring to all communications inside the system, and with other networked devices. It encompasses different transmission protocols and wireless technologies;
- *Processes* - as data is gathered in bulk by the system, there needs to be preprocessing, followed by actual processing, to first slim the data volume, and then

transform data so that information can be extracted from it. This pertains both to the processing and the equipment required to conduct that processing;

- *Application* - the application layer usually refers to traditional and mobile computing, as the means through which users interact with the system;
- *User* - users can be described as anyone or anything that communicates with or controls the system. These can be either human, or other systems or digital entities, external to the system, that are able to communicate with it.

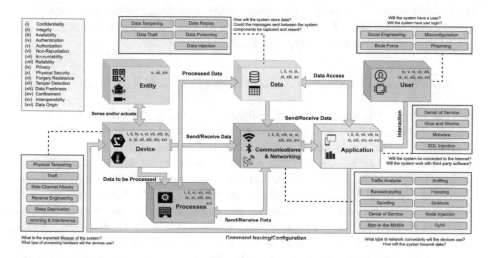

Fig. 1. High-level scheme for the proposed IoT architecture illustrating the several components, their dependencies and connections, and the proposed corresponding attack taxonomy.

Figure 1 shows a high-level representation of the components described above and how they interact with each other. Furthermore, it contains a mapping of the security requirements that can be applied to each one of them. This set of components partially reminisces those found in the traditional proposals for IoT architectures. However, these components are being considered from a security perspective, where a set of requirements will be defined for each component, which will lead to a more granular identification of the parts of the system or software that need attention or a given control, security mechanism or technology.

To further flesh out the proposed architecture, a total of eight different ecosystems were considered, for which the architecture can be applied to: healthcare, wearables, smart cities, environmental monitoring, smart grids, transportation and logistics, smart environments and smart manufacturing. Specifically, the recommendation of the presented requirements on a given component, for a given ecosystem, are defined in the SRE tool. This allows for each system to have its own set of requirements, that can then be applied to each system component. The next section defines an attack taxonomy that elaborates on this architecture and on the several proposed IoT ecosystems.

3.1   Attack Taxonomy for the Proposed Architecture

Taking into consideration the architecture proposed in the previous section, a corresponding attack taxonomy for the IoT is proposed in this section. This taxonomy, already depicted in Fig. 1, attempts to correlate attacks with the more relevant parts of the architecture. Since the IoT is subject to a wide panoply of different attacks (due to its complexity and heterogeneity), it is important to provide categories that are simultaneously descriptive, cover all the ecosystems and exhibit low intersections. These features guided the development of the taxonomy, resulting in a total of 29 categories/attack scenarios. Notice that the proposed architecture favors the orthogonal definition of the categories. Five of the total seven components were considered for this taxonomy: *Device*, *Data*, *Communications and Networking*, *User*, and *Application*.

4   Automated Creation of Attack Trees for an IoT System

This section presents the tool (and the underlying approach) devised for the automatic creation of attack trees. For a given description of an IoT system (composed of the identification of the ecosystem it belongs to and of several of its characteristics), the ATIoT generates a set of trees on the most relevant attacks that may affect that system. These trees will describe the different possibilities and targets for that specific attack. To further aid developers, each tree is accompanied by a description of each node, where some examples on how an attacker may proceed or what they may gain access to is presented. As ATIoT targets users with little to no security expertise, its purpose is to automatically generate the trees starting from descriptive information provided by the user regarding the system under analysis. The tool has four main components, the *Input Gatherer*, the *System Analyzer*, and the *Tree Generator*. Figure 2 presents the three main components and the overall framework. The components are briefly presented as follows:

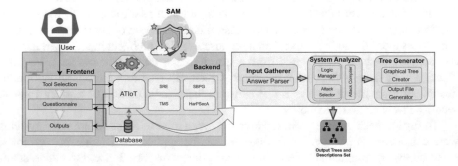

Fig. 2. High-level architecture of the framework and ATIoT tool.

– **Input Gatherer** - Information about a system is gathered through a questionnaire, to guide users into providing the right data in a structured way. This questionnaire has a total of 24 questions (included in Table 1). The questions are multiple choice, so that the user inputs are kept in scope. The *Input Gatherer* will communicate with SAM to obtain previously provided answers from the database, to avoid having the user answering the same questions when using different tools of the framework;

– **System Analyzer** - All the logic used to map the description (the answers) to the applicable attacks is codified in this component. From the answers, a set of attacks is selected, from which a set of nodes and their descriptions is then inputted into the tree generator component (see below), creating the corresponding attack tree and description. This attack selection is done through a set of rules, where each rule considers a given answer, and selects a potential attack that the system may be susceptible to. The system analyzer rule logic takes into account the results from the *Input Gatherer*, and decides on which trees should be presented to the user, e.g., if the user answers positively to the question *"Can someone gain physical access to the system or components and access and/or perform some type of modification to its hardware?"*, and the selected system belongs to an ecosystem where physical sensors are in accessible locations, the System Analyzer will generate attack trees such as *Physical Tampering*, *Theft* and *Reverse Engineering*. Though most of the rules are intricate, some of them can be depicted as a series of conditional statements, such as the examples included below:

- IF the system is from the Environmental Monitoring ecosystem, THEN select Physical Tampering, Theft, Reverse Engineering, and Eavesdropping;
- IF the system communications can be eavesdropped, THEN select Eavesdropping, Sniffing.
- IF the system does not have a user, THEN do not select Brute Force and Social Engineering;
- IF someone can attempt to impersonate a user to gain access to private information, THEN select Spoofing and Social Engineering;
- IF someone is able to gain physical access to the machine, THEN select Physical Tampering, Side-Channel Attacks, Device Theft, and Reverse Engineering.

– **Tree Generator** - This component receives information on the different selected attacks from the *system analyzer*, subsequently generating the corresponding attack tree. The information is given to the component in the form of a key-value store, where each key represents a node, and the values its direct children in the tree. Similarly, the node descriptions are also passed as a key-value store. The *Tree Generator* is then able to generate two representations of the tree: a visual representation, and a text-based representation. The visual representation of the tree is followed by text descriptions of each of the tree nodes. After completing its task, an output is generated and shown to the user;

To ensure that the tool remains viable, and its outputs relevant, system administrators can, through SAM, access and edit all information and logic, so that new trees, descriptions, and analysis can be inputted into the framework, whenever an update is necessary.

5 Functionality Testing with Use Cases

Two use cases are presented herein to provide a basic validation and demonstration on how ATIoT works. Validation is performed by describing potential real IoT systems, using the tool to generate the respective attack trees and then critically analyze the outputs. This approach was used for many use cases but, to keep the explanation manageable, we are only mentioning two scenarios below.

– **Smart Irrigation System** - This scenario proposes a smart (agriculture) irrigation system. The system should measure soil moisture and temperatures and control irrigation from these parameters. The sensors will send their data to a receiver node, that will then process the data. This data is then sent to be stored in the cloud. A graphical web interface and mobile application allow the operator to monitor data in real-time and be notified when certain threshold levels are hit, which then actuate upon the irrigation motor to change irrigation as needed. The system is schematized in Fig. 3(2). Considering this scenario, the SRE tool was run to collect the set of answers that represent the application scenario. Table 1 presents the questions and the chosen answers. After that, the ATIoT tool was run, which parsed those answers from the database, and its logic selected the following set of attacks: *Physical Tampering, Theft, Reverse Engineering, Data Theft, Side-Channel Attacks, Jamming & Interference, Eavesdropping, Sinkhole, Man-in-the-Middle, Sleep Deprivation, Traffic Analysis, Denial of Service, Node Injection, Brute Force, Social Engineering, Sniffing, and Misconfiguration*. For illustrative purposes, Fig. 4 presents an excerpt of the ATIoT output for the physical tampering attack. The outputted report presents the tree and a description for each node.

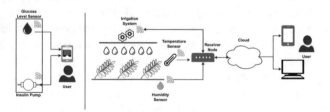

Fig. 3. Example of the logical architectures for the (1) glucose monitoring and insulin administration system and (2) smart irrigation system.

– **A Smart Glucose Monitoring and Insulin Administration System** - This scenario refers to a glucose monitoring system that measures the glucose level in the blood of diabetic patients. The sensor communicates with a smartphone, which will then decide whether insulin administration is needed. The smartphone will instruct an insulin pump to inject the correct dosage when needed. A log of the levels and administered dosages is also kept. A schematic for this system is presented in Fig. 3(1). The SRE tool was run for this scenario and fed with the answers shown in the right column of Table 1. In this case, the execution of the ATIoT tool outputted the following set of attacks: *Sleep Deprivation, Jamming & Interference,*

Physical Tampering

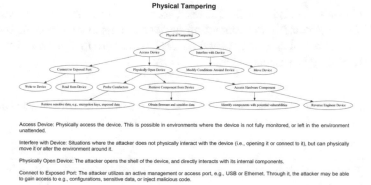

Access Device: Physically access the device. This is possible in environments where the device is not fully monitored, or left in the environment unattended.

Interfere with Device: Situations where the attacker does not physically interact with the device (i.e., opening it or connect to it), but can physically move it or alter the environment around it.

Physically Open Device: The attacker opens the shell of the device, and directly interacts with its internal components.

Connect to Exposed Port: The attacker utilizes an active management or access port, e.g., USB or Ethernet. Through it, the attacker may be able to gain access to e.g., configurations, sensitive data, or inject malicious code.

Fig. 4. Example of the ATIoT generated output for the physical tampering attack.

Data Tampering, Data Replay, Data Poisoning, Data Theft, Data Injection, SQL Injection, Traffic Analysis, Man-in-the-Middle, Eavesdropping, Denial of Service, Sniffing, Spoofing, Social Engineering, Brute Force, and Misconfiguration. Figure 5 presents a part of the ATIoT output for the *Data Tampering Attack*.

– **Output Analysis** - Focusing on the trees presented in Figs. 4 and 5, the *Physical Tampering* and *Data Poisoning* attacks show how an attacker may, for the first case, and given physical access to the system (the temperature and the moisture sensors), or, for the second case, given the possibility of tampering with data (so that the insulin pump either does not administer, or administers the incorrect dosage), proceed and attempt different techniques, which steps they can take, and the potential reward they may reap from conducting said attack. By identifying them, developers can then ensure their system design and implementation can close that attack avenue (*e.g.*, given the possibility of a management or access port being accessible, developers may choose to disable, remove or configure strong authentication mechanisms, so that an attacker cannot use this particular path, or developers can incorporate data integrity mechanisms, to ensure that sensor data is correct and not maliciously injected).

Data Tampering

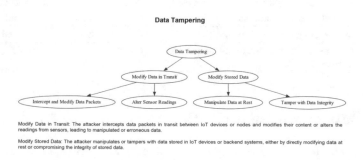

Modify Data in Transit: The attacker intercepts data packets in transit between IoT devices or nodes and modifies their content or alters the readings from sensors, leading to manipulated or erroneous data.

Modify Stored Data: The attacker manipulates or tampers with data stored in IoT devices or backend systems, either by directly modifying data at rest or compromising the integrity of stored data.

Fig. 5. Example of the ATIoT generated output for the data tampering attack.

The remaining selected attacks, for each scenario, encompass, when examined in detail, and considered in the context of the descriptions of each system, a complete and coherent set which addresses and covers a strong group of potential attacks that the systems may be susceptible to.

Table 1. Answers fed to the framework for both use case scenarios (the *irrigation* and the *glucose and insulin administration* systems).

Question	Irrigation	Glucose
State the domain type for your IoT system:	Smart Environment	Smart Healthcare
Will the system have a user?	Yes	Yes
Will the system have user login?	Yes	Yes
Will the system hold any user information?	Yes	Yes
Will the system store any kind of information?	Yes	Yes
What will the level of information stored be?	Normal	Critical
Will this information be sent to an external entity?	No	No
How will the system transmit data?	Device-gateway-cloud	Device-application
Will the system be connected to the internet?	Yes	No
Will it send its data to a cloud?	Yes	No
Will it store data in a database?	Yes	Yes
How will the system store data?	Hybrid storage	Application storage
Will the system receive regular updates?	No	Yes
How will the system receive software updates?	N/A	Manual updates
Will the system work with third-party software?	No	No
Could the messages sent between the system components be captured and resent?	Yes	No
Can someone try to impersonate a user to gain access to private information?	No	Yes
Can someone gain physical access to the system or components and access and/or perform modifications to it?	Yes	No
What type of processing hardware will the devices use?	Microcontroller	Microcontroller
What type of operating system will the devices run?	RTOS	RTOS
What type of network connectivity will the devices use?	Wi-fi	Bluetooth
What is the expected lifespan of the system?	Long (5+ years)	Medium (3–5 years)
What is the expected usage pattern of the system?	Continuous use	Continuous use
What are the expected communication protocols and data formats used by the system?	CoAP	MQTT

6   Conclusions and Future Work

Many companies are placing security in the backseat while giving priority to the development of new IoT systems, features and functionalities, especially in smaller teams and mostly motivated by fast growing markets and technology penetration. This is not expected to change in the coming years, further contributing to the poor cybersecurity state and to the skills gap on that area. This paper proposed a new comprehensive IoT architecture that builds on existing architectures while focusing on security requirements, to then propose a corresponding attack taxonomy and, subsequently, a tool for automated generation of attack trees for IoT systems (ATIoT). This tool targets developers with low security expertise, presenting them with the trees already constructed and described, to aid in the integration of security in the design and development process. ATIoT was tested through use case scenarios that mimicked the design of IoT systems, through which the outputs of the tool were analyzed for how appropriate they were for the given scenarios. As future work, the expansion and further definition of the proposed architecture will be performed, connecting the different system components and security requirements to the different IoT ecosystems, and the preponderance of each component and attack on these systems. The development of accompanying tools to ATIoT, that target the design of attack models based on other approaches, such as those presented in Sect. 2, and the development of tools targeting system modeling and data flows on the system, are also future research directions, to completely cover the entire system or software development cycle. Finally, future development will also investigate the integration of AI into the presented tools to broaden their scope and enhance their capabilities, including the development and automatic update of recommendations, and facilitating user input through freeform text.

Acknowledgements. This work was performed under the scope of Project SECURIoTESIGN, with funding from FCT/COMPETE/FEDER (projects with reference numbers UIDB/50008/2020 and POCI-01-0145-FEDER-030657), and FCT research and doctoral grants BIM/n°32/2018-B00582 and SFRH/BD/133838/2017, respectively, and also supported by operation Centro-01-0145-FEDER-000019 - C4 - Centro de Competências em Cloud Computing, co-financed by the European Regional Development Fund (ERDF) through the Programa Operacional Regional do Centro (Centro 2020), in the scope of the Sistema de Apoio à Investigação Científica e Tecnológica - Programas Integrados de IC&DT.

References

1. Al-Qaseemi, S.A., Almulhim, H.A., Almulhim, M.F., Chaudhry, S.R.: IoT architecture challenges and issues: lack of standardization. In: 2016 Future Technologies Conference (FTC), pp. 731–738 (2016). https://doi.org/10.1109/FTC.2016.7821686
2. Al-Qaseemi, S.A., Almulhim, H.A., Almulhim, M.F., Chaudhry, S.R.: IoT architecture challenges and issues: lack of standardization. In: 2016 Future Technologies Conference (FTC), pp. 731–738. IEEE (2016)
3. Caltagirone, S., Pendergast, A., Betz, C.: The diamond model of intrusion analysis. Technical report, Center For Cyber Intelligence Analysis and Threat Research Hanover Md (2013)
4. Cisco: IoT reference model (2014). http://cdn.iotwf.com/resources/72/IoT_Reference_Model_04_June_2014.pdf
5. ENISA: Good practices for security of IoT: secure software development lifecycle. Technical report (2019)
6. ETSI: M2M architecture (2011). https://docbox.etsi.org/workshop/2011/201110_m2mworkshop/02_m2m_standard/m2mwg2_architecture_pareglio.pdf
7. Hassan, R., Qamar, F., Hasan, M.K., Aman, A.H.M., Ahmed, A.S.: Internet of things and its applications: a comprehensive survey. Symmetry 12(10), 1674 (2020)

8. Kordy, B., Kordy, P., Mauw, S., Schweitzer, P.: ADTool: security analysis with attack–defense trees. In: Joshi, K., Siegle, M., Stoelinga, M., D'Argenio, P.R. (eds.) QEST 2013. LNCS, vol. 8054, pp. 173–176. Springer, Heidelberg (2013). https://doi.org/10.1007/978-3-642-40196-1_15

9. Mishra, N., Pandya, S.: Internet of things applications, security challenges, attacks, intrusion detection, and future visions: a systematic review. IEEE Access **9**, 59353–59377 (2021). https://doi.org/10.1109/ACCESS.2021.3073408

10. Mohsin, M., Anwar, Z., Husari, G., Al-Shaer, E., Rahman, M.A.: IoTSAT: a formal framework for security analysis of the internet of things (IoT). In: 2016 IEEE Conference on Communications and Network Security (CNS), pp. 180–188. IEEE (2016)

11. Samaila, M.G., José, M.Z., Sequeiros, J.B., Freire, M.M., Inácio, P.R.: IoT-HarPSecA: a framework for facilitating the design and development of secure IoT devices. In: Proceedings of the 14th International Conference on Availability, Reliability and Security, pp. 1–7 (2019)

12. Schneier, B.: Attack trees. Dr. Dobb's J. **24**(12), 21–29 (1999)

13. Sequeiros, J.B., Chimuco, F.T., Samaila, M.G., Freire, M.M., Inácio, P.R.: Attack and system modeling applied to IoT, cloud, and mobile ecosystems: embedding security by design. ACM Comput. Surv. (CSUR) **53**(2), 1–32 (2020)

14. Instituto de Telecomunicações: Securiotesign (2022). https://lx.it.pt/securIoTesign/

15. Yadav, T., Rao, A.M.: Technical aspects of cyber kill chain. In: Abawajy, J.H., Mukherjea, S., Thampi, S.M., Ruiz-Martínez, A. (eds.) SSCC 2015. CCIS, vol. 536, pp. 438–452. Springer, Cham (2015). https://doi.org/10.1007/978-3-319-22915-7_40

16. Yathuvaran, A.: AT-AT (attack tree analysis tool) (2022). https://github.com/yathuvaran/AT-AT

A Distributed Platform for Cycle Detection and Analysis in Cyber-Physical Systems

Gabriel Iuhasz[1,2(✉)], Silviu Panica[2,3], Florin Fortis[1,2], and Alecsandru Duma[4]

[1] West University of Timişoara, 300223 Timişoara, Romania
{iuhasz.gabriel,florin.fortis}@e-uvt.ro
[2] Research Institute e-Austria Timişoara, 300223 Timişoara, Romania
silviu@innoqube.ro
[3] Innovation Qube SRL, Timisoara, Romania
[4] ETA2U Computers, Gh. Dima Nr 1, Timisoara, Romania
aduma@eta2u.ro

Abstract. This paper presents a distributed platform that collects and processes data streams from Industrial IoT/Cyber-Physical Systems. We demonstrate the design and performance of our platform with an emphasis on accuracy and scalability. We validate our platform using three use cases: production cycle detection, cycle analysis, and anomaly detection.

Keywords: Cyber-Physical Systems · Services · Analysis · Machine Learning

1 Introduction

The Industrial IoT (IIoT) refers to the network of interconnected devices and sensors that gather, monitor, and analyze data from industrial operations. This technological approach can enhance productivity, efficiency, safety, and quality across various sectors, including manufacturing, transportation, energy, and mining.

It is enabled by a plethora of technologies such as cyber-physical systems, cloud computing, edge computing, big data analytics, and artificial intelligence. Industrial IoT facilitates real-time communication, data processing, and decision-making at the network's edge. It also allows for centralized data storage and analysis in the cloud. Moreover, Industrial IoT can utilize machine learning (ML) and artificial intelligence techniques to optimize processes, predict failures, and automate tasks.

As a part of the broader digital transformation of industry, Industry 4.0, IIoT aims to contribute to the creation of smart factories and smart cities. These entities can adapt to changing demands, reduce costs, and minimize environmental impact. The adoption of Industrial IoT is expected to significantly impact the global economy by creating new business models, revenue streams, and competitive advantages for organizations. In this context, a definition of IIoT can

© The Author(s), under exclusive license to Springer Nature Switzerland AG 2024
L. Barolli (Ed.): AINA 2024, LNDECT 200, pp. 437–450, 2024.
https://doi.org/10.1007/978-3-031-57853-3_37

be found in [1], "Industrial IoT (IIoT) is the network of intelligent and highly connected industrial components that are deployed to achieve high production rate with reduced operational costs through real-time monitoring, efficient management and controlling of industrial processes, assets and operational time".

While the Internet of Things (IIoT) represent one of the many facets of IoT, Industry 4.0 is defined in the context of IIoT, focusing on safety and production efficiency. Thus, Industry 4.0 can be set at the intersection of technologies such as Cloud Computing, the Internet of Things, Machine Learning and Artificial Intelligence methods, particularly within the manufacturing sector, with a specific goal to establish interconnected systems that automate industrial processes, leading to the creation of intelligent factories. Such factories will host machines that are monitored and controlled by computer systems, aiming at developing systems that are capable to create virtual machine models, which can then facilitate the incorporation of predictive methods and enhance production process optimization.

Digital Twins, a fundamental concept in the context of Industry 4.0, enable the creation of these virtual models. They offer digital replicas of physical products, allowing companies to identify physical issues more rapidly, predict outcomes with greater precision, and manufacture superior products. Digital Twins can provide companies with a digital footprint of products, providing a robust link to the product for diagnostics and design modifications.

Predictive maintenance is another crucial concept within the context of Industry 4.0. It involves identifying anomalies and failure patterns, learning from the patterns that lead to machine anomalies and failures, and predicting mechanical issues before they manifest. Predictive maintenance can manage asset maintenance proactively and preemptively based on predicted failures, significantly enhancing operation uptime and optimizing productivity.

Anomaly detection involves identifying unusual elements, events, or observations in data sets, whether in raw form or as streams. This is particularly important for sensors used in the industry. Anomaly detection techniques can be categorized into three types based on the nature of the trainable models used: unsupervised, supervised, and semi-supervised. However, it's important to note that anomaly detection is a side effect for both unsupervised and semi-supervised methods, as these techniques are generally not designed to account for anomalies. Furthermore, some methods force data allocation into groups, which makes anomaly detection impossible.

The remainder of this paper will first present some background information in the context of Industrial IoT and IoT-based Predictive Maintenance approaches, followed by a short description of the SCAMP-ML project[1] and the solution that was selected for the SCAMP-ML platform, targeting stream-processing and ML activities.

---

[1] Advanced computational statistics for planning and tracking production environments.

2   Background Information

2.1   Industrial IoT Architectures

The IoT has received considerable attention in the last ten years due to various initiatives and projects focused on it, intending to harness its potential to create and deliver value to multiple sectors and fields, such as smart cities, agriculture, health, energy, and the environment. However, IoT also involves issues and threats like privacy, security, ethics, and governance.

The European Union has become a critical player in this process as one of the main drivers of IoT research and innovation agenda, providing funding and coordination for several projects and initiatives that address the technical, social, and economic aspects. IoT is a diverse and rapidly growing field, encompassing various domains, stakeholders, and technologies, requiring a shared vision, architecture, and platform to enable interoperability, scalability, and innovation.

With the emergence of IIoT and the concept of Industry 4.0, new architectural models have been proposed to capitalize on the unique features of enabling technologies, such as cyber-physical systems (CPS), wireless sensor networks, cloud and edge computing, or big data, coupled with methods of artificial intelligence, quantum computing, and others for the development of the next generation in the context of the expected Industry 5.0.

RAMI 4.0 (Reference Architectural Model Industrie 4.0) is a framework for designing and implementing IoT solutions in the industrial sector. It was initially developed by the German Electrical and Electronic Manufacturers' Association (ZVEI) to support Industrie 4.0 initiatives. Its core features are explained in [2,3]. According to the *DIN SPEC 91345:2016-4* specification, RAMI is "a three-dimensional layer model, which depicts a basic architecture for Industry 4.0 using a sophisticated coordinate system." Additionally, the model introduces the concept of Industry 4.0 component and "shows on which level and in which phase each *Industry 4.0 component* plays a role, and where that component interacts with other components at other levels" [3].

The IIRC (Industrial Internet Reference Architecture), developed by the Industry IoT Consortium[2], defines a "standards-based open architecture for IIoT systems", an architecture framework, where four different viewpoints are defined, the implementation viewpoint offering specifications for a large number of architectural and Design Patterns for cross-industry use[3] [4].

The Internet of Things (IoT) is a rapidly evolving field that requires various architecture approaches to address its challenges and opportunities. Some of these approaches are provided by cloud service providers, such as the Azure IoT Reference Architecture [5], which defines a set of best practices and design patterns for building scalable and secure IoT solutions on Azure.

The AWS Event-driven reference architecture [6] demonstrates how to use AWS-based services to ingest, process, and analyze IoT data in real-time, illustrated by some relevant IoT sensor-based use cases, showing how an event-driven

---

[2] https://www.iiconsortium.org/.

[3] https://www.iiconsortium.org/patterns/.

architecture can be applied to different domains, such as medical device data collection, industrial IoT, connected vehicle, and sustainability and waste reduction.

Other approaches are proposed by industry consortia or research groups, such as the IoT Architectures and the related software stacks, as defined in an Eclipse whitepaper [7], which presents a comprehensive overview of the IoT landscape, including use cases, architectural patterns, and software stacks. These approaches offer essential references for further developments for IoT-based applications, as they provide guidance, insights, and recommendations for designing and implementing IoT systems.

2.2 Predictive Maintenance

In [8], R. Keith Mobley defines predictive maintenance (PdM) as "a philosophy or attitude that uses the actual operating condition of plant equipment and systems to optimize total plant operation." This definition reflects the evolution of PdM from preventive maintenance and condition-based maintenance practices. However, in the Internet of Things (IoT) era, PdM has gained a new dimension, as it leverages data mining, artificial intelligence, and ML to automatically analyze and predict the health and performance of the equipment and to optimize maintenance actions. Moreover, considered an important application of IoT, especially in the manufacturing industry, PdM benefits from the integration of cloud computing, edge computing, and digital twin technologies, which enable scalable, distributed, and real-time analysis and decision-making [8,9].

An overview of the different ML approaches considered for PdM was conducted in [10], showing that Support Vector Machines, Artificial Neural Networks, Random Forests, Deep Learning, and k-means have been considered for different PdM applications. The paper [11] presents a predictive maintenance methodology based on a PdM ML approach (Random Forest) on a cutting machine, where, based on inputs from a data management tool thus, dynamic decision rules can be adopted for maintenance management.

An overview of ML approaches for predictive maintenance in the context of the automotive industry is offered in [12], showing that Artificial Neural Networks, with fully labeled data sets, are the preferred approach. Another evaluation performed in [13] concluded that RandomForest (bagging ensemble) and XGBoost (boosting) outperformed individual algorithms in their experiments.

3   The SCAMP-ML Approach

SCAMP-ML is a research project that aims to move towards Industry 4.0 of heterogeneous production environments by using smart sensors and human-machine interfaces to collect and analyze data from different types of production processes. The project proposes a non-invasive approach, with minimal costs and logistical efforts, that does not require modifying the existing equipment or software. The project also uses advanced computational statistics and ML techniques to provide insights, predictions, and recommendations for improving production efficiency and quality. There are three significant objectives defined in the context of this project.

First, the development of a generic framework for data acquisition, processing, and analysis in heterogeneous production environments. This framework uses smart sensors and human-machine interfaces to collect data from the various components involved in the production process. It transforms the data into generic models that can be processed and analyzed unitarily, regardless of the specifics of the industry or the equipment.

Second, investigate the use of advanced computational statistics and ML methods to the data collected to provide insights, predictions, and recommendations for improving production efficiency and quality. These methods include, for example, anomaly detection, fault diagnosis, predictive maintenance, optimization, etc.

Third, validate and demonstrate the applicability and effectiveness of the proposed solution in real-world industrial scenarios involving different types of production processes and domains.

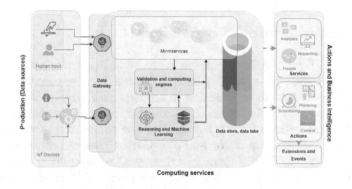

Fig. 1. A Simplified SCAMP-ML Architecture

Thus, it is in SCAMP-ML's intention to approach several challenges, like:

- Achieving interoperability between different types of equipment and software, especially those from older generations, which do not have standard data acquisition interfaces or require substantial integration effort and additional licensing costs.
  Therefore, the SCAMP-ML investigated a unitary and homogeneous approach that essentializes the production process and structures the data collected in a generic way, applying the same principles and standards regardless of the specifics of the industry in which the solution is applied. This approach enables the compatibility and integration of various kinds of equipment and software and facilitates data collection and processing;
- Overcoming data fragmentation, caused by the heterogeneous structure of the production environment, which makes it difficult to correlate and analyze

the data collected from various components involved in production processes. SCAMP-ML proposal is to enable the transformation of the acquired data at the time of acquisition into generic models that can be processed and analyzed unitarily. This proposal allows the consolidation and standardization of the data collected and enhances the data quality and usability;

– The lack of adequate analysis tools, especially in less technologized environments, but also in those where technologization took place in stages, the analysis tools are missing, are rudimentary, or are particularized to the stage of technologization that introduced them.

This challenge limits the ability and opportunity to perform a complex and consistent analysis that correlates the entire set of co-dependent activities carried out in the enterprise and to provide insights, predictions, and recommendations for improving production efficiency and quality. In contrast, the SCAMP-ML project uses advanced computational statistics and ML techniques to analyze the collected data and provide valuable information and feedback for the production process.

4   The Architecture of SCAMP-ML

The SCAMP-ML architecture that was depicted in Fig. 1 was partly inspired by one of Azure solution ideas for PdM for IIoT[4]. Even if native communication with the Data Gateway and data provision in the expected stream format is desired by the IIoT devices, the deployed solution will consider either supporting device data aggregation prior to feeding this data to the computing services or offering the means for raw data acquisition and preparation, near the computing services.

At the core of SCAMP-ML's architecture are the computing services, implementing a series of execution scenarios, including: a) Data pipelines for sensor data analysis using stream-processing components, handler for Time Series databases or similar solutions, supporting event detection services – such as event detection engine (EDE) Training and EDE Prediction –, and Event Handler components; b) Data ingestion, which is done in a streaming fashion using microbatches, which can handle high-speed data; c) Data analysis involves training predictive models on historical data, detecting anomalies, events, or production cycles on live data, and generating reports for distribution.

Data ingestion is based on data generated by the IIoT devices (sensors) and fed into specialized stream-processing software, to push raw or filtered data toward specialized handlers, such as a Time Series handled when a Time Series data repository is included in the deployed solution. A detailed view of the services required for EDE support is offered in Fig. 2.

In the Scamp-ML framework, obtaining data from devices is a challenging task. For example, sensors can provide monitoring data in various formats, either

---

[4] https://learn.microsoft.com/en-us/azure/architecture/solution-ideas/articles/iot-predictive-maintenance.

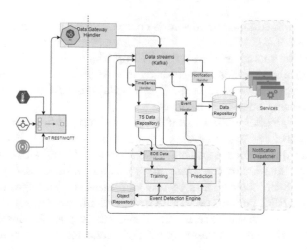

Fig. 2. Data Flows and Data Ingestion in the SCAMP-ML platform

structured, semi-structured, or unstructured, and use different protocols to transmit it (such as MQTT, AMQP, REST-based, and others). Specialized services like the *Data Gateway Handler* from Fig. 2 are necessary to handle, transfer, and prepare this data.

Data Gateway Handler communicates this data (raw or preprocessed) to a distributed and scalable platform of the publish-subscribe type, which supports data pipelines and streaming. This enables the data to be distributed to all the other components (usually specified as handlers) that are responsible for processing information, such as *TimeSeries Handler* that stores data persistently or *Event Handler* that retrieves data. Moreover, this component can also support exposing various data and products generated by the Scamp-ML components.

Data provided by sensors, even if compliant with a specific protocol, requires additional processing to support activities carried out in *Event Detection Engine*'s context. This prevents possible problems related to data incompatibility with the format assumed by the ML methods used. Therefore, in addition to activities associated with ingestion mechanisms, a data adaptation level (*EDE Data handler*) is provided, which ensures data compatibility with formats required for training and prediction activities.

This data transformation does not consider the data source, allowing the presentation of data frames to functional ML mechanisms. However, if data incompatible with data frame transformation are provided, raw data exploitation, collected through *Data Gateway Handler*, is considered. The preprocessing can be based on the usual types of operators (scaling/normalization, one-hot encoding, augmentation such as SMOTE) or user-defined operators to allow better flexibility of this component.

The *EDE Training* component is used for training predictive models. It has implicit support for typical algorithms from integrated libraries, allowing the user to define the analysis methods and their parameters. It also offers the possibility of adding user-defined analysis methods compatible with the main API (mainly, the scikit-learn[5] ones).

The prediction model trained by *EDE Training* is sent to *Object Repository*, a specialized database dedicated to storing models and metadata associated with them, such as optimization reports, XAI (explainable AI, if activated at EDE level), preprocessing operators used, etc.

Finally, *EDE Predict* is used to instantiate trained prediction models to attain anomaly detection and production cycles in near real-time. Prediction model results generate JSON messages that describe the detected situations, delivered through *Event Handler*, to the *Data Streams* and *Data Repository* facilities to allow other platform components to use and integrate these results quickly. In essence, as each EDE subcomponent is independent, we can scale each of them as required. We can instantiate a new detection or training instance for each new device or data source. Various execution scenarios were included in [14].

5   Validation and Experiments

We aimed to validate our proposed solution in a realistic production environment. In order to achieve this, we deployed all relevant components on a custom Kubernetes instance to achieve this. We then gathered power consumption data from at least ten hydro-forming machines. In a previous work, we outlined the problem domain and some of the experimental outcomes [14]. The problem domain for validation remains the same: detecting production cycles from power measurements. Several ML techniques have been tested. A detailed description of predictive model stability and hyper-parameter choices were at the basis of the methodology described in [14]. This methodology stands at the basis of the analysis of detected cycles for finding anomalies that can be useful for predictive maintenance scenarios.

Real data streams of power consumption from hydro-forming machines were used to validate our findings. We focused on a specific time interval of 3 months (June to September 2023), where manual annotations for a set of five machines were performed. These annotations include the number of cycles, scrapped items, cycle length, and unplanned production stops. With a high polling rate of 10 measurements per second, each machine was able to generate about 5 GB of data, resulting in more than 25 GB of annotated time-series data.

As one of our intentions was to have a validation of the analytics platform in terms of detection and performance, we exploited the annotations to score the predictive performance, and measured the execution times of selected functions. These functions included data loading, windowing, distance calculation, heuristic and ML method training, and inference. By profiling the execution times, one can assess the analytics solution's performance on constrained hardware, which is typical for Edge/Fog solutions and CPS.

---

[5] https://scikit-learn.org/stable/modules/classes.html.

It is worth mentioning that solutions like Intel RAPL[6] can be employed to measure both power consumption and execution time. However, these two metrics are directly proportional because we used no special acceleration or optimization techniques.

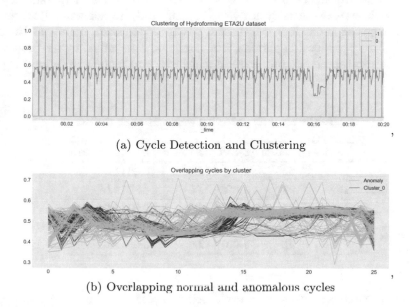

(a) Cycle Detection and Clustering

(b) Overlapping normal and anomalous cycles

Fig. 3. Detected cycles annotated with clusters (a) and overlapping cycles (b)

5.1   Results

Firstly, a template cycle for each dataset to detect cycles was defined. The only input needed for this was the approximate cycle length given in the annotations. Recurring motifs in a sub-sample of the data were identified, usually corresponding to about 2 h of monitoring data. Thus, five candidate templates were obtained. However, users can define their own templates.

Next, the distance between the template and the possible cycles in the data was computed, using Dynamic Time Warping (DTW). DTW was considered as it can handle misalignment between cycles, which makes it suitable for further ML-based analysis. Other methods, such as Euclidean distance, are too strict and may miss anomalous cycles that are relevant to our case. Figure 3(a) shows the detected cycles with vertical lines.

For cycle type detection, our choice was for HDBSCAN, while for anomaly detection, IsolationForest [14]. The minimum cluster size was set to 100, the distance metric to DTW, and the minimum samples to 1 for HDBSCAN. Also,

---

[6] https://web.eece.maine.edu/~vweaver/projects/rapl/.

the number of estimators was set to 100 and used all features for Isolation-Forest. These parameters were chosen based on the available data character-istics and our previous findings, from [14]. Figure 3(a) shows a sub-sample of the detected cycles, with green lines indicating clusters and red lines indicating noise. Figure 3(b) shows the overlap of all detected cycles in the sub-sample. One may observe that the cycles varied significantly in length and amplitude. Additionally, the anomalous cycles tend to cluster together, mainly at the start and end of long production periods, which may be determined by the production line startup and shutdown processes. Each cluster was mapped to a specific piece type.

Table 1. Cycle detection scores for each device

Device	precision	recall	specificity	f1	geometric
168	0.92	1.00	1.00	0.96	1.00
169	0.90	1.00	1.00	0.95	1.00
170	0.82	1.00	1.00	0.90	1.00
172	0.61	1.00	1.00	0.76	1.00
321	0.91	1.00	1.00	0.95	1.00

The probable root cause of a cycle being anomalous is presented in Figs. 4, while Fig. 4(a) shows global feature impact, where each feature represents a data point from a cycle. This highlights which part of anomalous cycles is most effective in determining anomalies. Figure 4(b) illustrates a comparison of the DTW distance calculations between the template cycle (top) and an abnormal cycle (left). The combination of these two visualizations shows how we can easily determine what part of a cycle is most anomalous and, in theory, can lead to aid in PdM-related issues as each part of the production cycle can indicate which component might require maintenance.

The overall scores obtained for all devices used for testing are included in Table 1. Although these are unsupervised methods, we used the F1 score for validation as access to labeled data was provided. This was performed to gauge the overall effectiveness of the predictive models and was not used during train-ing. Supervised ML methods give better results in most scenarios where labeled data is available; however, it is not typical for the scenario evaluated here. Most devices have relatively high F1 scores. Device 172 is an outlier from these scores, which can be explained by the fact that it had extended periods of inactivity with a high number of short production cycles. Likely, there were insufficient data points to train a reliable predictive model for this production line.

Finally, Fig. 5(a) shows cycle detection in case of high and low thresholds used for DTW distance calculations for the entire annotated period for all mon-itored devices. As expected, a high minimum distance leads to fewer correctly detected cycles. This is partly due to the high variability of cycle phasing and

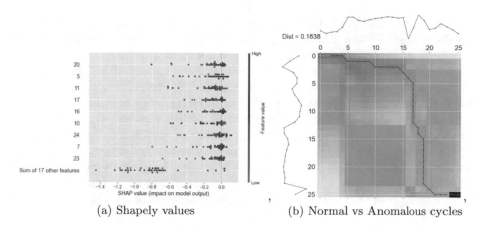

(a) Shapely values   (b) Normal vs Anomalous cycles

Fig. 4. Cycle Shapely value computation (a) and DTW distance visualization

the presence of anomalous cycles. Figure 5(b) shows the number of detected anomalies when using the contamination estimates hyperparameter for Isolation Forest. Performance was relatively good. However, more anomalous events were incorrectly identified, either as a result of overdetection or underdetection. The reason for this issue can be explained by the tendency of the selected detection method to overfit, leading to poor out-of-sample performance. The solution to this is to expand the available data and try different detection methods that are less susceptible to overfitting.

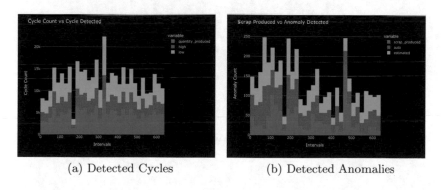

(a) Detected Cycles   (b) Detected Anomalies

Fig. 5. Detected Cycles (a) Anomalies detected (b) entire period

We measured then the execution times of each operation in our experiments. Figure 6 shows some of the execution times with different settings. We re-sampled the data to one second in experiments $e2$ and $e5$, and used different heuristics and thresholds in experiments $e1$, $e3$, $e6$ and $e9$. The heuristics, that were previously explained in [14], aimed to avoid overlapping cycles, which could occur due to our interest in detecting anomalous cycles, not just the best-fitting ones.

Training of the predictive models were the most time-consuming operation, in all situations. We also may notice the execution time of the windowing function in Fig. 6(a). The initial implementation was a simple loop that created windows with the same length as the target cycle and a step size of 1. This was fast for small datasets, taking less than one second. However, it took over 1,500 s for the 25 GB of data. A parallelization of the windowing function, using eight threads, eventually reduced the time to around 250 s. However, this method had some drawbacks: first, limited hardware could not parallelize the operation as much; second, this solution was hard to scale. We used the *sliding_window_view* specialized function from NumPy[7], which is a tailored algorithm for our use case. This method was able to achieve a linear scaling of execution time with respect to the input size $\mathcal{O}(N)$. This resulted in less than 1 s execution time, as seen in Fig. 6(b).

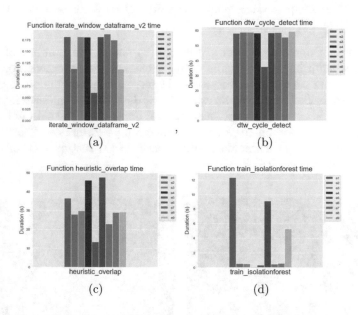

Fig. 6. Execution time (x-axis) for cycle detection and analytical operators

6   Conclusion

This paper offers an overview of the SCAMP-ML platform, with an emphasis on the computing services of the platform, which aims to achieve two main goals of the project: the development of a generic framework for data acquisition, processing, and analysis in heterogeneous production environments, and the use of ML methods applied on the collected data.

---

[7] https://numpy.org/.

Our contribution focuses on the data ingestion and analysis aspect and highlights the following features: (a) The capability to collect (ingest) data from various sources/sensors and store them in a time-series or similar database; (b) The use of a "publish-subscribe" component to enable a scalable and distributed system; (c) The incorporation, in the analysis and detection component (the Event Detection Engine - EDE) of both data preprocessing and predictive modeling to support different scenarios, such as anomaly detection, event classification, production cycle detection, and more; (d) The possibility to enrich EDE with user-defined features; (e) The scalability of EDEs execution system, both vertically and horizontally, due to the subcomponents that generate and implement the predictive models. (f) The ability to submit detected events to different platform components so that they can be utilized by users of SCAMP-ML and other components. (g) Validation on real-world data and execution optimizations.

Acknowledgement. The Romania Competitiveness Operational Programme partially supports this paper under project number SMIS 120725 - SCAMP-ML (Advanced computational statistics for planning and tracking production environments) and UEFISCDI COCO research project PN III-P4-ID-PCE-2020-0407.

References

1. Khan, W.Z., Rehman, M.H., Zangoti, H.M., Afzal, M.K., Armi, N., Salah, K.: Industrial internet of things: recent advances, enabling technologies and open challenges. Comput. Electr. Eng. **81**, 106522 (2020)
2. Lydon, B.: RAMI 4.0 reference architectural model for Industrie 4.0. *InTech*, April 2019
3. Reference architecture model Industrie 4.0 (RAMI4.0). Technical report DIN SPEC 91345:2016-04, Deutsches Institut fur Normung E.V. (DIN) (2016)
4. The industrial internet reference architecture. Technical report, Industry IoT Consortium (2022)
5. Cosner, M., Garcia, A.B.: Azure IoT reference architecture. White paper, Microsoft (2023)
6. Sodabathina, R., Shan, J., Ulloa, M.: Building event-driven architectures with IoT sensor data. AWS architecture blog, Amazon Web Services (2022)
7. Eclipse IoT Working Group. The three software stacks required for IoT architectures. White paper (2016)
8. Keith Mobley, R.: Introduction to Predictive Maintenance. Plant Engineering, 2nd edn. Elsevier Science & Technology, Oxford (2002). Description based on publisher supplied metadata and other sources
9. Passlick, J., Dreyer, S., Olivotti, D., Grützner, L., Eilers, D., Breitner, M.H.: Predictive maintenance as an internet of things enabled business model: a taxonomy. Electron. Mark. **31**(1), 67–87 (2020)
10. Carvalho, T.P., Soares, F.A.A.M.N., Vita, R., Francisco, R.P., Basto, J.P., Alcalá, S.G.S.: A systematic literature review of machine learning methods applied to predictive maintenance. Comput. Ind. Eng. **137**, 106024 (2019)
11. Paolanti, M., Romeo, L., Felicetti, A., Mancini, A., Frontoni, E., Loncarski, J.: Machine learning approach for predictive maintenance in industry 4.0. In: 2018 14th IEEE/ASME International Conference on Mechatronic and Embedded Systems and Applications (MESA), July 2018. IEEE (2018)

12. Theissler, A., Pérez-Velázquez, J., Kettelgerdes, M., Elger, G.: Predictive maintenance enabled by machine learning: use cases and challenges in the automotive industry. Reliab. Eng. Syst. Saf. **215**, 107864 (2021)
13. Ayvaz, S., Alpay, K.: Predictive maintenance system for production lines in manufacturing: a machine learning approach using IoT data in real-time. Exp. Syst. Appl. **173**, 114598 (2021)
14. Iuhasz, G., Panica, S., Duma, A.: Cycle detection and clustering for cyber physical systems. In: Barolli, L. (eds.) Advanced Information Networking and Applications, AINA 2023. LNNS, vol. 655, pp. 100–114. Springer, Cham (2023). https://doi.org/10.1007/978-3-031-28694-0_10

From Fault Tree Analysis to Runtime Model-Based Assurance Cases

Luis Nascimento[1]($\boxtimes$), Andre L. de Oliveira[1], Regina Villela[1], Ran Wei[2],
Richard Hawkins[3], and Tim Kelly[1]

[1] Universidade Federal de Juiz de Fora, Juiz de Fora, Brasil
`{luis.felipe.almeida,andre.oliveira,regina.braga}@ice.ufjf.br`
[2] University of Cambridge, Cambridge, UK
`rw741@cam.ac.uk`
[3] The University of York, Cambridge, UK
`richard.hawkins@york.ac.uk`

Abstract. Cyber-physical systems (CPSs) demand the justification of system safety and component reliability. Assurance cases provide an explicit means for justifying/assessing confidence in system dependability with explicit and implicit references to design, safety, and reliability artifacts. Fault Tree Analysis (FTA) is one of the most popular safety analysis techniques, which is an integral part of Model-Based Safety Analysis. Conversely, the open and adaptive nature of CPSs, demands a paradigm shift from design-time to runtime system assurance. Although the Structured Assurance Case Metamodel (SACM) provides the foundations for runtime system assurance, enabling the traceability between assurance cases and FTA results, which is part of Executable Digital Dependability Identities (EDDIs) of CPS components, it is still a barrier. In this paper, we introduce a model-driven methodology to support the synthesis of assurance cases from FTA as another step towards demonstration of CPS safety and component reliability at runtime.

1 Introduction

Cyber-physical systems (CPS) integrate physical processes and computer systems that contain sensors to observe the environment and actuators to influence physical processes [1]. They have vast economic potential and societal impact, enabling new types of promising applications in different embedded systems domains such as automotive, avionics, railway, healthcare, and home automation [4]. These systems perform safety-critical functions, and if they fail may harm people or lead to the collapse of important infrastructures with catastrophic consequences to industry, and/or society [5]. Therefore, safety-critical CPS domains such as automotive require the justification and demonstration of systems dependability. Assurance cases provide an explicit means for justifying and assessing confidence in safety, security, and other dependability properties of interest [6]. An assurance case supported by evidence is the key artifact for safety and security acceptance of systems before their release for operation.

The required evidence of system assurance can be provided by applying safety and reliability analysis techniques. Fault Tree Analysis (FTA) [3] and Failure Modes and

L. Barolli (Ed.): AINA 2024, LNDECT 200, pp. 451–464, 2024.
https://doi.org/10.1007/978-3-031-57853-3_38

Effects Analysis (FMEA) are among the most popular techniques. They support engineers in determining the ways a system can fail and the likelihood of those failures. FTA and FMEA form the integral parts of Model-Based Safety Analysis (MBSA). However, due to the highly open and adaptive nature of CPSs, i.e., systems can connect at runtime, and are capable of adapting their behaviors to changing contexts, the evidence and justification of system assurance need to change from design time to runtime. With this shift, we will have a transition from the current design time assurance cases produced from manually created artifacts to assurance case models that can be automatically synthesized and evaluated at runtime [9]. To achieve this goal, the CPS or a CPS component is equipped with all information that uniquely describes its dependability characteristics (design, analysis, and process models) within a Digital Dependability Identity (DDI) [4].

The Structured Assurance Case Metamodel (SACM) [14] issued by the Object Management Group (OMG) provides the foundations for runtime model-based system assurance and in order to ensure the dependability of CPSs the assurance cases are expected to be exchanged, integrated, and verified at runtime [7]. SACM defines a standardized metamodel and visual notation for representing structured assurance cases, which was developed to support interoperability between different assurance case approaches (e.g., GSN [6], CAE [8]). SACM was used in the DEIS project [4] as a backbone for its Open Dependability Exchange (ODE) metamodel, which defines the appropriate format of a DDI, in the first step towards runtime assurance of CPSs. Assurance case patterns are useful in capturing good practice in system argumentation for re-use, by defining the required system information to instantiate abstract assurance claims and evidence to support those claims [10].

Although SACM and ODE metamodels provide the backbone for the assurance of CPSs at runtime, it is still a barrier enabling the traceability between the assurance case and FTA results, which is part of the Executable Digital Dependability Identity (EDDI) of open and adaptive CPS components, at runtime. To fill this gap, in this paper, we introduce a model-driven methodology to support the automated synthesis of assurance cases from Fault Tree Analysis results to demonstrate CPS safety and CPS component reliability at runtime. We applied the proposed methodology to generate assurance case cases from FTA results of an automotive wheel braking system to evaluate its effectiveness. In Sect. 2, we provide an overview of ISO 26262 safety life-cycle, Fault Tree Analysis, Digital Dependability Identity, SACM, and its pattern extensions needed for the reader to understand the contributions of this paper. In Sect. 3, we introduce a novel model-driven methodology to synthesize assurance cases from FTA results. In Sect. 4, we illustrate the evaluation in the automotive domain. In Sect. 5, we discuss the related work, and we present the conclusions and sketch future work in Sect. 6.

2 Background

In this section, we provide an overview of the ISO 26262 safety standard and its lifecycle, and Fault Tree Analysis. We describe the concept of Digital Dependability Identity (DDI), and the Structured Assurance Case Metamodel (SACM) needed for the reader to understand the contributions of this paper.

2.1 ISO 26262 Safety Lifecycle

Standards define the concept of Safety Integrity Levels (SILs) to allocate requirements to mitigate hazard, subsystem, and/or component failure effects on the overall system safety. The IEC 61508 [25] industry standard defines four different levels of safety integrity, ranging from the least stringent SIL 1 to the most stringent SIL 4. The most stringent SILs demand significant verification and validation effort such as performing control-flow and data-flow testing to be achieved. On the other hand, achieving the least stringent SIL demands less rigorous verification activities such as software inspection. Standards also define a risk-based approach, e.g., based on severity and likelihood levels, to classify the risk posed by system hazards and component failures in a given SIL, and they prescribe requirements and provide guidance to achieve safety per SIL. It also defines rules for decomposing safety requirements allocated to mitigate hazard effects (in the form of SILs) into requirements to mitigate architectural subsystems and component failures.

In the automotive domain, the ISO 26262 [12] standard defines the requirements for ensuring functional safety in electrical/electronic systems of small and medium-sized general-purpose road vehicles (up to 3.5 tons). This standard introduces a comprehensive safety lifecycle, a risk-based approach to determine Automotive Safety Integrity Levels (ASILs), requirements for validation, and confirmation measures to ensure an acceptable level of safety. At the concept phase (ISO 26262 Part 3), firstly, functions are allocated to systems, subsystems, and components, which can be software, electric, or electronic at the item definition (Sec. 3.5). Then, item dependencies and their interactions with the environment at the vehicle level are specified to initiate the safety lifecycle (Sec. 3.6). After the initiation of the safety lifecycle, Hazard Analysis and Risk Assessment (HARA) is performed to identify and categorize the hazards that malfunctions the item (Sec. 3.7) at the system level, e.g., using HAZard and OPerability studies [2]. Safety goals, in the form of ASILs, are then allocated to eliminate or minimize hazard effects, and decomposed into functional and technical safety requirements, at the Functional Safety Concept (Sec. 3.8), e.g., using Fault Tree Analysis (FTA) and Failure Modes and Effects Analysis (FMEA) safety analysis techniques. System safety requirements, e.g., fault prevention, detection and removal, tolerance, and/or forecasting, assigned to mitigate hazard effects are preliminary allocated to architectural elements (i.e., Functional Safety Requirements - FSRs) of the item to mitigate hazardous behaviors into subsystems. FSRs allocated to architectural subsystems are finally decomposed into technical safety requirements (TSR) to mitigate component failure effects on safety.

2.2 Fault Tree Analysis

Fault Tree Analysis (FTA) is a widely used, top-down reasoning technique for safety and reliability analysis [3]. FTA is based on a visual notation to support the analysis of combinations of fault propagation paths (i.e., using AND, OR, NOT gates), comprising intermediate and basic events, that can lead the system to fail (top event), causing harm to people, environment, or property. The starting point of this analysis is a given top event (i.e., a system failure). Then, the possible combinations of causes that could trigger the top event (system failure) are progressively explored until individual component failures (basic events) have been reached. The FTA provides support for qualitative (logical) and quantitative (probabilistic) analysis. FTA requires the use of extensions to model dynamic relationships between causes and component failures (i.e., modeling the dynamic failure behavior). Fault trees have been used for diagnosis purposes in runtime applications. In [13], sensor readings are connected to "complex basic events" (CBE) as inputs to the fault tree analysis. The failure model is updated in real-time to provide diagnoses, predictions of failures, and reliability evaluation at runtime. Runtime FTA is the first step towards the integration of fault trees into the system Executable Digital Dependability Identity.

2.3 Digital Dependability Identity and ODE Metamodel

A Digital Dependability Identity (DDI) comprises all the information that uniquely describes the dependability characteristics (design, analysis, and process models) of a CPS or a CPS component [4]. DDIs provide the basis for automated and dynamic integration of independent systems into systems of systems in the field at runtime. The Open Dependability Exchange (ODE) metamodel defines the appropriate format of a DDI. Therefore, the ODE metamodel provides the basis for the representation and exchange of safety information (e.g., FME(D)A, FTA, and Markov Chains) between open-adaptive CPSs and CPS components. The SACM metamodel is the backbone of the ODE, providing the formal traceability between assurance cases and DDI information. Therefore, SACM assurance case patterns can be instantiated based on the information provided by CPS and CPS component DDIs.

2.4 SACM and Its Pattern Extensions

The SACM [14] was developed to support interoperability with well-established assurance case frameworks such as GSN [6] and CAE [8]. It defines a metamodel and a visual notation for representing assurance cases. The SACM metamodel has five packages. The **Structured Assurance Case Base Classes** captures the foundational concepts of SACM elements throughout the rest of the metamodels. **Structured Assurance Case Terminology package** defines the concepts of Term, Expression, and external interface, which support the definition of a controlled and reusable vocabulary referred to in the argumentation of an assurance case. **Assurance Case Packages** define the concept of modularity in assurance cases. **Argumentation Metamodel** defines the abstractions for representing structured arguments. **Artefact Metamodel** specifies the concepts for structuring evidence in assurance cases.

SACMElement contains the basic properties shared among all elements and contains a + *isAbstract* Boolean flag to indicate whether or not it is considered abstract. The ModelElement refines SACMElement with references to Description and ImplementationConstraint utility elements. Thus, a ModelElement can contain a multi-language Description, which may contain Terms and Expressions that provide its content, e.g., a Description that provides the text of a Claim. ModelElement can also contain zero or more ImplementationConstraints, which specify the details of a constraint that must be satisfied to convert a referencing ModelElement from isAbstract = true to isAbstract = false. The abstract Terms referenced in ModelElement's Descriptions have an + externalReference property with the *url* to external artifacts (files, models), which are part of the

CPS component DDI. In addition, ImplementationConstraints with queries, written using computer languages such as Object Constraint Language (OCL) and Epsilon Object Language (EOL), to DDI models (e.g., fault trees) provide the traceability between the assurance case and the evidence (in this case, an ODE-compliant model). A structured SACM argument is expressed through Claims, citations of artifacts, or ArtifactReferences (e.g., Evidence or Context for Claims), and relationships between these elements, expressed via AssertedRelationships. A detailed description of SACM metamodel and its visual notation is available in [14].

3 Proposed Methodology

In this section, we introduce a new SACM pattern extension, named Children implementation constraint subtype (Subsect. 3.1), and a methodology to support the generation of assurance cases from FTA results (Subsects. 3.2 and 3.3).

3.1 SACM Pattern Extension Add-On

The SACM pattern extensions presented in [15] provide the basis for the specification of executable assurance case patterns by adopting four ImplementationConstraint subtypes. **Mapping (p)** constraint is used to link the + *value* slot of an abstract Term to the information from an external artifact. **Multiplicity (m)** denotes zero-or-more n-ary cardinality of an abstract SACM ModelElement used for pattern instantiation. **Optional (o)** denotes zero-or-one n-ary cardinality of an abstract ModelElement. **Choice (c)** is used to denote possible alternatives (choices) in asserted relationships. However, these constraints do not support the recursive instantiation of abstract terms. Thus, this gap needs to be fulfilled to enable the recursive instantiation of abstract terms based on the hierarchical structure of FTA results. To address this issue, we introduce the Children implementation constraint subtype detailed in the following.

Children (s): This constraint subtype can be attached to abstract Terms. *Semantics*: It has similar semantics of the mapping (p) constraint, i.e., mapping abstract Terms to an element from an external artifact via model query. Children are used to indicate hierarchical relationships (i.e., parent and child) between instances of the same abstract Term retrieved from an external artifact. A children constraint is used to obtain the child elements of each model element retrieved from an external artifact by executing the queries stored into mapping (p) and children (s) constraints associated with the given abstract Term t. A children (s) ImplementationConstraint manipulates the result of a mapping (p) constraint. For this reason, a children constraint should be used in conjunction with a mapping constraint.

Example: consider an abstract *Term*, and recursive hierarchical relationships between components (parent) and sub-components (child) in the design model (i.e., an external artifact), a query that captures hierarchical relationships between components and sub-components of a design model can be specified as *Component.all.selectOne(c| c.name = '$parent').subcomponents.collect(s| s.name)*. This type of query is attached to the + *content* slot of a children (s) ImplementationConstraint subtype. The *$parent* query parameter could be an element retrieved from an external artifact via execution of a mapping (p) query, or a child element of the tree data structure (i.e., a child element of a component that could be a leaf node, or having other child sub-components). This is similar to the relationships between the top, intermediate, and basic events of a fault tree.

3.2 Assurance Case Pattern Specification

The inputs of this phase are the Hazard avoidance [10] and Hazardous Software Failure Mode (HSFM) [16] assurance case pattern catalog. In this phase, engineers specify the structure of the assurance case pattern using the SACM visual notation [14]. SACM assurance case modeling tools such as ACME [7] and ACEditor [15] can be used to support the specification of assurance case patterns with explicit links to evidence. This phase encompasses six steps detailed in the following:

Step 1: Specify the vocabulary. *Description:* In this step, we specify the placeholders using abstract SACM terminology elements (i.e., Terms and Expressions), and the textual information using concrete SACM Expression elements, which constitute the vocabulary of the targeted Hazard Avoidance and HSFM argumentation patterns used to build the assurance case pattern structure. Still in this step, we specify the relationships between abstract and concrete terminology elements to define expressions to be used as descriptions of SACM Claims, Reasoning, and Artifact Reference argumentation elements. Abstract SACM Terms and Expressions, via + *origin* and + *elements* properties respectively, provide the context to enable the automated instantiation of the pattern. SACM abstract Term and Expression elements enable an assurance case pattern specification to be machine-readable, i.e., it enables the specification of mappings between abstract terminology elements to the concrete information from design, safety assessment, and process models needed to instantiate abstract assurance case argumentation elements. *Output:* assurance case pattern vocabulary comprising concrete and abstract Term and Expression elements.

Step 2: Specify the argumentation elements. *Input*: the assurance case pattern vocabulary, i.e., abstract (non-instantiated) and concrete SACM terminology elements. *Description*: in this step, we specify the claims, artifact references, reasoning elements, and their relationships, using SACM AssertedRelationships, to define the hierarchical structure of the Hazard Avoidance and HSFM assurance case patterns. *Output*: the hierarchical structure of the pattern.

Step 3: Specify mappings between argumentation and terminology elements. *Input:* the assurance case pattern vocabulary and argumentation structure. *Description*: here, we define the description of each argumentation element (Claim, Reasoning, and Artifact Reference) specified in the pattern. A Description of an argumentation element includes explicit references to one or several SACM ExpressionElements. An expression element can be composed of both abstract and concrete SACM Term and Expression elements. *Output*: the assignment of descriptions to each SACM Claim, Reasoning, and Artifact Reference from the assurance case pattern.

Step 4: Specify implementation constraints associated with argumentation elements. *Inputs*: the assurance case pattern vocabulary, structure, and SACM argumentation elements (Claim, Reasoning, and Artifact Reference) enriched with SACM Description elements. *Description*: here, we assign Multiplicity and Optional SACM Implementation Constraint subtypes, defined into the SACM pattern extensions [15], to abstract Claims, Reasoning, and/or Artifact Reference assurance case pattern argumentation elements. In this step, we also assign SACM Choice implementation constraint to abstract SACM Asserted Relationship elements from the assurance case pattern. *Output*: assurance case pattern specification enriched with implementation constraints assigned to argumentation elements.

Step 5: Specify implementation constraints associated with terminology. *Inputs*: the assurance case pattern vocabulary, structure, and SACM argumentation elements (Claim, Reasoning, and Artifact Reference) enriched with SACM Description and implementation constraints elements. *Description*: here, we assign Multiplicity, Mapping, and **Children** ImplementationConstraint subtypes, defined into the SACM pattern extensions, to abstract Terms. In this step, we also assign SACM Multiplicity implementation constraint to SACM Expression elements. *Output*: assurance case pattern specification enriched with implementation constraints assigned to terminology elements.

Step 6: Specify mappings between abstract Terms and FTA results. *Inputs*: the assurance case pattern vocabulary with implementation constraints, structure, and SACM argumentation elements (Claim, Reasoning, and Artifact Reference) enriched with SACM Description, implementation constraint elements, and the ODE metamodel. *Description*: in this step, we define model-based queries for the Mapping and **Children** implementation constraint subtypes assigned to abstract Terms. These queries provide traceability links between abstract terms of the pattern and ODE FTA package metamodel using a computer language such as EOL to map the values of these terms to FTA results. *Output*: an assurance case pattern specification enriched with traceability links to FTA results.

3.3 Assurance Case Pattern Instantiation

This phase encompasses three steps that should be performed to generate an assurance case for a target system with references to FTA results.

Step 1: Performing system safety analysis. *Input*: system design. *Description*: in this step, the engineers must conduct safety analysis at both system, function, and component levels, e.g., using HAZOP at the system level to identify the potential hazards that malfunction the system, their safety risks, and safety goals; and fault tree analysis at function and component levels to identify how architectural subsystems and components may fail and contribute to the occurrence of hazards that may cause harm, and allocate functional and technical safety requirements to subsystems and components respectively. Model-Based Safety Assessment (MBSA) tools such as HiP-HOPS [17], OSATE AADL [19], CHESS framework [23], or EMFTA [24] can be used to support this step. *Outputs*: fault trees describing the fault propagation paths for each identified system hazard.

Step 2: Integrate the FTA results into the EDDI. *Inputs*: fault trees of each identified system hazard. *Description*: in this step, engineers execute a model transformation algorithm, produced by the authors, to convert the input fault tree models, e.g., specified using HiP-HOPs, into the Open Dependability Exchange (ODE) metamodel compliant fault tree format. For fault tree models produced using third-part MBSA tools other than HiP-HOPs, e.g., Component Fault Trees [18] OSATE AADL Error Annex [19], the engineers need to specify a model transformation to map the tool metamodel elements to the ODE elements. *Output*: ODE fault tree compliant models.

Step 3: Execute the instantiation program. *Inputs*: an assurance case pattern, and the ODE fault tree compliant models. In this step, engineers provide the pattern and the fault tree models to the assurance case pattern instantiation program, developed by the authors in [15], to synthesize the system safety argument based on the FTA results. *Output*: a product safety argument for the targeted system with references to FTA results (evidence).

4   Evaluation

In this section, we demonstrate the feasibility of our methodology in supporting the synthesis of a system assurance case from FTA results through executable SACM argument patterns for an automotive wheel braking system. However, due to space limitations, the results have not been fully described in this paper.

4.1   Braking System

HBS is a hybrid brake-by-wire system (Fig. 1) for electric vehicles propelled by four In-Wheel Motors (IWMs) taken from [11]. Hybrid means that braking is achieved through the combined action of electrical IWMs, and frictional Electro-Mechanical Brakes (EMBs). While braking, IWMs transform the vehicle's kinetic energy into electricity, which charges the power train battery, increasing the vehicle's range.

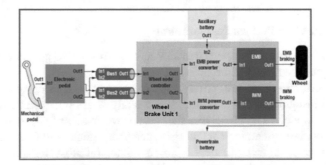

Fig. 1. Hybrid Braking System Architecture [11].

The system is activated when the driver presses the mechanical pedal. The Electronic Pedal component senses the driver's action, and it sends the braking forces, via a duplex bus system, to the Wheel Node Controller (WNC) of each wheel brake module. Each WNC generates commands to the power converters to activate EMB and IWM braking actuators. While braking, the power flows from the auxiliary battery to the EMB and from IWM to the powertrain battery. In the HBS system, due to interactions between wheel-braking system components, different hazards with different causes and criticality (i.e., ISO 26262 Automotive Safety Integrity Level - ASILs) can arise.

4.2 Assurance Case Pattern Specification

Here, we describe the application of the Assurance Case Pattern Specification Phase. Table 1 illustrates each step performed in this phase and its reflection on the SACM assurance case pattern model. In **step 1**, we built the pattern vocabulary by creating terms and expressions. The HSFMType abstract term is the abstraction for the top and intermediate events of a fault tree. The abstract term HSFMEvent represents the basic events of a fault tree, and it has a reference to the HSFMType term via its + *origin* property. Expressions are created referencing these terms through their + *element* property using the $ < ExpressionElement.value > $ as the production rule for the expression value. In **step 2**, we created claims with no description. In **step 3**, the mappings between claims (7, 8) and the terminology elements were created by inserting a description within them. These descriptions contain ExpressionLangString elements referencing the abstract expressions 5 and 6 respectively. In **step 4**, multiplicity implementation constraint subtypes were attached to claims 7 and 8 due to the multiplicity of fault tree event types (i.e., top, intermediate, and basic events). In **step 5**, we added multiplicity implementation constraints to all terminology elements, also due to the multiplicity of fault tree events. Mapping constraints were assigned to both HSFMType and HSFMEvent abstract terms to map their values to fault tree events. Due to the hierarchical (parent and child) structure of FTA results, we added a **Children(s)** constraint to the HSFMType abstract term. Finally, in **step 6**, we specified ODE-compliant fault tree model queries (q1, q2, and q3) inside mapping and children implementation constraints subtypes. Query q1 was assigned to a mapping constraint to map the values of the abstract term HSFMType to the top-events of fault trees. The query q2 was assigned to a children constraint to map the values of the abstract term HSFMType to intermediate events of a fault tree. The query q3 was assigned to a mapping constraint to map the values of the abstract term HSFMEvent to the basic events of a fault tree.

Figure 2 shows an excerpt of Hazardous Software Failure Mode (HSFM) Assurance Case Pattern in SACM visual notation. This pattern decomposes the claim *ABSHSFM-Type* into the *AbsTypeSecondary* subclaim arguing the failure modes of other components that contribute to the current failure mode are acceptable. The *AbsTypeSecondary* is further decomposed into *HSFMAccept* fault mitigation sub-claims arguing that all causes of each failure event specified in fault tree leaf nodes are acceptable, i.e., they do not lead the system to an unsafe state. For each fault tree non-leaf node, the *AbsHSFM* is decomposed into other "Absence Hazardous Software Failure Mode" (HSFM) fault mitigation citation claims referencing top-claims within other argument packages (i.e. recursive pattern instantiation).

4.3 Assurance Case Pattern Instantiation

In this section, we describe the Assurance Case Pattern Instantiation process. In **step 1**, we performed safety analysis using Fault Tree Analysis to identify and classify the risk

Table 1. An Excerpt of Assurance Case Pattern Specification.

Step	Type	Gid	isAbstract	Value/Description	Constraints
1	Term	1	True	HSFMType	
1	Term	3	True	HSFMEvent	
1	Expression	5	True	{ HSFMType} is absent	
1	Expression	6	True	{ HSFMEvent} is acceptable	
2	Claim	7/8	True		
3	Claim	7	True	{ HSFMType} is absent	
3	Claim	8	True	{ HSFMEvent} is acceptable	
4	Claim	7	True	{ HSFMType} is absent	M
4	Claim	8	True	{ HSFMEvent} is acceptable	M
5	Term	1	True	HSFMType	m, p, s
5	Term	3	True	HSFMEvent	m, p
5	Expression	3	True	{ HSFMType} is absent	M
5	Expression	3	True	{ HSFMEvent} is acceptable	M
6	Term	1	True	HSFMType	m, p^{q1}, s^{q2}
6	Term	3	True	HSFMEvent	m, p^{q3}

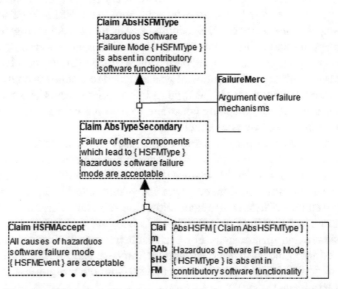

Fig. 2. An Excerpt of Hazardous Software Failure Mode Assurance Case Pattern in SACM.

posed by hazards, and their failure propagation paths. We identified the following hazards
for the wheel-braking system: No braking four wheels, and no braking three wheels, both

classified as ASIL D. Figure 3 shows an excerpt of the FTA for the no-braking four-wheel hazard using HiP-HOPs. In **step 2** of this phase, the input fault.

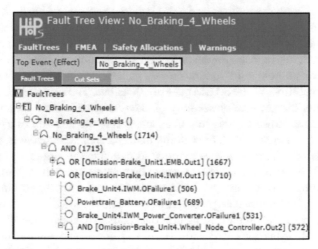

Fig. 3. No Braking Four Wheels HiP-HOPs Faut Tree Excerpt.

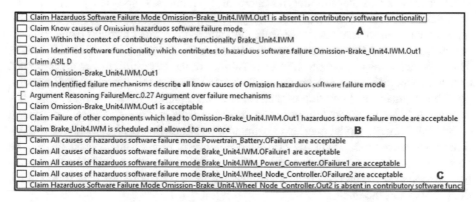

Fig. 4. HBS Hazardous Software Failure Mode Assurance Case.

tree models were converted into the Open Dependability Exchange (ODE) meta-model compliant fault tree format. Finally, in **step 3,** the FTA results were synthesized into an assurance case for the wheel-braking system using the assurance case pattern instantiation program developed by the authors. Figure 4 illustrates the product safety argument resulting from **step 6** where the abstract claims have now been instantiated referencing FTA results (evidence). ODE Gates (i.e., top and intermediate events) are represented by claims A and C, and ODE Output Events (i.e., basic events) are represented by claims within the B set.

5   Related Works

[15] propose a methodology for assurance case pattern instantiation based on implementation constraint subtypes defined as SACM pattern extensions. However, those pattern extensions do not support the specification and instantiation of recursive assurance case pattern elements. In this paper, we propose an add-on to SACM pattern extensions for supporting the recursive instantiation of SACM argumentation elements. We also introduced a novel methodology for assurance case pattern specification and instantiation to enhance the capabilities of the SACM pattern extensions. The AdvoCATE tool proposed by [20] supports the automatic assembly of safety arguments and the instantiation of argument patterns. However, this tool does not enable the creation of assurance arguments directly from models with traceability to one or more metamodels. It requires the creation of a table with data entries required to populate the assurance case pattern to instantiate it. Therefore, a data table needs to be created for each system to instantiate a specific pattern. Thus, the usage of data tables does not fully support interoperability and integration of assurance case patterns and evidence (i.g., design or analysis models). We address this issue by implementing the concept of Executable Digital Dependability Identities (EDDI) using the ODE and SACM metamodels to support interoperability between evidence and assurance cases. The ACME tool support for model-based assurance cases developed in SACM was presented in [7]. Although ACME supports model transformations and other model management capabilities, it does not yet support the automatic instantiation of assurance case patterns. [21] proposes an approach for automated model-based assurance case validation and management through a new concept called Constrained Natural Language (CNL) using SACM implementation constraints. The approach aims to promote comprehensibility of the traceability from an assurance case to its supporting engineering model elements. However, this technique does not support the automatic instantiation of an assurance case pattern from evidence (e.g., FTA results). In [22], the authors propose a methodology to support the automatic execution of SACM elements model queries to validate the engineering artifacts, and the evaluation of assurance cases. It also describes a development time assurance case that can be converted to a dynamic assurance case, with traceability to runtime data. Our methodology supports the traceability from assurance case patterns to runtime data stored in the ODE-compliant models.

6   Conclusion

In this paper, we introduced a model-driven methodology to support the specification and synthesis of executable SACM argument patterns with traceability links between claims and ODE-compliant FTA results within an Executable Digital Dependability Identity. We propose the SACM **Children** implementation constraint subtype as an add-on to the SACM Pattern extension to support the automated recursive instantiation within assurance case patterns. We describe a systematic process to support engineers during assurance case patterns specification and instantiation with references to FTA results (evidence). We evaluated the feasibility of our solution in a medium-sized illustrative automotive brake-by-wire system. Our methodology may improve argument-evidence traceability, and reduce the complexity of maintaining SACM argument structures. The medium size and complexity of the models used in the evaluation are a threat to the validity of our study, thus, not supporting the generalization of the results. Case studies using systems and cyber-physical systems from other safety and security-critical domains need to be conducted to properly evaluate the feasibility of our SACM argument pattern specification and instantiation process. Moreover, experimental studies in the industry still need to be conducted to evaluate usability aspects of the traceability between executable argument patterns enriched with Implementation Constraints and FTA results. Although ODE model-based queries are simpler than metamodel-specific queries, it is still a challenge to simplify them to enable property chains while increasing the understanding and usability of the specified queries. In future work, we also intend to use ontologies to simplify the queries into external artifacts within SACM assurance case patterns.

Acknowledgments. This work was supported by CAPES, finance code 001, FAPEMIG under Grant APQ-00743–22, and CNPq Brazilian research funding agencies, by the Secure and Safe Multi-Robot Systems (SESAME) H2020 Project under Grant 101017258, and by the Assuring Autonomy International Programme (https://www.york.ac.uk/assuring-autonomy).

References

1. Kopetz, H., Bondavalli, A., Brancati, F., Frömel, B., Höftberger, O., Iacob, S.: Emergence in Cyber-Physical Systems-of-Systems (CPSoSs). In: Bondavalli, A., Bouchenak, S., Kopetz, H. (eds.) Cyber-Physical Systems of Systems, pp. 73–96. Springer International Publishing, Cham (2016). https://doi.org/10.1007/978-3-319-47590-5_3
2. Kletz, Trevor A. "Specifying and designing protective systems." Loss prevention 6 (1972)
3. NASA. "Fault Tree Analysis Handbook for Aerospace Applications". WA, USA, 2002
4. Wei, T. P. Kelly, R. Hawkins, and E. Armengaud, DEIS: dependability engineering innovation for cyber-physical systems. In: Seidl, M., Zschaler, S. (eds) Software Technologies: Applications and Foundations - STAF 2017 Workshops, Marburg, Germany, July 17–21, vol. 10748 of Lecture Notes in Computer Science, pp. 409–416, Springer, Cham (2017). 10.1007/978-3-319-74730-9_37
5. Trapp, M., Schneider, D., Liggesmeyer, P.: A safety roadmap to cyber-physical systems. In: Münch, J., Schmid, K. (eds.) Perspectives on the Future of Software Engineering: Essays in Honor of Dieter Rombach, pp. 81–94. Springer Berlin Heidelberg, Berlin, Heidelberg (2013). https://doi.org/10.1007/978-3-642-37395-4_6
6. Hawkins R. Kelly, T.: A systematic approach for developing software safety arguments. In: 27th International System Safety Conference, pp. 25–33 (2010)
7. Wei, R., Kelly, T.P., Dai, X., Zhao, S., Hawkins, R.: Model based system assurance using the structured assurance case metamodel. J. of Syst. and Soft. **154**, 211–233 (2019)
8. Bloomfield, R., Bishop, P.: Safety and Assurance Cases: Past, Present and Possible Future – an Adelard Perspective. In: Dale, C., Anderson, T. (eds.) Making Systems Safer: Proceedings of the Eighteenth Safety-Critical Systems Symposium, Bristol, UK, 9-11th February 2010, pp. 51–67. Springer London, London (2010). https://doi.org/10.1007/978-1-84996-086-1_4
9. Wei, R. Kelly, T. Reich, J. Gerasimou, S.: On the transition from design time to runtime model-based assurance cases. in MoDELS (Workshops), pp. 56–61, 2018

10. Kelly, T. P. McDermid, J. A.: Safety case construction and reuse using patterns, in Safe Comp 97, pp. 55–69, Springer, 1997
11. Azevedo, L., Parker, D., Walker, M., Papadopoulos, Y., Araújo, R.: Assisted assignment of automotive safety requirements. IEEE Softw. **31**(1), 62–68 (2014)
12. ISO, 2018. ISO 26262: Road Vehicles Functional Safety
13. Koorosh, A.,et al. "Safedrones: Real-time reliability evaluation of uavs using executable digital dependable identities." International Symposium on Model-Based Safety and Assessment. Springer International Publishing, Cham (2022) https://doi.org/10.1007/978-3-031-15842-1_18
14. OMG: Structured Assurance Case Metamodel (sacm) Version 2.2.
15. <https://www.omg.org/spec/SACM/2.2/About-SACM/>. Access on: Nov. 19th, 2023
16. Luis, N., et al.: Runtime model-based assurance of open and adaptive cyber-physical systems. In: International Conference on Advanced Information Networking and Applications. Springer International Publishing, Cham (2023). 10.1007/978-3-031-29056-5_46
17. Weaver, R.A.: The safety of software: Constructing and assuring arguments. University of York, Department of Computer Science (2003)
18. Papadopoulos, Y., et al.: Engineering failure analysis and design optimisation with HiP-HOPS. Eng. Fail. Anal. **18**(2), 590–608 (2011). https://doi.org/10.1016/j.engfailanal.2010.09.025
19. Marc, Z., et al.: Open dependability exchange metamodel: a format to exchange safety information. In: 2023 Annual Reliab and Maintain Symposium (RAMS). IEEE, (2023)
20. Delange, Julien, and Peter H. Feiler: Supporting the ARP4761 safety assessment process with AADL. Embedded real time software and systems (ERTS2014) (2014)
21. E. Denney, G. Pai, and J. Pohl: Advocate: an assurance case automation toolset. In: International Conference on Computer Safety, Reliability, and Security, pp. 8–21, Springer (2012). 10.1007/978-3-642-33675-1_2
22. Wei, Ran, et al.: Designing critical systems with iterative automated safety analysis. In: Proceedings of the 59th ACM/IEEE Design Automation Conference (2022)
23. Wei, Ran, et al. "Automated Model Based Assurance Case Management Using Constrained Natural Language." IEEE Transactions on CAD (2023)
24. CHESS: Composition with guarantees for High-integrity Embedded Software components aSsembly. http://www.chess-project.org. Accessed 15 Nov 2023
25. CMU-SEI: EMFTA: EMF-based Fault-Tree Analysis Tool, 20-Sep-2017. https://github.com/cmu-sei/emfta. Accessed 20 Nov 2023
26. IEC, 2010. IEC 61508 Functional safety of electrical/electronic/programmable electronic safety-related systems, Parts 1 to 7, Edition 2.0. International Electrotechnical Commission

An Energy Consumption Model to Change the TBFC Model of the IoT

Dilawaer Duolikun[1]([✉]), Tomoya Enokido[2], and Makoto Takizawa[1]

[1] RCCMS, Hosei University, Tokyo, Japan
dilewerdolkun@gmail.com, makoto.takizawa@computer.org
[2] Faculty of Business Administration, Rissho University, Tokyo, Japan
eno@ris.ac.jp

Abstract. The IoT (Internet of Things) consumes plenty of electric energy since tremendous amount of sensor data is transmitted and processed. Fog nodes support application processes of sensor data in the FC (Fog Computing) model. Fog nodes are tree-structured to interconnect with devices as leaves and a server cloud as a root in the TBFC (Tree-Based FC) model for realizing the energy-efficient FC model. We discuss how to design a homogeneous TBFC (HTBFC) model composed of homogeneous fog nodes given application processes and size of sensor data from each device. In our approach, starting from a tree composed of a pair of a root server node and one leaf device node, i.e. CC (Cloud Computing) model, the HTBFC tree is obtained by adding fog nodes through applying three types of operations. If an operation is applied on a node, not only the node but also other nodes like the child nodes are also changed, which are affected nodes. First, a node most consuming energy is selected as a target node. Then, an operation by which the total energy consumption of all the affected nodes is most reduced is applied on the target node. In the evaluation, we show the energy consumption and delivery time of a CC model can be reduced in the HTBFC model obtained in the proposed algorithm.

Keywords: Green computing · Fog computing (FC) model · TBFC model · HTBFC model

1 Introduction

In the FC model [3] of the IoT, fog nodes support application processes of sensor data. Fog nodes are tree-structured to interconnect with devices as leaves and a server cloud as a root in the TBFC (Tree-Based FC) model [13,19–23] for realizing the energy-efficient FC model. Variants of the TBFC model [14–16,19, 23–25] are so far discussed.

In this paper, a *homogeneous* TBFC (HTBFC) model is considered, where each fog node is homogeneous. A CC model [1,2] is an HTBFC model composed of a pair of a root server cloud node and a leaf device node, i.e. a set of devices. In

© The Author(s), under exclusive license to Springer Nature Switzerland AG 2024
L. Barolli (Ed.): AINA 2024, LNDECT 200, pp. 465–475, 2024.
https://doi.org/10.1007/978-3-031-57853-3_39

this paper, an algorithm to design an energy-efficient HTBFC model is proposed by including fog nodes through three types of operations *migrate* (*MG*), *replicate* (*RP*), and *expand* (*EP*) starting from an CC model. A node most consuming energy is a *target* node in a tree. First, one operation of *MG*, *RP*, and *EP* is applied on a target node. Application processes migrate from a target node to every child node in the *MG* operation. Replicas of the target node are created as the sibling nodes in the *RP* operation. In the *EP* operation, new nodes are created as child nodes of a target node. In addition, processes migrate from a target node to the created nodes. A target node, nodes newly created, and nodes to which processes migrate are *affected* nodes of the operation. Based on the estimation formulas [18], mew estimation formulas are proposed to give the total energy consumption of all the affected nodes in the HTBFC model. By using the estimation formulas, one operation on a target node is selected and applied to change the tree, by which the total energy consumption of all the affected nodes is most reduced. Until no operation can be found for a target node, the procedure is iterated. In the evaluation, we show a lower-energy HTBFC tree is obtained from the CC model in the proposed algorithm.

In Sect. 2, we present the HTBFC model and the energy consumption in Sect. 3. In Sect. 4, we propose the algorithm to obtain the energy-efficient HTBFC tree. We evaluate the proposed algorithm in Sect. 5.

2   An HTBFC Model

In the TBFC (Tree-Based FC) model [13, 19–23], fog nodes are interconnected in a tree structure, where a *root* node denotes a cloud of servers and each leaf *device* node shows a set of devices. A *DU* (Data Unit) is a unit of data exchanged among processes. On receipt of DUs from the child nodes, the DUs are processed by application processes on a node and a processed DU is sent to the parent node. In this paper, we consider an *homogeneous* TBFC (HTBFC) model where every fog node is homogeneous. Here, "f_I" stands for a node where I is a sequence $i_1 i_2 \ldots i_{h-1} i_h (h \geq 1)$ of numbers and "f" stands for a root node. Each node $f_I (= f_{i_1 \ldots i_{h-1} i_h})$ is a parent node of $b_I (\geq 0)$ child nodes $f_{Ii} (= f_{i_1 i_2 \ldots i_{h-1} i_h i}) (i = 1, \ldots, b_I)$ and has one parent node $f_{i_1 \ldots i_{h-1}}$. A root node f has $b (\geq 1)$ child nodes. An *edge* node f_I is a fog node which has one child device node d_I. Here, d_I is a collection of $nd_I (\geq 1)$ devices. The HTBFC tree includes nd devices.

We consider a linear application AP of sensor data, i.e. AP is a sequence $\langle p_1, \ldots, p_{np} \rangle (np > 0)$ of application processes [14,15]. First, the *tail* process p_{np} takes an input DU of sensor data from devices. Each process p_h receives an input DU od_{h+1} from p_{h+1} and passes an output DU od_h to $p_{h-1} (h = np - 1, np - 2, \ldots, 2)$. The *top* process p_1 finally delivers an output DU od_1 to devices. A subsequence $P_I = \langle p_{tp_I}, p_{tp_I+1}, \ldots, p_{tl_I} \rangle$ $(1 \leq tp_I \leq tl_I \leq np)$ of $m_I (= tl_I - tp_I + 1 \geq 1)$ processes of AP is installed on a node f_I. For every child node f_{Ii} of f_I, $tp_{Ii} = tl_I + 1$. Let P show a process subsequence $\langle p_1, p_2, \ldots, p_{tl} \rangle (1 \leq tl \leq np)$ and m be the number tl of processes of a root node f. For a path $\langle f, f_{i_1}, \ldots, f_{i_1 i_2 \ldots i_h} \rangle$ from f to each edge node $f_{i_1 i_2 \ldots i_h}$, the process subsequences P, P_{i_1}, $\ldots$, $P_{i_1 i_2 \ldots i_h}$ are the application process AP.

$PCA(s)$ is the amount [bit] of the computation of each process p_h to obtain an output DU od from an input DU of size s and is $a_1 s^2 + a_2 s + a_3$ where a_1, a_2, and a_3 are constants. The size $|od|$ is $\rho_h \cdot s$ where ρ_h is the *output ratio*. In this paper, ρ_h is ρ for every process p_h. The reduction ratio π_I of a node f_I is ρ^{m_I} and π shows the reduction ratio ρ^m of a root node f. id_{Ii} is an input DU of f_I from each child node f_{Ii}. ID_I is a set $\{id_{I1}, \ldots, id_{Ib_I}\}$ of input DUs of f_I. Let s be the size $|ID_I|$. $FCA(s, tl_I, tp_I)$ [bit] is the total computation amount of each node f_I to obtain an output DU od_I of size $\pi^{m_I} \cdot s$ by executing m_I processes $p_{tl_I}, p_{tl_I-1}, \ldots, p_{tp_I}$ for input DUs ID_I, which is $\Sigma_{h=tl_I}^{tp_I} PCA(\rho^{tl_I-h} \cdot s)$:

$$FCA(s, tl, tp) = \begin{cases} a_1 \dfrac{1 - \rho^{2v}}{1 - \rho^2} s^2 + a_2 \dfrac{1 - \rho^v}{1 - \rho} s + a_3 v \\ \text{where } v = 1 + tl - tp \text{ and } tl \geq tp. \end{cases} \tag{1}$$

The communication amounts $RA(s)$ and $SA(s)$ [bit] of each node f_I to receive and send a DU of size s are $rc_1 \cdot s + rc_2$ and $sc_1 \cdot s + sc_2$, respectively, where rc_1, rc_2, sc_1, and sc_2 are constants.

3 Energy Consumption of the HTBFC Model

3.1 Nodes to be Affected

Each fog node f_I consumes the maximum power FXP [W] if some application process is executed, else the minimum power FMP $(< FXP)$ as discussed in the SPC model [4–12]. Each fog node f_I also consumes the power FRP $(= Fr \cdot FXP)$ and FSP $(= Fs \cdot FXP)$ [W] to receive and send a DU where $Fr \leq 1$ and $Fs \leq 1$. A root node f consumes the maximum and minimum power SXP and SMP [W] and the communication power SRP $(= Sr \cdot SXP)$ and SSP $(= Ss \cdot SXP)$ [W] to receive and send a DU.

FCR and SCR are *computation rates* [bps], i.e. how many bits of an input DU ID_I a fog node f_I and a root node f process for one second, respectively. The *energy-bit ratios* (EB) FEB and SEB of f_I and f are FXP/FCR and SXP/SCR [W sec (J)/bit], respectively. Let s be the size $|ID_I|$ [bit] of input DUs ID_I. It takes time $FPT(s) = PCA(s)/FCR$ and $SPT(s) = PCA(s)/SCR$ [sec] to execute a process p_h on f_I and f for input size s, respectively. The energy $FPE(s) = PCA(s) \cdot FEB$ and $SPE(s) = PCA(s) \cdot SEB$ [J] are consumed by f_I and f, respectively.

A fog node f_I and a root node f support process subsequences $\langle p_{tp_I}, \ldots, p_{tl_I} \rangle$ and $\langle p_1, \ldots, p_{tl} \rangle$, respectively. The total execution time $FCT(s, tl_I, tp_I)$ and $SCT(s, tl)$ [sec] for f_I and f to process input DUs ID_I of size s are $FCA(s, tl_I, tp_I)/FCR$ and $FCA(s, tl, 1)/SCR$, respectively. Here, f_I and f consume the total energy $FCE(s, tl_I, tp_I)$ $=$ $FCA(s, tl_I, tp_I) \cdot FEB$ and $SCE(s, tl)$ $=$ $FCA(s, tl, 1) \cdot SEB$, respectively. It takes time $FRT(s)$ $(= RA(s)/FCR)$ and $FST(s)(= SA(s)/FCR)$ [sec] for each f_I to receive and send a DU du of size s, respectively. For a root node f, the receiving time $SRT(s)$

is $RA(s)/SCR$ and the sending time $SST(s)$ is $SA(s)/SCR$ [sec]. The energy $FRE(s) = RA(s) \cdot Fr \cdot FEB$ and $FSE(s) = SA(s) \cdot Fs \cdot SEB$ [J] are consumed by f_I and f to receive and send du, respectively.

A fog node f_I and a root node f totally consume the energy $FE(s, tl_I, tp_I) = (RA(s) \cdot Fr + FCA(s, tl_I, tp_I) + SA(\pi_I \cdot s) \cdot Fs) \cdot FEB$ and $SE(s, tl) = (RA(s) \cdot Sr + SCA(s, tl) + SA(\pi \cdot s) \cdot Ss) \cdot SEB$ [J] for an input DU of size s, respectively. The total execution time $FT(s, tl_I, tp_I)$ of f_I and $ST(s, tl)$ of f are $(RA(s) + FCA(s, tl_I, tp_I) + SA(\pi_I \cdot s))/FCR$ and $(RA(s) + SCA(s, tl) + SA(\pi \cdot s))/SCR$ [sec], respectively.

Let f_P and f_C denote the parent node and a child node f_{Ii} of a node f_I, respectively. The node f_I receives an input DU id_C of size $\frac{s}{b_I}$ from each child node f_C where $s = |ID_I|$. Each f_C consumes the energy $FSE(\frac{s}{b_I})$ to send a DU id_C of size $\frac{s}{b_I}$ to f_I. Then f_I consumes the energy $FRE(\frac{s}{b_I})$ to receive the DU from each f_C, $FCE(s, tl_I, tp_I)$ to obtain an output DU od_I of size $\pi_I \cdot s$ from ID_I, and $FSE(\pi_I s)$ to send od_I to f_P. The energy $FRE(\pi_I s)$ is consumed by f_P to receive od_I if f_P is a fog node, otherwise, i.e. root node, $SRE(\pi_I \cdot s)$. Thus, the energy $TFE(s, tl_I, tp_I)$ is totally consumed by f_I, every f_C, and f_P:

$$TFE(s, tl_I, tp_I) = \begin{cases} FSE(\frac{s}{b_I}) \cdot b_I + \\ FRE(\frac{s}{b_I}) \cdot b_I + FCE(s, tl_I, tp_I) + FSE(\pi_I s) + \\ FRE(\pi_I \cdot s) \text{ if } f_P \text{ is not a root, else } SRE(\pi_I \cdot s). \end{cases} \quad (2)$$

A root node f and every child node f_C consume the energy $TSE(s, tl)$ as follows:

$$TSE(s, tl) = \begin{cases} FSE(\frac{s}{b}) \cdot b + \\ SRE(\frac{s}{b}) \cdot b + SCE(s, tl, 1) + SSE(\pi \cdot s). \end{cases} \quad (3)$$

3.2   Affected Nodes

We consider following operations MG, RP, and EP on a target node f_I to obtain an energy-efficient TBFC tree from a CC model. Here, a target node f_I and each child node f_C support process subsequences $\langle p_{tp_I}, p_{tp_I+1}, \ldots, p_{tl_I} \rangle$ and $\langle p_{tp_C}, p_{tp_C+1}, \ldots, p_{tl_C} \rangle$, respectively.

1. Migrate: $MG(f_I, k)$.
2. Replicate: $\langle f_{N_1}, \ldots, f_{N_{n-1}} \rangle = RP(f_I, n)$.
3. Expand: $\langle f_{N_1}, \ldots, f_{N_n} \rangle = EP(f_I, n, k)$.

Here, n is the number of replica fog nodes to be created in the RP and EP operations and k is the number of processes to migrate from f_I in the MG and EP operations. Even if the energy consumption of a target node f_I can be smaller, some affected nodes may consume more energy. Hence, we have to reduce the total energy consumption of all the affected nodes. Let s be the size of input data ID_I of a target node f_I.

In the $MG(f_I, k)$ operation, $k(< m_I)$ processes $p_{tl_I}, \ldots, p_{tl_I-k+1}$ migrate from a target node f_I to every child node f_C through the live migration of virtual machines [11–13, 17]. Here, all the affected nodes consume the total energy $FME(s, tl_I, tp_I, k)$:

$$FME(s, tl_I, tp_I, k) = \begin{cases} [FCE(\frac{s}{b_I}, tl_I, tl_I-k+1) + FSE(\frac{\rho^k s}{b_I})]b_I+ \\ FRE(\frac{\rho^k s}{b_I})b_I + FCE(\rho^k s, tl_I-k, tp_I) + FSE(\pi_I s)+ \\ FRE(\pi_I \cdot s) \text{ if } f_P \text{ is not a root node, else } SRE(\pi_I \cdot s). \end{cases} \quad (4)$$

If a target node is a root node f, all the affected nodes totally consume the energy $SME(s, tl, k)$:

$$SME(s, tl, k) = \begin{cases} [FCE(\frac{s}{b_I}, tl, tl-k+1) + FSE(\frac{\rho^k \cdot s}{b})] \cdot b+ \\ SRE(\frac{\rho^k \cdot s}{b}) \cdot b + SCE(\rho^k \cdot s, tl-k, 1) + SSE(\pi \cdot s). \end{cases} \quad (5)$$

In the $RP(f_I, n)(1 < b_I < n)$ operation, $(n-1)$ replica nodes $f_{N_1}, \ldots, f_{N_{n-1}}$ of the target fog node f_I are created as the sibling nodes. All the affected nodes consume the total energy $FRE(s, tl_I, tp_I, n)$:

$$FRE(s, tl_I, tp_I, n) = \begin{cases} FSE(\frac{s}{b_I})b_I + \\ (FRE(\frac{s}{b_I})\frac{b_I}{n} + FCE(\frac{s}{n}, tl_I, tp_I) + FSE(\frac{\pi_I s}{n}))n+ \\ FRE(\frac{\pi_I s}{n})n \text{ if } f_P \text{ is not a root node, else } SRE(\frac{\pi_I s}{n})n. \end{cases} \quad (6)$$

If a target node is a root f, all the affected nodes totally consume the energy $SRE(s, tl, n)$:

$$SRE(s, tl, n) = \begin{cases} FSE(\frac{s}{b}) \cdot b + \\ (FRE(\frac{s}{b})\frac{b}{n} + SCE(\frac{s}{n}, tl) + SSE(\frac{\pi s}{n})) \cdot n. \end{cases} \quad (7)$$

In the $EP(f_I, n, k)$ operation, $n (\geq b_I)$ child fog nodes $f_{N_1}, \ldots, f_{N_n}$ of a target node f_I are created and $k(\leq m_I)$ processes $p_{tl_I}, p_{tl_I-1}, \ldots, p_{tl_I-k+1}$ migrate to each child replica node. All the affected nodes totally consume the energy $FEE(s, tl_I, tp_I, n, k)$:

$$FEE(s, tl_I, tp_I, n, k) = \begin{cases} FSE(\frac{s}{b_I}) \cdot b_I+ \\ (FRE(\frac{s}{b_I})\frac{k}{n} + FCE(\frac{s}{n}, tl_I, tl_I-k+1) + FSE(\frac{\rho^k \cdot s}{n})) \cdot n + \\ FRE(\frac{\rho^k \cdot s}{n}) \cdot n + FCE(\rho^k s, tl_I-k, 1) + FSE(\pi s)+ \\ FRE(\pi_I \cdot s) \text{ if } f_P \text{ is not a root node, else } SRE(\pi_I \cdot s). \end{cases} \quad (8)$$

If a target node is a root node f, all the affected nodes consume the total energy $SEE(s, tl, n, k)$:

$$SEE(s, tl, n, k) = \begin{cases} FSE(\frac{s}{b}) \cdot b + \\ (FRE(\frac{s}{b})\frac{b}{n} + FCE(\frac{s}{n}, tl, tl-k+1) + FSE(\frac{\rho^k \cdot s}{n})) \cdot n + \\ SRE(\frac{\rho^k s}{n}) \cdot n + SCE(\rho^k \cdot s, tl-k, 1) + SSE(\pi \cdot s). \end{cases} \quad (9)$$

4   An Energy-Efficient TBFC Tree

Let f_C, f_P, and f_N denote a child node f_{Ii}, a parent node, a node f_{N_i} newly created in the operations $RP(f_I, n)$ and $EP(f_I, n, k)$ on a node f_I, respectively. A target node f_I is a node whose receipt queue is not empty and which most consumes the energy. Let TN be an ordered set of target nodes, where target nodes are ordered in the descending order of levels in the tree T. By using a function $f_I = deQ(TN)$, a top node f_I is dequeued from TN.

First, for a target node $f_I = deQ(TN)$, the energy consumption TE of f_I is calculated as $TFE(s, tl_I, tp_I)$ if f_I is not a root node, else $TSE(s, tl_I)$. Here, s is the size of input DUs ID_I of f_I. Then, in $\langle op, n, k \rangle = tope(f_I, s)$, a tuple $\langle op, n, k \rangle$ is obtained, where n is the number of nodes to be newly created and k is the number of processes to migrate from f_I Here, let $ME_I(s, tl_I, tp_I, km)$, $RE_I(s, tl_I, tp_I, nr)$, and $EE_I(s, tl_I, tp_I, ne, ke)$ be the smallest energy consumption of all the affected nodes for MG, RP, and EP operations, respectively. That is, $ME_I(s, tl_I, tp_I, km) \leq ME_I(s, l_I, tp_I, ck) < TE$ for every $ck = 1, \ldots, m_I - 1$, $RE_I(s, tl_I, tp_I, nr) \leq RE_I(s, tl_I, tp_I, cn) < TE$ for every $cn = 1, \ldots, b_I - 1$, and $EE_I(s, tl_I, tp_I, ne, ke) \leq EE_I(s, tl_I, tp_I, cn, ck) < TE$ for every $cn = 1, \ldots, b_I$, $ck = 1, \ldots, m_I - 1$. Then, MG is taken as an operation op and $k = km$ if $ME_I(s, tl_I, tp_I, km)$ is the smallest, $op = RP$ and $n = nr$ if $RE_I(s, tl_I, tp_I, nr)$ is the smallest, and $op = EP, n = ne$, and $k = ke$ if $EE_I(s, tl_I, tp_I, ne, ke)$ is the smallest. The operation $op(f_I, n, k)$ is applied on the target node f_I and a tree which consumes smaller energy is obtained.

[**Algorithm to select an operation**]

$tope(f_I, s)$ {
    XE = maximum energy;
    $E = TFE_I(s, tl_I, tp_I)$ **if** f_I is not a root, **else** $TSE(s, tl)$;
    $k = n = 0$; $op = 0$:
    for $ck = 1, \ldots, m_I - 1$ /* Migrate */ {
        $E_{MG} = FME_I(s, tl_I, tp_I, ck)$ **if** f_I is not a root, **else** $SME(s, tl)$;
        if $E_{MG} < E$, $\{op = MG;\ E = E_{MG};\ n = 0;\ k = ck;\}$;
    };   **for** $cn = 1, \ldots, b_I$ /* Replicate */ {
    $E_{RP} = FRE_I(s, tl_I, tp_I, cn)$ **if** f_I is not a root, **else** XE;
    if $E_{RP} < E$, $\{op = RP;\ E = E_{RP};\ n = cn;\ k = 0;\ \}$;
    for $ck = 1, \ldots, np$ /* Expand */ {
        $E_{EP} = FEE_I(s, tl_I, tp_I, cn, ck)$ **if** f_I is not a root, **else** $SEE(s, tl)$;
        if $E_{EP} < E$, $\{op = EP;\ E = E_{EP};\ n = cn;\ k = ck;\}$;
    };
    };
    return $(\langle op, n, k \rangle)$; };

The following procedure is iterated as long as op is found, i.e. $op \neq 0$ for every target node. For every target node f_I in TN, a most-energy efficient operation $op \in \{MG, RP, EP\}$ is selected by the function $tope(f_I, s)$. The tree T is changed by applying op on the target node f_I by the function $ex(T, f_I, op, n, k)$ [18].

[Procedure to change a tree]
  TN = set of target nodes in a tree T;
  while ($TN \neq \phi$) {
      $f_I = deQ(TN)$; /* target node f_I */
      $s = |ID_I|$ for f_I;
      $\langle op, n, k \rangle = tope(f_I, s)$;
      if op is found, $T = ex(T, f_I, op, n, k)$; /* change T */
  };

5 Evaluation

We make the following assumptions on the application $AP = \langle p_1, \ldots, p_{np} \rangle$.

1. The output ratio ρ_h of each process p_h is ρ. ρ^{np} is 0.1. For example, ρ is 0.562 for $np = 4$.
2. The process computation amount $PCA(s)$ is $a_1 \cdot s^2 + a_2 \cdot s + a_3$ where $a_1 = 0.1$, $a_2 = 1$, and $a_3 = 10$.
3. The communication amount $RA(s)$ is $r_1 \cdot s + r_2$ and $SA(s)$ is $s_1 \cdot s + s_2$ [bit] where $r_1 = s_1 = 0.01$ and $r_2 = s_2 = 0.02$.

We also make the following assumptions on nodes:

1. The computation rate SCR is 800 [Kbps], maximum power consumption SXP is 301.3 [W], minimum power consumption SMP is 126.1 [W], $Ss = 0.68$, and $Sr = 0.73$ for a root node f. Here, the energy-bit ratio SEB is 0.37 [J/Kbit].
2. FCR is 148 [Kbs], FXP is 3.7 [W], FMP is 2.1 [W], Fs is 0.68, and $Fr = 0.73$ for every fog node f_I [20,22,26]. Here, the energy-bit ratio FEB is 0.025 [J/Kbit].
3. Each of 1,000 devices, i.e. $nd = 1,000$ sends a DU of sensor data of size 8 [Kbit] every 10 [sec]. Hence, totally 8,000 [Kbit] are sent to the edge nodes.

Let IT be an initial tree composed of one root node f and one device node d. The device node d shows a collection of nd (= 1,000) devices. IT denotes a CC model. All the target nodes in a tree are collected in a set TN. If $TN \neq \phi$, for each target node f_I in TN and input size $s = ID_I$, an operation op on f_I is found by $\langle op, n, k \rangle = tope(f_I, s)$. If such an operation op is found, the total energy consumption TE_T of the tree T can be most reduced. T is changed by applying op on f_I. The procedure is applied as long as TE_T can be reduced as presented in the preceding section.

For each number np of processes ($2 \leq np \leq 24$) in AP, the HTBFC tree T is obtained from the initial tree IT by the algorithm. The delivery time shows how long it takes to deliver sensor DUs sent by device nodes to a root node f in a tree T. Here, ET_T is the total energy consumption of nodes and DT_T is the delivery time for an HTBFC tree T obtained for each np in the proposed algorithm. The ratios TE_T/TE_{IT} and DT_T/DT_{IT} are calculated.

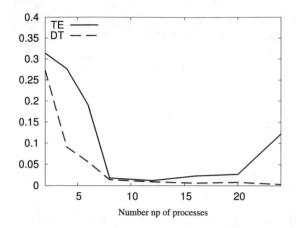

Fig. 1. Ratios of total energy consumption (TE) and delivery time (DT)

Figure 1 shows the ratios TE_T/TE_{IT} and DT_T/DT_{IT} for number np of processes of an application AP. In the evaluation, $2 \leq np \leq 24$. As shown in Fig. 1, starting from a CC model, an energy-efficient HTBFC tree T can be obtained in the proposed algorithm. Especially, for $np = 12$, the total energy consumption of the HTBFC tree is only 1.1 [%] of the CC model. For $np > 20$, the energy ratio TE_T/TE_{IT} increases.

Figure 2 shows the number of nodes in the HTBFC tree for number np of processes in AP. The more processes an application AP includes, the more fog nodes are created. The number of nodes drastically increases for the umber $np \geq 16$. For $np < 16$, child fog nodes of a root node are replicated in the RP operation. For $np \geq 16$, fog nodes are expanded by the EP operation, i.e. the height of the tree T gets higher.

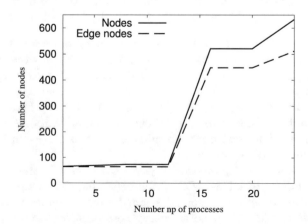

Fig. 2. Number of nodes in a tree

6   Concluding Remarks

In this paper, the HTBFC model is considered, where every fog node is implemented by a same computer like Raspberry Pi. We discussed how to design an energy-efficient HTBFC model starting from the CC model of the IoT. In our approach, a target node whose energy consumption is the largest is changed by one of MG, RP, and EP operations. By applying an operation on a target node, the energy consumption of another node is also changed. Nodes whose energy consumption changes are affected nodes. We proposed formulas to estimate the total energy consumption of all the affected nodes in the HTBFC model. By using the estimation formulas, we proposed the algorithm to obtain an HTBFC tree starting from a CC model. In the evaluation, we showed that a HTBFC tree whose total energy consumption of the tree and the delivery time of sensor data are smaller than the CC model is obtained in the proposed algorithm.

In this evaluation, even if a pair of applications AP_1 and AP_2 include different numbers np_1 and $np_2(np_1 \neq np_2)$ of processes, the output ratios are the same 0.1, i.e. $\rho^{np_1} = \rho^{np_2} = 0.1$ but the total amounts of computation of the np_1 and np_2 processes are different. The total amount of computation of an application of np processes increase as the number np increases while $\rho^{np} = 0.1$. For the number np of processes and size s of input DUs, we are now getting the parameters $a_{np.1}$, $a_{np,2}$, and $a_{np.3}$ of each process of an application of np processes, which satisfies the equation $FCA(s, np, 1) = FCA(s, 1, 1)$, i.e. $a_{np,1} \cdot (1-\rho^{2np})/(1-\rho^2) \cdot s^2 + a_{np,2} \cdot (1-\rho^{np})/(1-\rho) \cdot s + a_{np,3} \cdot np = a_1 \cdot s^2 + a_2 \cdot s + a_3$. That is, $a_{np,1} = a_1 \cdot (1 - \rho^2)/(1 - \rho^{2np})$, $a_{np,2} = a_2 \cdot (1 - \rho)/(1 - \rho^{np})$, and $a_{mp,3} = a_3/np$. The total computation amount $FCA(s, np, 1)$ of an application of np (≥ 1) processes is the same as $FCA(s, 1, 1)$ of an application of one process. By using the parameters $a_{np,1}$, $a_{np,2}$, and $a_{np,3}$ of each process obtained for each np, we are now evaluating the algorithm.

Acknowledgment. This work is supported by Japan Society for the Promotion of Science (JSPS) KAKENHI Grant Number 22K12018.

References

1. Dayarathna, M., Wen, Y., Fan, R.: Data center energy consumption modeling: a survey. IEEE Commun. Surv. Tut. **18**(1), 732–787 (2016)
2. Qian, L. Luo, Z., Du, Y.,Guo, L.: Cloud computing: an overview. In: Proceedings of the 1st International Conference on Cloud Computing, pp. 626–631 (2009)
3. Rahmani, A.M., Liljeberg, P., Preden, J.-S., Jantsch, A. (eds.): Fog Computing in the Internet of Things, 1st edn., p. 172. Springer, Cham (2018). https://doi.org/10.1007/978-3-319-57639-8
4. Enokido, T., Aikebaier, A., Takizawa, M.: Process allocation algorithms for saving power consumption in peer-to-peer systems. IEEE Trans. Ind. Electron. **58**(6), 2097–2105 (2011)
5. Enokido, T., Aikebaier, A., Takizawa, M.: A model for reducing power consumption in peer-to-peer systems. IEEE Syst. J. **4**(2), 221–229 (2010)

6. Enokido, T., Aikebaier, A., Takizawa, M.: An extended simple power consumption model for selecting a server to perform computation type processes in digital ecosystems. IEEE Trans. Ind. Inform. **10**(2), 1627–1636 (2014)
7. Enokido, T., Takizawa, M.: Integrated power consumption model for distributed systems. IEEE Trans. Ind. Electron. **60**(2), 824–836 (2013)
8. Kataoka, H., Duolikun, D., Enokido, T., Takizawa, M.: Energy-efficient virtualisation of threads in a server cluster. In: Proceedings of the 10th International Conference on Broadband and Wireless Computing, Communication and Applications, BWCCA 2015, pp. 288–295 (2015)
9. Kataoka, H., Duolikun, D., Sawada, A., Enokido, T., Takizawa, M.: Energy-aware server selection algorithms in a scalable cluster. In: Proceedings of the 30th International Conference on Advanced Information Networking and Applications, pp. 565–572 (2016)
10. Kataoka, H., Nakamura, S., Duolikun, D., Enokido, T., Takizawa, M.: Multi-level power consumption model and energy-aware server selection algorithm. Int. J. Grid Util. Comput. **8**(3), 201–210 (2017)
11. Duolikun, D., Enokido, T., Takizawa, M.: Energy-efficient dynamic clusters of servers. In: Proceedings of the 8th International Conference on Broadband and Wireless Computing, Communication and Applications, pp. 253–260 (2013)
12. Duolikun, D., Enokido, T., Takizawa, M.: Static and dynamic group migration algorithms of virtual machines to reduce energy consumption of a server cluster. In: Nguyen, N.T., Kowalczyk, R., Xhafa, F. (eds.) Transactions on Computational Collective Intelligence XXXIII. LNCS, vol. 11610, pp. 144–166. Springer, Heidelberg (2019). https://doi.org/10.1007/978-3-662-59540-4_8
13. Duolikun, D., Nakamura, S., Enokido, T., Takizawa, M.: Energy-consumption evaluation of the tree-based fog computing (TBFC) model. In: Proceedings of the 22nd International Conference on Broadband and Wireless Computing, Communication and Applications, pp. 66–77 (2022)
14. Duolikun, D., Enokido, T., Barolli, L., Takizawa, M.: A flexible fog computing (FTBFC) model to reduce energy consumption of the IoT. In: Proceedings of the 10th International Conference on Emerging Internet, Data and Web Technologies, pp. 256–262 (2022)
15. Duolikun, D., Enokido, Takizawa, M.: An energy-aware algorithm for changing tree structure and process migration in the flexible tree-based fog computing model. In: Proceedings of the 37th International Conference on Advanced Information Networking and Applications, pp. 268–278 (2023)
16. Duolikun, D., Enokido, T., Takizawa, M.: An energy-aware dynamic algorithm for the FTBFC model of the IoT. In: Proceedings of the 17th International Conference on Complex, Intelligent and Software Intensive Systems, CISIS-2023, Toronto, ON, Canada, pp. 38–47 (2023)
17. Duolikun, D., Enokido, T., Takizawa, M.: An energy-aware negotiation protocol for live migration of virtual machines. Int. J. Web Grid Serv. **19**(4), 446–462 (2023)
18. Duolikun, D., Enokido, T., Takizawa, M.: An algorithm to change the TBFC model to reduce the energy consumption of the IoT. In: Proceedings of the 12th International Conference on Emerging Internet, Data and Web Technologies (2024, accepted)
19. Mukae, K., Saito, T., Nakamura, S., Enokido, T., Takizawa, M.: Design and implementing of the dynamic tree-based fog computing (DTBFC) model to realize the energy-efficient IoT. In: Proceedings of the 9th International Conference on Emerging Internet, Data and Web Technologies, pp. 71–81 (2021)

20. Oma, R., Nakamura, S., Duolikun, D., Enokido, T., Takizawa, M.: An energy-efficient model for fog computing in the Internet of Things (IoT). IoT J. **1–2**, 14–26 (2018)
21. Oma, R., Nakamura, S., Enokido, T., Takizawa, M.: A tree-based model of energy-efficient fog computing systems in IoT. In: Proceedings of the 12th International Conference on Complex, Intelligent, and Software Intensive Systems, pp. 991–1001 (2018)
22. Oma, R., Nakamura, S., Duolikun, D., Enokido, T., Takizawa, M.: Evaluation of an energy-efficient tree-based model of fog computing. In: Barolli, L., Kryvinska, N., Enokido, T., Takizawa, M. (eds.) NBiS 2018. LNDECT, vol. 22, pp. 99–109. Springer, Cham (2019). https://doi.org/10.1007/978-3-319-98530-5_9
23. Oma, R., Nakamura, S., Duolikun, D., Enokido, T., Takizawa, M.: A fault-tolerant tree-based fog computing model. Int. J. Web Grid Serv. **15**(3), 219–239 (2019)
24. Oma, R., Nakamura, S., Duolikun, D., Enokido, T., Takizawa, M.: Energy-efficient recovery algorithm in the fault-tolerant tree-based fog computing (FTBFC) model. In: Barolli, L., Takizawa, M., Xhafa, F., Enokido, T. (eds.) AINA 2019. AISC, vol. 926, pp. 132–143. Springer, Cham (2020). https://doi.org/10.1007/978-3-030-15032-7_11
25. Oma, R., Nakamura, S., Enokido, T., Takizawa, M.: A dynamic tree-based fog computing (DTBFC) model for the energy-efficient IoT. In: Proceedings of the 8th International Conference on Emerging Internet, Data and Web Technologies, pp. 24–34 (2020). pp. 268–275 (2020)
26. Raspberry Pi 3 Model B (2016). https://www.raspberrypi.org/products/raspberry-pi-3-model-b

Automotive User Interface Based on LSTM-Grid Search Deep Learning Model for IoT Security Change Request Classification

Zaineb Sakhrawi[1(✉)], Taher Labidi[2], Asma Sellami[1], and Nadia Bouassida[1]

[1] MIRACL Laboratory, Higher Institute of Computer Science and Multimedia,
University of Sfax, Sfax, Tunisia
{zaineb.sakhraoui,asma.sellami,nadia.bouassida}@isims.usf.tn
[2] MIRACL Laboratory, Higher Institute of Computer Science of Medenine,
University of Gabes, Gabes, Tunisia
taherlabidi@gmail.com

Abstract. The field of the Internet of Things (IoT) is rapidly growing in significance, with roughly fifty billion devices being used in technology for computing by the end of 2020. However, the interdependence of IoT devices, as well as the variety of components used in their implementation, has caused a variety of issues, such as insufficient testing of change requests (CR) that affect security requirements. One way to address these security issues is to provide a deeper classification of security requirements that have an impact on the overall software process, such as software testing. Thus, the primary goal of this study is to assist software testers in prioritizing changes requested to enhance software security on IoT-based devices. Therefore, a deep learning-based approach to security CR classification is proposed. In this study, the Long Short Term Memory (LSTM) model is used and enhanced through the grid search tuning method. The LSTM-based grid search deep learning classifier identifies the class (*i.e.,* sub-characteristics defined by the ISO 25010 quality model) of a given security CR from IoT-based devices with an average classification accuracy of 79%. Finally, to validate the robustness of our proposed research methodology, we developed an automated classification user interface for software testers.

Keywords: IoT-based devices · software regression testing · LSTM deep learning · Grid Search tuning method · Security change request

1 Introduction

The IoT is one of the fastest-growing computing areas, with approximately fifty billion gadgets by the end of 2020 [1]. As IoT devices become more prevalent in our daily activities, guaranteeing their security is ever more crucial. Security issues have emerged due to the interdependence of IoT systems and the variety of components used in their implementation [1]. They may be related to one of the security sub-characteristics (integrity, confidentiality, non-repudiation, accountability, and authenticity) as described by ISO 25010 [2]. Conventional security testing methods frequently

L. Barolli (Ed.): AINA 2024, LNDECT 200, pp. 476–486, 2024.
https://doi.org/10.1007/978-3-031-57853-3_40

fall short of keeping up with the ever-changing threat landscape and the dynamic nature of IoT-based devices [3]. Testing the security of IoT based-devices when change requests (CR) are frequent is becoming increasingly important [3].

When performing regular changes, the security of IoT based-devices may be degraded by a lack of understanding and the wrong selection of adequate test cases (TC) [4]. One method of addressing these security issues is by implementing a security CR. When security CR occurs, several TCs are usually generated over time that could be relevant or irrelevant to a particular CR [5]. CR may occur within one of the IoT standard devices, such as a sensor, an API and terminal, a user system, or a data processor [5]. For IoT devices, security CR could include software updates, configuration changes, or other modifications to enhance security [3]. Accordingly, there is an increasing demand for a tool that ensures the effectiveness of security testing in IoT-based devices, detects security issues, and selects the relevant TC [6]. Security issues can be treated as a classification problem [7]. Thus, a deeper classification of the quality characteristic (*i.e.,* non-functional requirement), such as security, can speed up the software development process. This classification has an impact on other aspects of software development, such as test prioritization [8]. For each CR, a large number of TCs can be generated. As a result, testers frequently become confused about which TC to begin with. In other words, which TC will be given top priority? As a result, to respond appropriately and quickly to security CRs, it is necessary to automatically classify changes.

This paper has three main contributions. The first contribution consists of collecting data related to security requirements in the context of IoT systems. The security requirements fall into five sub characteristics, as described in the ISO 25010 quality model. Second, we propose a deep learning based approach for classifying security CR into authenticity, integrity, accountability, confidentiality, or non-repudiation. When reviewing security CR, we discovered that most of them are represented by lengthy sentences. To deal with this situation, one of the key reasons for using the LSTM is its ability to classify both short and long sentences, as well as accomplish sentiment analysis [9]. Given a large number of possible variable combinations, the grid search tuning method is employed to optimize the LSTM parameters. To the best of our knowledge, our study is the first research that used the ISO 25010 security decomposition to classify security CR for IoT-based devices. Because manual classification of software quality requirements requires time, particularly on large projects with a large number of requirements [10], we propose, as a third contribution, an automated classification of the individual security CR using the LSTM-based grid search tuning model classifier. This classification can benefit software testers in selecting TC and doing test regression.

The remainder of this paper is organized as follows: Sect. 2 discusses the background and related work. Section 3 describes our proposed research methodology. Section 4 discusses the experimental results. Section 5 presents the threats to the study's validity. Section 6 concludes the paper and outlines some of its future works.

2   Background and Related Work

2.1   LSTM Deep Learning Architecture

The LSTM model is a type of recurrent neural network (RNN) architecture that has recently emerged as the dominant RNN structure [11]. It can learn long-term dependencies and overcome RNN restrictions [11]. Many researchers have investigated the effectiveness of the LSTM for particular sequence problems, such as text classification [11]. In addition, the LSTM model can classify both short and long sentences as well as perform sentiment analysis [9]. In [12], Yahya et al., have indicated that the LSTM deep learning architecture provides the most accurate NFR classification when compared to CNN, and RNN gated recurrent units (GRU). In [13], Rahimi et al. indicated that the use of a two-stage automatic classification system-based LSTM is more reliable than a single-stage classification system.

2.2   Grid Search Tuning Method

To demonstrate the superiority of the LSTM model, we also used the Gird Search tuning method, showing good performance in various classification problems [16]. There are several approaches to tuning hyperparameters [14]. In this research, we used the grid search method, which entails exhaustively searching through a subset of the algorithm's hyperparameter space, followed by a performance metric [14]. Although several other more robust optimization methods have been investigated, grid search remains among the most frequently employed because its computational needs can be met easily in most cases [14]. Many approaches used the grid search tuning method to optimize the LSTM deep learning architecture hyperparameters [15, 16].

2.3   Term Frequency-Inverse Document (TF-IDF)

In our work, using unstructured data, determining term frequency and inverse document frequency is an essential weighting technique for feature extraction. Using the TF-IDF vectorization technique, the frequency of a term in a document is offset by the frequency of the term across a set of documents [17]. Thus, the numeric matrix can be used to train machine learning and deep learning models. The TF-IDF technique is increasingly being used by many researchers in the pre-processing phase [17].

3   Research Methodology

The overall process of our research methodology is depicted in Fig. 1.

3.1   Data Collection

Although we face numerous challenges in collecting data directly from industries due to confidentiality concerns, the dataset was collected from IoT application developers

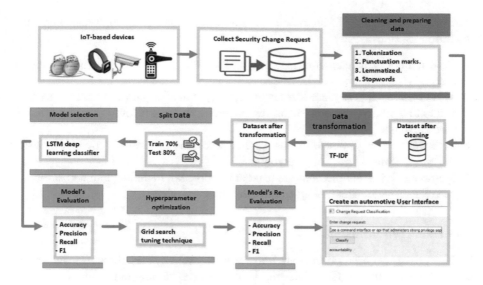

Fig. 1. Proposed approach

using Google Forms. Some of the data is collected from websites describing IoT applications. All the data collected is used to secure IoT devices. The collected dataset contains four features, which are the security Change Requests (CR) that can be described in the form of user stories, the related IoT devices, the acceptance criteria, and their corresponding classes. As mentioned in the introduction section, the class is generated based on the security goals identified as either confidentiality, integrity, authenticity, or availability (*i.e.,* based on the ISO 25010 sub-characteristics of security). The collected dataset contains 1100 security CRs for IoT devices (such as security cameras, smart locks, medical pacemakers, fitness trackers, smart consumer drones, in-car entertainment systems, smart TVs, smart refrigerators, smartwatches, smart beds, and MRI machines). We gathered 100 security CRs for each device.

Changes requested for security cover a wide range of security issues, including privacy, authorization, software enhancements, accessibility, privacy on networks, and authentication. The dataset includes security CRs such as implementing end-to-end encryption to secure transmitted data, enforcing strong password policies to avoid bruteforce attacks, allowing intrusion detection systems in place, turning off unused services and ports, and regularly updating the firmware to fix security vulnerabilities and bugs. The collected data are checked for any bias or inconsistencies. We asked a cybersecurity expert to examine the security CR and their assigned security sub-characteristics.

3.2   Dataset Cleaning/Preparation

This step involves converting an input requirement into a format that can be handled and implemented into deep learning models via algorithms. Preparing data means transforming text into a form that is foreseeable and measurable for the task [15]. This step involves converting the "Change Request" raw into vectors. To do so, we accomplished the following steps:

- Case folding: This step contributes to unifying the cases of letters throughout the text.
- Tokenization: This step contributes to dividing input data into tokens. In this paper, texts are tokenized by employing the Natural Language Toolkit (NLTK) library's word_tokenize() method.
- Stop words: The generated tokens are then processed to remove stopwords by using the NLTK library's set() of stopwords.
- Stemming: Different methods are used in this process, including lemmatization, the use of semi-automatic lookup tables, and suffix stripping.

3.3   Data Transformation

In this section, we perform the data transformation step by step through word vectorization which is the TF-IDF. This step is required for training our model on the corresponding textual data, which must be converted into numerical format. The TF-IDF vectorizes or scores a word by multiplying the Term Frequency (TF) of the word by the Inverse Document Frequency (IDF), where:

- The term frequency (TF) measures how frequently a term or word appears in a document relative to all the words.
- Inverse Document Frequency (IDF) of a term: The proportion of documents in the corpus that contain the term is reflected in the IDF.

3.4   Building a Classification Model for Change Requests

In this section, we conducted a series of experimental assessments examining the use of the LSTM deep learning model. We split the vectorized dataset into training (70%) and test (30%) sets. Two LSTM deep learning classifier models are built, one with and one without the grid search tuning method. These models were used to achieve a high level of accuracy in classifying security CRs for IoT-based devices. Experiments were carried out using Jupyter Python programming[1].

3.4.1   LSTM Deep Learning Architecture Classifier Construction
This section presents the classification outcomes for the LSTM deep learning classifier.

Parameter Setting of LSTM Deep Learning Model
The most important factor influencing model performance is correctly choosing the value of the model's parameters. Certainly, it is necessary for good results. The main parameters taken into account when building the LSTM deep learning model are batch size, epochs, optimization, neurons in the hidden layer, and embedding dimension. These parameters are detailed in Table 1.

---

[1] https://jupyter.org/.

Table 1. Parameters of the GS

Parameters	Description
Batch size	specifies the number of samples to be considered for tuning before updating the internal model parameters
Epcohes	estimates how many times the learning algorithm will run on the training dataset
Optimizer	reduce the gap between enhancement model parameters and the loss function
Neurons in hidden layers	indicate the number of neurons in a layer
Embedding dimension	specifies how many features or dimensions the embedding space has

LSTM Deep Learning Classifier Model Performance Assessment
We assessed the constructed models using a variety of evaluation metrics, including accuracy, precision, recall, and F1 score, which were measured as follows:

$$Accuracy = \frac{TP+TN}{TP+TN+FP+FN} \tag{1}$$

$$Precision = \frac{TP}{TP+FP} \tag{2}$$

$$Recall = \frac{TP}{TP \mid FN} \tag{3}$$

$$F1 = \frac{2*Precision*Recall}{Precision+Recall} = \frac{2*TP}{2*TP+FP+FN} \tag{4}$$

The accuracy of the built model is equal to 0.33%, indicating the insufficiency of the LSTM deep learning model.

3.4.2 LSTM-Based Grid Search Deep Learning Classifier Construction

Figure 2 depicts the steps required to build the security CR classification model using the LSTM-based grid search tuning method.

Parameter Setting of LSTM-Based Grid Search Deep Learning Classifier
Using the grid search tuning method, each combination of an established set of hyperparameter values is investigated. The best combination is subsequently selected based on a 10-fold cross-validation test. For tuning hyperparameters, we considered batch size, epochs, optimization, neurons in the hidden layer, and embedding dimension. We also used the grid search to adjust the LSTM deep learning architecture and improve the accuracy of the classification results (see Fig. 2) by analyzing each possible combination of hyperparameters. Table 2 shows the best hyperparameter combinations that improve the accuracy of the classification model.

LSTM-Based Grid Search Deep Learning Classifier Model Performance Assessment
To evaluate the LSTM-based grid search deep learning classifier, we evaluate the confusion matrix. A confusion matrix, as the name implies, is a matrix of numbers that indicates where a model becomes confused [7]. The confusion matrix is obtained when

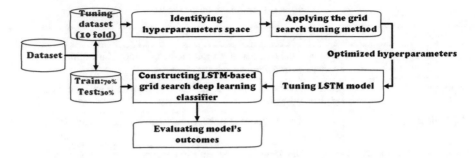

Fig. 2. LSTM-based grid search deep learning classifier Process

Table 2. Parameters of the GS

Parameters	values
Batch size	10
Epoch	50
Optimization	Adam
Neurons in hidden layer	2 (*i.e.,* two input features "change request", "Class")
embedding dimension	50

all four classes are used (see Fig. 3). The true class labels are listed along the x-axis, and the class predictions are listed along the y-axis. The correct classifications are shown along the first diagonal, while all of the other entries show misclassifications. A total of 173 samples were correctly predicted out of the total of 220 samples. Thus, the overall accuracy is 78.6%. Based on the curve of the validation accuracy shown in Fig. 4, we concluded that our constructed model produces accurate results.

3.5   Automatically Classifying IoT Security Change Request Through an Automotive User Interface

Regarding the experimental results of this study, we discovered that LSTM deep learning based on the grid search tuning method yielded the most accurate model. However, manually employing this model is time-consuming [10]. As a result, we proposed to use an automotive user interface for our model. The automotive user interface is designed for testers to easily and automatically make the security CR classification. It is developed using Jupyter Python programming. The user interface is intended to meet the requirements of agile project testers. It will allow testers to express the security CR in both formal and informal descriptions. Based on the description of the CR, its class is generated as authenticity, integrity, accountability, or confidentiality. The user interface depicted in Fig. 5 includes two sessions: the training session built on the back end and the classification session built on the front end for the testers. As shown in Fig. 5, when the "classify" button is pressed, the security CR class is automatically displayed.

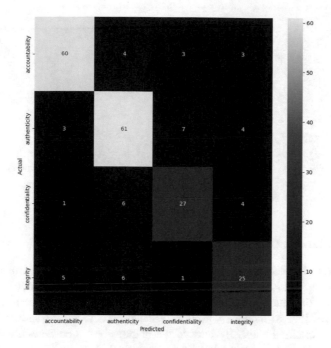

Fig. 3. Results

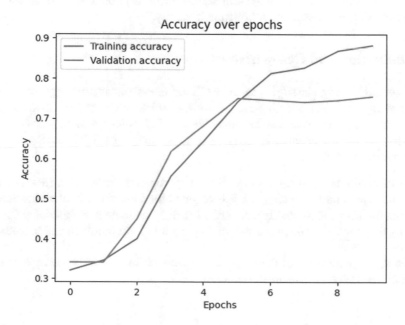

Fig. 4. Accuracy curve of the LSTM-based grid search deep learning classifier

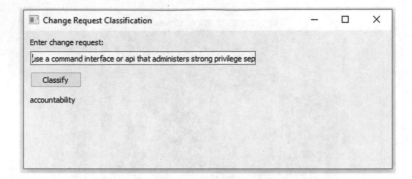

Fig. 5. Classify security change request outcome

3.6   User Interface Demonstrate

To demonstrate the accuracy of our automotive user interface using an LSTM-based grid search deep learning classifier, we created a demonstration video[2]. Through the user interface, for instance, the user starts by choosing a CR from our collected dataset: "access control logs to monitor who is accessing the device." As shown in the video, when the user clicks the classify button, the output gives a correct classification (*i.e., as classified in the dataset*), which is "accountability." Another CR chosen was to "use secure coding standards and practices to prevent buffer overflow and other common software vulnerabilities." When the user selects the classify button, the interface's output shows a correct classification, which is "integrity."

4   Discussion and Comparison

In this research study, classifying security CR in IoT systems is primarily used to assist testers in appropriately selecting and prioritizing security CR and the associated test cases, resulting in the successful management of IoT software system projects. The following are the major factors influencing the accuracy of LSTM-based grid search deep learning classifier:

- Gathering/Preparing dataset: Selecting the appropriate model input (dataset) while having a good understanding of CR is required for more accurate classification.
- Using the ISO 25010 quality standards: Formally defining non-functional CR (*i.e., using the ISO 25010 quality standards*) is required for a thorough understanding of customer needs.
- Setting the parameters of the deep learning model: In this study, we approved that the grid search tuning method is efficient.

---

[2] https://drive.google.com/file/d/17Yvj3olZPlpawMG2SmoBZMGYs4kTTUJX/view?usp=sharing.

5   Threats to Validity

In this section, we identify and summarize some threats that affect the validity of this study, along with the actions taken to mitigate them.

- External Validity: The first threat could be the choice of the dataset, where only one dataset, privately gathered, was used. The second threat was the use of only one tuning method, grid search. As a result, other tuning methods should have been used to improve the generalisability of the current study's findings.
- Threats of construct: The new challenge in this study is the use of ISO 25010 quality standards when identifying the different sub-classes of the security CR. In terms of measurement validity, the current study used the three most used evaluation metrics which are accuracy, precision, recall, and F-1 score.

6   Conclusion

In summary, we proposed a classifier model based on LSTM deep learning classifier. The models' hyperparameters are adjusted using the grid search tuning method. The built model represents text data in low-dimensional vector space and classifies security CR. It is then used by the users (testers) to classify a specific security CR in IoT-based devices without prior knowledge of deep learning through an automotive user interface. Future work could include investigating providing a dynamic selection of TC for security after an appropriate classification for the tester to know which TC to prioritize, experimenting with different neural network architectures, and testing the model on larger datasets.

References

1. Al-Garadi, M.A., et al.: A survey of machine and deep learning methods for internet of things (IoT) security. IEEE Commun. Surv. Tutor. **22**(3), 1646–1685 (2020)
2. Haoues, M., Sellami, A., Ben-Abdallah, H., Cheikhi, L.: A guideline for software architecture selection based on ISO 25010 quality related characteristics. Int. J. Syst. Assur. Eng. Manag. **8**, 886–909 (2017)
3. Medhat, N., Moussa, S., Badr, N., Tolba, M.F.: Testing techniques in IoT-based systems. In: 2019 Ninth International Conference on Intelligent Computing and Information Systems (ICICIS), December 2019, pp. 394–401. IEEE (2019)
4. Ahmad, W., Rasool, A., Javed, A.R., Baker, T., Jalil, Z.: Cyber security in IoT-based cloud computing: a comprehensive survey. Electronics **11**(1), 16 (2021)
5. Medhat, N., Moussa, S.M., Badr, N.L., Tolba, M.F.: A framework for continuous regression and integration testing in IoT systems based on deep learning and search-based techniques. IEEE Access **8**, 215716–215726 (2020)
6. Feng, X., Zhu, X., Han, Q.L., Zhou, W., Wen, S., Xiang, Y.: Detecting vulnerability on IoT device firmware: a survey. IEEE/CAA J. Automatica Sin. **10**(1), 25–41 (2022)
7. AlDhafer, O., Ahmad, I., Mahmood, S.: An end-to-end deep learning system for requirements classification using recurrent neural networks. Inf. Softw. Technol. **147**, 106877 (2022)

8. Zhu, S., Yang, S., Gou, X., Xu, Y., Zhang, T., Wan, Y.: Survey of testing methods and testbed development concerning internet of things. Wirel. Pers. Commun. **123**, 165–194 (2021). https://doi.org/10.1007/s11277-021-09124-5

9. Johnson, R. Zhang, T.: Supervised and semi-supervised text categorization using LSTM for region embeddings. In: International Conference on Machine Learning, pp. 526–534. PMLR (2016)

10. Navarro-Almanza, R., Juarez-Ramirez, R., Licea, G.: Towards supporting software engineering using deep learning: a case of software requirements classification. In: 2017 5th International Conference in Software Engineering Research and Innovation (CONISOFT), pp. 116–120. IEEE (2017)

11. Han, J., Pak, W.: Hierarchical LSTM-based network intrusion detection system using hybrid classification. Appl. Sci. **13**(5), 3089 (2023)

12. Yahya, A.E., Gharbi, A., Yafooz, W.M., Al-Dhaqm, A.: A novel hybrid deep learning model for detecting and classifying non-functional requirements of mobile apps issues. Electronics **12**(5), 1258 (2023)

13. Rahimi, N., Eassa, F., Elrefaei, L.: One-and two-phase software requirement classification using ensemble deep learning. Entropy **23**(10), 1264 (2021)

14. Bacanin, N., Stoean, C., Zivkovic, M., Rakic, M., Strulak-Wójcikiewicz, R., Stoean, R.: On the benefits of using metaheuristics in the hyperparameter tuning of deep learning models for energy load forecasting. Energies **16**(3), 1434 (2023)

15. Kim, T.Y., Cho, S.B.: Particle swarm optimization-based CNN-LSTM networks for forecasting energy consumption. In: 2019 IEEE Congress on Evolutionary Computation (CEC), pp. 1510–1516. IEEE (2019)

16. Priyadarshini, I., Cotton, C.: A novel LSTM-CNN-grid search-based deep neural network for sentiment analysis. J. Supercomput. **77**(12), 13911–13932 (2021)

17. Tiun, S., Mokhtar, U.A., Bakar, S.H., Saad, S.: Classification of functional and non-functional requirement in software requirement using Word2vec and fast Text. J. Phys. Conf. Ser. **1529**(4), 042077 (2020)

Author Index

L. Barolli (Ed.): AINA 2024, LNDECT 200, pp. 487–489, 2024.
https://doi.org/10.1007/978-3-031-57853-3

Printed in the United States
by Baker & Taylor Publisher Services